How to Listen to the Audio Files

The audio files that accompany this book are available streaming online at
www.petersonbirdsounds.com

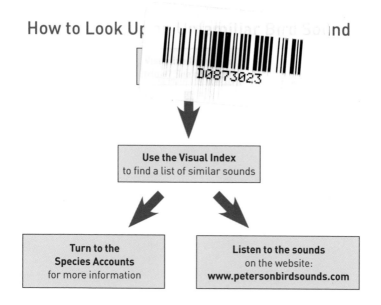

How to Look U̶p̶ nd

Use the Visual Index
to find a list of similar sounds

**Turn to the
Species Accounts**
for more information

Listen to the sounds
on the website:
www.petersonbirdsounds.com

How to Use the Visual Index

The index organizes sounds by pattern, providing a quick reference
to soundalikes and a way to look up unfamiliar sounds.

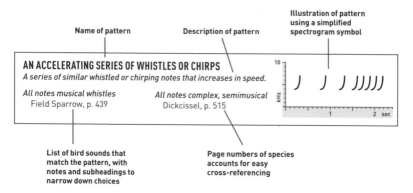

Name of pattern

Description of pattern

Illustration of pattern
using a simplified
spectrogram symbol

AN ACCELERATING SERIES OF WHISTLES OR CHIRPS
A series of similar whistled or chirping notes that increases in speed.

All notes musical whistles
Field Sparrow, p. 439

All notes complex, semimusical
Dickcissel, p. 515

List of bird sounds that
match the pattern, with
notes and subheadings to
narrow down choices

Page numbers of species
accounts for easy
cross-referencing

PETERSON FIELD GUIDE TO

BIRD SOUNDS
of Western North America

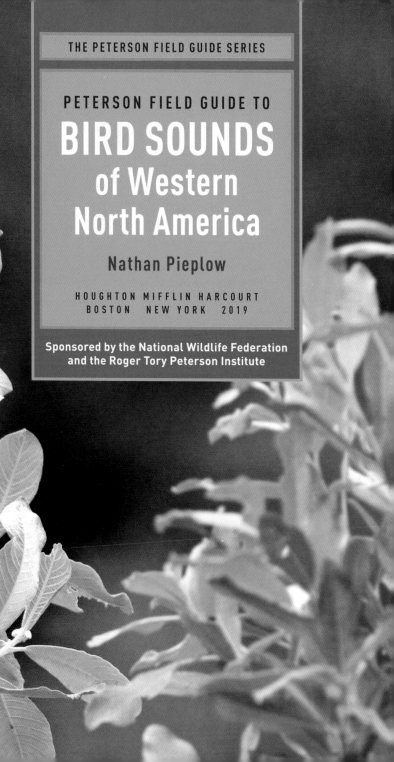

PETERSON FIELD GUIDE TO

BIRD SOUNDS
of Western
North America

Nathan Pieplow

HOUGHTON MIFFLIN HARCOURT
BOSTON   NEW YORK   2019

Sponsored by the National Wildlife Federation
and the Roger Tory Peterson Institute

Address requests for permission to make copies of Houghton Mifflin Harcourt
material to trade.permissions@hmhco.com or Permissions, Houghton Mifflin Harcourt
Publishing Company, 3 Park Avenue, 19th Floor, New York, NY 10016.

www.hmhco.com

Library of Congress Cataloging-in-Publication Data is available.
ISBN 978-0-547-90557-0

Book design by Eugenie S. Delaney

Printed in China

SCP 10 9 8 7 6 5 4 3 2 1

*This book is dedicated to
my mother*

ROGER TORY PETERSON INSTITUTE
OF NATURAL HISTORY

Continuing the work of Roger Tory Peterson through Art, Education, and Conservation

In 1984, the Roger Tory Peterson Institute of Natural History (RTPI) was founded in Peterson's hometown of Jamestown, New York, as an educational institution charged by Peterson with preserving his lifetime body of work and making it available to the world for educational purposes.

RTPI is the only official institutional steward of Roger Tory Peterson's body of work and his enduring legacy. It is our mission to foster understanding, appreciation, and protection of the natural world. By providing people with opportunities to engage in nature-focused art, education, and conservation projects, we promote the study of natural history and its connections to human health and economic prosperity.

Art—Using Art to Inspire Appreciation of Nature
The RTPI Archives contains the largest collection of Peterson's art in the world—iconic images that continue to inspire an awareness of and appreciation for nature.

Education—Explaining the Importance of Studying Natural History
We need to study, firsthand, the workings of the natural world and its importance to human life. Local surroundings can provide an engaging context for the study of natural history and its relationship to other disciplines such as math, science, and language. Environmental literacy is everybody's responsibility—not just experts and special interests.

Conservation—Sustaining and Restoring the Natural World
RTPI works to inspire people to choose action over inaction, and engages in meaningful conservation research and actions that transcend political and other boundaries. Our goal is to increase awareness and understanding of the natural connections between species, habitats, and people—connections that are critical to effective conservation.

For more information, and to support RTPI, please visit rtpi.org.

CONTENTS

INTRODUCTION

A New Approach to Bird Sounds

Bird sounds fill the wilderness, echo between city skyscrapers, and penetrate the windows of speeding cars. Few other natural phenomena are as ubiquitous, and few are as rich in beauty, variety, and meaning.

They have been called "nature's music," but bird sounds are perhaps better described as nature's language. Like human speech, they carry complex messages from singer to listener. This book is a dictionary for the language of the birds.

Identifying the species of a bird by sound, also called "birding by ear" or "ear-birding," has long been important to naturalists. In fact, some say that experienced birders detect and identify ten times as many birds with their ears as with their eyes. But listening to a bird can reveal far more than simply the identity of the singer. It can also reveal what the bird is doing and why, what it is communicating and to whom.

Listening to birds is one of the best ways to connect to the natural world. Whether out in the wilderness or in between the skyscrapers, birds can help remind us that we humans are not the only organisms on Earth capable of complex social interaction, communication, drama, and even artistry.

This book, the most comprehensive reference to North American bird sounds ever produced, takes a new approach to the language of the birds based on visualizing and indexing sounds.

Learning to visualize sounds takes only a short time, but revolutionizes listening and hearing. Picturing sounds makes it possible to understand at a glance what a bird will sound like, even before hearing the accompanying recording. It makes it possible to search this book visually for a bird song heard in the field.

Thus, this book makes it possible, for the first time, to look up an unfamiliar sound, just as one can look up an unfamiliar word in the dictionary.

Whenever possible, this book also seeks to illuminate the meanings of sounds—that is, their behavioral context, including whether they are made by males, females, or both; whether by adults, juveniles, or birds of all ages; whether in association with alarm, courtship, territorial defense, or the need to keep a flock together. Our knowledge of the sounds in this book is incomplete, and readers might add to that knowledge with their own careful observations of bird sounds and bird behaviors.

We have much else to learn about birds as well. Perhaps the most pressing question is how the birds are faring, and how they will continue to fare, on a planet increasingly repurposed and redesigned for human ends. This book cannot answer that question, but it does aim to bring bird sounds out of the background noise, making them more comprehensible, more meaningful, more enjoyable, and thereby more beautiful, and more difficult to ignore.

Organization of the book

The introductory section of this book explains how to listen to and visualize bird sounds, and discusses some aspects of their production, function, and development. The species accounts, arranged family-by-family in taxonomic order, depict the sounds of each species spectrographically. The visual index lists similar sounds in groups by pattern, with reference to the page number of the species account in which each sound is discussed.

Scope of the book

This book covers 537 species of birds regularly found in the western United States and southwestern Canada (hereafter, "the West"), south of about the 50th parallel and east to roughly the 100th meridian. Certain species that occur regularly only at the fringes of this area are omitted. Most pelagic species are omitted.

The book primarily treats sounds known to be given in the West. Thus, the breeding-season sounds of some species that nest in the Arctic,

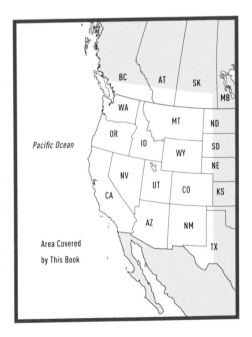

such as certain sandpipers, have been omitted, unless they are known to be given in migration or winter.

Extinct species are included when recordings of their voices exist. Extinct species for which no recordings exist are mentioned in the family introductions.

A number of exotic species are included in the guide, when wild or feral individuals can regularly be encountered in the West. Whenever possible, recordings of these species come from within the geographic scope of this book, because in some cases wild or feral birds may sound different than their wild ancestors.

Notes on the audio files

This book is accompanied by more than 7,500 audio files of bird sounds, which are available streaming on the Web at www.petersonbirdsounds.com.

The recordings were made by nearly 350 different recordists. Many of the recordings have been edited, sometimes extensively, in order to amplify the target sound, reduce interference from background sounds, and/or shorten intervals between vocalizations. Some recordings are of natural vocalizations, while others are of birds responding to playback. A few recordings are of captive birds.

The website lists information about each audio file, including the name of the recordist and the date and location of the recording, sometimes with additional notes on the behavioral context of the sounds and the circumstances of the recording. The website also hosts links to additional resources as well as video tutorials on how to read spectrograms.

Notes on the spectrograms

All the spectrograms in this book were created from the audio files on the accompanying website using the Raven Pro software developed by the Cornell Laboratory of Ornithology. The spectrograms were generated using Hann windows, typically with a window size of 512 samples and an overlap of 90 percent. Window size was sometimes adjusted for clarity. All spectrograms were then modified using graphics editing software, to remove visible traces of echoes and background sounds and to improve contrast.

In most cases the spectrogram shown is of the first vocalization on the sound file, but some spectrograms were generated from later sections of a recording. The website indicates which recordings are illustrated in the book.

How Birds Produce Sound

Sounds produced by a bird's vocal tract are called **vocal sounds** or **phonations**, while those produced in other ways are called **nonvocal sounds, mechanical sounds,** or **sonations.** Most of the sounds in this book are vocal sounds.

Vocal sounds

The vocal organ of birds is called a **syrinx** (plural **syringes**). The birds with the most complex vocalizations—hummingbirds, parrots, and passerines—tend to be those with the most complex musculature around the syrinx. Black and Turkey Vultures lack most or all syringeal muscles, so their vocal repertoires are limited to hisses and grunts.

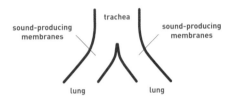

FIG. 1. *The avian syrinx*

In many species, the syrinx can produce two independently controlled voices at once, one from each lung. The result is called a **polyphonic** sound. Many birds have this capability but do not use it; they may produce all their sounds with one side of the syrinx, or they may use both sides, but not at the same time—for example, to produce different notes in the same song.

Nonvocal sounds

1. Beating the bill against a hard surface: Perhaps the most familiar type of mechanical bird sound is the drumming that woodpeckers make by striking hard surfaces with their bills. The initial sounds made by a displaying male Ruddy Duck may also qualify, as the bird apparently makes those sounds by slapping its bill rapidly downward against its own chest feathers.

2. Snapping the bill shut: In most birds, the sound of the bill snapping shut is not very loud, but some species, particularly owls, flycatchers, and gnatcatchers, use it in close-range aggressive displays. A few birds, notably Greater Roadrunner and the Wood Stork of the East, make clattering sounds by beating their mandibles together many times in quick succession.

3. Clapping the wings together: Among North American birds, this type of sound production typically takes place in flight. Some species, such as Long-eared and Short-eared Owls, clap their wings together below their bodies, while others, such as pigeons, clap their wings together above their backs.

4. Moving feathers through air: The wings of most birds can be heard in flight, at least at close range, but some species are specialized for sound production with their wings or tail. Sounds produced by feathers include the drumming of Ruffed Grouse, the winnowing of Wilson's Snipe, the high-pitched whistles of Mourning Doves flushing from a perch, and many hummingbird sounds.

5. Inflating body cavities with air: All birds have air sacs inside their body, but only a few species have specially modified air sacs that are used in sound production. Displaying Greater and Gunnison Sage-Grouse make popping sounds by expelling air from their air sacs explosively. The bizarre song of the American Bittern apparently involves the inflation of the esophagus, although the mechanism of production of this sound, as with many bird sounds, is not completely understood.

In some cases, an inflated air sac or body cavity may serve as a resonating chamber for a vocal sound. Birds such as Trumpeter Swans and Sandhill Cranes have tracheas that are elongated or even looped inside the body, apparently to amplify or modify the voice (see p. 126).

Visualizing Sound

Visualizing bird sounds makes it easier to identify them, because the aspects of bird sound that are important for visualization are the same ones that are important for identification: pitch pattern, speed, repetition, pauses, and tone quality. Creating a mental image of the sound makes it possible to look up the sound in the visual index of this book (p. 529), where similar sounds are grouped by their visual pattern.

How to see it

The best way of depicting bird sounds visually is the spectrogram (often called the Sonagram), a computer-generated graph of sound frequencies across time. Below, compare how the same familiar tune appears in two different visual depictions: standard musical notation on the left, and a spectrogram on the right:

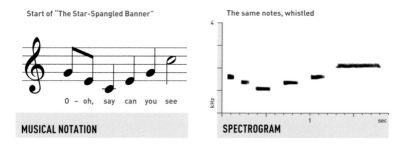

Start of "The Star-Spangled Banner"

The same notes, whistled

O – oh, say can you see

MUSICAL NOTATION

SPECTROGRAM

Musical notation places notes on a staff, while spectrograms measure the frequency of sounds in kilohertz (kHz), but the basic principle is the same: they both read from left to right, with high notes near the top of the chart and low notes near the bottom. On a spectrogram, the more horizontal space a note takes up, the longer it lasts in time.

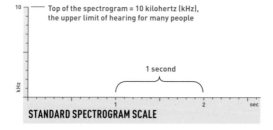

Top of the spectrogram = 10 kilohertz (kHz), the upper limit of hearing for many people

1 second

STANDARD SPECTROGRAM SCALE

Most spectrograms in this book conform to the scale above, with the top of the spectrogram at 10 kHz (near the upper limit of hearing in most adults) and numbers across the bottom marking intervals of one second.

Real spectrograms vs. spectrogram symbols

This book uses two types of visualizations: real spectrograms in the species accounts, where accuracy and detail are important; and spectrogram symbols in the visual index, where basic patterns and similarities are more important than detail.

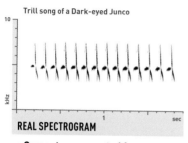

Trill song of a Dark-eyed Junco

REAL SPECTROGRAM

- **Computer-generated from an audio recording**
- **Shows fine details, even some not audible to the human ear**
- **Used in the species accounts**

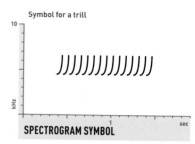

Symbol for a trill

SPECTROGRAM SYMBOL

- **Artistic approximation of a real spectrogram**
- **Emphasizes basic patterns rather than details**
- **Used in the visual index**

The five basic pitch patterns

Unlike music, bird sound identification does not require attention to the precise pitch of notes; more important is *how the pitch changes*. All bird sounds can be described with just five basic pitch patterns (or combinations thereof), which can be visualized this way:

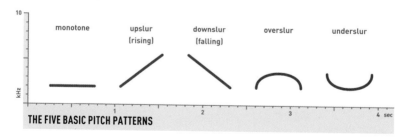

monotone   upslur (rising)   downslur (falling)   overslur   underslur

THE FIVE BASIC PITCH PATTERNS

- **Monotone sounds** do not change in pitch, and appear horizontal on the spectrogram.
- **Upslurred sounds** rise in pitch, and appear tilted upward.
- **Downslurred sounds** fall in pitch, and appear tilted downward.
- **Overslurred sounds** rise and then fall in pitch, appearing and sounding highest in the middle.
- **Underslurred sounds** fall and then rise, appearing and sounding lowest in the middle.

The four basic patterns of repetition and speed

The key to hearing and visualizing the repetition and speed of bird sounds lies in two questions:

1. Does the bird ever sing the same note twice?
2. Are the notes slow enough to count, or too fast to count?

Together, these two questions make it possible to identify four basic patterns of bird sound: **phrases, series, warbles,** and **trills.** *Phrases* and *series* are slower sounds, with individual notes slow enough to count; phrases contain unique notes that are not repeated, while series consist of one note repeated over and over. *Warbles* and *trills* are faster versions of phrases and series, with notes too fast to count (faster than about eight notes per second). At this speed, unique notes run together into a single warbled sound, and similar notes merge into a trill.

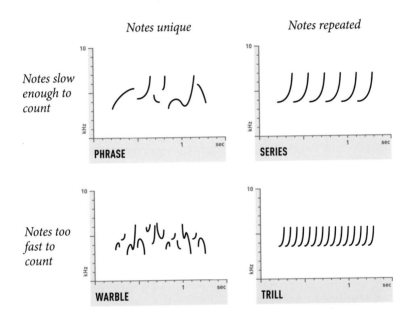

Even tremendously long and complex bird songs can be described as a combination of phrases, series, warbles, and trills. These four patterns form the most important building blocks of the visual index in this book.

The examples on the following page illustrate these four basic patterns with real spectrograms, generated from recordings that you can listen to on the accompanying website **www.petersonbirdsounds.com.**

Examples of phrases (unique notes, slow enough to count)

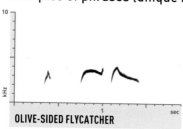

OLIVE-SIDED FLYCATCHER

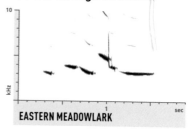

EASTERN MEADOWLARK

Examples of series (repeated notes, slow enough to count)

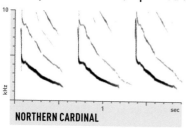

NORTHERN CARDINAL

AMERICAN GOLDFINCH

Examples of warbles (unique notes, too fast to count)

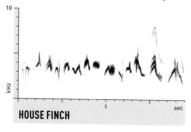

HOUSE FINCH

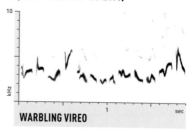

WARBLING VIREO

Examples of trills (repeated notes, too fast to count)

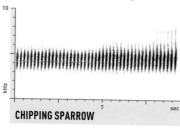

CHIPPING SPARROW

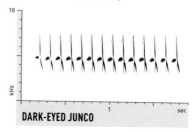

DARK-EYED JUNCO

Changes in speed and pitch of whole songs

Some bird sounds consist of series that change in speed. If the elements in a series are more closely spaced on the spectrogram as you move from left to right, then they are growing more closely spaced in time, which means that the series accelerates. If the elements grow farther apart, the series decelerates.

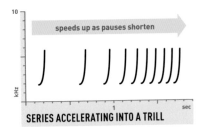

SERIES ACCELERATING INTO A TRILL

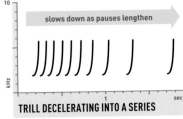

TRILL DECELERATING INTO A SERIES

Phrases, series, warbles, and trills can also change in pitch. For example, a warble might sound upslurred if it shows an overall trend toward higher notes. Similarly, a series might fall in pitch if each note starts slightly lower than the last, regardless of the pitch pattern of the individual notes.

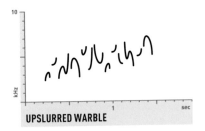

UPSLURRED WARBLE

DOWNSLURRED SERIES (OF UPSLURS)

The song of the Canyon Wren is a downslurred, decelerating series. Notice that the individual notes in this particular song are not downslurred, however; most of them are underslurred. As they progress from left to right, they slow down. They also become fainter on the spectrogram, meaning they are not as loud.

Canyon Wren Male Song

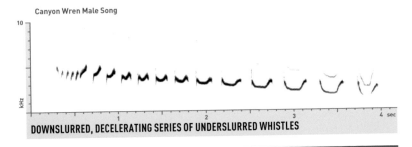

DOWNSLURRED, DECELERATING SERIES OF UNDERSLURRED WHISTLES

These songs both accelerate. The Field Sparrow's song begins as a series and speeds up into a trill; in this example the series is downslurred. The Horned Lark's song begins as a phrase and becomes a warble, rising in pitch at the end.

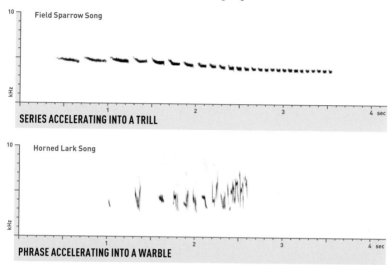

Complex series

Sometimes the repeated elements in a series themselves consist of multiple notes. A **couplet series** sounds like a 2-syllable word repeated, such as "peter peter peter"; a **triplet series** sounds like a 3-syllable word repeated, such as "teakettle teakettle teakettle." Series of 1-syllabled notes can be called **simple series**.

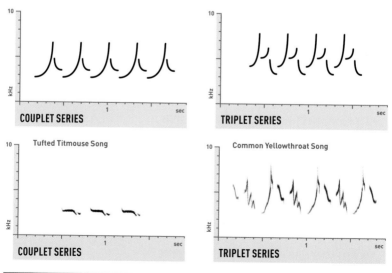

Pauses

A very important question to ask about bird sounds is whether and when the bird stops to "take a breath." Some birds can sing for 30 seconds or more without any noticeable pause. The song of the American Goldfinch, for example, often consists of several short series strung seamlessly together:

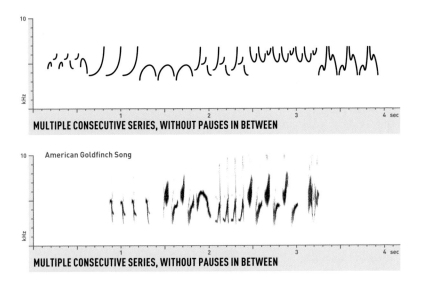

MULTIPLE CONSECUTIVE SERIES, WITHOUT PAUSES IN BETWEEN

American Goldfinch Song

MULTIPLE CONSECUTIVE SERIES, WITHOUT PAUSES IN BETWEEN

Compare the song pattern of this Northern Mockingbird, in which the bird takes short "breaths" between almost all of the series in its song:

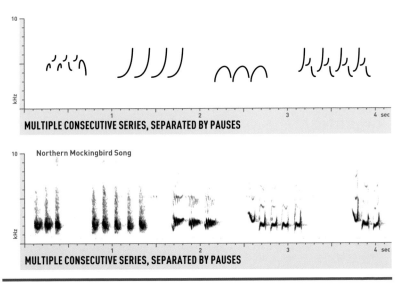

MULTIPLE CONSECUTIVE SERIES, SEPARATED BY PAUSES

Northern Mockingbird Song

MULTIPLE CONSECUTIVE SERIES, SEPARATED BY PAUSES

Some birds, especially many vireos and flycatchers, sing very short unique phrases separated by relatively long pauses:

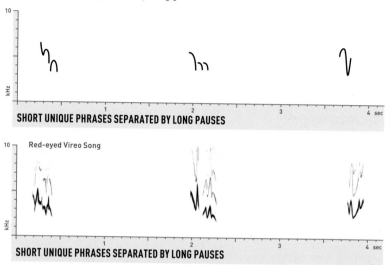

SHORT UNIQUE PHRASES SEPARATED BY LONG PAUSES

Red-eyed Vireo Song

SHORT UNIQUE PHRASES SEPARATED BY LONG PAUSES

Other singers, including the American Robin, Rose-breasted and Black-headed Grosbeaks, and many tanagers, group short unique phrases into clusters that are separated by longer pauses:

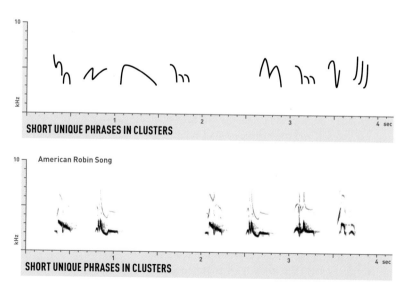

SHORT UNIQUE PHRASES IN CLUSTERS

American Robin Song

SHORT UNIQUE PHRASES IN CLUSTERS

The seven basic tone qualities

The distinctive voice of a bird sound, also called its tone quality, can be an important clue to its identification. Many species have unique tone qualities, and although the differences are easy to perceive, many people have found them difficult to describe. However, all bird voices are composed of variations or combinations of just seven basic tone qualities, illustrated below and explained in more detail on the following pages.

Whistled sounds p. 16

Whistles are the most basic sounds, appearing on the spectrogram as simple nonvertical lines. Examples include typical human whistling and the sounds of flutes and piccolos. Birds with whistled songs include Black-capped Chickadee, Yellow Warbler, and Northern Cardinal.

Hooting and cooing sounds p. 16

The hoots of large owls and the coos of doves are just extremely low-pitched whistles, less than 1 kHz in frequency. Because they are so low, they are illustrated in special spectrograms with a vertical scale of just 0 to 2 kHz. These low-pitched spectrograms are marked with an owl symbol.

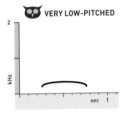

VERY LOW-PITCHED

Ticking sounds p. 17

Instantaneous bursts of noise sound like ticks, snaps, or knocks, and appear on the spectrogram as vertical lines. The ticking of a clock, the drumming of a woodpecker's bill against a tree, the bill snap of an angry flycatcher, and the ticking song of Yellow Rail all fall into this category.

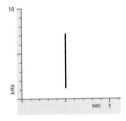

Burry and buzzy sounds p. 19

When a sound rises and falls very rapidly in pitch, it forms a squiggly line on the spectrogram and sounds trilled, like a referee whistle. If the squiggles are tall and fast enough, they sound less musical, more like an electric buzzer. These sounds are common among birds such as warblers and sparrows.

Noisy sounds   p. 20

Noisy sounds contain noise—that is, random sound at multiple frequencies, which looks like television static on the spectrogram and sounds like static to the ear. Very noisy bird sounds have a rough or harsh quality, like the alarm chatters of wrens and the hissing of angry swans and geese.

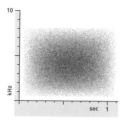

Nasal sounds   p. 20

Nasal sounds are composed of whistles stacked vertically that the human ear interprets as a single sound. Examples include police sirens, the whine of mosquito wings, and the sounds of oboes and violins. Birds with nasal voices include Red-breasted Nuthatch, Black-billed Magpie, and Pinyon Jay.

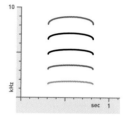

Polyphonic sounds   p. 23

Birds can make two whistled or nasal sounds simultaneously, creating a distinctive sound with a spectrographic pattern of lines that cross and/or stack without exactly matching. Such sounds may be dissonant and "whiny," like the calls of goldfinches, or metallic, like the songs of many thrushes.

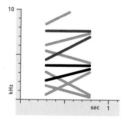

These seven tone qualities interact and combine in various ways. A few of the more common combinations are listed here.

- **Nasal and noisy sounds** include shrieks and screeches.
- **Nasal, noisy, and burry sounds** include the caws of crows and the quacks of ducks.
- **Polyphonic and noisy sounds** include the agitated whines of vireos and the harsh metallic cries of scrub-jays. Many of these sounds are also burry.

Whistles and hoots

Tone quality changes with pitch

The pitch of a sound has a significant impact on its tone quality. This is true of all sounds, but it is particularly true of whistles.

Hoots and coos are low-pitched whistles

The hooting of owls and the cooing of doves are just whistles that are lower than about 1 kHz. Similar low whistles can be made by blowing across the top of a large bottle. Extremely low-pitched sounds like these are illustrated using a different vertical scale than the rest of the spectrograms in this book, and marked with the symbol of an owl.

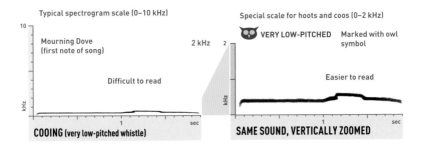

Pitch and whistle quality

As whistles rise to a pitch above hooting and cooing sounds, they first take on a mellow quality. As they continue to rise, above 3 kHz or so, they gradually become thinner and more penetrating. Whistles above 6 kHz tend to sound sibilant, almost like the hiss of air escaping a tire.

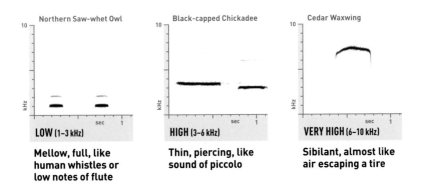

Ticking sounds and sharp whistles

As explained on page 14, nonvertical lines on the spectrogram represent whistles, and vertical lines represent ticks. As whistles become **sharper** (more vertical on the spectrogram), they gradually lose their musical quality and become more like ticks.

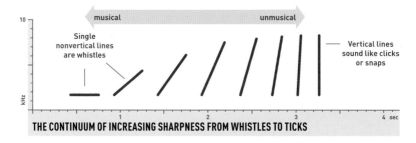

THE CONTINUUM OF INCREASING SHARPNESS FROM WHISTLES TO TICKS

The song of the Black-chinned Sparrow provides an excellent example of this phenomenon. The song is an accelerating series of whistles in which each successive note is slightly sharper and therefore less musical. The first half of the song is musical because the whistles are closer to the horizontal, but as the song continues, the whistles approach the vertical and become decidedly ticklike. The last few notes just sound like a toneless buzz.

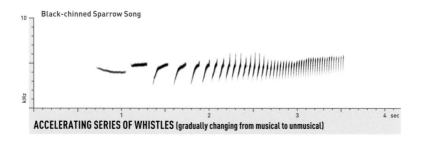

Black-chinned Sparrow Song

ACCELERATING SERIES OF WHISTLES (gradually changing from musical to unmusical)

The Whit calls of *Empidonax* flycatchers and the Chip calls of warblers are examples of very sharp, unmusical whistles. Whits are sharp upslurs; Chips are downslurs.

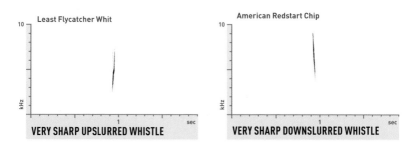

Least Flycatcher Whit

American Redstart Chip

VERY SHARP UPSLURRED WHISTLE

VERY SHARP DOWNSLURRED WHISTLE

The quality of a trill depends on the individual note

The principles explained on the previous page apply to trills as well as individual notes. A trill of musical notes sounds musical, while a trill of unmusical notes sounds unmusical.

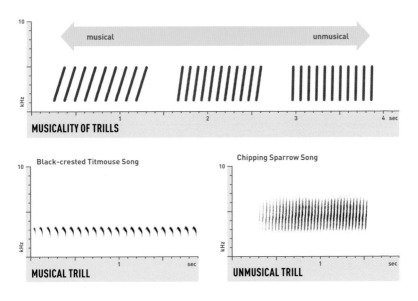

MUSICALITY OF TRILLS

Black-crested Titmouse Song

MUSICAL TRILL

Chipping Sparrow Song

UNMUSICAL TRILL

Musical trumps unmusical

Our perception of a sound's musicality tends to be determined by its most musical elements. For example, when a whistle on a spectrogram has both steep and flat sections, we tend to hear the flat, musical sections more strongly. This applies to trills as well.

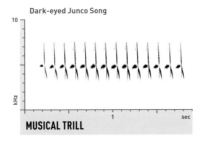

Dark-eyed Junco Song

MUSICAL TRILL

This junco's song consists of two alternated notes—one musical, one unmusical—repeated so fast that we hear only a single trill. Because musical sounds tend to dominate, the trill sounds rather musical to the ear.

Burry and buzzy sounds

These sounds rise and fall very rapidly in pitch, creating up-and-down squiggles, or **beats,** on the spectrogram. **Burrs** are more musical, and tend to take up less vertical space on the spectrogram; **buzzes** are less musical, and tend to take up more vertical space. In **coarse** burrs and buzzes, the squiggles are farther apart; in **fine** burrs and buzzes, they are closer together.

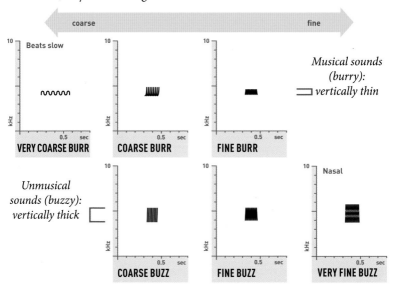

Musical sounds (burry): vertically thin

VERY COARSE BURR **COARSE BURR** **FINE BURR**

Unmusical sounds (buzzy): vertically thick

COARSE BUZZ **FINE BUZZ** **VERY FINE BUZZ**

Tremolos are burrs so coarse that the individual beats are countable, or nearly countable. **Peents** are buzzes so fine that they take on a nasal quality.

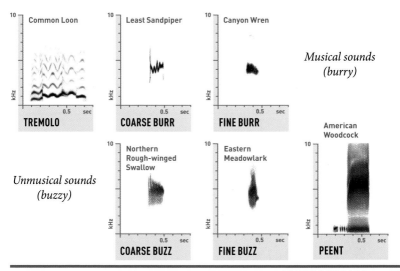

Musical sounds (burry)

TREMOLO **COARSE BURR** **FINE BURR**

Unmusical sounds (buzzy)

COARSE BUZZ **FINE BUZZ** **PEENT**

Noisy sounds

Darker is louder

The darker a note appears on the spectrogram, the louder it is.

In noisy sounds, the pitch and pitch pattern are determined by the placement and shape of the darkest (and thus the loudest) parts of the noise.

If the darkest part of the sound is closer to the top of the spectrogram, the noise will sound higher in pitch. If the darkest part rises toward the end of the spectrogram, the noise will sound upslurred; if it falls, the noise will sound downslurred.

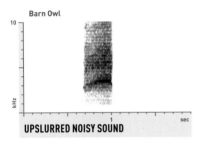

Nasal sounds

Many bird sounds are actually combinations of multiple simultaneous whistles on different pitches that the human brain typically perceives as a single sound. The individual whistles that make up a complex sound of this kind are called **partials.** The first (lowest) partial is often called the **fundamental.**

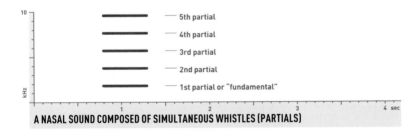

All nasal sounds are stacks of partials, but not all stacks of partials sound nasal. The tone quality of a complex sound depends on which partials are loudest—that is, darkest on the spectrogram.

The higher the darkest line in a stack, the more nasal the sound

If the fundamental or the second partial is the loudest (darkest), then the call will not sound nasal at all, but whistled. Once the darkest band climbs as high as the third partial, the call will start to sound nasal, and the higher it climbs, the more nasal the tone quality, as the following graph illustrates.

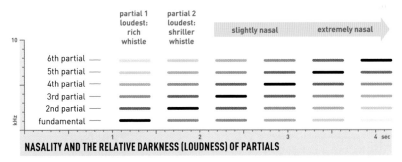

NASALITY AND THE RELATIVE DARKNESS (LOUDNESS) OF PARTIALS

The six bird sounds below are arranged in order of increasing nasality; the higher the loudest (darkest) partial, the more nasal the sound. Note that the counting of partials starts at the fundamental even if it is too faint to appear on the spectrogram. This is because in nasal sounds, the numbering of the partials is determined by the mathematical relationship between their frequencies: the frequency of the second partial is twice that of the fundamental, the frequency of the third partial is three times that of the fundamental, and so on.

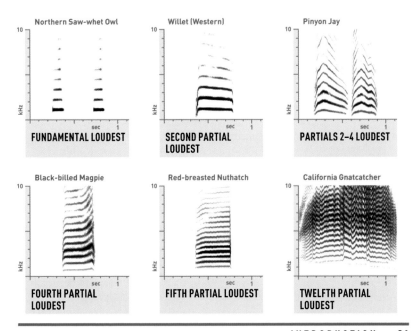

FUNDAMENTAL LOUDEST

SECOND PARTIAL LOUDEST

PARTIALS 2–4 LOUDEST

FOURTH PARTIAL LOUDEST

FIFTH PARTIAL LOUDEST

TWELFTH PARTIAL LOUDEST

The farther apart the stacked lines, the higher the pitch

The key to gauging the pitch of a nasal sound is **the spacing between the partials.** The farther apart the partials, the higher the pitch of the sound. If the partials stretch to the top and bottom of the spectrogram, counting them is helpful in estimating pitch. High sounds with three to five visible partials have a distinctive seminasal tone quality.

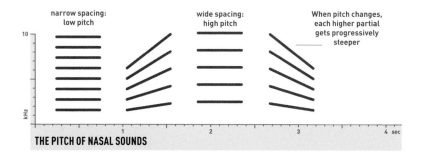

THE PITCH OF NASAL SOUNDS

The voices of large birds often break

The voices of large birds, such as geese, hawks, gulls, shorebirds, and woodpeckers, may jump suddenly to a higher or lower pitch. In nasal sounds, voice breaks appear on the spectrogram as vertical "fault lines" along which the stripes of the partials don't match up.

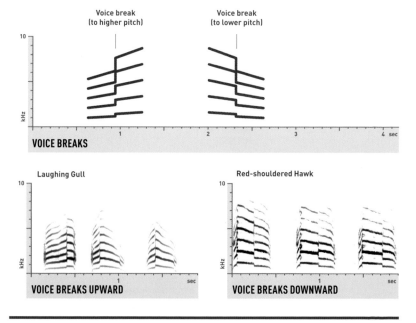

VOICE BREAKS

VOICE BREAKS UPWARD

VOICE BREAKS DOWNWARD

Polyphonic sounds

Many birds can produce two separate sounds simultaneously, one from each lung. When birds use this ability, the two original sounds blend into one polyphonic sound. Polyphonic sounds generally look similar to nasal sounds on the spectrogram, but they usually show one or more of these telltale signs, illustrated below:

1. Simultaneous rising and falling partials;
2. Stacks of partials with dissimilar shapes;
3. Partials that cross each other;
4. Partials that are irregularly spaced.

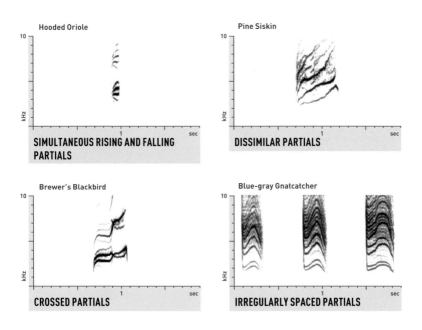

Polyphonic sounds are diverse, but with practice, they can be consistently distinguished from all other types of sounds. Most often, they sound either distinctively metallic or distinctively whiny.

If the polyphonic notes are very brief or contain monotone segments, they tend to sound metallic, like the Hooded Oriole call above, certain calls of Red-winged Blackbird, and the jangling melodies of thrushes like the Veery.

If the polyphonic notes do not contain any monotone segments, they tend to have a whiny quality, like the Pine Siskin and Blue-gray Gnatcatcher calls above, as well as the common calls of House Finch.

The Naming of Bird Sounds

Traditionally, bird sounds have been named either for their biological function ("alarm call," "flight call," "begging call," "song") or for their sound ("chewink," "chip," "buzz," "rattle").

Naming sounds for their function is often difficult, because little is known about the functions of many bird sounds. Even more important, one sound frequently serves several different functions, and different sounds can serve the same function. But assigning names based on sound can also be difficult, especially for sounds that are complex, highly variable, or plastic (see p. 28).

No method of naming sounds is perfect. This book seeks to strike a balance by using the name "Song" for primary breeding-season vocalizations that likely serve in mate attraction and territorial defense, and using sound-based names in most other cases. Mnemonic phrases are listed for some songs, when they are deemed useful.

In this book, the names of vocalizations are capitalized.

Transcribing bird sounds into human speech

One way to describe bird sounds is by transcribing them into the English alphabet. Bird voices and human voices are vastly different, and no standard method of transcription exists, but people do tend to follow certain patterns when imitating bird sounds in speech.

Vertical = consonant; horizontal = vowel

On the spectrogram, many bird sounds start and end with brief steep sections. The shape of the steep part tends to determine the consonant sound that many people hear, as follows:

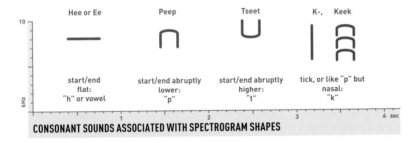

CONSONANT SOUNDS ASSOCIATED WITH SPECTROGRAM SHAPES

Listeners do not always agree on consonants, especially at the end of a sound. There tends to be more agreement on vowels.

Certain consonants correspond to tone qualities

In addition to the consonants that correspond to spectrogram shapes, some consonants are often used to indicate the quality of a sound, as follows:

consonant	often denotes a sound that is
s	very high-pitched
ch, sh	noisy
r	burry (but see below)
z	buzzy or polyphonic
l	broken

Different vowels represent different pitches

Linguists describe vowels as "high" or "low" based on the height of the tongue in the mouth during pronunciation. Low-pitched bird sounds are often transcribed using low vowels, and high-pitched bird sounds are often transcribed using high vowels.

The letters r, w, and y are vowel-like consonants (or "semivowels") that also correspond to pitches: **r** and **w** are low, while **y** is high.

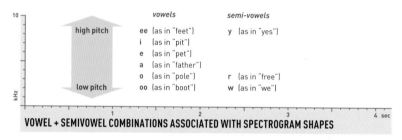

VOWEL + SEMIVOWEL COMBINATIONS ASSOCIATED WITH SPECTROGRAM SHAPES

Vowel combinations represent changes in pitch

Monotone bird sounds are typically written with a single vowel sound. Bird sounds that change in pitch are typically written with a combination of vowels and semivowels that mirrors the change in pitch. For example, upslurred sounds are usually transcribed with a low vowel or semivowel, followed immediately by a higher one.

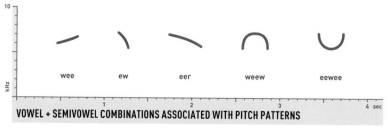

VOWEL + SEMIVOWEL COMBINATIONS ASSOCIATED WITH PITCH PATTERNS

The Biology of Bird Sounds

Innate vs. learned sounds

Some sounds are unlearned, or innate, meaning that birds can produce them without ever having heard them before. The ability to produce innate sounds is inborn and genetically controlled.

Some bird sounds are learned, meaning that in order to produce them properly, a young bird must hear them from an adult tutor of the same species. Tutors do not actively "teach"; instead, young birds learn by listening from a distance. The tutors are often the neighboring males on a young bird's first summer territory.

As far as we know, the only North American birds that can learn sounds are the hummingbirds, the parrots, and the passerines (excluding the flycatchers). Even in these groups, many sounds are innate.

Songs vs. calls

In 1961, W. H. Thorpe, the pioneering British ornithologist who originated the spectrographic analysis of bird sounds, laid out a set of criteria to distinguish between the songs and the calls of birds. According to his scheme, songs tend to be complex, learned, and given principally by males in prolonged bouts in the breeding season to establish a territory and attract a mate. Calls tend to be simple, innate, and used by both sexes in more general contexts, such as to raise an alarm, maintain contact between flock members, or beg for food.

The traditional distinction between songs and calls is blurred in many species. For example, the "Che-bek" of Least Flycatcher is innate, rather simple, and regularly given by females—but because it is given incessantly by territorial males and resembles the more complex songs of related flycatcher species, it is traditionally considered a song.

Ultimately, drawing the line between songs and calls is a matter of personal judgment. In drawing that line, this book seeks to strike a balance between traditional descriptions of bird sounds and the most recent research into their structure and function.

Mislearned songs

In captivity, some young birds will learn the songs of other species if they do not hear their own species' song. In a very few cases, birds in the wild will do the same. This happens most frequently among closely related species like Eastern and Western Meadowlarks, but even then it is rare.

Abnormalities may affect a whole song or just part of a song. One Song Sparrow from British Columbia sang a song that began with the notes of a Northern Waterthrush but ended with those of a typical Song Sparrow. In the wild, birds that sing abnormally are usually considered less likely to attract mates and pass their genes on to the next generation, meaning that most serious song abnormalities are selected out of the gene pool.

The learning process: subsong, plastic song, and crystallization

Young birds learning to sing usually start by producing **subsong,** a soft, disorganized jumble of notes that sometimes bears little resemblance to adult song. Subsong is typically heard between late summer and early spring, often from birds hidden in dense vegetation. Many subsongs include imitations of other species.

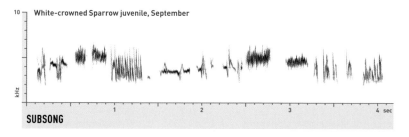

Eventually, young birds begin singing versions of their own species' songs that are recognizable, but variable and poorly stereotyped. This **plastic song** is typical of many young birds in their first spring. Adults in some species also give plastic song from fall through spring, as vocal centers in the brain undergo seasonal atrophy to reduce energy needs.

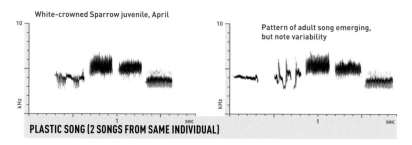

The final stage of song development is marked by a process called **crystallization.** At this stage, changes in the bird's brain "lock in" stereotyped versions of the song or songs that it has learned. In many species, after crystallization, no further learning ever occurs.

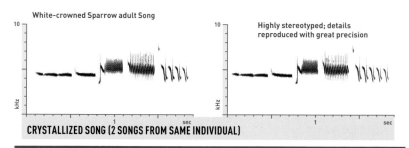

Song differences within individuals

Plastic vs. stereotyped sounds

Plastic sounds differ slightly each time they are produced, even by the same bird. Stereotyped sounds are always the same when made by the same bird. Plastic songs are typical of young birds, but some sounds of certain species remain plastic into adulthood, like several calls of the House Finch.

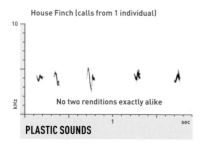

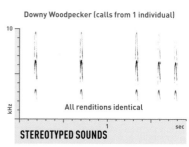

In adult birds, plasticity may vary with situation and season. Highly stereotyped versions of song tend to be associated with courtship. Plastic versions may be heard in winter, when levels of breeding hormones are lower.

Repertoires

In some species, individual birds learn only a single version of their species' song. In other species, individual birds learn multiple versions. These versions are called **songtypes**, and the set of all the songtypes that an individual bird can sing is called its **song repertoire.**

Birds with multiple songtypes may cycle through their repertoire, singing each song just once, or they may repeat one songtype multiple times before switching to another. If consecutive songtypes differ, a bird is said to be singing with **immediate variety.** If consecutive songtypes tend to be the same, it is said to be singing with **eventual variety.** Many intermediate patterns occur, and some birds change pattern depending on the situation.

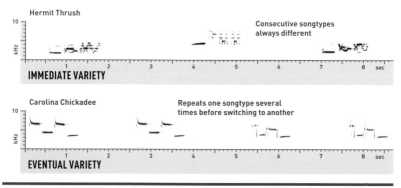

Song differences between individuals

Innovation, dialects, and individual recognition

Not all young birds copy songs from a tutor. Some invent their own songtypes according to the typical pattern of their species. In some species, like Song Sparrows, this is a rare behavior shown by just a few individuals. In other species it is a common strategy. Young American Robins may innovate up to 80 percent of their song phrases, learning the remaining 20 percent from tutors. Sedge Wrens may invent all their songtypes.

In species that learn from tutors, some young birds typically end up with slightly different versions than their tutors sang. The difference may be as minor as a slight modification to one note, or as major as a mixing of two previously distinct songtypes. As these birds get older they may serve as tutors for other young birds, establishing the modified song as a local variant. When birds breeding in a particular area sound similar to each other but different from members of the same species elsewhere, they are said to have a regional **dialect.**

Even within a dialect group, songs of individual birds may vary slightly. This variation can serve a biological purpose, allowing individual members of some species to recognize one another by song. Females may also use subtle distinctions between male songs to determine which males might make the fittest mates.

Variable vs. uniform sounds

This book uses the word **variable** to describe sounds that show high levels of individual variation, and **uniform** to describe sounds with low levels of individual variation. Variable sounds differ from one bird to the next; uniform sounds are similar in all members of a species.

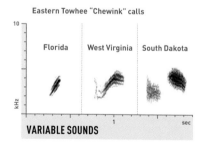

Eastern Towhee "Chewink" calls

Florida West Virginia South Dakota

kHz

VARIABLE SOUNDS

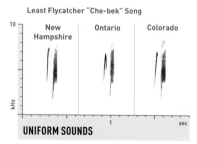

Least Flycatcher "Che-bek" Song

New Hampshire Ontario Colorado

kHz

UNIFORM SOUNDS

Imitation

Some bird species incorporate the sounds of other bird species, or even manmade sounds, into their own songs. This practice has often been called "mimicry," but many biologists prefer the terms "imitation" or "appropriation."

Why some birds appropriate others' songs is not well understood. However, the purpose is likely not to impersonate the imitated species. Most birds that imitate always signal their true identity by arranging the notes of other species in diagnostic patterns, mixing them with distinctive call notes, or both. Blue Jays that imitate hawks may be an exception. It has been hypothesized that jays may impersonate hawks as a way of drawing attention to the presence of a silent hawk or in order to frighten other birds, but more study is needed.

Species that regularly imitate include some vireos, corvids, gnatcatchers, mimids, European Starling, mynas, Phainopepla, Yellow-breasted Chat, western forms of Fox Sparrow, some orioles, and several species of cardueline finch, particularly Pine Grosbeak and Lesser Goldfinch. Many other birds imitate during the song-learning process.

Duets and choruses

Some bird species regularly perform duets, in which two birds sing together. Duetting birds are often mated pairs. In synchronized duets, the notes of the two birds are carefully timed, and the performance stereotyped. In unsynchronized duets, the timing and the performance are variable.

Duets occur in several North American families, including cranes, grebes, loons, chachalacas, and icterids. When more than two birds participate, the result is called a chorus. In this book, duets and choruses are indicated on the spectrograms by contrasting colors.

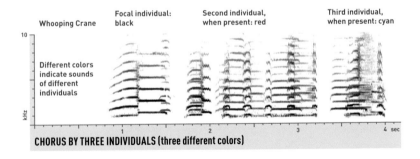

CHORUS BY THREE INDIVIDUALS (three different colors)

Sounds and species boundaries

Because females in many species choose mates by listening to the precise characteristics of each male's vocalizations, vocal differences can be a key factor keeping members of different species from pairing with one another—a reproductive isolating mechanism. In other words, differences in vocalizations can be not merely a symptom of being different species, but actually a mechanism that drives populations apart and keeps them separate. For this reason, taxonomists consider differences in song to be one important indicator of species boundaries.

Vocal differences can reproductively isolate populations that seem otherwise nearly identical. Such populations are often called **cryptic species.** For example, Willow and Alder Flycatchers were thought to be one species until the 1960s, when fieldwork showed that they often nest in the same areas without interbreeding, owing in large part to differences in their songs and calls.

Not all vocal differences are taxonomically important. Demonstrating that different-sounding populations are actually cryptic species requires careful fieldwork and genetic analysis to provide convincing evidence of reproductive isolation.

Sounds of hybrid birds

When birds of two different species pair and produce offspring, the resulting hybrid birds often give songs and calls that show characteristics of both parent species. This is generally true of birds with innate sounds, such as flycatchers, but can also occur in birds with learned sounds, such as sparrows.

In a few species pairs, such as Blue-winged and Golden-winged Warblers, hybrid offspring typically sound like one parent species or the other, not a mixture of both.

Insects and mammals that sound like birds

Some animals can sound quite similar to birds, especially frogs, toads, chipmunks, squirrels, and singing insects such as crickets, katydids, and cicadas. Such species are beyond the scope of this guide, but a few of the nonbird sounds most likely to cause confusion are mentioned in parentheses in the visual index alongside the bird sounds they resemble, so as to warn listeners of the possibility of misidentification.

Listeners should also be aware that recordings of birds in distress and/or birds of prey are sometimes used to repel pigeons, starlings, and gulls from roosting on buildings. Such recordings can sometimes be mistaken for the sounds of wild birds.

Recording Bird Sounds and Making Spectrograms

Creating one's own recordings and spectrograms is an excellent way to learn about bird sounds, and it has never been easier.

Many people can record audio with devices they already own, such as smartphones and digital cameras. Many different pocket voice recorders are now available, some for less than $100, that work quite well to record sounds that are reasonably close and loud.

Creating high-quality audio recordings of birds requires a high-quality microphone and a dedicated audio recorder. Many recordists use a shotgun microphone or a parabolic reflector dish to amplify the sound of a target bird relative to the background noise. The price of excellent audio recording equipment is comparable to the price of excellent photography equipment, but far fewer people own recording equipment. Good recordings have enormous potential to advance our knowledge of bird behavior, identification, and taxonomy.

From the 1960s through the 1990s, the only way to make spectrograms was using a Sona-Graph machine, manufactured by the Kay Elemetrics Company. The resulting "Sonagrams" were printed on paper, often in high-contrast black-and-white, which was not always ideal for showing the fine details of the sound.

Modern computer software makes it much easier to generate detailed and comprehensible pictures of bird sounds. Free software such as Raven Lite, available for download from the Cornell Laboratory of Ornithology website (**http://birds.cornell.edu**), allows anyone to create spectrograms like those used in this book and customize them in myriad ways.

Playback and birding ethics

Improvements in technology have made it increasingly easy to play bird sound recordings back to birds in the field, usually in an attempt to draw them closer so that people can get a better view. This practice has long been controversial.

Playback changes the behavior of wild birds. It simulates a territorial challenge, or in some cases a threat from a predator, creating a situation to which birds must respond aggressively. What birders may call "cooperative" individual birds may in fact be extremely agitated.

Some have argued that playback may negatively impact the breeding success of the responding birds. The evidence on this point is scanty and mixed. However, playback that is extremely loud, long, and/or disruptive to the birds could be considered harassment, which is unethical in all cases, and illegal in the case of threatened and endangered species.

In deciding whether to use playback, consider these questions:

- **Can I achieve my purpose without playback?** If you can, perhaps you should.
- **Would the use of playback violate any regulations at this site?** Many state and national parks, national wildlife refuges, and other sites prohibit playback. Know and follow the relevant rules.

- **How often will these same individual birds likely be the target of playback?** If you think the answer is more than two to three times per season, consider moving to another area to target different individuals, or avoiding playback altogether.
- **Would playback disturb other people?** If you are birding with others, or near others, seek consensus on the use of playback before beginning.
- **How loud is too loud, and how long is too long?** The answers may vary, but consider erring on the side of quietude and brevity.

When in doubt about the appropriateness of playback or any other behavior, check the American Birding Association Code of Ethics, **www.aba.org/about/ethics.html**, for general guidelines.

Additional resources

Online bird sound libraries

In addition to the sounds that accompany this book online at **www.peterson birdsounds.com**, these websites host streaming audio of vast numbers of bird sounds:

Xeno-Canto: **www.xeno-canto.org**
Macaulay Library: **www.macaulaylibrary.org** and eBird: **ebird.org/media**
Borror Laboratory of Bioacoustics: **drc.ohiolink.edu/handle/2374 .OX/30658**
Florida Museum of Natural History: **www.floridamuseum.ufl.edu/ bird-sounds**

Nocturnal flight call resources

Flight Calls of Migratory Birds (CD-ROM), by Bill Evans and Michael O'Brien, available online at **www.oldbird.org**

Books

Two books in particular provide good comprehensive introductions to the biology of bird sounds and the reading of spectrograms:

The Singing Life of Birds by Donald Kroodsma, 2007, Houghton Mifflin, Boston, MA
The Sound Approach to Birding by Mark Constantine and the Sound Approach, 2006, The Sound Approach, Poole, Dorset, UK

Blog

Earbirding.com: Nathan Pieplow (author of this book) and Andrew Spencer blog about recording, identifying, and interpreting bird sounds.

BIRDS BY
SONG:
SPECIES
ACCOUNTS

WATERFOWL (Family Anatidae)

This family includes the geese, the swans, and the ducks, all of which are adapted to floating on the surface of the water. Many species are brightly and distinctively colored, and some have been domesticated.

WHISTLING-DUCKS (Family Anatidae, Subfamily Dendrocygninae)

In size and shape, these species are intermediate between ducks and geese, with long legs and long necks. Formerly called "tree ducks," they are partly arboreal. Their vocalizations, consisting mostly of breathy whistles, are apparently innate.

FULVOUS WHISTLING-DUCK

Dendrocygna bicolor

A tropical species that expanded into our area in the nineteenth century. U.S. population has fluctuated ever since; closely tied to flooded rice fields.

Flocks often give Twitter of wheezy Pips, punctuated by Ki-wheers

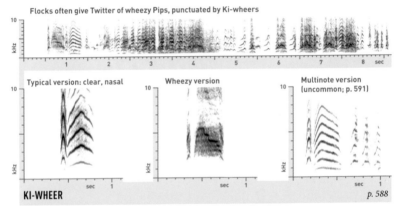

Typical version: clear, nasal

Wheezy version

Multinote version (uncommon; p. 591)

KI-WHEER   *p. 588*

All year; most common and distinctive call, likely by both sexes, often in flight. Fairly variable and plastic. Clear and wheezy versions may correspond to differences in sex, age, or function; more study needed. Multinote version recalls Black-bellied Whistling-Duck, but lower and more nasal.

Longer, downslurred versions (Wheer)

Shorter Whips and upslurred Wheeps

Here, followed by short Twitter (p. 571)

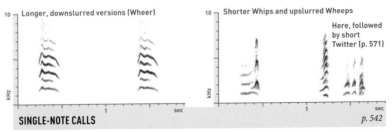

SINGLE-NOTE CALLS   *p. 542*

A set of variable calls that may intergrade; given all year, in alarm and in flight, sometimes in long Twitters. Most versions are 1-noted, like single syllables of Ki-wheer; same note often given twice in quick succession. Quality nasal to slightly polyphonic to fairly noisy. Usually given with Ki-wheers.

BLACK-BELLIED WHISTLING-DUCK

Dendrocygna autumnalis

Highly social and vocal, often gathering in loud flocks in shallow wetlands or adjacent fields. Range is expanding in the United States.

Flocks often give constant Twitter punctuated by Kip-whee-witters

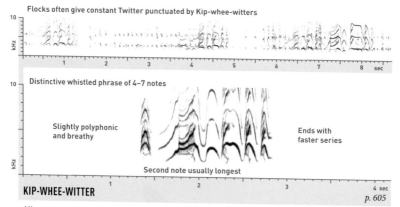

Distinctive whistled phrase of 4–7 notes

Slightly polyphonic and breathy

Ends with faster series

Second note usually longest

KIP-WHEE-WITTER   *p. 605*

All year; most common and distinctive call, likely by both sexes, often in flight. Fairly variable and plastic; individuals seem to have distinctive versions. Function little known.

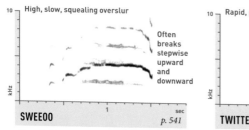

High, slow, squealing overslur

Often breaks stepwise upward and downward

SWEEOO   *p. 541*

Given by birds in groups, often with Twitters and Kip-whee-witters; function unknown.

Rapid, plastic series

High, seminasal notes

TWITTER   *p. 571*

Given by birds in groups, perhaps in contact. Highly plastic.

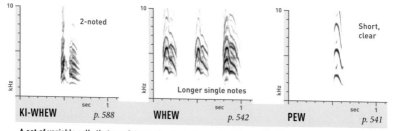

2-noted

KI-WHEW   *p. 588*

Longer single notes

WHEW   *p. 542*

Short, clear

PEW   *p. 541*

A set of variable calls that may intergrade; given all year, in alarm and in flight, sometimes in long series. Whew and Pew are most common; Ki-whew recalls Ki-wheer of Fulvous Whistling-Duck, but shorter. Always at least slightly polyphonic, but some short versions are nearly whistled.

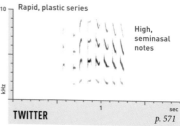

GEESE (Family Anatidae, Subfamily Anserinae)

Geese are familiar to all as the larger, longer-necked cousins of ducks. The voices and displays of the North American species are surprisingly little studied, but on average, their vocal communication appears to be simpler than that of ducks.

All goose sounds are apparently innate; most are plastic variations on a basic Honk, Yap, or Grunt. High levels of variation and plasticity make it difficult to determine the degree to which different sounds serve different functions, but at least some species have several discrete types of calls and displays.

Domestic and Feral Geese and Ducks

Many farms and urban parks host flocks of geese and ducks of domestic origin, which come in a variety of shapes, sizes, and colors. Many are entirely or partially white, with orange bills and legs; others may be mostly or partly black. Domestic geese are mostly derived from Eurasian species: the Greylag Goose (*Anser anser*), the Swan Goose (*Anser cygnoides*), or hybrids between them. Domestic ducks are mostly derived from the Mallard (p. 52) or the Muscovy Duck.

These feral birds can be highly vocal; their voices generally resemble those of their wild ancestors, but generations of selective breeding and hybridization can alter their sounds.

Emperor Goose

Anser canagicus

Breeds only in the Bering Sea region; rare in fall and winter along the West Coast, usually singly or in pairs in large flocks of other geese.

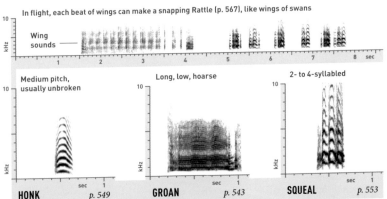

In flight, each beat of wings can make a snapping Rattle (p. 567), like wings of swans

Wing sounds

Medium pitch, usually unbroken

HONK *p. 549*

Long, low, hoarse

GROAN *p. 543*

2- to 4-syllabled

SQUEAL *p. 553*

All sounds variable and plastic. Honk usually unbroken; lower, longer than Honk of Snow Goose. Rarely shortened to a brief Yap (p. 549). Groans given in agitation; also sometimes in flight. Distinctive Squeal is most common call; like Squeal of Greater White-fronted Goose, but lower, often clearer.

Snow Goose

Anser caerulescens

Frequents marshes and farmland. Most are white with black wingtips. Dark morph, called "Blue Goose" (shown), is most common mid-continent.

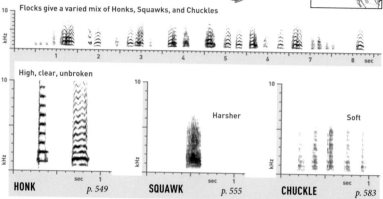

Flocks give a varied mix of Honks, Squawks, and Chuckles

High, clear, unbroken

Harsher

Soft

HONK *p. 549*　　**SQUAWK** *p. 555*　　**CHUCKLE** *p. 583*

All sounds variable and plastic, heard year-round. Sound of large flocks giving high clear Honks in alarm is distinctive, higher than Canada Goose and lower than Greater White-fronted. Soft Chuckles, often given along with other sounds by flying flocks, nearly unique to this species and Ross's Goose.

Ross's Goose

Anser rossii

Smaller than Snow Goose, with shorter neck and stubby, triangular bill with blue warty base, lacking Snow's black "grinning patch."

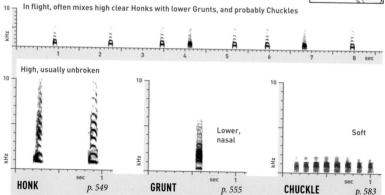

In flight, often mixes high clear Honks with lower Grunts, and probably Chuckles

High, usually unbroken

Lower, nasal

Soft

HONK *p. 549*　　**GRUNT** *p. 555*　　**CHUCKLE** *p. 583*

Voice little studied, but very similar to Snow Goose; averages lower-pitched with shorter notes, but probably not reliably separable. All sounds variable and plastic, likely given all year. Sole available recording of Chuckles is from nest site, but flying flocks probably also give a version, as in Snow Goose.

GREATER WHITE-FRONTED GOOSE

Anser albifrons

Medium-sized, with conspicuous orange legs. In migration and winter, forms flocks with other geese in wetlands and agricultural fields. Vocally distinctive.

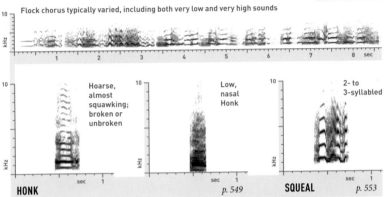

Flock chorus typically varied, including both very low and very high sounds

Hoarse, almost squawking; broken or unbroken

HONK

Low, nasal Honk

p. 549

2- to 3-syllabled

SQUEAL *p. 553*

All sounds variable and plastic. Honks average lower and harsher than in any other goose except Brant. Most distinctive sound is a 2- to 3-syllabled Squeal, high and hoarse, that can give away even a single individual of this species in a large flock of other geese. Rarely gives Chuckle like Snow Goose.

BRANT

Branta bernicla

Winters in flocks in salt water. Eastern birds have pale bellies and western birds dark bellies; a gray-bellied form breeds in the central Canadian Arctic.

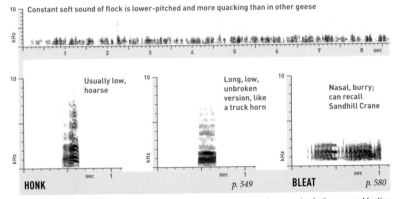

Constant soft sound of flock is lower-pitched and more quacking than in other geese

Usually low, hoarse

HONK

Long, low, unbroken version, like a truck horn

p. 549

Nasal, burry; can recall Sandhill Crane

BLEAT *p. 580*

All sounds variable and plastic. Generally lower-pitched and hoarser than sounds of other geese; bleating sounds are distinctive. If vocal differences exist between paler Atlantic population (top and right) and western "Black Brant" population (left and center), they are subtle.

Canada Goose

Branta canadensis

Abundant and familiar; now resident in many areas where it formerly occurred only in winter or migration. Subspecies differ greatly in size.

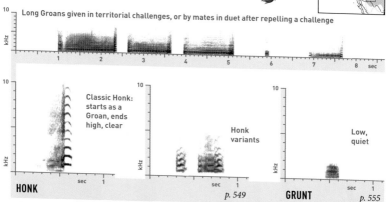

Long Groans given in territorial challenges, or by mates in duet after repelling a challenge

Classic Honk: starts as a Groan, ends high, clear

HONK

Honk variants

p. 549

Low, quiet

GRUNT

p. 555

All sounds highly variable and plastic; larger-bodied individuals have lower-pitched voices. Dominant birds in group are far more vocal than submissive birds. Subtle differences in Honks associated with different behaviors and displays; Grunts given in close contact. When threatened, adults give Hiss (p. 556).

Cackling Goose

Branta hutchinsii

Like a small Canada Goose, and formerly considered part of that species. Subspecies vary in size, with larger birds nearly identical to small Canadas.

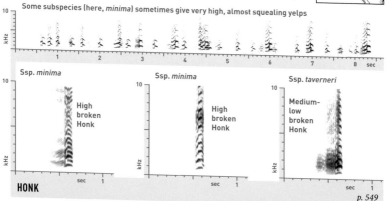

Some subspecies (here, *minima*) sometimes give very high, almost squealing yelps

Ssp. *minima*

High broken Honk

HONK

Ssp. *minima*

High broken Honk

Ssp. *taverneri*

Medium-low broken Honk

p. 549

All sounds highly variable and plastic, varying between subspecies, individuals, and situations. Larger subspecies (like Alaskan *taverneri*) can sound like Canada Goose; smaller subspecies (like Pacific Northwest *minima*) make some sounds that are higher than any Canada. Full repertoires poorly known.

EGYPTIAN GOOSE

Alopochen aegyptiacus

Native to Africa; now feral in parts of Florida and Texas, and escapes occur elsewhere. Nests in large trees or on ground, often in urban parks.

Display, often by duetting pairs: Grunt Series, sometimes accelerating into rapid Chatter (p. 582)

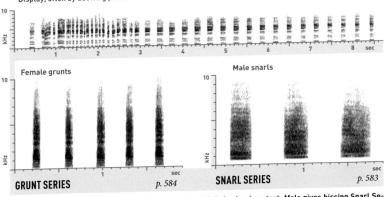

GRUNT SERIES p. 584

Female grunts

Male snarls

SNARL SERIES p. 583

All year. Only slightly variable, but calls vary subtly with behavioral context. Male gives hissing Snarl Series, while female gives louder Grunt Series, sometimes nasal or quacking. Series often continue for many minutes. Pairs and groups display loudly, often bowing heads or spreading wings.

SWANS (Family Anatidae, Subfamily Anserinae)

Currently classified in the same subfamily as geese, the swans are a distinctive group. The largest of the waterfowl, they dip their extremely long necks under the water to forage on submerged plants. Our species have primarily white plumage as adults and gray plumage as juveniles.

Sounds are apparently innate and tend to be relatively simple and gooselike; pair and group displays may be accompanied by loud, unsynchronized series of calls in duet or chorus. In flight, the wings of all three species make distinctive hums or rattles that can carry for considerable distances, possibly serving a communicative function.

The Mute Swan (*Cygnus olor*), native to Europe, has become established along the Mid-Atlantic coast of the U.S. and much of the Great Lakes region. It is widely kept as an exotic in the West, and escapes are frequently seen. It resembles native species, but has an orange bill with a large black knob, and often swims with its wings lifted slightly above its back. It is not mute at all; calls of adults include a Wee-Snarl, a Rattle, and a loud aggressive Hiss. Juveniles give various whooping calls.

TRUMPETER SWAN

Cygnus buccinator

Our largest swan. Hunted to near-extinction in the nineteenth century and extirpated from the East. Wild and re-introduced populations now increasing.

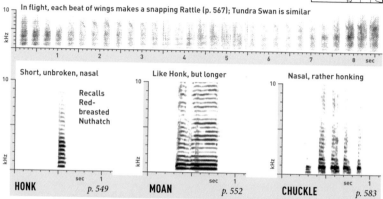

In flight, each beat of wings makes a snapping Rattle (p. 567); Tundra Swan is similar

Short, unbroken, nasal

Recalls Red-breasted Nuthatch

HONK *p. 549*

Like Honk, but longer

MOAN *p. 552*

Nasal, rather honking

CHUCKLE *p. 583*

All sounds variable and plastic, mostly variations on a low, nasal Honk, with quality recalling a paper party horn. Lower and more nasal than Tundra Swan, with almost no overlap. Slow Chuckles are typical at the end of Long Calls given during displays by pairs or families when reuniting or after aggressive encounters.

TUNDRA SWAN

Cygnus columbianus

Breeds in the Arctic; winters in large flocks on coastal estuaries and fresh-water lakes. At a distance, best distinguished from Trumpeter by voice.

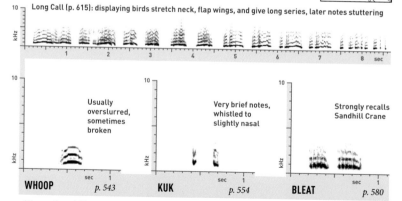

Long Call (p. 615): displaying birds stretch neck, flap wings, and give long series, later notes stuttering

Usually overslurred, sometimes broken

WHOOP *p. 543*

Very brief notes, whistled to slightly nasal

KUK *p. 554*

Strongly recalls Sandhill Crane

BLEAT *p. 580*

All sounds variable and plastic, mostly variations on a high clear Whoop, with quality that recalls Snow Goose. More varied than Trumpeter; rare low Moans, perhaps given only on breeding grounds, most similar to Trumpeter. Long Call (top) given by pairs and families when reuniting or after aggressive encounters.

DUCKS (Family Anatidae, Subfamily Anatinae)

The "true" ducks in the subfamily Anatinae are a diverse group. As far as is known, all duck sounds are innate. Many sounds are sex-specific.

In many species, courtship occurs primarily on the wintering grounds, starting as early as October. Especially in the dabbling ducks, pairs are thought to migrate northward together in spring, and the pair bond typically lasts only until the eggs hatch, with new pairs forming the following winter.

Courtship generally takes place in groups on the water, with several males simultaneously courting one or more females. Many species have multiple courtship displays, some of which include vocalizations. In this context, some behaviors that may seem casual—such as preening, flapping the wings, and/or turning away from the female—can actually represent highly stereotyped courtship behaviors.

Courtship in many ducks is associated with a high degree of aggression, especially by males toward females. Many female ducks direct soft Chuckles or Growls (sometimes termed "inciting calls") at males they prefer, and louder Screeches (sometimes termed "repulsion calls") at males they are trying to fight off. Females may take to the air with males following in pursuit flights, one or both sexes often vocalizing loudly.

Females of all dabbling duck species give Long Calls (or "Decrescendo Calls"), a series of quacks that usually accelerates. Given in long-distance contact, especially to a bird's mate, Long Calls often incite other females to respond in kind. The Long Call of Mallard is by far the most frequently heard; in many species it is rare. In Cinnamon Teal and both wigeons, no definitively identified recordings are available.

MANDARIN DUCK

Aix galericulata

Native to East Asia. Related to Wood Duck, with similar vocalizations. A small feral population breeds in Sonoma County, California, and escapes are frequently seen elsewhere, especially in southern California.

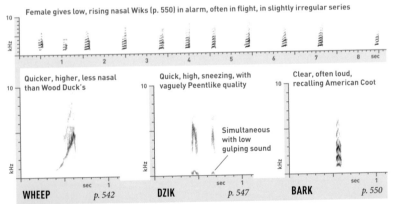

Female gives low, rising nasal Wiks (p. 550) in alarm, often in flight, in slightly irregular series

Quicker, higher, less nasal than Wood Duck's

WHEEP *p. 542*

Quick, high, sneezing, with vaguely Peentlike quality

Simultaneous with low gulping sound

DZIK *p. 547*

Clear, often loud, recalling American Coot

BARK *p. 550*

Wheep and Dzik are most common sounds of male in display, audible mostly at close range. Some versions of Wheep are more grating than shown. Other calls include soft peeping Twitter (p. 571). Bark reportedly by female, in display and when separated from mate.

Wood Duck

Aix sponsa

Found on secluded, wooded rivers and
ponds, usually in pairs. Nests in tree
cavities or nest boxes. Generally quiet
and fairly inconspicuous.

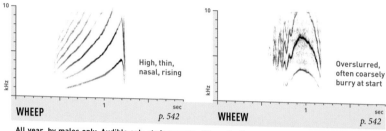

WHEEP High, thin, nasal, rising *p. 542*

WHEEW Overslurred, often coarsely burry at start *p. 542*

All year, by males only. Audible only at close range. Wheep is given frequently in a variety of situations.
Wheew is associated mostly with alarm and the urge to fly. In courtship, males give a variety of plastic
notes, all with similar distinctive thin, nasal quality, easily mistaken for polyphonic.

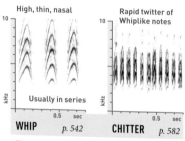

WHIP High, thin, nasal. Usually in series *p. 542*

CHITTER Rapid twitter of Whiplike notes *p. 582*

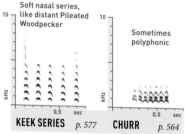

KEEK SERIES Soft nasal series, like distant Pileated Woodpecker *p. 577*

CHURR Sometimes polyphonic *p. 564*

These and similar calls given mostly Sept.–May,
by males in courtship, along with a very soft, low,
coughing Whup (not shown).

These and similar calls given all year by females, in courtship and in family contact. Somewhat variable and plastic; never loud.

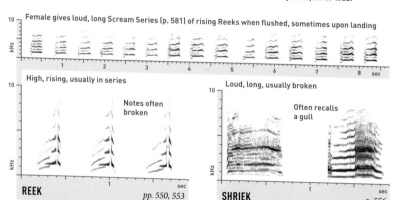

Female gives loud, long Scream Series (p. 581) of rising Reeks when flushed, sometimes upon landing

REEK High, rising, usually in series. Notes often broken *pp. 550, 553*

SHRIEK Loud, long, usually broken. Often recalls a gull *p. 556*

All year, by female only, in alarm and in pair contact. Plastic and variable but distinctive; by far
the species' most common call.

Mostly Aug.–May, by females only, in courtship
and in long-distance pair contact. Highly plastic.
Usually given singly at long intervals.

BLUE-WINGED TEAL

Spatula discors

A small duck of shallow ponds and wetlands. Blue shoulder patches, visible on both sexes in flight, are shared with Cinnamon Teal and Northern Shoveler.

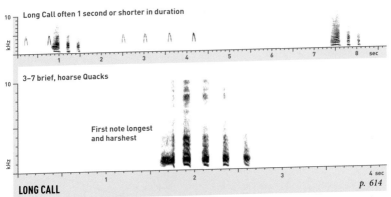

Long Call often 1 second or shorter in duration

3–7 brief, hoarse Quacks

First note longest and harshest

LONG CALL p. 614

All year, by females only, likely in pair contact. Individual females have fairly distinctive versions, but these are plastic, especially in number of notes. Shortness of notes can be useful in identification; Long Calls of Gadwall and Northern Shoveler can be quite similar.

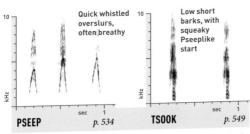

Quick whistled overslurs, often breathy

Low short barks, with squeaky Pseeplike start

PSEEP p. 534

TSOOK p. 549

Pseep given all year, by males, in courtship and other contexts, sometimes in series. Tsook apparently by males, possibly during territorial or courtship conflicts; may grade into Pseep.

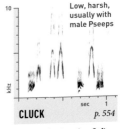

Low, harsh, usually with male Pseeps

CLUCK p. 554

By courting females. Soft and rather plastic; never particularly fast.

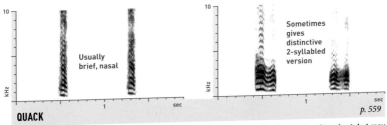

Usually brief, nasal

Sometimes gives distinctive 2-syllabled version

QUACK p. 559

All year, by females only, in contact and alarm. Somewhat variable and quite plastic; often short, but may become longer and rougher in agitation. Long series of Quacks sometimes given by females in spring, especially at dusk.

Cinnamon Teal

Spatula cyanoptera

Found on shallow ponds with emergent vegetation. Closely related to Blue-winged. Females similar, but Cinnamon usually has larger bill and plainer face.

Long Call (p. 614): brief, nasal, with stuttering, syncopated rhythm

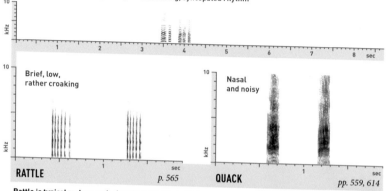

Brief, low, rather croaking

RATTLE

p. 565

Nasal and noisy

QUACK

pp. 559, 614

Rattle is typical male sound, given in courtship, in flight, and in alarm; completely unlike any other dabbling duck sound. Quacks, by females in alarm, are variable and plastic, but average medium-harsh. Females give Long Call infrequently; only available recording shown.

Northern Shoveler

Spatula clypeata

Found on shallow ponds. Named for its very large bill, which is immediately distinctive in both sexes. Closely related to Blue-winged and Cinnamon Teal.

Long Call (p. 614): 3-8 notes, first one usually longest, the rest in rapid series

Quick, swallowed squeaky bark

TSOOK

p. 549

Generally long, nasal, slightly harsh

Sometimes distinctively 2-syllabled

QUACK

p. 559

Tsook is typical male sound, given in courtship, in flight, and in alarm; some versions more nasal and barking. In flight, male wings make a low beating sound. Female Quacks rather variable and plastic, ranging from nasal to harsh; sometimes given in long series in spring, especially at dusk. Long Call infrequent.

GADWALL

Mareca strepera

Breeds in dense wetland vegetation or on islands; widespread in migration and winter. Note white patch on trailing edge of wing.

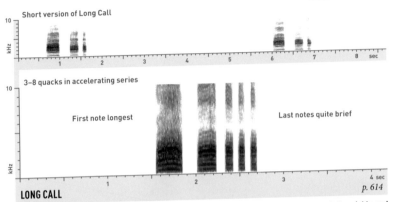

Short version of Long Call

3–8 quacks in accelerating series

First note longest

Last notes quite brief

LONG CALL *p. 614*

Apparently all year, in long-distance contact by females only, but rarely given. Somewhat variable and plastic. Quality typically quite nasal, slightly harsh, without much change in pitch between notes. When 5 or more notes are present, note rapid pace at end.

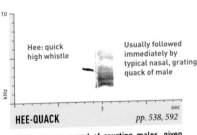

Hee: quick high whistle

Usually followed immediately by typical nasal, grating quack of male

HEE-QUACK *pp. 538, 592*

Most common sound of courting males, given while rearing up out of water with bill pressed to chest. Somewhat plastic.

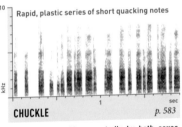

Rapid, plastic series of short quacking notes

CHUCKLE *p. 583*

Given in courtship, reportedly by both sexes. Much like Mallard Chuckle.

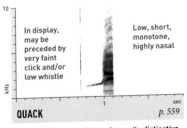

In display, may be preceded by very faint click and/or low whistle

Low, short, monotone, highly nasal

QUACK *p. 559*

Most common sound by males; quite distinctive.

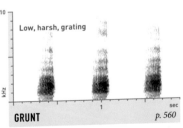

Low, harsh, grating

GRUNT *p. 560*

Given by females in alarm. Generally harsher than Mallard female Quack.

American Wigeon

Mareca americana

Found in shallow ponds, bays, estuaries, and city parks, sometimes in large flocks. Sometimes called "Baldpate" because of the male's white forehead.

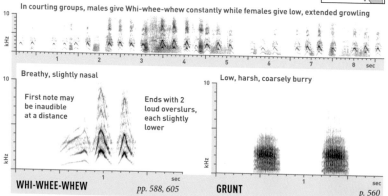

In courting groups, males give Whi-whee-whew constantly while females give low, extended growling

Breathy, slightly nasal

First note may be inaudible at a distance

Ends with 2 loud overslurs, each slightly lower

WHI-WHEE-WHEW *pp. 588, 605*

Low, harsh, coarsely burry

GRUNT *p. 560*

Whi-whee-whew given all year by males, in alarm, contact, flight, and courtship. Slightly plastic, but highly distinctive. Most versions are 3-noted; some have 2 or 4 notes. Grunts given only by females, in alarm or flight. Female rarely gives a 3- to 5-note Long Call like other ducks; quality apparently like typical Grunts.

Eurasian Wigeon

Mareca penelope

The Old World counterpart of American Wigeon, and closely related to it; the two occasionally hybridize. Rare in winter, usually with American Wigeons.

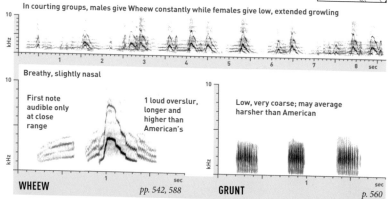

In courting groups, males give Wheew constantly while females give low, extended growling

Breathy, slightly nasal

First note audible only at close range

1 loud overslur, longer and higher than American's

WHEEW *pp. 542, 588*

Low, very coarse; may average harsher than American

GRUNT *p. 560*

Wheew given all year by males, in alarm, contact, flight, and courtship. Slightly individual and plastic, but highly distinctive. Grunts given only by females, in alarm or flight. Female rarely gives a 3- to 5-note Long Call like other ducks; quality apparently like typical Grunts.

NORTHERN PINTAIL

Anas acuta

A slender, elegant duck, named for the male's elongated central tail feathers. Common on shallow water. Note all-dark bill and unmarked face of female.

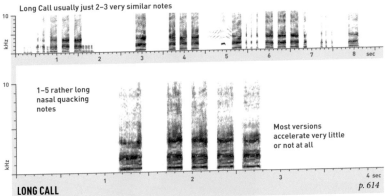

Long Call usually just 2–3 very similar notes

1–5 rather long nasal quacking notes

Most versions accelerate very little or not at all

LONG CALL *p. 614*

All year, by females only, likely in pair contact. Slightly variable and somewhat plastic; in some versions the individual notes are squeakier, vaguely 2-noted. Most versions are distinctively slow. Little distinction in this species between brief Long Calls and short series of Quacks.

Wheew: breathy, often barely audible

Hee: loud

HEE / WHEEW *pp. 538, 542*

By males in display. Hee given halfway through Wheew. Hee lower than Green-winged Teal's.

Grunt: faint, possibly mechanical

Hee: breathy, slightly higher than Green-winged Teal's

GRUNT-HEE *pp. 538, 592*

By males in display, with a flick of the head. Often follows Hee / Wheews; does not carry as far.

Given in short rapid churring bursts

CHUCKLE *p. 583*

By courting females, often at great length. Variable; ranges from nasal and purring to slightly noisy.

Hoarse nasal version

QUACK *p. 559*

Harsher version

Most versions noticeably upslurred

QUACK *p. 559*

All year, by females only, in contact and alarm. Somewhat variable and quite plastic. Generally less frequent than Chuckle in this species, but long series of Quacks sometimes given by females in spring. Most versions as rough as or rougher than Mallard Quack.

GREEN-WINGED TEAL

Anas crecca

Our smallest dabbling duck. Eurasian form, rare on coasts, has horizontal white stripe on wing, no vertical white bar on side; sounds similar.

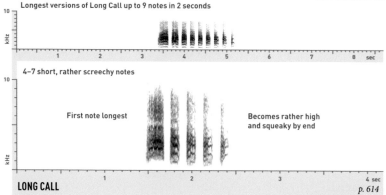

Longest versions of Long Call up to 9 notes in 2 seconds

4–7 short, rather screechy notes

First note longest

Becomes rather high and squeaky by end

LONG CALL   *p. 614*

All year, by females only, likely in pair contact. Slightly variable and somewhat plastic. On average the highest Long Call of any duck in our area; squeaky tone of last few notes distinctive when present.

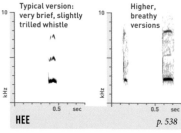

Typical version: very brief, slightly trilled whistle

Higher, breathy versions

HEE   *p. 538*

Mostly Sept.–Mar.; plastic. Given by courting males; different versions apparently accompany different courtship displays.

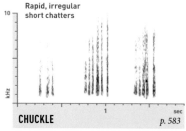

Rapid, irregular short chatters

CHUCKLE   *p. 583*

Mostly Sept.–Mar.; plastic. Given by courting females, usually with male Hees. Male gives a similar call during bill-up display.

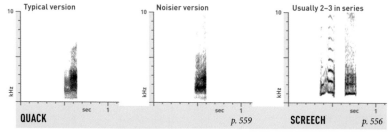

Typical version

Noisier version

Usually 2–3 in series

QUACK   *p. 559*   **SCREECH**   *p. 556*

All year, by females only, in alarm and contact. Somewhat plastic; may become rougher in agitation. Sometimes vaguely 2-syllabled. Long series of Quacks sometimes given by females in spring. Screeches given in aggression against males; may grade into Long Call.

MALLARD

Anas platyrhynchos

The world's most widespread duck. As in many ducks, females tends to be more vocal than males. Late summer males resemble females.

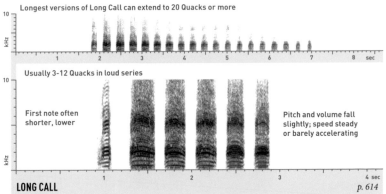

Longest versions of Long Call can extend to 20 Quacks or more

Usually 3-12 Quacks in loud series

First note often shorter, lower

Pitch and volume fall slightly; speed steady or barely accelerating

LONG CALL — p. 614

All year, especially Sept.–Apr., by females only. Somewhat variable and plastic, especially in number of notes. Given in long-distance contact with distant mate or with other groups of Mallards. Tone quality variable, but usually strongly nasal and only slightly harsh.

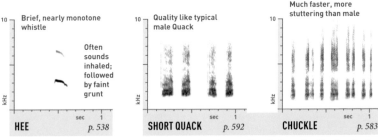

Brief, nearly monotone whistle

Often sounds inhaled; followed by faint grunt

HEE — p. 538

Quality like typical male Quack

SHORT QUACK — p. 592

Much faster, more stuttering than male

CHUCKLE — p. 583

Sept.–Mar., by males displaying in small groups. Slightly variable and plastic.

Mostly Sept.–Mar., in courtship; similar soft calls when feeding. Short Quacks given by males in plastic series. Chuckle given by females courted or pursued by males, including on the wing.

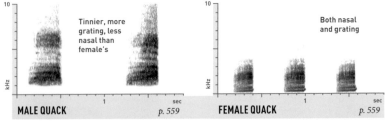

Tinnier, more grating, less nasal than female's

MALE QUACK — p. 559

Both nasal and grating

FEMALE QUACK — p. 559

All year; most common call, given loudly in alarm and contact, often upon takeoff or in flight. Usually in slow series. Male and female versions differ, often quite noticeably. Females Quack loudly and persistently for long periods in early spring when searching for nest sites.

MEXICAN DUCK

Anas [platyrhynchos] diazi

Currently a subspecies of Mallard, but may deserve species status. Male like female Mallard, but darker, with brown tail, olive bill. Hybridizes with Mallard.

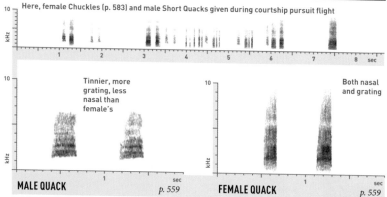

Here, female Chuckles (p. 583) and male Short Quacks given during courtship pursuit flight

Tinnier, more grating, less nasal than female's

MALE QUACK   *p. 559*

Both nasal and grating

FEMALE QUACK   *p. 559*

All vocalizations and displays apparently similar to those of Mallard, but few recordings available.

RING-NECKED DUCK

Aythya collaris

Often found on wooded ponds. Note distinctive bill markings, peaked head shape, and, in males, bright white vertical stripe on side of breast.

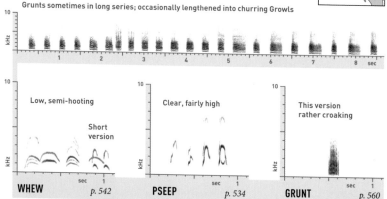

Grunts sometimes in long series; occasionally lengthened into churring Growls

Low, semi-hooting

Short version

WHEW   *p. 542*

Clear, fairly high

PSEEP   *p. 534*

This version rather croaking

GRUNT   *p. 560*

Several different sounds apparently by males in courtship, but more study needed. Whew like those of scaup, but sometimes longer. Pseeps recall Blue-winged Teal, but lower. Grunts, by females, vary from croaking to fairly clear, nasal, and quacking.

LESSER SCAUP

Aythya affinis

Very similar to Greater Scaup, but note slight peak on crown just behind eye. Generally more common than Greater, especially inland.

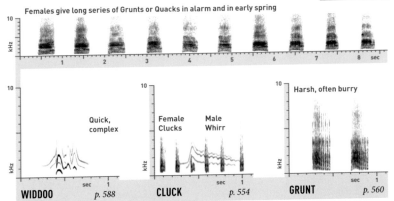

Females give long series of Grunts or Quacks in alarm and in early spring

WIDDOO — Quick, complex — *p. 588*

CLUCK — Female Clucks / Male Whirr — *p. 554*

GRUNT — Harsh, often burry — *p. 560*

Widdoo given mostly Feb.–May, by males displaying with backward toss of head. Whirr (p. 542) is likely one of several male courtship sounds; more study needed. Courting females give Cluck, sometimes run into slow Chuckles (p. 583). All sounds soft except for Grunt, often more of a Quack (p. 559), given by females.

GREATER SCAUP

Aythya marila

Resembles Lesser Scaup, but slightly larger, with more rounded head shape. In flight, wing stripe more extensively white. Most winter on salt water.

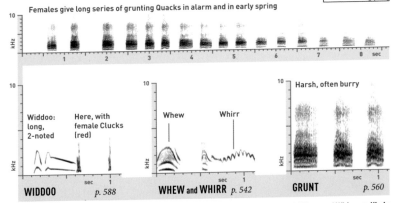

Females give long series of grunting Quacks in alarm and in early spring

WIDDOO — Widdoo: long, 2-noted. Here, with female Clucks (red) — *p. 588*

WHEW and WHIRR — Whew / Whirr — *p. 542*

GRUNT — Harsh, often burry — *p. 560*

Widdoo given mostly Feb.–May, by males displaying with backward toss of head. Whew and Whirr are likely other male courtship sounds, but more study needed. Females give Clucks (p. 554) in courting groups. All sounds fairly soft except for Grunt, given by females.

REDHEAD

Aythya americana

Breeds mostly in prairie marshes. As in many other ducks, females often lay eggs in nests of other Redheads, or even other duck species.

Females give long series of quacking Grunts in alarm and in early spring

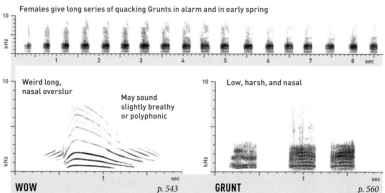

Weird long, nasal overslur

May sound slightly breathy or polyphonic

WOW *p. 543*

Low, harsh, and nasal

GRUNT *p. 560*

Wow is male display sound, given frequently Feb.–May, with a backward toss of the head. Distinctive, uniform, and stereotyped; longer than any similar sound by other *Aythya* ducks. Male may give some other sounds in display; more study needed. Female Grunts plastic, but average very rough.

CANVASBACK

Aythya valisineria

Larger than Redhead, with an elegant, sloping bill profile. Male has a striking white back. Population has fluctuated in recent years.

Females give long series of quacking Grunts in alarm and in early spring

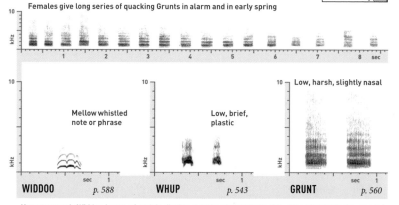

Mellow whistled note or phrase

WIDDOO *p. 588*

Low, brief, plastic

WHUP *p. 543*

Low, harsh, slightly nasal

GRUNT *p. 560*

Not very vocal. Widdoo is one of male's display sounds, given mostly Feb.–May. Highly plastic, possibly grading into other, poorly known types of display sounds. Highly plastic Whups also given during courtship; unknown whether by males or females. Female Grunts plastic, but average very rough.

Surf Scoter

Melanitta perspicillata

Our most strikingly patterned scoter. Nests on lakes in boreal forest; winters in flocks along seacoasts. Rare inland. Rarely vocalizes.

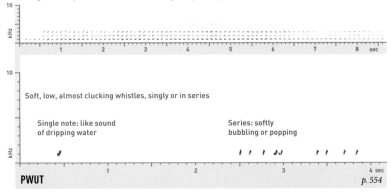

Wing Whistle (p. 574) audible at times in flight, reportedly from adult males

Soft, low, almost clucking whistles, singly or in series

Single note: like sound of dripping water

Series: softly bubbling or popping

PWUT

p. 554

Pwut given at least Mar.–June, by displaying males, but does not carry far. Displaying male stretches neck out, sometimes bringing bill to chest. Single notes and series likely accompany different displays. Females reportedly give crowlike Croaks in display and in nest defense; no recordings available.

White-winged Scoter

Melanitta fusca

Our largest scoter. White wing patch distinctive. Nests on lakes in boreal forest; winters in flocks along seacoasts. Rare inland. Rarely vocalizes.

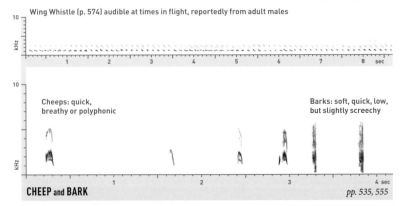

Wing Whistle (p. 574) audible at times in flight, reportedly from adult males

Cheeps: quick, breathy or polyphonic

Barks: soft, quick, low, but slightly screechy

CHEEP and BARK

pp. 535, 555

Vocalizations rarely heard; sources differ on which sex gives which sounds. Cheeps and Barks known only from group interactions on breeding grounds; both sounds audible only at close range. Cheep may accompany chin-lift display. Females may croak or gurgle in nest defense; no recordings available.

BLACK SCOTER

Melanitta americana

Our smallest scoter. Nests on lakes in boreal forest; winters in flocks along seacoasts. Rare inland. Males vocalize frequently.

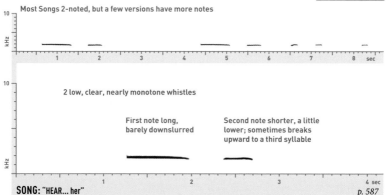

Most Songs 2-noted, but a few versions have more notes

2 low, clear, nearly monotone whistles

First note long, barely downslurred

Second note shorter, a little lower; sometimes breaks upward to a third syllable

SONG: "HEAR... her"

p. 587

Song given at least Nov.–June, by males; carries for long distances. Plaintive quality distinctive. Sometimes given in flight by flushed birds. Females reportedly give nasal or growling calls; may give harsh Quacks like scaup. Wing Whistle (p. 539) audible at times in flight, reportedly from adult males.

BUFFLEHEAD

Bucephala albeola

Our smallest diving duck. Nests in tree cavities, usually old flicker nests, in mixed woods. Winters on fresh and salt water. Rarely vocalizes.

In courtship, female Barks sometimes strung into chuckling series

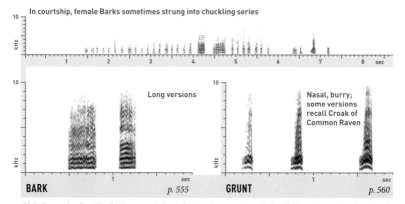

Long versions

Nasal, burry; some versions recall Croak of Common Raven

BARK   p. 555

GRUNT   p. 560

Male is nearly silent in display, except for soft slapping sounds made by flicking wings. Courting female gives plastic Barks, often in series, while swimming behind a displaying male. In alarm, female gives Grunts, some versions harsh and grating. Wings do not whistle in flight.

Common Goldeneye

Bucephala clangula

Breeds in tree cavities in boreal forests; winters on both fresh and salt water. Forehead slopes back from bill to peak above eye; bill of female mostly dark.

Wing Whistle (p. 574): in flight, male's wings make loud, clear whistled series

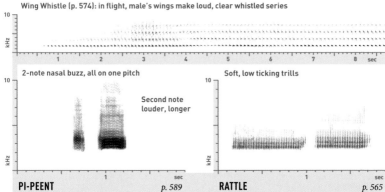

2-note nasal buzz, all on one pitch

Second note louder, longer

PI-PEENT *p. 589*

Soft, low ticking trills

RATTLE *p. 565*

Mostly Dec.–May, by males in courting groups on the water. Pi-peent is loudest and most distinctive sound; accompanies dramatic backward head throw. Rattles given with head stretched forward, nearly touching water.

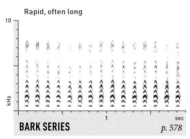

Rapid, often long

BARK SERIES *p. 578*

Mostly Apr.–June, by female, possibly in courtship, sometimes in flight. Recalls Pileated Woodpecker Keek Series, but lower, more nasal.

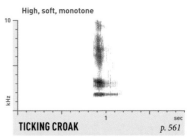

High, soft, monotone

TICKING CROAK *p. 561*

All year, by males, in alarm and in flight; may also play a role in courtship displays. Audible mostly at close range.

Harsh, nasal, burry

GRUNT *p. 560*

Low, cootlike

CROAK *p. 561*

By females. Grunts all year, in alarm and in flight; much like calls of *Aythya* ducks. Croaks not well known; possibly given mostly in nest defense. Both calls apparently plastic; may grade into one another and into Bark Series.

Barrow's Goldeneye

Bucephala islandica

Breeds in tree cavities in boreal and montane forests. Winters along coast. Peak of head is forward of eye; female's bill is mostly yellow.

Wing Whistle (p. 574): slightly lower than Common Goldeneye's

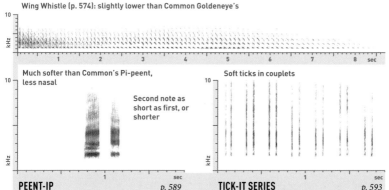

Much softer than Common's Pi-peent, less nasal

Second note as short as first, or shorter

PEENT-IP   p. 589

Soft ticks in couplets

TICK-IT SERIES   p. 593

Mostly Dec.–May, by males in courting groups on the water. Peent-ip accompanies dramatic backward head throw; sometimes preceded by soft short Hiss. Tick-it Series apparently given with head stretched forward, nearly touching water, or during head throw. Both sounds difficult to hear from a distance, especially Tick-it Series.

Only available example is slower than Common Goldeneye's

BARK SERIES   p. 578

Mostly Apr.–June, by females, possibly in courtship, sometimes in flight.

High, soft, monotone

TICKING CROAK   p. 561

All year, by males, in alarm and in flight; may also play a role in courtship displays. Audible mostly at close range.

Harsh, nasal, burry

GRUNT   p. 560

Low, cootlike

Here, followed by short nasal barks

CROAK   p. 561

By females. Grunts all year, in alarm and in flight; much like calls of *Aythya* ducks. Croaks not well known; possibly given mostly in nest defense. Both calls apparently plastic; may grade into one another and into Bark Series.

Long-tailed Duck

Clangula hyemalis

Breeds in the Arctic; winters primarily along seacoasts. Rare inland away from the Great Lakes. Molt schedule complex, resulting in a variety of plumages.

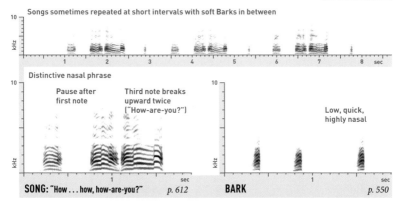

Songs sometimes repeated at short intervals with soft Barks in between

Distinctive nasal phrase

Pause after first note

Third note breaks upward twice ("How-are-you?")

Low, quick, highly nasal

SONG: "How . . . how, how-are-you?"   *p. 612*   BARK   *p. 550*

Song given all year, especially Jan.–June, reportedly only by males. Rather uniform; variation between males is mostly in final syllable. Sometimes given in flight. Barks generally soft, given in alarm or contact. Very soft Crackle (p. 557), possibly nonvocal, sometimes given with Barks in alarm.

Harlequin Duck

Histrionicus histrionicus

Strikingly patterned and social. Breeds along clear, fast-flowing streams; winters on rocky seacoasts, in or near the crashing surf. Rare inland.

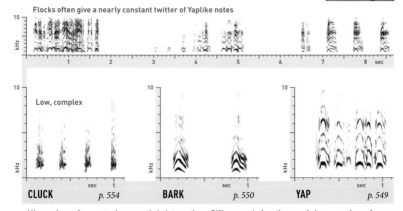

Flocks often give a nearly constant twitter of Yaplike notes

Low, complex

CLUCK   *p. 554*   BARK   *p. 550*   YAP   *p. 549*

All sounds are frequent, given mostly in interactions. Differences in function poorly known; voices of sexes may differ. Yaps are most frequent sounds; highly plastic, grading into low Barks and Clucks. Also gives soft, low Moans (not shown; p. 552).

COMMON MERGANSER

Mergus merganser

Nests on lakes and along clear rivers; winters on both fresh and salt water. Like other mergansers, dives to catch fish with serrated bill.

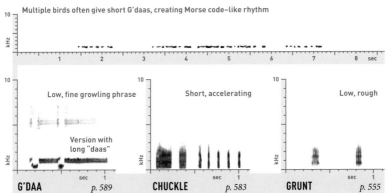

Multiple birds often give short G'daas, creating Morse code–like rhythm

| Low, fine growling phrase | Short, accelerating | Low, rough |

Version with long "daas"

G'DAA   *p. 589*   **CHUCKLE**   *p. 583*   **GRUNT**   *p. 555*

G'daa given mostly Dec.–May by males in courting groups on the water, while stretching neck and/or tossing head. Several different variants may correspond to different displays. Chuckle given infrequently by females during courtship. Grunts given all year, at least by females, in alarm, often in flight.

RED-BREASTED MERGANSER

Mergus serrator

Breeds on lakes and rivers; winters primarily on salt water, but some winter inland. Slightly smaller than Common; both sexes have ragged crest.

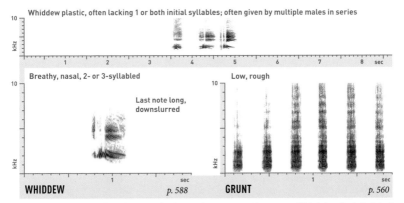

Whiddew plastic, often lacking 1 or both initial syllables; often given by multiple males in series

Breathy, nasal, 2- or 3-syllabled

Last note long, downslurred

Low, rough

WHIDDEW   *p. 588*   **GRUNT**   *p. 560*

Whiddew given mostly Dec.–May by males in courting groups on the water, with brief vertical stretch of head and bill. Distinctive; from a distance, may sound catlike. Females reported to growl in courtship; no recordings available. Grunts given all year, at least by females, in alarm, often in flight.

Hooded Merganser

Lophodytes cucullatus

Nests in tree cavities around wooded ponds. Male's white head patch is a large semicircle when crest is raised, a thin stripe when it is lowered.

Wing Whistle (p. 574): in flight, male's wings make soft whistles, much higher than those of other ducks

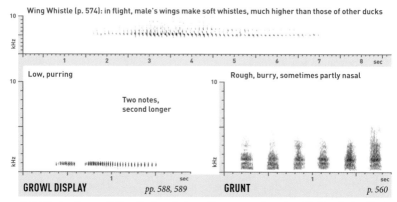

Low, purring

Two notes, second longer

GROWL DISPLAY   *pp. 588, 589*

Rough, burry, sometimes partly nasal

GRUNT   *p. 560*

Growl Display given mostly Dec.–May by males in courting groups on the water, while throwing head backward to touch the back. Does not carry far. Females reported to make low Croaks in courtship; no recordings available. Grunts given all year, at least by females, in alarm, often in flight.

Ruddy Duck

Oxyura jamaicensis

Breeds in marshes and on ponds with a mix of emergent vegetation and open water. Aggressive in behavior; highly aquatic, almost unable to walk on land.

In rushing display flight, males loudly splash water with feet, between Bubble Displays

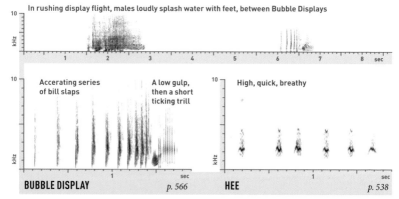

Accerating series of bill slaps

A low gulp, then a short ticking trill

BUBBLE DISPLAY   *p. 566*

High, quick, breathy

HEE   *p. 538*

Unlike other ducks, displays only on breeding grounds. Bubble Display mostly May–July, by courting males with tail vertical; bill slapped rapidly down against chest, frothing the water. Whistles given by courting females. Females also reportedly give nasal calls in alarm; no recordings available.

CHICKENLIKE BIRDS (Order Galliformes)

This diverse order contains several families of terrestrial birds ranging from medium-sized to very large. Often known as "game birds," they tend to fly only short distances, usually when flushed by a predator.

Most birds in this group have well-developed vocal repertoires. All vocalizations appear to be innate. Many species, especially peacocks, turkeys, and grouse, have striking courtship displays that involve fanned tails, wing flaps, or elaborate dances, often with distinctive sounds. Some species, including chachalacas and some quail, perform synchronized duets or choruses.

Domestic and feral game birds

Several species in this group have been domesticated, and most are hunted; many have been introduced to North America for hunting purposes, including Chukar, Himalayan Snowcock, Ring-necked Pheasant and Gray Partridge. In several species, some populations continue to be maintained or supplemented by released birds. Even some native species, such as Northern Bobwhite, are sometimes released far from their native range in order to be hunted.

The Domestic Chicken, Indian Peafowl, and Helmeted Guineafowl are among the most common bird species kept in captivity, and all three sometimes roam free in fields, yards, and even towns. Although they have not established self-sustaining feral populations in North America, their vocalizations are a significant component of the soundscape in many areas.

DOMESTIC CHICKEN

Gallus gallus

Ubiquitous and familiar. Domestic birds derived from the Red Junglefowl of southeast Asia, but show huge variation in plumage; vocalizations apparently still strongly resemble those of wild forms.

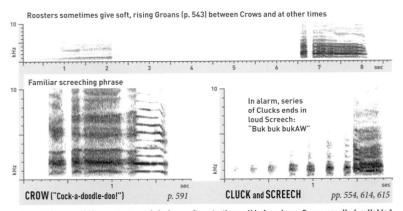

Roosters sometimes give soft, rising Groans (p. 543) between Crows and at other times

Familiar screeching phrase

CROW ("Cock-a-doodle-doo!")   *p. 591*

In alarm, series of Clucks ends in loud Screech: "Buk buk bukAW"

CLUCK and SCREECH   *pp. 554, 614, 615*

Crows given mostly by roosters, rarely by hens, often starting well before dawn. Crows usually 4-syllabled, sometimes 2- or 3-syllabled, ranging from very screechy to nearly clear. Clucks given in alarm, sometimes in long series, often culminating in loud Screech.

Indian Peafowl

Pavo cristatus

Common in zoos, often wandering into adjacent neighborhoods. Some apparently breed in the wild in parts of urban southern California and possibly south Florida.

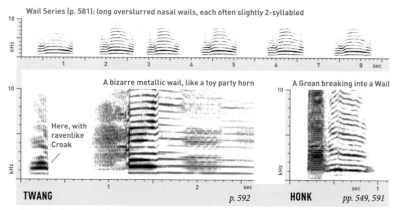

Wail Series (p. 581): long overslurred nasal wails, each often slightly 2-syllabled

A bizarre metallic wail, like a toy party horn

Here, with ravenlike Croak

TWANG   *p. 592*

A Groan breaking into a Wail

HONK   *pp. 549, 591*

Wail Series (top) is most familiar call; loud, frequent, and distinctive. Male with tail spread in display often silent, but may give faster, lower version of Wail Series, and less frequently Twang, sometimes with a soft Croak. Honks and similar calls may be given in alarm or long-distance contact.

Helmeted Guineafowl

Numida meleagris

Native to Africa; widely kept in captivity, sometimes roaming free. Generally does not breed in the wild in the West. Domestic birds may have white or pied plumage.

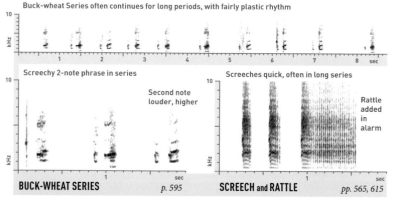

Buck-wheat Series often continues for long periods, with fairly plastic rhythm

Screechy 2-note phrase in series

Second note louder, higher

BUCK-WHEAT SERIES   *p. 595*

Screeches quick, often in long series

Rattle added in alarm

SCREECH and RATTLE   *pp. 565, 615*

All calls loud and generally unmusical. Buck-wheat Series given all year, reportedly only by female; variable and plastic, tone varying from screechy to nearly whistled. Screeches and Rattles reportedly given by both sexes; male often gives long series of quick Screeches. Rattles given by agitated birds.

WILD TURKEY

Meleagris gallopavo

Found in wooded areas and adjacent fields. Range shrank in nineteenth century owing to hunting; now larger than ever thanks to reintroductions.

Some versions of Yelp Series are clear, almost whistled

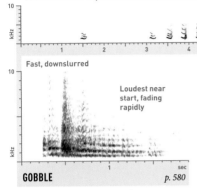

Fast, downslurred

Loudest near start, fading rapidly

GOBBLE *p. 580*

Unmistakable. By males, often from a tree, to attract a mate; often echoed by other males. Male in strutting display nearly silent.

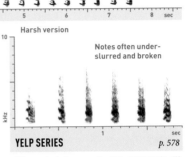

Harsh version

Notes often under-slurred and broken

YELP SERIES *p. 578*

All year, by both sexes. Variable and plastic; different versions may serve functions from alarm to reassembly of scattered flocks.

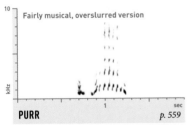

Fairly musical, overslurred version

PURR *p. 559*

All year. Variable, ranging from musical trills to harsh rattles; often nasal. Used to maintain spacing in flocks, in threats, and in alarm.

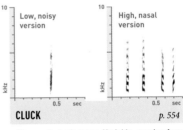

Low, noisy version

High, nasal version

CLUCK *p. 554*

All year, by both sexes. Variable, ranging from soft, noisy versions in contact to high, fast, nasal Keklike versions in high alarm.

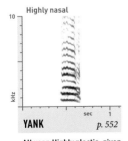

Highly nasal

YANK *p. 552*

All year. Highly plastic; given during flock interactions, usually with Purrs.

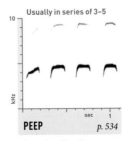

Usually in series of 3–5

PEEP *p. 534*

As early as May, by young birds separated from hen; grades into Yelp by fall.

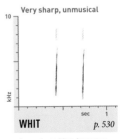

Very sharp, unmusical

WHIT *p. 530*

As early as May, by young birds in alarm, often with Peeps.

MOUNTAIN QUAIL

Oreortyx pictus

Our largest quail. Fairly common but shy resident in dense mountain scrub, far more often heard than seen. Has declined in northeastern part of range.

Song usually consists of single cries at wide intervals, but can run into series in high excitement

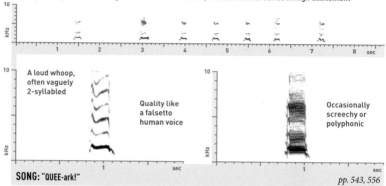

A loud whoop, often vaguely 2-syllabled

Quality like a falsetto human voice

Occasionally screechy or polyphonic

SONG: "QUEE-ark!"

pp. 543, 556

Mostly Jan.–June, by both mated and unmated males. Unmistakable in range; audible from a tremendous distance. Female sometimes responds to male Song with a soft variation on the Keek Series. In aggressive encounters, Songlike notes can run into faster series, often with clucking or Keeklike notes in between.

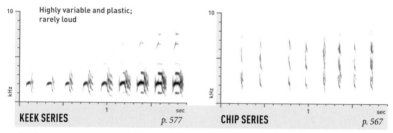

Highly variable and plastic; rarely loud

KEEK SERIES *p. 577*

CHIP SERIES *p. 567*

A variety of series calls given; two typical versions are shown. Most are rather soft and plastic. Keek Series and similar sounds given all year, when members of a flock are separated, and at dawn or dusk. Chip Series may represent higher agitation; sometimes given upon flushing.

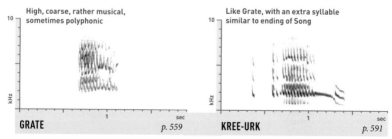

High, coarse, rather musical, sometimes polyphonic

Like Grate, with an extra syllable similar to ending of Song

GRATE *p. 559*

KREE-URK *p. 591*

All year, in high excitement and alarm. Highly plastic; rarely loud. Kree-urk is standard alarm call; grades into Grate, associated with higher alarm. Both calls often given with Chip Series. As birds calm down after a disturbance, Kree-urk may become soft nasal mewing sounds without a grating portion.

SCALED QUAIL

Callipepla squamata

Found in dry, grassy habitats with some cover, sometimes including xeriscaped residential areas. Quite social, forming coveys outside the breeding season.

Song consists of single Screeches repeated at widely spaced intervals

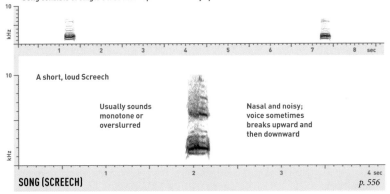

A short, loud Screech

Usually sounds monotone or overslurred

Nasal and noisy; voice sometimes breaks upward and then downward

SONG (SCREECH)   *p. 556*

Mostly Apr.–Aug., by unmated males, usually from a conspicuous perch at mid-height, such as a fencepost. Slightly variable and plastic. Kuk-curr calls sometimes interspersed. Can resemble notes from Yelp Series of Northern Bobwhite, but harsher and given at longer intervals.

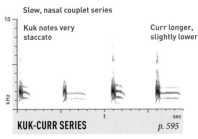

Slow, nasal couplet series

Kuk notes very staccato

Curr longer, slightly lower

KUK-CURR SERIES   *p. 595*

All year, by both sexes, when separated from other quail. Series often starts very softly and grows louder; can continue for long periods.

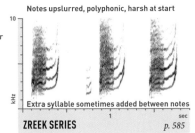

Notes upslurred, polyphonic, harsh at start

Extra syllable sometimes added between notes

ZREEK SERIES   *p. 585*

All year, but mostly during breeding season, by males in aggressive display, with head toss; rare from females.

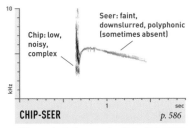

Chip: low, noisy, complex

Seer: faint, downslurred, polyphonic (sometimes absent)

CHIP-SEER   *p. 586*

All year, by both sexes, in alarm. Sometimes given upon flushing. Whistled portion often audible only at close range; see Chip (p. 531).

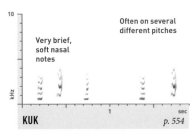

Very brief, soft nasal notes

Often on several different pitches

KUK   *p. 554*

All year, by both sexes, in close contact. Quite plastic; never loud. Low, rough versions sometimes given upon flushing.

CALIFORNIA QUAIL

Callipepla californica

Common in brushy habitats, including residential areas. Native to California and parts of adjacent states, but widely introduced elsewhere in the West.

Hu-HAA-how: usually given in series; first few renditions often partial, occasionally 4-noted

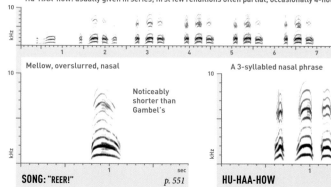

Mellow, overslurred, nasal

Noticeably shorter than Gambel's

SONG: "REER!"
p. 551

Mostly Feb.–July, by males, from fencepost or other low perch; rarely by females. Stereotyped, uniform. Soft nasal Pew calls often interspersed.

A 3-syllabled nasal phrase

Middle note highest, loudest, longest

HU-HAA-HOW
p. 591

All year, by both sexes, when separated or when covey is about to move. Highly distinctive, but geographically variable and slightly plastic.

Very quick, unmusical, overslurred

Here, mixed with soft Kuks

WIP
p. 550

Mostly Feb.–July, by males, especially dominant ones, in aggression toward other males. Also given during copulation. Rather plastic.

Quick, loud, overslurred

Downslurred nasal ending

SQUEAK
p. 548

Mostly Feb.–July, by males only, in aggression toward other males; also in duet with separated mate, overlapping last note of Hu-haa-how.

High, extremely sharp notes

Notes often start with slight nasal upslur

SPIT SERIES
p. 567

All year, by both sexes, in alarm; most common call. Plastic, becoming faster in high alarm; some versions are more Wiplike.

Mostly low, nasal notes

Grading into other calls

KUK
p. 554

All year, by both sexes, in close contact. Quite plastic; never loud. A large variety of other soft calls given in coveys at close range.

Gambel's Quail

Callipepla gambelii

Closely related to California Quail; the two rarely hybridize in narrow zone of overlap. Gambel's prefers slightly drier, grassier habitats.

Hu-HAA-ha-how: usually given in series; first few renditions often partial, frequently 3-noted

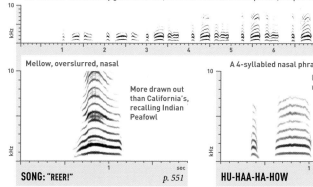

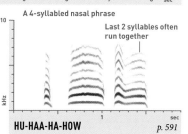

Mellow, overslurred, nasal

More drawn out than California's, recalling Indian Peafowl

SONG: "REER!" *p. 551*

Mostly Feb.–July, by unmated males, from fence-post or other low perch. Stereotyped, rather uniform. Can be vaguely 3-syllabled ("Ra-EE-uh").

A 4-syllabled nasal phrase

Last 2 syllables often run together

HU-HAA-HA-HOW *p. 591*

All year, by both sexes, when separated or when covey is about to move. Distinctive, but somewhat variable and plastic.

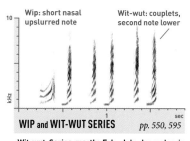

Wip: short nasal upslurred note

Wit-wut: couplets, second note lower

WIP and WIT-WUT SERIES *pp. 550, 595*

Wit-wut Series mostly Feb.–July, by males in close courtship or in aggression toward other males. Plastic; often shortened to single Wips.

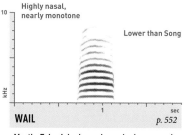

Highly nasal, nearly monotone

Lower than Song

WAIL *p. 552*

Mostly Feb.–July, by males only, in aggression toward other males; also in duet with separated mate, overlapping last note of Hu-haa-ha-how.

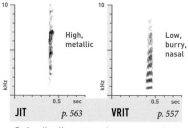

High, metallic

JIT *p. 563*

Low, burry, nasal

VRIT *p. 557*

Both calls all year, together or separately, by both sexes, in alarm. Jit grades into Witlike notes; often in series, including upon flushing.

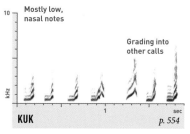

Mostly low, nasal notes

Grading into other calls

KUK *p. 554*

All year, by both sexes, in close contact. Quite plastic; never loud. A large variety of other soft calls given in coveys at close range.

NORTHERN BOBWHITE

Colinus virginianus

A social quail of forest edges, shelter-belts, open woods, and savannas. In many areas, large numbers are raised and released to be hunted.

Duet (p. 605): "Hoyp-woo . . . Bob . . . WHITE?," possibly by mated pairs

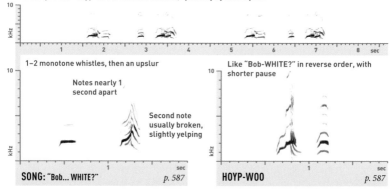

1–2 monotone whistles, then an upslur

Notes nearly 1 second apart

Second note usually broken, slightly yelping

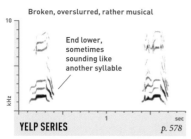

SONG: "Bob... WHITE?" *p. 587*

Highly distinctive. Mostly Apr.–Aug., by males, those either without mates or separated from mates. Females sing rarely and softly.

Like "Bob-WHITE?" in reverse order, with shorter pause

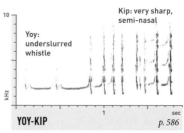

HOYP-WOO *p. 587*

All year, in series of 4–5, by flockmates reassembling or by females responding to male Song. Quite variable, somewhat plastic.

Broken, overslurred, rather musical

End lower, sometimes sounding like another syllable

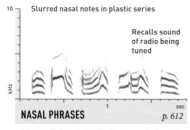

YELP SERIES *p. 578*

All year, in similar situations to Hoyp-woo; also at dawn by waking coveys. Variable, somewhat plastic.

Kip: very sharp, semi-nasal

Yoy: underslurred whistle

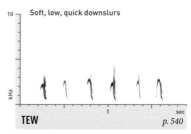

YOY-KIP *p. 586*

All year, in alarm. Yoy notes soft, sometimes nasal; Kips multiply as agitation rises, sometimes running into a sharp, chipping chatter.

Slurred nasal notes in plastic series

Recalls sound of radio being tuned

NASAL PHRASES *p. 612*

All year. Soft and plastic. Given mostly by males in aggressive encounters; also possibly in alarm.

Soft, low, quick downslurs

TEW *p. 540*

All year, in close contact, often with short wailing or squealing calls. Plastic; not very loud.

Montezuma Quail

Cyrtonyx montezumae

An elusive quail, most common in open, grassy oak woodlands, but ranging locally from shrubby deserts all the way up to pine forests.

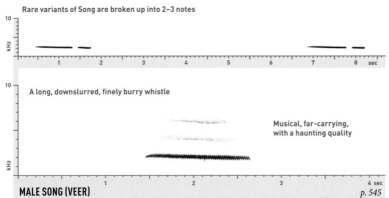

Rare variants of Song are broken up into 2–3 notes

A long, downslurred, finely burry whistle

Musical, far-carrying, with a haunting quality

MALE SONG (VEER) *p. 545*

Mostly Feb.–Sept., by males. May be heard at any time of day, but most frequently at dawn and dusk. May serve a territorial function; also sometimes given as birds attempt to reunite after covey is flushed.

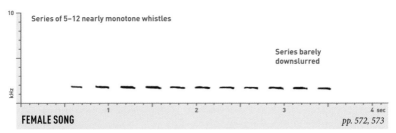

Series of 5–12 nearly monotone whistles

Series barely downslurred

FEMALE SONG *pp. 572, 573*

Medium-loud version given mostly Feb.–Sept., primarily at dawn and dusk, probably by unmated females. Causes nearby males to sing and approach. Quieter version given year-round by scattered birds, including immatures, seeking to reunite.

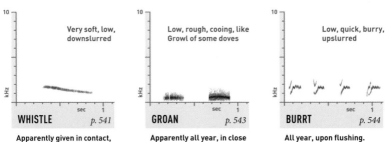

Very soft, low, downslurred

WHISTLE *p. 541*

Apparently given in contact, possibly by females. Some versions wheezy, nasal.

Low, rough, cooing, like Growl of some doves

GROAN *p. 543*

Apparently all year, in close contact.

Low, quick, burry, upslurred

BURRT *p. 544*

All year, upon flushing. Highly plastic, but generally fairly musical.

CHUKAR

Alectoris chukar

Native to Asia. Prefers dry, rocky slopes and canyons with little vegetation. Often escaped or released from captivity well away from established range in West.

Long Call: accelerating series of low Chuk notes, becoming louder and more syncopated in rhythm

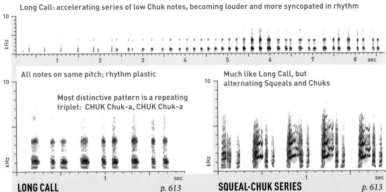

All notes on same pitch; rhythm plastic

Most distinctive pattern is a repeating triplet: CHUK Chuk-a, CHUK Chuk-a

LONG CALL   *p. 613*

Much like Long Call, but alternating Squeals and Chuks

SQUEAL-CHUK SERIES   *p. 613*

Vocal repertoire is large, but poorly represented in available recordings. Long Call is most frequently heard sound, given all year by both sexes, but especially Feb.–June by males. Some versions remain steady, but most develop characteristic stuttering rhythm. Squeal-Chuk Series apparently given in higher excitement.

HIMALAYAN SNOWCOCK

Tetraogallus himalayensis

Native to Asia. A small population of perhaps 1,000 birds is established above treeline in the Ruby Mountains of Nevada, in meadows near cliffs.

Long Call (p. 613): accelerating, rising series of Clucks, an overslurred whistle, then a few more Clucks

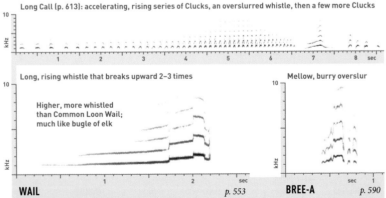

Long, rising whistle that breaks upward 2–3 times

Higher, more whistled than Common Loon Wail; much like bugle of elk

WAIL   *p. 553*

Mellow, burry overslur

BREE-A   *p. 590*

Wail is most frequently heard call in the wild; loud and far carrying. Long Call also loud and frequent, given by both sexes, possibly to unite distant birds. Softer Bree-a apparently by male in excitement or in display. In contact, gives rather high, peeping Clucks, singly or in series (not indexed).

Ring-necked Pheasant

Phasianus colchicus

Native to Asia. Prefers agricultural areas with some brushy or wooded cover nearby. In many areas, large numbers are raised and released to be hunted.

Flushing birds give loud, rapid series of Kuttuks, along with loud wingbeats

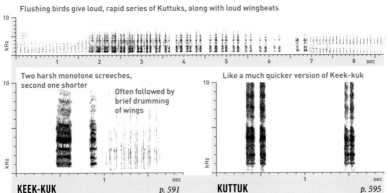

Two harsh monotone screeches, second one shorter

Often followed by brief drumming of wings

KEEK-KUK   *p. 591*

Like a much quicker version of Keek-kuk

KUTTUK   *p. 595*

Keek-kuk given all year, especially Mar.–June, by males only, from ground. Loud and far-carrying; some individual variation. Kuttuk given all year, by males only, in alarm. Rather plastic. Flushing females give only soft calls, or are silent. Chicks give high Seerlike whistles (not indexed).

Gray Partridge

Perdix perdix

Native to Europe. Found in grasslands and agricultural fields, especially of wheat and other grasslike crops. In flight, note short, rusty-colored tail.

Territorial males give Kee-raw at regular intervals

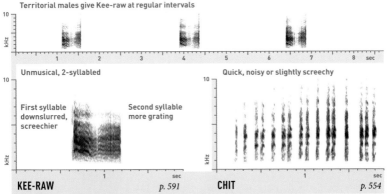

Unmusical, 2-syllabled

First syllable downslurred, screechier

Second syllable more grating

KEE-RAW   *p. 591*

Quick, noisy or slightly screechy

CHIT   *p. 554*

Kee-raw given all year, especially Feb.–June, by males. Somewhat variable. Often given in flight after flushing; these versions often shorter. Chit given all year, in alarm; given singly from ground, but often in series in high alarm, especially right before or after flushing.

SHARP-TAILED GROUSE

Tympanuchus phasianellus

Local in grasslands and shrub steppes, sometimes in bogs. Range has shrunk and numbers have declined, but still our most widespread prairie grouse.

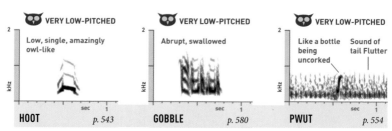

🦉 VERY LOW-PITCHED

Low, single, amazingly owl-like

HOOT   *p. 543*

🦉 VERY LOW-PITCHED

Abrupt, swallowed

GOBBLE   *p. 580*

🦉 VERY LOW-PITCHED

Like a bottle being uncorked   Sound of tail Flutter

PWUT   *p. 554*

Hoot given by lekking males, with neck sacs fully inflated and tail raised. Gobbles given in aggression toward other lekking males. Pwuts given with Tsooks in high-intensity dancing display for female, with wings spread; male shakes tail and stamps feet in rapid unison, creating a rattling Flutter (p. 566).

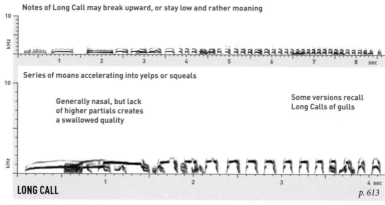

Notes of Long Call may break upward, or stay low and rather moaning

Series of moans accelerating into yelps or squeals

Generally nasal, but lack of higher partials creates a swallowed quality

Some versions recall Long Calls of gulls

LONG CALL   *p. 613*

Given during lekking displays by males, mostly during aggressive face-offs or fights with other males. Highly plastic, but usually not divided into distinct segments; and short snippets or single notes are frequently heard.

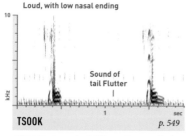

Loud, with low nasal ending

Sound of tail Flutter

TSOOK   *p. 549*

Given by aggressive lekking males, usually during or after dancing, possibly to interfere with sound of other males' Pwuts.

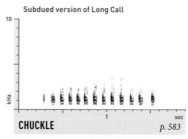

Subdued version of Long Call

CHUCKLE   *p. 583*

All year, upon flushing. Nearly identical to Greater Prairie-Chicken Chuckle.

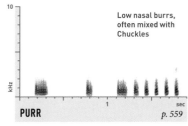

Low nasal burrs,
often mixed with
Chuckles

kHz

1   sec

PURR   *p. 559*

One of a variety of similar notes given by
alarmed hens with broods; prairie-chickens
likely give similar calls.

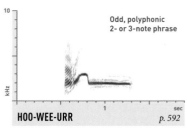

Odd, polyphonic
2- or 3-note phrase

kHz

1   sec

HOO-WEE-URR   *p. 592*

Poorly known. Apparently Aug.–Sept., by nearly
independent young separated from parents.
Prairie-chickens likely give similar calls.

LEKS AND LEKKING

All five North American species of "prairie
grouse" (Sharp-tailed Grouse, Greater and
Lesser Prairie-Chickens, and Greater and
Gunnison Sage-Grouse) exhibit a specialized
mating behavior called lekking. Males gather
each spring on traditional display sites, called
leks. Beginning before first light, the male
grouse perform elaborate display dances, inflating their throat sacs, spreading their tails,
and making a variety of unique sounds. These
displays are given for a few hours around
dawn, and sometimes in the evening. They
sometimes occur in fall, as well as in spring.

By displaying and fighting, males establish
a strict dominance hierarchy, and each claims
a small part of the lek as his territory, with the
highest-ranking males occupying the largest
territories near the center. After the male
hierarchy is established, females arrive to
mate primarily with the high-ranking males.

All North American prairie grouse have
declined dramatically in population over the
past century, primarily because of human-
caused changes in habitat. Lek mating may
make these species particularly vulnerable to
disturbances, and the tendency for a few
high-ranking males to do most of the mating
can result in low genetic diversity. The Heath
Hen, an East Coast subspecies of Greater
Prairie-Chicken, has been extinct since 1932.
The Gulf Coast subspecies, "Attwater's"
Prairie-Chicken, is critically endangered.

Typical display
postures of
prairie-chickens

GREATER PRAIRIE-CHICKEN

Tympanuchus cupido

Uncommon and local in mid- to tall-grass prairie and oak savanna; populations have disappeared from 12 states and provinces. Neck sac yellowish.

Ooh-loo-woo: low, 3-syllabled moaning coo, all on nearly the same pitch

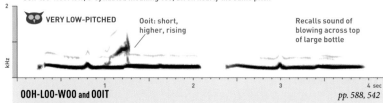

VERY LOW-PITCHED

Ooit: short, higher, rising

Recalls sound of blowing across top of large bottle

OOH-LOO-WOO and OOIT

pp. 588, 542

Ooh-loo-woo is main display sound of lekking males, given with neck sacs fully inflated, tail raised, and pinnae feathers erected into a pointed crest behind the head, often while stamping feet. Ooit given infrequently, when females present, almost always during another male's Ooh-loo-woo.

Some notes of Long Call may break upward to high, polyphonic whines

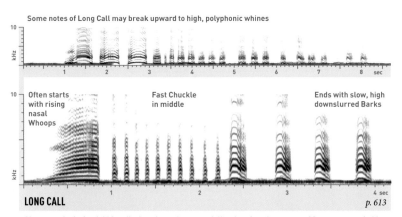

Often starts with rising nasal Whoops

Fast Chuckle in middle

Ends with slow, high downslurred Barks

LONG CALL

p. 613

Given mostly during lekking displays by males, especially when females appear. Often accompanied by a flutter-jump display, in which the bird leaps into the air briefly, flapping its wings. Quite plastic; often 3-parted, as shown above, but each part of the Long Call also given separately at times.

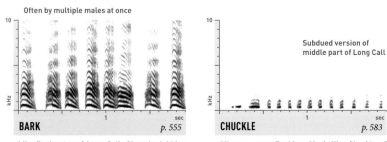

Often by multiple males at once

BARK

p. 555

Like final notes of Long Call. Given by lekking males, possibly in antiphonal duets with other males; rising nasal cries often mixed in.

Subdued version of middle part of Long Call

CHUCKLE

p. 583

All year, upon flushing. Much like Chuckle of Lesser Prairie-Chicken, but pitch slightly lower.

LESSER PRAIRIE-CHICKEN

Tympanuchus pallidicinctus

Very uncommon and local in sand sage prairie and oak shrublands; range has contracted by over 75 percent since the 1960s. Note pinkish neck sac.

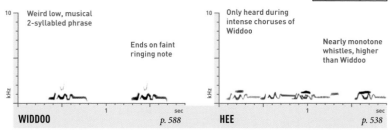

Weird low, musical 2-syllabled phrase

Ends on faint ringing note

WIDDOO *p. 588*

Only heard during intense choruses of Widdoo

Nearly monotone whistles, higher than Widdoo

HEE *p. 538*

Widdoo is main display sound of lekking males, given with neck sacs fully inflated, tail raised, and pinnae feathers erected into a pointed crest behind the head, often with a flick of the tail and a bob of the head. Sometimes males face each other and give Widdoos at high rate, with Hees interspersed.

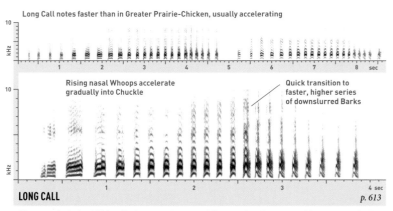

Long Call notes faster than in Greater Prairie-Chicken, usually accelerating

Rising nasal Whoops accelerate gradually into Chuckle

Quick transition to faster, higher series of downslurred Barks

LONG CALL *p. 613*

Given mostly during lekking displays by males, especially when females appear. Often accompanied by a flutter-jump display, in which the bird leaps into the air briefly, flapping its wings. Quite plastic; often 2-parted, as shown above, but each part of the Long Call also given separately at times.

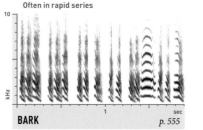

Often in rapid series

BARK *p. 555*

Like final notes of Long Call. Given rapidly by lekking males, possibly in antiphonal duets with other males; longer notes often mixed in.

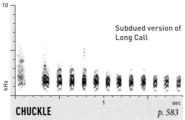

Subdued version of Long Call

CHUCKLE *p. 583*

All year, upon flushing. Much like Chuckle of Greater Prairie-Chicken, but slightly higher.

GREATER SAGE-GROUSE

Centrocercus urophasianus

Our largest grouse, a spectacular bird very closely tied to extensive sagebrush steppe. Still widespread, but most populations have declined.

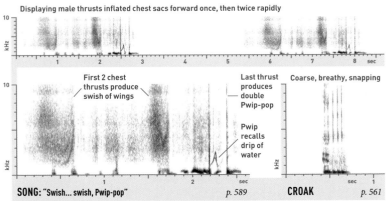

Displaying male thrusts inflated chest sacs forward once, then twice rapidly

First 2 chest thrusts produce swish of wings

Last thrust produces double Pwip-pop

Pwip recalls drip of water

Coarse, breathy, snapping

SONG: "Swish... swish, Pwip-pop"   *p. 589*

CROAK   *p. 561*

Song is main display sound of lekking males. Highly uniform and stereotyped. Dripping-water sound is produced separately from Pops, but often sounds like part of them. Low, faint Hoots occur during wing swishes; audible only at close range. Croak given occasionally between Songs with a quick exhalation, like a belch.

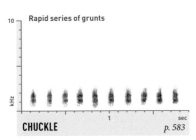

Rapid series of grunts

CHUCKLE   *p. 583*

All year, by female, in alarm and in agitation. Plastic; some versions are more like male Croak Series.

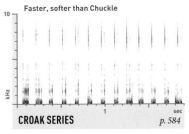

Faster, softer than Chuckle

CROAK SERIES   *p. 584*

At least Mar.–May, by males in aggression, often during standoffs and fights on leks. Sometimes directed at other species, including humans.

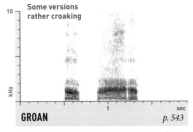

Some versions rather croaking

GROAN   *p. 543*

All year, by both sexes, in high agitation, often with Chuckle or Kuk. Quite plastic.

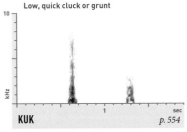

Low, quick cluck or grunt

KUK   *p. 554*

All year, by both sexes, in contact and mild alarm. Plastic.

GUNNISON SAGE-GROUSE

Centrocercus minimus

Distinctly smaller than Greater Sage-Grouse; display differs subtly. Ranges do not overlap. Small population survives in only 10 percent of former range.

Displaying male thrusts inflated chest sacs forward once, then twice rapidly

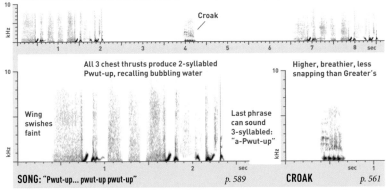

Croak

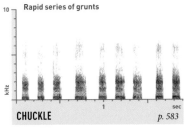

All 3 chest thrusts produce 2-syllabled Pwut-up, recalling bubbling water

Wing swishes faint

Last phrase can sound 3-syllabled: "a-Pwut-up"

SONG: "Pwut-up... pwut-up pwut-up" *p. 589*

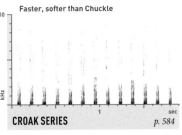

Higher, breathier, less snapping than Greater's

CROAK *p. 561*

Song is main display sound of lekking males. Highly uniform and stereotyped; differs consistently and audibly from Song of Greater Sage-Grouse. Croak given occasionally between Songs with a quick exhalation, like a belch; quality is fluttering, almost as though made by feathers.

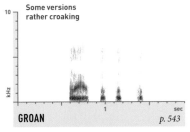

Rapid series of grunts

CHUCKLE *p. 583*

All year, by female, in alarm and in agitation. Plastic; some versions are more like male Croak Series.

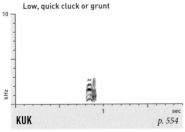

Faster, softer than Chuckle

CROAK SERIES *p. 584*

At least Mar.–May, by males in aggression, often during standoffs and fights on leks. Much like Croak Series of Greater Sage-Grouse.

Some versions rather croaking

GROAN *p. 543*

All year, by both sexes, in high agitation, often with Chuckle or Kuk. Quite plastic.

Low, quick cluck or grunt

KUK *p. 554*

All year, by both sexes, in contact and mild alarm. Plastic.

RUFFED GROUSE

Bonasa umbellus

A bird of dense forests, especially young mixed forests and aspen groves; drumming of males is a characteristic sound of spring throughout range.

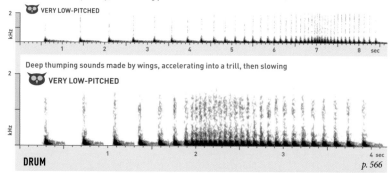

Drum starts very slowly; accelerating portion can last more than 10 seconds

🦉 VERY LOW-PITCHED

Deep thumping sounds made by wings, accelerating into a trill, then slowing

🦉 VERY LOW-PITCHED

DRUM *p. 566*

All year, especially Apr.–May and Sept.–Oct. By males only, atop preferred log or boulder, in courtship and territorial defense. Unmistakable, far-carrying. Loudest at low frequencies; often felt in the chest as much as heard. Poorly reproduced by most speakers. Bird also flushes with thumping wingbeats.

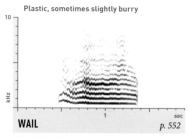

Plastic, sometimes slightly burry

WAIL *p. 552*

Mostly May–Sept., by females in high alarm, especially when chicks are threatened. Squealing versions sometimes given in distraction display.

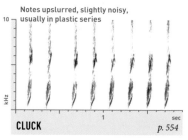

Notes upslurred, slightly noisy, usually in plastic series

CLUCK *p. 554*

Likely all year, in alarm. Can strongly recall alarm notes of Red Squirrel. Often given in conjunction with Purrs.

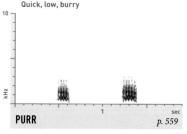

Quick, low, burry

PURR *p. 559*

All year, in alarm, often in conjunction with other calls. Fairly plastic.

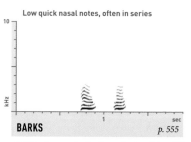

Low quick nasal notes, often in series

BARKS *p. 555*

Likely all year, in family contact and in mild alarm. Plastic, grading into Wails. Rather soft.

Spruce Grouse (Franklin's)

Falcipennis canadensis franklinii

A tame but quiet and elusive resident of spruce forests. "Franklin's" subspecies possibly a separate species from the birds in northern and eastern Canada.

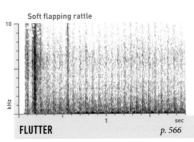

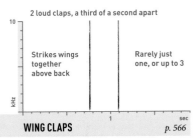

Soft flapping rattle

FLUTTER   *p. 566*

2 loud claps, a third of a second apart

Strikes wings together above back

Rarely just one, or up to 3

WING CLAPS   *p. 566*

Mostly Feb.–June. Male in display makes mechanical sounds only, inaudible beyond 150 feet (50 meters) except for loud Wing Claps of "Franklin's" population. Flutter and Wing Claps both given as bird stalls in flight down from tree; claps often follow Flutter. Also gives soft Whoosh (p. 557) by opening and closing tail.

Long Call often 2-parted, with accelerating Kuks followed by series of nasal squeals

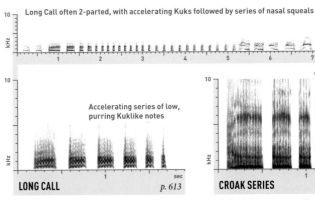

Accelerating series of low, purring Kuklike notes

LONG CALL   *p. 613*

Mostly Mar.–May, by females only, in songlike fashion from trees at dawn and dusk, as well as in territorial encounters.

Accelerating

CROAK SERIES   *p. 584*

Distinctive; harsh, sometimes inflected like human speech. Apparently by males only, in close-range aggression. Eastern population shown.

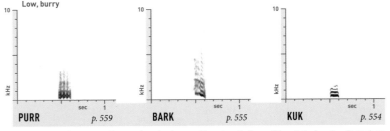

Low, burry

PURR   *p. 559*

BARK   *p. 555*

KUK   *p. 554*

A variety of quick low nasal notes given in alarm and brood contact, possibly only by females. Purr given singly or followed by Kuks. Barks and Kuks given singly; quite plastic, possibly grading into other calls.

DUSKY GROUSE

Dendragapus obscurus

One of our largest grouse, uncommon and shy, found in open coniferous forests and adjacent shrublands. Formerly lumped with Sooty Grouse.

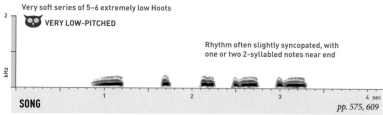

Very soft series of 5–6 extremely low Hoots

VERY LOW-PITCHED

Rhythm often slightly syncopated, with one or two 2-syllabled notes near end

SONG
pp. 575, 609

Mostly Mar.–May, by males in courtship display with tail spread and inflatable neck pouch exposed; neck pouch swells with each note of Song. A short-range signal; soft, unlikely to be heard beyond 50 feet. Usually given from the ground. Probably the lowest-pitched vocal bird sound in North America.

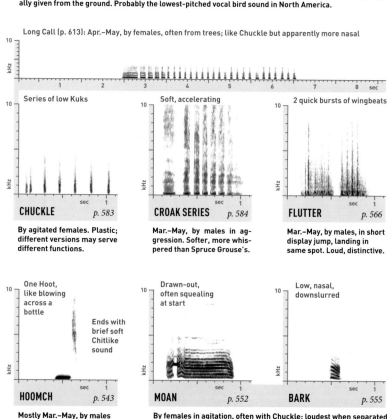

Long Call (p. 613): Apr.–May, by females, often from trees; like Chuckle but apparently more nasal

Series of low Kuks

CHUCKLE
p. 583

By agitated females. Plastic; different versions may serve different functions.

Soft, accelerating

CROAK SERIES
p. 584

Mar.–May, by males in aggression. Softer, more whispered than Spruce Grouse's.

2 quick bursts of wingbeats

FLUTTER
p. 566

Mar.–May, by males, in short display jump, landing in same spot. Loud, distinctive.

One Hoot, like blowing across a bottle

Ends with brief soft Chitlike sound

HOOMCH
p. 543

Mostly Mar.–May, by males in close courtship, at end of rush toward female. Loud.

Drawn-out, often squealing at start

MOAN
p. 552

Low, nasal, downslurred

BARK
p. 555

By females in agitation, often with Chuckle; loudest when separated from chicks. Females may also give Moanlike sounds in courtship. Moan and Bark quite plastic, grading into one another.

SOOTY GROUSE

Dendragapus fuliginosus

Very similar to Dusky Grouse but male plumage is darker, and inflatable throat sac is yellow in most of range (Dusky's is wine red). Some sounds differ.

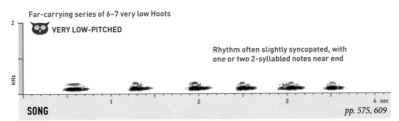

Far-carrying series of 6–7 very low Hoots

VERY LOW-PITCHED

Rhythm often slightly syncopated, with one or two 2-syllabled notes near end

SONG *pp. 575, 609*

Mostly Mar.–May, by males in courtship display with tail spread and inflatable neck pouch exposed; neck pouch swells with each note of Song. A long-range signal; can be heard from several hundred yards away. Usually given from high in a tree. Averages slightly higher than Dusky Grouse Song, often with 1 more note.

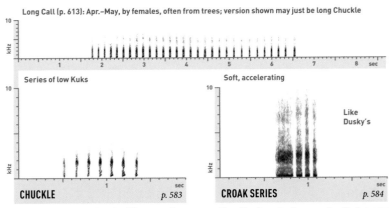

Long Call (p. 613): Apr.–May, by females, often from trees; version shown may just be long Chuckle

Series of low Kuks

Soft, accelerating

Like Dusky's

CHUCKLE *p. 583*

CROAK SERIES *p. 584*

By agitated females. Plastic; different versions may serve different functions. May grade into Long Call.

Mar.–May, by males in aggression. Sooty seems to lack exact equivalent of Dusky's Flutter display, but both species can take off or land loudly.

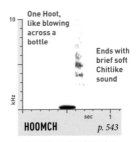

One Hoot, like blowing across a bottle

Ends with brief soft Chitlike sound

HOOMCH *p. 543*

Mostly Mar.–May, by males in close courtship, at end of rush toward female. Loud.

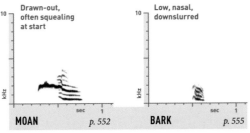

Drawn-out, often squealing at start

Low, nasal, downslurred

MOAN *p. 552*

BARK *p. 555*

By females in agitation, often with Chuckle; loudest when separated from chicks. Females may also give Moanlike sounds in courtship. Moan and Bark average slightly higher than Dusky's.

White-tailed Ptarmigan

Lagopus leucura

Elusive resident of high tundra, mottled in summer and white in winter. Often unnoticed until nearly stepped on, but surprisingly loud in spring courtship.

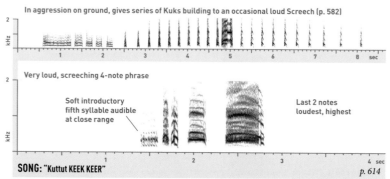

In aggression on ground, gives series of Kuks building to an occasional loud Screech (p. 582)

Very loud, screeching 4-note phrase

Soft introductory fifth syllable audible at close range

Last 2 notes loudest, highest

SONG: "Kuttut KEEK KEER" *p. 614*

Mostly May–July, by both sexes, but especially by males in low, horizontal display flight. Audible from very long distances; usually repeated every 2 seconds or so during flight, often preceded and followed by series of clucks. Also given from ground in high excitement. Highly distinctive; uniform and stereotyped.

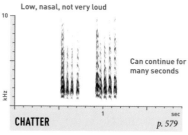

Low, nasal, not very loud

Can continue for many seconds

CHATTER *p. 579*

All year, by both sexes, in agitation and alarm. Variable and plastic; different versions given in different situations.

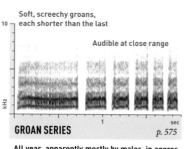

Soft, screechy groans, each shorter than the last

Audible at close range

GROAN SERIES *p. 575*

All year, apparently mostly by males, in aggression or in agitation. May also function in close courtship.

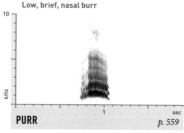

Low, brief, nasal burr

PURR *p. 559*

All year, by both sexes, in contact and in alarm. Slightly variable; usually monotone, but some versions are overslurred. Not very loud.

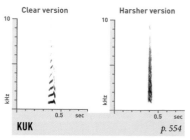

Clear version   Harsher version

KUK *p. 554*

All year, by both sexes. Soft versions given in contact as birds forage; louder versions given in territorial encounters, often in series (p. 615).

GREBES (Family Podicipedidae)

Highly aquatic, grebes have lobed feet attached far back on the body, limiting their mobility on land. They dive for food such as fish and arthropods, and build floating nests of plant matter. Young chicks often ride on the backs of their parents in the water. They are not closely related to any other birds, but surprisingly, their closest living relatives may be the flamingos.

Vocalizations of grebes are likely innate. Vocal repertoires are often complex; some species engage in synchronized duets. Courtship displays of some species include elaborately choreographed water dances.

CLARK'S GREBE

Aechmophorus clarkii

Formerly considered a color morph of Western Grebe, but the two breed side-by-side with little hybridization. More common to the west and south.

Songs sometimes mixed with Dziks, especially when calling to chicks

Typical version: a single grate

Monotone or slightly rising

Clearer, more whistled version

SONG: "Kreeek"   *p. 559*

All year, to attract potential mates and to maintain contact with distant family members; sometimes repeated in slow series. Variable and plastic; female versions average slightly higher. In areas where both species breed, Clark's and Western generally do not respond to one another's Songs.

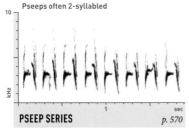

Pseeps often 2-syllabled

PSEEP SERIES   *p. 570*

Mostly July–Nov., by begging juveniles. Varies individually and with age. Loud and often incessant, carrying far across the water.

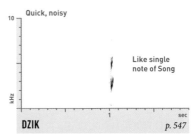

Quick, noisy

Like single note of Song

DZIK   *p. 547*

All year, in alarm and family contact. Highly plastic. Similar notes, singly or in series, accompany certain courtship displays.

WESTERN GREBE

Aechmophorus occidentalis

Shares many displays with Clark's, including the famous rushing courtship display: 2–3 birds rise out of water, run silently across it together, then dive.

Consecutive songs may differ slightly, but individuals often recognizable by Song

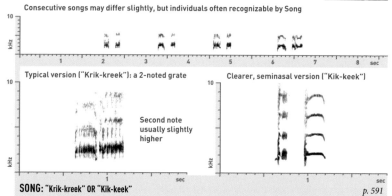

Typical version ("Krik-kreek"): a 2-noted grate

Second note usually slightly higher

Clearer, seminasal version ("Kik-keek")

SONG: "Krik-kreek" OR "Kik-keek"   *p. 591*

All year, to attract potential mates and to maintain contact with distant family members; sometimes repeated in slow series. Variable and plastic; beware occasional 1-note versions similar to Song of Clark's Grebe. Female versions average slightly higher. Occasionally heard at night.

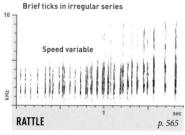

Brief ticks in irregular series

Speed variable

RATTLE   *p. 565*

Mostly May–June, in ratchet-pointing display: birds face off with necks low, occasionally shaking bill in water, often before rushing.

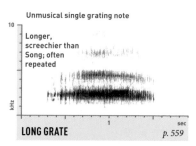

Unmusical single grating note

Longer, screechier than Song; often repeated

LONG GRATE   *p. 559*

May–June, in barge-trill display: two competing males rise out of water as if to rush, but move slowly, giving ritual head turns.

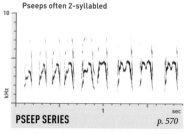

Pseeps often 2-syllabled

PSEEP SERIES   *p. 570*

July–Nov., by begging juveniles. Varies individually and with age. Loud and often incessant, carrying far across the water.

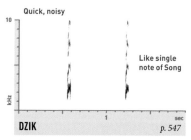

Quick, noisy

Like single note of Song

DZIK   *p. 547*

All year, in alarm and family contact. Highly plastic. Similar notes, singly or in series, accompany certain courtship displays.

Eared Grebe

Podiceps nigricollis

Nests on shallow ponds with emergent vegetation, often in dense colonies. In migration and winter, gathers in vast numbers on saline lakes in the West.

Song often given in series; here, answered by Twitter of another bird

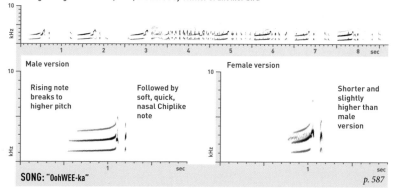

Male version

Rising note breaks to higher pitch

Followed by soft, quick, nasal Chiplike note

Female version

Shorter and slightly higher than male version

SONG: "OohWEE-ka"

p. 587

Most common vocalization; given mostly Mar.–July, by solo birds seeking mates, or by pairs in courtship or when feeding young. Male and female versions similar but often separable by ear. At start of breeding season, Song may be followed by elaborate dancing courtship displays by pair.

Often a short burr followed by a series of shrill Pips

Some versions faster, more trilling

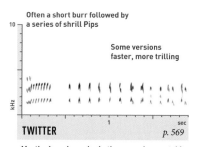

TWITTER

p. 569

Mostly Apr.–June, by both sexes, in courtship. Aggressive versions faster, trilled. Rarely downslurred and decelerating, recalling Sora.

Long breathy, burry whistle

Monotone or slightly rising

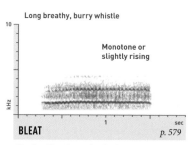

BLEAT

p. 579

Mostly May–June, by both sexes, in close courtship. Does not carry far. Often followed by copulation, accompanied by clearer Wail.

Slightly nasal; usually sounds upslurred

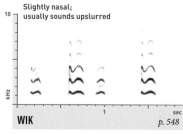

WIK

p. 548

Likely mostly Apr.–June. Not well understood, but likely given in agitation or alarm. Begging juveniles give Pseep Series (p. 570).

Soft, abrupt

Sometimes in 2-note pattern: Pik-up

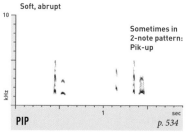

PIP

p. 534

Likely all year, in alarm and contact. Does not carry far. Like single note of Twitter; grades into both Wik and Twitter.

Red-necked Grebe

Podiceps grisegena

Large, strikingly patterned in breeding season, with long yellowish bill. Breeds on ponds with emergent vegetation; winters mostly along coasts.

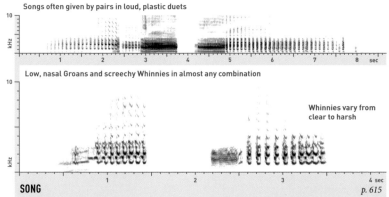

Songs often given by pairs in loud, plastic duets

Low, nasal Groans and screechy Whinnies in almost any combination

Whinnies vary from clear to harsh

SONG *p. 615*

All year, but especially Jan.–Aug., by lone birds of either sex to attract a mate, and by duetting pairs in courtship or after successful defense of territories. Little difference between male and female versions. Extremely plastic, but distinctive. Long, burry versions of Groan given during copulation.

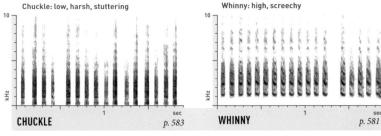

Chuckle: low, harsh, stuttering

Whinny: high, screechy

CHUCKLE *p. 583*

WHINNY *p. 581*

A set of similar vocalizations that intergrade. All year, but especially Jan.–Aug., by both sexes. Some versions resemble Whinnies from Song; other versions are noisier, more chattering. Some versions may be given during elaborate courtship displays prior to pairing.

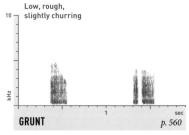

Low, rough, slightly churring

GRUNT *p. 560*

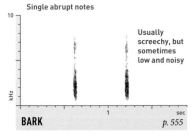

Single abrupt notes

Usually screechy, but sometimes low and noisy

BARK *p. 555*

Function not entirely clear; may be given in high alarm. Plastic. Begging juveniles likely give Pseep Series (no recordings).

All year, in mild alarm and possibly contact. Like single note of Whinny from Song; sometimes given in slow, irregular series.

Horned Grebe

Podiceps auritus

Breeds on shallow ponds with emergent vegetation, usually not in colonies. Winters on both salt and fresh water. Head flatter, less peaked than Eared's.

In greeting, pairs duet with plastic, trilled Kwirrs and wailing sounds (p. 615)

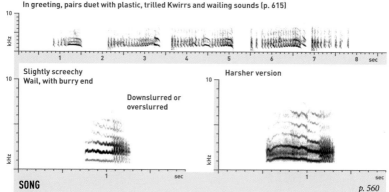

Slightly screechy Wail, with burry end

Downslurred or overslurred

Harsher version

SONG

p. 560

All year, but especially Jan.–Aug., by lone birds seeking a mate, and by courting pairs, starting in late winter and extending through spring migration. Little difference between male and female versions. Variable and plastic.

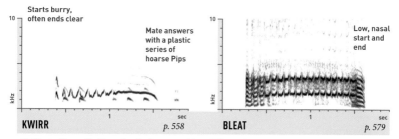

Starts burry, often ends clear

Mate answers with a plastic series of hoarse Pips

Low, nasal start and end

KWIRR

p. 558

BLEAT

p. 579

Highly plastic vocalizations given in close courtship and high excitement. Kwirrs often precede copulation; Bleat given during copulation. Both may grade into Song. Often given by pairs in unsynchronized duets. Hoarse Pips given by male near nest, grading into Screech Series.

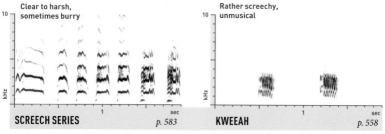

Clear to harsh, sometimes burry

Rather screechy, unmusical

SCREECH SERIES

p. 583

KWEEAH

p. 558

Mostly May–Aug., in high alarm. Extremely plastic. Usually in series. Some versions clear, not screechy. Juveniles give Pseep Series (p. 570).

Likely mostly May–July, in alarm near nest or young. Fairly uniform. May grade into Screeches.

PIED-BILLED GREBE

Podilymbus podiceps

Prefers marshy ponds. Outlandish voice is often heard but frequently given from dense cover, and casual observers often do not connect it with this species.

Three-part version of Song: third part alternates gulps with rising Whoops

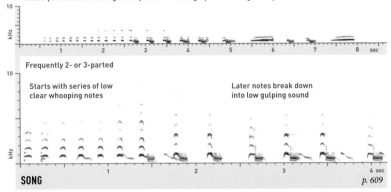

Frequently 2- or 3-parted

Starts with series of low
clear whooping notes

Later notes break down
into low gulping sound

SONG   *p. 609*

All year, especially Apr.–June. Highly variable; usually contain 2–3 contiguous series, each slower than the last and with more complex notes. Any series may last up to 15–20 notes; first and second series often given by themselves. Both sexes sing, but apparently only males give 3-part songs.

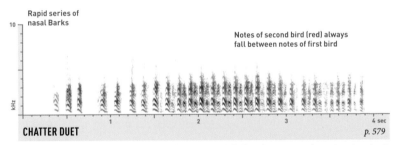

Rapid series of
nasal Barks

Notes of second bird (red) always
fall between notes of first bird

CHATTER DUET   *p. 579*

All year. Chatters sometimes given by solo birds, but mostly in duets between members of a pair on territory, resulting in a distinctive rapid laughing couplet series. One bird (female?) often sounds significantly higher than the other. Often introduces Song, or given simultaneously with mate's Song.

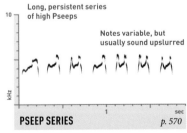

Long, persistent series
of high Pseeps

Notes variable, but
usually sound upslurred

PSEEP SERIES   *p. 570*

Mostly May–Aug., by begging juveniles. Often long and loud. Sometimes given while young ride on parent's back during brooding.

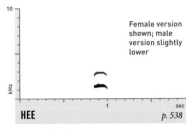

Female version
shown; male
version slightly
lower

HEE   *p. 538*

At least Nov.–May, by birds exploring occupied territories, or in mild alarm. Recalls Keek of Song. Like first notes of Song, but given singly.

LEAST GREBE

Tachybaptus dominicus

Our smallest grebe, widespread in the tropics on densely vegetated ponds, rare in southern Arizona. Black chin of breeding birds becomes white in winter.

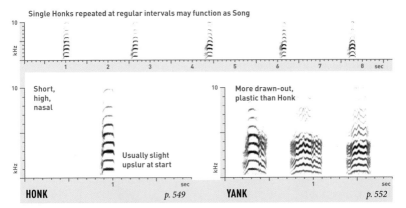

Single Honks repeated at regular intervals may function as Song

HONK   *p. 549*

Short, high, nasal

Usually slight upslur at start

YANK   *p. 552*

More drawn-out, plastic than Honk

Honk given all year; uniform but slightly plastic. Used by mated pairs when separated, or in alarm; possibly also to advertise for mates. Often accompanied by erect posture, bird sitting high in water, with neck stretched vertically and head feathers fluffed. Longer, plastic Yanks apparently given in alarm.

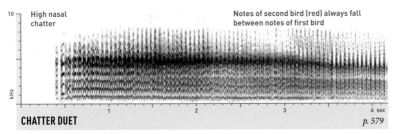

High nasal chatter

Notes of second bird (red) always fall between notes of first bird

CHATTER DUET   *p. 579*

All year; most commonly heard call. Often given by pairs in duet, alternating notes so rapidly that they sound like a Churr from a single bird; apparently to reinforce pair bond and/or maintain territory. Plastic; speed can change irregularly.

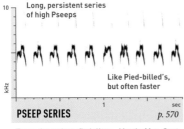

Long, persistent series of high Pseeps

Like Pied-billed's, but often faster

PSEEP SERIES   *p. 570*

From dependent fledglings. Mostly May–Sept., but the species can breed at any time of year if weather is mild.

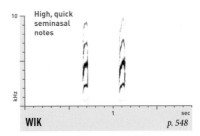

High, quick seminasal notes

WIK   *p. 548*

All year, in contact or alarm. Variable; some versions much higher and more piercing.

PIGEONS AND DOVES (Family Columbidae)

Often common even in cities, pigeons and doves are some of our most widespread and familiar birds. Many have adapted well to human changes to the landscape; some species from the dry Southwest, such as White-winged and Inca Doves, have expanded their ranges north and east in the past century. Species from Europe and Asia, such as the Rock Pigeon and the Eurasian Collared-Dove, have spread across the continent. Not every species has fared well. The Passenger Pigeon (*Ectopistes migratorius*), which likely numbered in the billions when Europeans began colonizing North America, became extinct in 1914.

The sounds of doves are innate, varying little between or within individuals. Most are soft and low-pitched, with a characteristic cooing or hooting quality that makes them easy to mistake for owls. Low growling or purring sounds are typical of courtship. Voice breaks are common.

Many dove species also make sounds with their wings, including wing whistles on takeoff that may warn other birds of danger (see p. 574) and wing claps or wing rattles that may serve territorial or courtship functions (see p. 566).

ROCK PIGEON

Columba livia

The familiar pigeon of cities and parks, native to Europe but now widespread in North America. Extremely variable in plumage color and pattern.

Song lower than those of other doves; not very loud

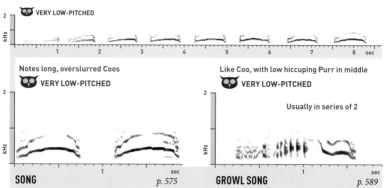

VERY LOW-PITCHED

Notes long, overslurred Coos
VERY LOW-PITCHED

SONG   p. 575

Like Coo, with low hiccuping Purr in middle
VERY LOW-PITCHED

Usually in series of 2

GROWL SONG   p. 589

Song given all year, as the species nests year-round, by both sexes near nest or potential nest site. Single Coos given singly in alarm. Growl Song given all year, by males bowing with puffed-out throats and fanned tails, to court females and to repel other males. Does not carry far. Wings often slap loudly on takeoff (p. 567).

Band-tailed Pigeon

Patagioenas fasciata

A large pigeon that inhabits dry pine-oak forests inland, and wet coniferous forests along the West Coast. Common in places; sometimes visits feeders.

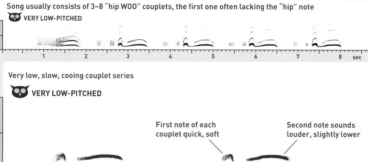

Song usually consists of 3–8 "hip WOO" couplets, the first one often lacking the "hip" note

🦉 VERY LOW-PITCHED

Very low, slow, cooing couplet series

🦉 VERY LOW-PITCHED

First note of each couplet quick, soft

Second note sounds louder, slightly lower

SONG: "Hip WOO, hip WOO..."   *pp. 588, 594*

Mostly Apr.–Aug, by males. Uniform and sterotyped; sounds soft, but carries well. Unmated males reportedly sing from high exposed perches, while mated males choose hidden perches inside the canopy. At close range, Gruntlike notes often audible between Song couplets. Often precedes wing-clapping display flight.

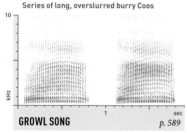

Series of long, overslurred burry Coos

GROWL SONG   *p. 589*

Little known. Possibly given in close courtship. In some examples, typical Song leads into Growl Song. Quality like Grunt, but more musical.

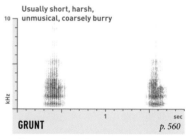

Usually short, harsh, unmusical, coarsely burry

GRUNT   *p. 560*

All year, in altercations. Variable, plastic. Similar calls given in series in wing-clapping display flight; these may be related to Growl Song.

Spotted Dove   *Streptopelia chinensis*

Native to Asia. Introduced to urban southern California; once abundant there, but now only a few remain in Los Angeles, on Catalina Island, in Bakersfield, and in Fresno. Song burry, rather variable: usually a 3-note phrase, but some versions have 2 or 4 notes. Sometimes rapidly repeats song phrases. Apparently lacks a Wheezelike call.

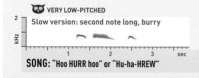

🦉 VERY LOW-PITCHED

Slow version: second note long, burry

🦉 VERY LOW-PITCHED

Faster version: last note long, burry

SONG: "Hoo HURR hoo" or "Hu-ha-HREW"   *pp. 589, 594*

EURASIAN COLLARED-DOVE

Streptopelia decaocto

Native to Eurasia, but since the 1980s
has spread rapidly across North American cities, towns, and farmsteads. In
many areas, nests nearly year-round.

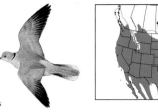

Songs given singly or strung together in long triplet series

🦉 VERY LOW-PITCHED

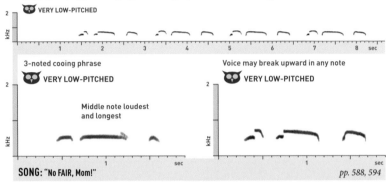

3-noted cooing phrase

🦉 VERY LOW-PITCHED

Middle note loudest
and longest

Voice may break upward in any note

🦉 VERY LOW-PITCHED

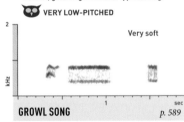

SONG: "No FAIR, Mom!"   *pp. 588, 594*

Nearly all year. Highly variable, but triplet pattern of Coos is distinctive. All notes typically on the same
pitch, the triplets either given singly or strung together in long series. Rare 2-noted variant omits the final
note.

Hoarse, groaning version of typical Song

🦉 VERY LOW-PITCHED

Very soft

GROWL SONG   *p. 589*

Nearly all year; from courting birds near potential nest sites, often well hidden. Never in series.
Does not carry far.

Nasal, finely burry overslur

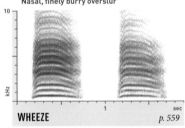

WHEEZE   *p. 559*

All year. Distinctive; usually given on the wing,
including in wing-clapping display flight. Often
in series of 2–3.

AFRICAN COLLARED-DOVE   *Streptopelia roseogrisea*

Domesticated form, sometimes called
"Ringed Turtle-Dove" (*S. risoria*), escapes
widely. Resembles Eurasian Collared-
Dove; hybrids sound intermediate.

🦉 VERY LOW-PITCHED

Repeated 2-noted burry phrase

SONG: "WHIP brooOOm"   *pp. 589, 594*

Falling, slowing nasal series

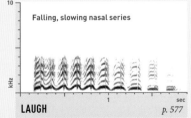

LAUGH   *p. 577*

INCA DOVE

Columbina inca

Originally restricted to deserts, but has become a familiar bird of residential areas throughout the Southwest, often nesting in yards and visiting feeders.

Song repeated at short intervals for long periods

🦉 VERY LOW-PITCHED

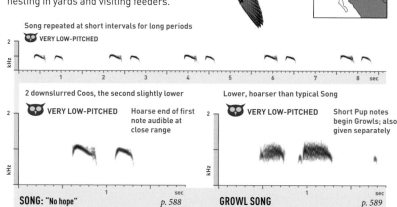

2 downslurred Coos, the second slightly lower

🦉 VERY LOW-PITCHED   Hoarse end of first note audible at close range

SONG: "No hope"   *p. 588*

Lower, hoarser than typical Song

🦉 VERY LOW-PITCHED   Short Pup notes begin Growls; also given separately

GROWL SONG   *p. 589*

Song given all year, but mostly Feb.–Aug., by males. Repeated incessantly, even in the heat of the day; highly uniform. Growl Song mostly Feb.–Oct., in close courtship or aggressive interactions. Soft; rhythm varies with excitement, but above pattern is typical. Wing Flutter (p. 566) given on takeoff and in aggression.

COMMON GROUND-DOVE

Columbina passerina

Fairly common in dry scrub and second growth. Also found in residential areas, but has declined as suburbs expand, unlike some other dove species.

Song repeated in slow steady series for long periods

🦉 VERY LOW-PITCHED

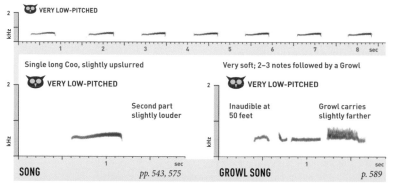

Single long Coo, slightly upslurred

🦉 VERY LOW-PITCHED   Second part slightly louder

SONG   *pp. 543, 575*

Very soft; 2–3 notes followed by a Growl

🦉 VERY LOW-PITCHED   Inaudible at 50 feet   Growl carries slightly farther

GROWL SONG   *p. 589*

Song given all year, especially Jan.–Sept., by males. Sings even in the heat of the day; highly uniform. Growl Song given mostly Jan.–Sept., in close courtship and possibly some territorial encounters; rarely noticed because of low volume. Wing Flutter (p. 566) given on takeoff.

Mourning Dove

Zenaida macroura

Our most familiar native dove, common in a huge variety of habitats across the continent. One of the first birds that many people learn to identify by sound.

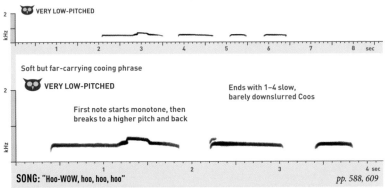

Number of Hoo notes at the end varies, but usually 3

👁 **VERY LOW-PITCHED**

Soft but far-carrying cooing phrase

👁 **VERY LOW-PITCHED**

First note starts monotone, then breaks to a higher pitch and back

Ends with 1–4 slow, barely downslurred Coos

SONG: "Hoo-WOW, hoo, hoo, hoo"   *pp. 588, 609*

Mostly Apr.–Aug. A well-known and evocative song, the sad sound of which gives the bird its common name. Sometimes mistaken for an owl, but generally given only during the day. Highly uniform.

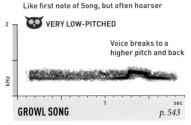

Like first note of Song, but often hoarser

👁 **VERY LOW-PITCHED**

Voice breaks to a higher pitch and back

GROWL SONG   *p. 543*

Mostly Mar.–Aug. Given by male at nest or prospective nest site, often well concealed.

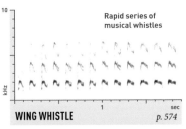

Rapid series of musical whistles

WING WHISTLE   *p. 574*

All year. Often mistaken for a vocal sound. Mostly heard at takeoff, but takeoff can be silent, and whistle can occur in sustained flight.

PIGEON AND DOVE WING SOUNDS

All species of North American doves make sounds with their wings, and many of these sounds serve communicative functions.

Whistles: Most familiar is the loud musical whistle of the Mourning Dove's wings. Mourning Doves may be able to control this sound, and it can provoke alarm responses in other birds, leading some to hypothesize that the Wing Whistle is an intentional alarm signal. The wings of Rock Pigeon, White-winged Dove, and Eurasian Collared-Dove sometimes whistle in a similar fashion.

Flutters: Several species of doves can produce a loose, slapping Flutter on takeoff, instead of or in addition to Wing Whistles. Inca Doves also give Flutters while perched in aggressive displays toward other members of their species.

Claps: Most dove and pigeon species in our area give Wing Claps in courtship flight displays. In all such displays, a male bird sallies out from a high perch, claps its wings together above its back 5–10 times, then goes into a stiff-winged downward glide, usually to a different perch.

WHITE-WINGED DOVE

Zenaida asiatica

Vocally and visually conspicuous in a variety of habitats. Native southwestern range has expanded, especially to the north and east.

Fewer repeating phrases than in other long dove songs

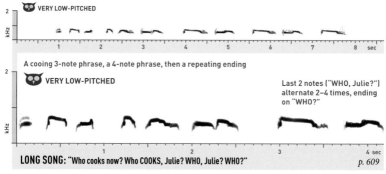

VERY LOW-PITCHED

A cooing 3-note phrase, a 4-note phrase, then a repeating ending

VERY LOW-PITCHED

Last 2 notes ("WHO, Julie?") alternate 2–4 times, ending on "WHO?"

LONG SONG: "Who cooks now? Who COOKS, Julie? WHO, Julie? WHO?" *p. 609*

Nearly all year, but mostly Jan.–Aug. Highly distinctive; uniform and stereotyped. Given by males upon first landing on a song perch, after courtship flights, and when rival males approach. If Long Song is not answered, male often switches to Short Song or alternates the two song forms.

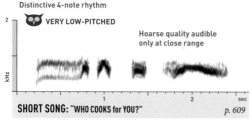

Distinctive 4-note rhythm

VERY LOW-PITCHED

Hoarse quality audible only at close range

SHORT SONG: "WHO COOKS for YOU?" *p. 609*

Nearly all year, but mostly Jan.–Aug. Never in series. Higher than Barred Owl's "Who cooks for you?" Burrier 2- and 3-note versions, possibly from female, given in some courtship duets.

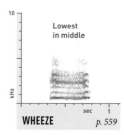

Lowest in middle

WHEEZE *p. 559*

All year, in encounters with conspecifics. Quality like air escaping balloon.

CUCKOOS AND RELATIVES (Family Cuculidae) *next pages*

The cuckoo family is a diverse, cosmopolitan group. Sounds are apparently innate. The familiar 2-note "COO-coo" song popularized by the cuckoo clock is not heard in North America; it is the song of the Common Cuckoo of Eurasia. The North American cuckoos in the genus *Coccyzus*, slender and secretive birds of dense woods, tend to give mellow coos and mechanical-sounding knocks and clucks. They usually build their own nests, but sometimes lay eggs in the nests of other cuckoos or other bird species. The roadrunner, a terrestrial desert species, also gives coos, among other sounds.

Yellow-billed Cuckoo

Coccyzus americanus

A striking but reclusive bird of dense deciduous woods; more often heard than seen. Eats mostly large insects, especially caterpillars.

Slow series of mellow, slightly nasal downslurred Coos or Barks

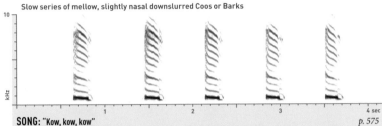

SONG: "Kow, kow, kow"   p. 575

Mostly Apr.–June, by unmated males seeking mates. Given with bill closed, throat inflating with each note. Quality varies from clear, rather dovelike cooing to low, swallowed barking; some versions polyphonic. Length of series plastic, but often 9–12 notes; last few often lower, softer, more 2-syllabled.

Initial rattle sometimes omitted

Classic version: rapid knocking Rattle, then slower, decelerating couplet or triplet series of Clucks

Loudest in middle

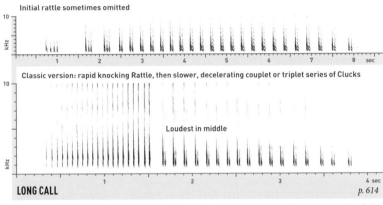

LONG CALL   p. 614

Mostly Apr.–July, possibly only by males. Most common vocalization, but usually given at very long intervals of 10 or more minutes. Often prompts other cuckoos to respond with Long Call. May function to reunite pairs and to maintain territorial boundaries. Very distinctive, but rhythm is variable and plastic.

Rattle: like first part of Long Call

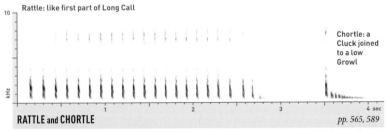

Chortle: a Cluck joined to a low Growl

RATTLE and CHORTLE   pp. 565, 589

Rattle given by both sexes in pair contact, mostly in breeding season. Can be given rather frequently near nest. Rattle may end in Chortle; Chortle also given separately, sometimes as short as 2 notes. Night migrants give all sounds, but mostly Chortle, often in series.

BLACK-BILLED CUCKOO

Coccyzus erythropthalmus

Generally uncommon near the edges of deciduous and mixed forests. Similar to Yellow-billed Cuckoo, but wing and tail patterns differ; eye-ring red, not yellow.

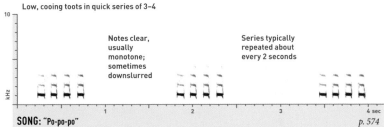

Low, cooing toots in quick series of 3–4

Notes clear, usually monotone; sometimes downslurred

Series typically repeated about every 2 seconds

SONG: "Po-po-po"   *p. 574*

Mostly May–July, reportedly by both sexes. Can be given at any time of day or night on the breeding grounds, sometimes in nocturnal song flights; also occasionally by night migrants. Sometimes gives up to 9 notes in a series, especially if excited. Recalls Least Bittern Song, but higher and clearer.

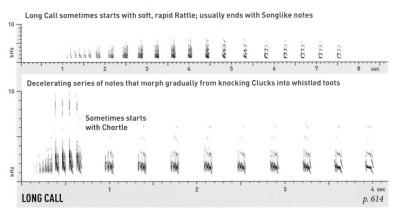

Long Call sometimes starts with soft, rapid Rattle; usually ends with Songlike notes

Decelerating series of notes that morph gradually from knocking Clucks into whistled toots

Sometimes starts with Chortle

LONG CALL   *p. 614*

Mostly May–July. Function little known, but likely similar to Long Call of Yellow-billed Cuckoo. Usually given at very long intervals of 10 or more minutes. May introduce bouts of Song. Pattern variable, especially at start. Gradual transition to clear, musical notes is highly distinctive.

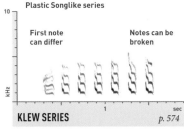

Plastic Songlike series

First note can differ

Notes can be broken

KLEW SERIES   *p. 574*

Much like Song, but series longer, more plastic. Perhaps a vocalization intermediate between Long Call and Song.

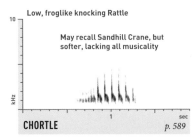

Low, froglike knocking Rattle

May recall Sandhill Crane, but softer, lacking all musicality

CHORTLE   *p. 589*

Distinctive. Given in night flights on breeding grounds and in migration; rarely during the day. Some versions shorter, more monotone.

GREATER ROADRUNNER

Geococcyx californianus

Unmistakable. Usually solitary; prowls arid places in search of snakes, lizards, and other prey. Can run up to 20 miles per hour; flies infrequently but well.

Song: number of notes in series may vary, but most often 5 or 6

🦉 VERY LOW-PITCHED

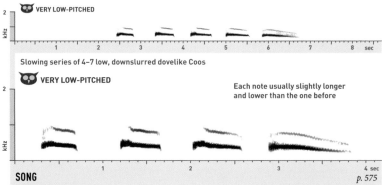

Slowing series of 4–7 low, downslurred dovelike Coos

🦉 VERY LOW-PITCHED

Each note usually slightly longer and lower than the one before

SONG   *p. 575*

Mostly Feb.–June, reportedly only by males. Uniform. Seems soft but carries far; often given from an exposed perch with bill pointed downward, head bobbing forward with each note. Final notes often distinctly 2-syllabled, with a downward voice break. Very soft 1- and 2-note versions sometimes given.

Polyphonic, finely burry, usually rising

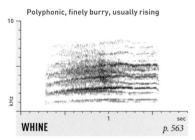

WHINE   *p. 563*

By both sexes, in alarm near nest and when pairs forage together; also in crouching display with complex head motions. Highly plastic.

Low, purring, audible only at close range

🦉 VERY LOW-PITCHED

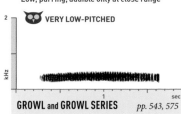

GROWL and GROWL SERIES   *pp. 543, 575*

Variable. Long single Growls given by courting males; short Growls or Grunts given in series by both sexes, in courtship and at nest.

Loud, rapid, snapping

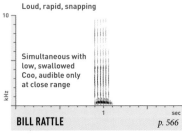

Simultaneous with low, swallowed Coo, audible only at close range

BILL RATTLE   *p. 566*

All year, by both sexes, in alarm and contact. Plastic; varies from 1 snap to many. Series of pops made with wings also reported.

Low, clear, nasal; slows slightly at end

First and last few notes lower

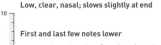
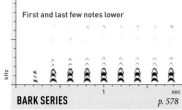

BARK SERIES   *p. 578*

Mostly Feb.–June, reportedly only by female, often in response to male Song or Growl.

NIGHTJARS (Order Caprimulgiformes, Family Caprimulgidae)

Active mostly at dawn and dusk, nightjars have large eyes and wide mouths lined with bristles to capture flying insects in low light. Cryptic plumage lets them avoid detection during the day, when they roost on the ground or lengthwise along tree branches.

Nightjars give simple, often repetitive vocalizations which appear to be innate. Many species vocalize for long periods from perches, usually at dawn or dusk, less often in the middle of the night, and very rarely during daylight hours. Some species are named for their evocative, repetitive calls; several species give wing-clapping displays.

The nighthawks (genus *Chordeiles*) are more aerial than other nightjars and more likely to be active by day.

Common Nighthawk

Chordeiles minor

Widespread; seen mostly at dusk and dawn, but also sometimes at midday. Often vocal on the wing. Nests on open sandy ground or flat urban rooftops.

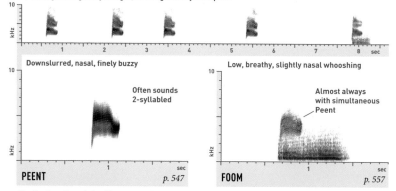

Peents given regularly in flight; Fooms generally infrequent

Downslurred, nasal, finely buzzy

Often sounds 2-syllabled

PEENT   sec   *p. 547*

Mostly Apr.–Aug. A familiar sound of summer evenings across North America; usually given in flight with a rapid wing flutter.

Low, breathy, slightly nasal whooshing

Almost always with simultaneous Peent

FOOM   sec   *p. 557*

May–Aug. Produced in flight at bottom of shallow dive in vicinity of nest; may be made by wings. Sounds soft, but carries far.

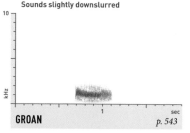

Sounds slightly downslurred

GROAN   sec   *p. 543*

May–Aug. Given from perch by male during courtship, usually in presence of female, and often in conjunction with Peent.

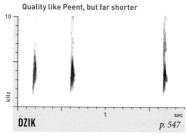

Quality like Peent, but far shorter

DZIK   sec   *p. 547*

May–Aug., in flight chases; infrequent. Buzzier, louder than Lesser Nighthawk Kuk.

Lesser Nighthawk

Chordeiles acutipennis

A desert bird, usually seen only at dusk and dawn. Often silent in flight, but can be quite vocal on predawn perches and during flight chases.

Trill may continue for long periods, but regularly punctuated by Chortle or by brief pauses

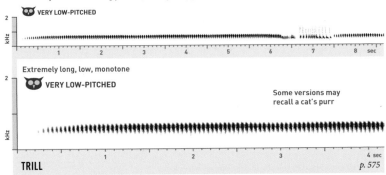

VERY LOW-PITCHED

Extremely long, low, monotone

VERY LOW-PITCHED

Some versions may recall a cat's purr

TRILL

p. 575

Mar.–Aug., from ground or bush, usually before dawn or after dusk; rarely in flight. Uniform and stereotyped. Like screech-owl Trills but much longer, sometimes continuing for many minutes; soft and easily overlooked or mistaken for a distant mechanical sound or the similar trilling of some toad species.

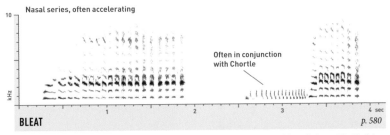

Nasal series, often accelerating

Often in conjunction with Chortle

BLEAT

p. 580

Mostly Mar.–Aug. Frequently given in flight chases; possibly also in courtship and alarm. Plastic; intergrades with Kuk, and sometimes appended to Trill. Distinctive in our area.

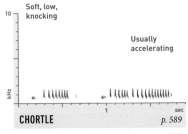

Soft, low, knocking

Usually accelerating

CHORTLE

p. 589

Mostly Mar.–Aug. Usually heard as part of Trill, but also given separately on the wing. Function unknown.

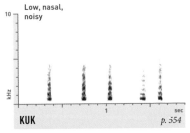

Low, nasal, noisy

KUK

p. 554

All year. Like single notes of Bleat, but usually lower-pitched, briefer, and noisier. Never grating or buzzy.

Mexican Whip-poor-will

Antrostomus arizonae

Locally common in pine-oak forests and brushy canyons, but nocturnal and rarely seen. Named for its distinctive, incessant nighttime call.

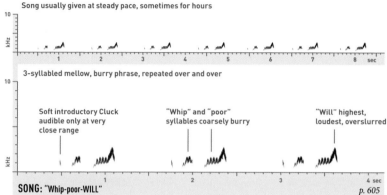

Song usually given at steady pace, sometimes for hours

3-syllabled mellow, burry phrase, repeated over and over

Soft introductory Cluck audible only at very close range

"Whip" and "poor" syllables coarsely burry

"Will" highest, loudest, overslurred

SONG: "Whip-poor-WILL"
p. 605

All year, but especially Apr.–Aug., from ground or tree, around dawn and dusk and on moonlit nights, but also on rare occasions during the day. Not known whether females sing. Quite uniform and stereotyped. Last bouts before sunrise often frantic in pace, without pauses.

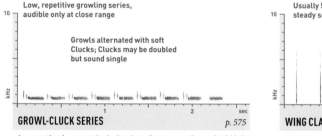

Low, repetitive growling series, audible only at close range

Growls alternated with soft Clucks; Clucks may be doubled but sound single

GROWL-CLUCK SERIES
p. 575

Apparently given mostly during breeding season by excited birds, often between Song and wing-clapping display. Also gives single Growls (p. 543) and Growl Series (p. 575) at nest.

Usually 5–10 in steady series

WING CLAPS
p. 566

Given in flight, often after Growl-Cluck Series; may be territorial in function.

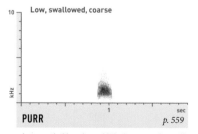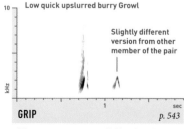

Low, swallowed, coarse

PURR
p. 559

Apparently May–June. Little known; sole available recording given with Grips by agitated bird in response to recordist near nest.

Low quick upslurred burry Growl

Slightly different version from other member of the pair

GRIP
p. 543

All year; most common call. Given in pair contact and in alarm, including distraction display near nest, along with hisses or soft Groans.

Buff-collared Nightjar

Antrostomus ridgwayi

A primarily Mexican species, rare and local north of the border in thorny desert canyons and washes. Vocally unique among North American nightjars.

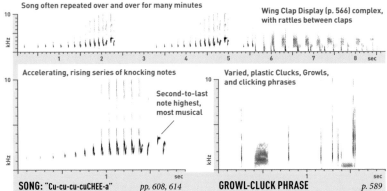

Song often repeated over and over for many minutes

Wing Clap Display (p. 566) complex, with rattles between claps

Accelerating, rising series of knocking notes

Second-to-last note highest, most musical

SONG: "Cu-cu-cu-cuCHEE-a"   *pp. 608, 614*

Varied, plastic Clucks, Growls, and clicking phrases

GROWL-CLUCK PHRASE   *p. 589*

Song given all year, but mostly Mar.–Aug., around dawn and dusk and on moonlit nights; also on rare occasions during the day. Uniform and stereotyped. Soft Growl-Cluck Phrase given in response to playback; sometimes introduces Song bouts. Also gives individual Clucks when flushed (p. 554).

Common Poorwill

Phalaenoptilus nuttallii

Our smallest nightjar. Fairly common in scrubby canyons, rocky areas, and open coniferous forests, but nocturnal and rarely seen.

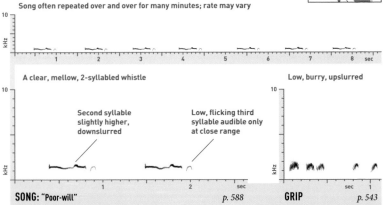

Song often repeated over and over for many minutes; rate may vary

A clear, mellow, 2-syllabled whistle

Second syllable slightly higher, downslurred

Low, flicking third syllable audible only at close range

SONG: "Poor-will"   *p. 588*

Low, burry, upslurred

GRIP   *p. 543*

All year, but mostly Mar.–Aug., around dawn and dusk and on moonlit nights, but also on rare occasions during the day. Highly uniform and stereotyped. Both sexes reported to sing. Song is easily imitated by human whistling. Grip given all year; most common call, but infrequent. Somewhat plastic.

HUMMINGBIRDS (Order Apodiformes, Family Trochilidae)

Hummingbirds are adapted to hovering in front of flowers in order to drink nectar. Our smallest birds, they live at a frenetic pace, their wings a blur. Combative and pugnacious, they sometimes drive away birds many times their size. They frequently visit backyard sugar water feeders.

Vocal repertoires vary from fairly simple to very well developed; some species sing exceedingly complex songs. At least some sounds in some species are learned.

Most of the North American species perform elaborate flight displays, often accompanied by mechanical sounds produced by the tail and/or wings.

BLUE-THROATED HUMMINGBIRD

Lampornis clemenciae

Our largest hummingbird. Uncommon in riparian woods in pine-oak forest. Large white tail corners, visible from below, are distinctive.

Seet Song (pp. 571, 596): Seetlike notes on subtly different pitches, given at steady pace for long periods

Complex Song: whispered scratchy warble with rasping rattles interspersed; patterns often repeat

COMPLEX SONG *p. 612*

Both types of song given by males all year, especially Mar.–Aug. Seet Song most frequent early in the morning or late in the evening, given from a perch. Complex Song infrequent, audible only at close range. Female Complex Song has been reported, but no complex flight display like those of smaller species.

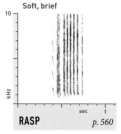

Soft, brief

RASP *p. 560*

All year, in aggression. Plastic; often finer, briefer, and more snarling than shown.

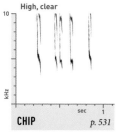

High, clear

CHIP *p. 531*

All year, in interactions. Infrequent. Highly plastic; usually in brief bursts.

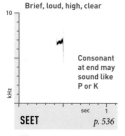

Brief, loud, high, clear

Consonant at end may sound like P or K

SEET *p. 536*

All year; most common call. Highly distinctive. Usually monotone or slightly rising.

RIVOLI'S HUMMINGBIRD

Eugenes fulgens

Very large for a hummingbird. Found in
pine-oak woodlands, usually nesting in
riparian areas along streams. Formerly
called Magnificent Hummingbird.

Repertoire size unknown; Songs apparently somewhat plastic

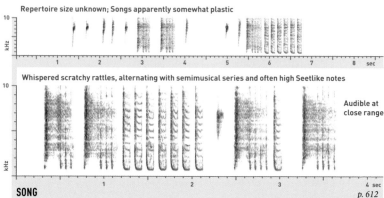

Whispered scratchy rattles, alternating with semimusical series and often high Seetlike notes

Audible at
close range

SONG   *p. 612*

Infrequent and little known; apparently mostly May–June, by males, usually from a perch. Song of young
birds less structured, just a few single well-spaced scratchy or squeaky sounds. Flight displays not reported in this species.

Rapid squeaky series, often upslurred

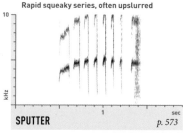

SPUTTER   *p. 573*

All year, in interactions and chases, often with
Snarl and other calls. Highly plastic, grading
smoothly into Twitter.

Soft, brief

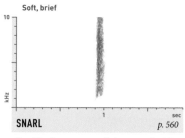

SNARL   *p. 560*

Possibly all year, apparently in aggressive interactions, usually with other calls. Infrequent.

Usually high, sputtering, variably squeaky

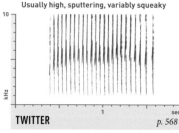

TWITTER   *p. 568*

All year, in interactions and chases, grading
smoothly into Chittip and Chip, but individual
notes often more Spitlike than Chiplike.

Like 2 notes of Rattle   Low, musical

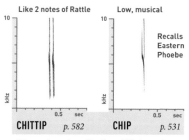

Recalls
Eastern
Phoebe

CHITTIP   *p. 582*   **CHIP**   *p. 531*

All year, in various situations. Chip is most common call. Both sexes give long steady Songlike
series of Chips from perch; function unknown.

LUCIFER HUMMINGBIRD

Calothorax lucifer

One of our smallest hummingbirds, rare and local north of Mexico on scrubby slopes in Chihuahuan Desert. Note distinctly curved bill.

Soft 2- to 3-syllabled buzzy phrase, repeated 4–6 times with slight pauses

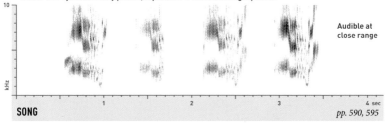

Audible at close range

SONG   *pp. 590, 595*

At least May–Aug., by males, including immature males, apparently in defense of territory. Usually given from a perch, but sometimes during Shuttle Display, in rhythm with the back-and-forth movements. Level of variation unknown; individual males generally repeat the same phrase.

Shuttle Display (p. 576): fast series of beelike buzzes with rattling component sometimes audible

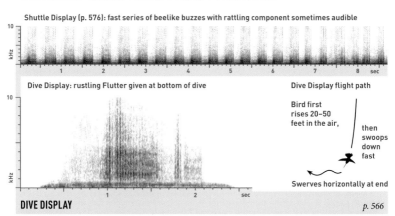

Dive Display: rustling Flutter given at bottom of dive

Dive Display flight path

Bird first rises 20–50 feet in the air,

then swoops down fast

Swerves horizontally at end

DIVE DISPLAY   *p. 566*

Shuttle Display given by male, apparently in courtship, including to female on nest: male buzzes back and forth along a short horizontal path, very close to the female; rate about 2–2.5 shuttles in 1 second. Dive Display often follows Shuttle: male climbs and dives once (rarely twice), swerving away with Flutter of tail.

Often starts with high chipping Twitter

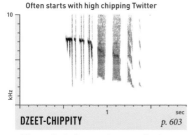

DZEET-CHIPPITY   *p. 603*

All year, by both sexes, in aggressive interactions. Highly plastic; the note types shown here combine with others into many patterns.

Sharp, medium-low

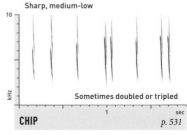

Sometimes doubled or tripled

CHIP   *p. 531*

All year, by both sexes. Most common call, given in many situations. May average slightly lower and more musical than Chips of *Selasphorus*.

BLACK-CHINNED HUMMINGBIRD

Archilochus alexandri

Very closely related to Ruby-throated Hummingbird of the East. Breeds in riparian woods, pinyon-juniper forests, and residential areas.

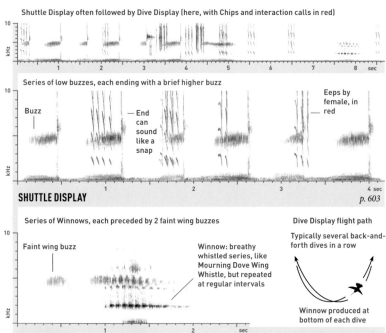

Shuttle Display often followed by Dive Display (here, with Chips and interaction calls in red)

Series of low buzzes, each ending with a brief higher buzz

Buzz

— End can sound like a snap

Eeps by female, in red

SHUTTLE DISPLAY *p. 603*

Series of Winnows, each preceded by 2 faint wing buzzes

Dive Display flight path

Faint wing buzz

Winnow: breathy whistled series, like Mourning Dove Wing Whistle, but repeated at regular intervals

Typically several back-and-forth dives in a row

Winnow produced at bottom of each dive

DIVE DISPLAY *p. 574*

Two types of flight display given during the breeding season by males. Displays of young males lack wing and tail sounds of adult. Shuttle Display given mostly in courtship: male buzzes back and forth along a short horizontal path, very close to the female; rate about 3 shuttles in 2 seconds. Dive Display given in aggression or courtship, occasionally directed at non-hummingbirds; rate about 1 dive every 3–4 seconds. All display sounds nonvocal: buzzes in Shuttle Display produced by wings, Winnow produced by tail.

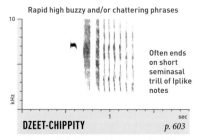

Rapid high buzzy and/or chattering phrases

Often ends on short seminasal trill of Iplike notes

DZEET-CHIPPITY *p. 603*

All year, by both sexes, in aggressive interactions. Highly plastic; the note types shown here combine into many patterns.

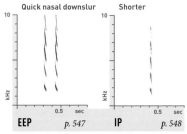

Quick nasal downslur

Shorter

EEP *p. 547*   **IP** *p. 548*

All year; most common calls. In interactions, often gives loud repeating series of 2–3 Eeps; solo birds give short, single Ips.

ANNA'S HUMMINGBIRD

Calypte anna

As late as 1930, restricted to southern coastal California chapparal and oaks; now uses residential areas extensively, and range has expanded manyfold.

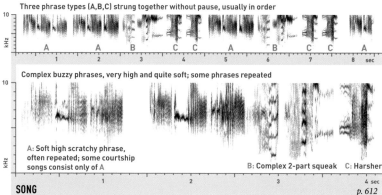

Three phrase types (A,B,C) strung together without pause, usually in order

A   A   B   C   C   A   B   C   C   A

Complex buzzy phrases, very high and quite soft; some phrases repeated

A: Soft high scratchy phrase, often repeated; some courtship songs consist only of A

B: Complex 2-part squeak   C: Harsher

SONG *p. 612*

All year, though many songs June–Nov. are plastic versions from first-year males. Song is learned, with local dialects; given frequently from perch or in flight, but never loud. Anna's apparently lacks true stereo-typed Shuttle Display, but males can challenge other males with plastic back-and-forth "strafing" flight.

A single loud 1- or 2-syllabled whistle

Higher first syllable often inaudible

Ends with a loud Hee

Faint trill rarely heard

Dive Display flight path

Bird first rises 20–50 feet in the air,

then swoops down fast

DIVE DISPLAY *pp. 538, 587*

Mostly Feb.–May, by males in aggression or courtship, occasionally directed at non-hummingbirds. Often dives just once, but sometimes repeats display many times, with dives 5–6 seconds apart. Dive is always oriented into sun, to maximize iridescence of gorget. Loud Hee is made by tail, preceding sounds by wings.

Most notes buzzy

Rhythm often stuttering

CHITTER *p. 582*

All year, by both sexes, in aggression. Highly plastic, grading into Twitter, but typical versions rather distinctive.

Very sharp, almost smacking

Often run into rapid Twitter (p. 568)

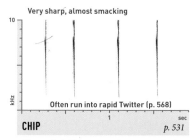

CHIP *p. 531*

All year, by both sexes; most common call, given in a variety of situations. Distinctive; higher, sharper than Chips of other hummingbirds.

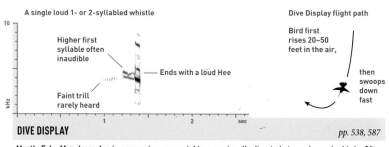

Costa's Hummingbird

Calypte costae

Breeds in dry desert, coastal chapparal, sage scrub, and some residential areas. Seasonal movements complex. Barely larger than Calliope Hummingbird.

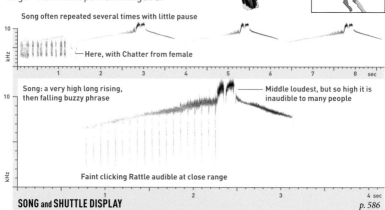

Song often repeated several times with little pause

Here, with Chatter from female

Song: a very high long rising, then falling buzzy phrase

Middle loudest, but so high it is inaudible to many people

Faint clicking Rattle audible at close range

SONG and SHUTTLE DISPLAY *p. 586*

Song, a vocal sound, given nearly all year, but especially Nov.–May, by males. Plastic song of young males often heard June–Oct. Uniform and stereotyped. When given from perch, often consists of a single phrase. During Shuttle Display (4 shuttles/second), phrases often repeated with little pause.

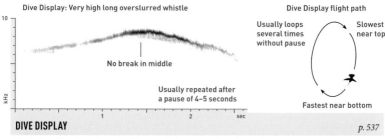

Dive Display: Very high long overslurred whistle

No break in middle

Usually repeated after a pause of 4–5 seconds

Dive Display flight path

Usually loops several times without pause

Slowest near top

Fastest near bottom

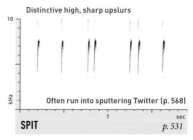

DIVE DISPLAY *p. 537*

Mostly Dec.–May, by males. Sound is very similar to vocal Song, but made by tail at bottom of display dive; not broken into 2 notes, and never repeated without a pause of several seconds as bird makes another loop. Display may begin or end with a zig-zagging horizontal flight by male. Diving bird can be hard to see.

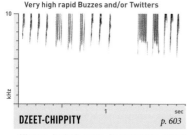

Very high rapid Buzzes and/or Twitters

DZEET-CHIPPITY *p. 603*

All year, by both sexes, in aggression. Highly plastic. Usually higher than other hummingbird interaction calls, but also gives lower Chatters.

Distinctive high, sharp upslurs

Often run into sputtering Twitter (p. 568)

SPIT *p. 531*

All year, by both sexes; most common call, given in a variety of situations. Subtly different from any other North American hummingbird call.

CALLIOPE HUMMINGBIRD

Selasphorus calliope

The smallest bird in North America. Tiny, with very short tail and bill. Nests along streams and in shrubby clearings in coniferous mountain forests.

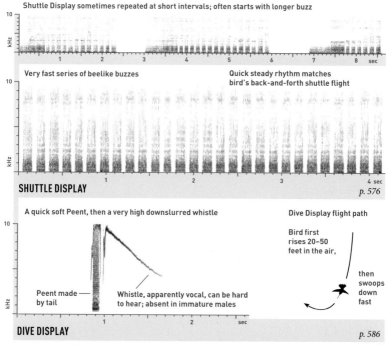

Shuttle Display sometimes repeated at short intervals; often starts with longer buzz

Very fast series of beelike buzzes

Quick steady rhythm matches bird's back-and-forth shuttle flight

SHUTTLE DISPLAY   *p. 576*

A quick soft Peent, then a very high downslurred whistle

Dive Display flight path

Bird first rises 20–50 feet in the air,

then swoops down fast

Peent made by tail

Whistle, apparently vocal, can be hard to hear; absent in immature males

DIVE DISPLAY   *p. 586*

Two types of flight display given during the breeding season by males. Displays of young males lack wing and tail sounds of adult. Shuttle Display given mostly in courtship: male buzzes back and forth along a short horizontal path, very close to the female; rate about 6–7 shuttles in 1 second. Dive Display given in aggression or courtship, occasionally directed at non-hummingbirds; often dives just once, but sometimes repeats display 3–4 times, with dives 5–6 seconds apart. Wings of male in flight make rather soft beelike buzz.

Rapid high buzzy and/or twittering phrases

DZEET-CHIPPITY   *p. 603*

All year, by both sexes, in aggressive interactions. Highly plastic; the note types shown here combine into many patterns.

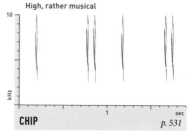

High, rather musical

CHIP   *p. 531*

All year, by both sexes; most common call, given in a variety of situations. Quite similar to Chip of Broad-tailed Hummingbird.

RUFOUS HUMMINGBIRD

Selasphorus rufus

Extensive orange color is distinctive. Common late-summer migrant well to east of breeding range. Males are very aggressive, defending feeders fiercely.

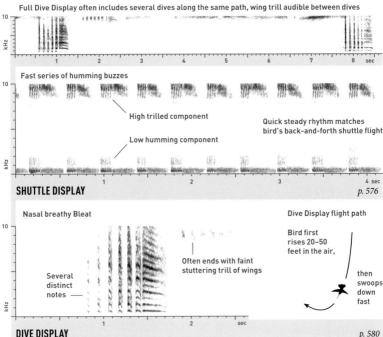

Full Dive Display often includes several dives along the same path, wing trill audible between dives

Fast series of humming buzzes

High trilled component

Low humming component

Quick steady rhythm matches bird's back-and-forth shuttle flight

SHUTTLE DISPLAY   *p. 576*

Nasal breathy Bleat

Dive Display flight path

Bird first rises 20–50 feet in the air,

Often ends with faint stuttering trill of wings

then swoops down fast

Several distinct notes

DIVE DISPLAY   *p. 580*

Two types of flight display given during the breeding season by males. Displays of young males lack wing and tail sounds of adult. Shuttle Display given mostly in courtship: male buzzes back and forth along a short horizontal path, very close to the female; rate about 2–3 shuttles in 1 second. Dive Display given in aggression or courtship, occasionally directed at non-hummingbirds; often dives just once, but sometimes repeats display 3–4 times, with dives 6–8 seconds apart. Wings of male in flight make high trill, like Broad-tailed but higher and softer.

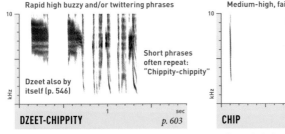

Rapid high buzzy and/or twittering phrases

Short phrases often repeat: "Chippity-chippity"

Dzeet also by itself (p. 546)

DZEET-CHIPPITY   *p. 603*

All year, by both sexes, in aggressive interactions. Highly plastic; the note types shown here combine into many patterns.

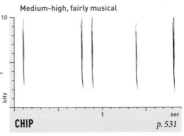

Medium-high, fairly musical

CHIP   *p. 531*

All year, by both sexes; most common call, given in a variety of situations. Averages slightly lower and sharper than Chip of Broad-tailed.

Allen's Hummingbird

Selasphorus sasin

Like Rufous Hummingbird, but males have green back and cap (as do some hybrids with Rufous). Vocalizations are nearly identical, but displays differ.

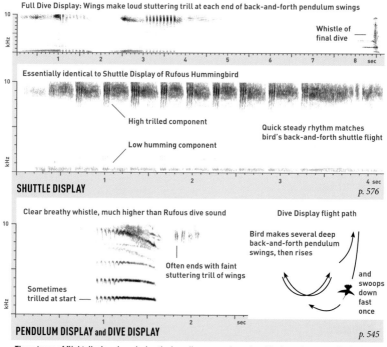

Full Dive Display: Wings make loud stuttering trill at each end of back-and-forth pendulum swings

Whistle of final dive

Essentially identical to Shuttle Display of Rufous Hummingbird

High trilled component

Low humming component

Quick steady rhythm matches bird's back-and-forth shuttle flight

SHUTTLE DISPLAY — p. 576

Clear breathy whistle, much higher than Rufous dive sound

Dive Display flight path

Bird makes several deep back-and-forth pendulum swings, then rises

Often ends with faint stuttering trill of wings

Sometimes trilled at start

and swoops down fast once

PENDULUM DISPLAY and DIVE DISPLAY — p. 545

Three types of flight display given during the breeding season by males. Displays of young males lack wing and tail sounds of adult. Shuttle Display given mostly in courtship: male buzzes back and forth along a short horizontal path, very close to the female; rate about 2–3 shuttles in 1 second. Pendulum Display followed by Dive Display given in aggression or courtship, usually to female hummingbirds. Pendulum Display can be performed without Dive Display, but Dive rarely occurs without preceding Pendulum Display. Wings of male in flight make high trill identical to that of Rufous.

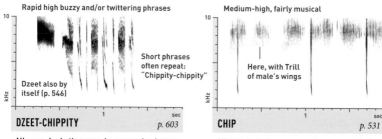

Rapid high buzzy and/or twittering phrases

Short phrases often repeat: "Chippity-chippity"

Dzeet also by itself (p. 546)

DZEET-CHIPPITY — p. 603

All year, by both sexes, in aggressive interactions. Highly plastic; the note types shown here combine into many patterns.

Medium-high, fairly musical

Here, with Trill of male's wings

CHIP — p. 531

All year, by both sexes; most common call, given in a variety of situations. Not known to be distinguishable from Chip of Rufous Hummingbird.

Broad-tailed Hummingbird

Selasphorus platycercus

Common as a breeder in foothill and mountain meadows and open forests. Slightly larger than other *Selasphorus* hummingbirds, with wider tail.

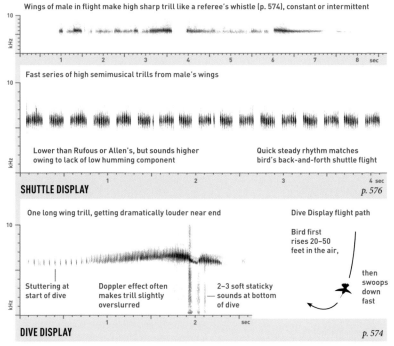

Wings of male in flight make high sharp trill like a referee's whistle (p. 574), constant or intermittent

Fast series of high semimusical trills from male's wings

Lower than Rufous or Allen's, but sounds higher owing to lack of low humming component

Quick steady rhythm matches bird's back-and-forth shuttle flight

SHUTTLE DISPLAY   *p. 576*

One long wing trill, getting dramatically louder near end

Dive Display flight path

Bird first rises 20–50 feet in the air,

then swoops down fast

Stuttering at start of dive

Doppler effect often makes trill slightly overslurred

2–3 soft staticky sounds at bottom of dive

DIVE DISPLAY   *p. 574*

Two types of flight display given during the breeding season by males. Displays of young males lack wing and tail sounds of adult. Shuttle Display given mostly in courtship: male buzzes back and forth along a short horizontal path, very close to the female; rate about 4 shuttles in 1 second. Dive Display given in aggression or courtship, occasionally directed at non-hummingbirds; often dives just once, but sometimes repeats display 3–4 times, with dives 5–6 seconds apart. All display sounds nonvocal.

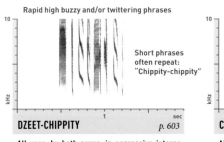

Rapid high buzzy and/or twittering phrases

Short phrases often repeat: "Chippity-chippity"

DZEET-CHIPPITY   *p. 603*

All year, by both sexes, in aggressive interactions. Highly plastic; the note types shown here combine into many patterns.

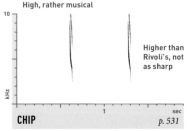

High, rather musical

Higher than Rivoli's, not as sharp

CHIP   *p. 531*

All year, by both sexes; most common call, given in a variety of situations. Quite similar to Chip of Calliope Hummingbird.

BROAD-BILLED HUMMINGBIRD

Cynanthus latirostris

A small hummingbird, locally common in riparian areas in its limited U.S. range. Bill is indeed broad at base, though this can be difficult to see.

Males apparently have a single Song, repeated with little variation

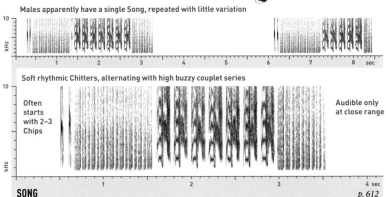

Soft rhythmic Chitters, alternating with high buzzy couplet series

Often starts with 2–3 Chips

Audible only at close range

SONG *p. 612*

Possibly all year, by males, usually from a perch. Infrequent. Details differ from one male to the next, but overall song pattern seems quite uniform rangewide. Male also performs display dives, accompanied by loud hum of wings, but these dives apparently lack specialized vocal or mechanical sounds.

High, complex, slightly buzzy couplet series

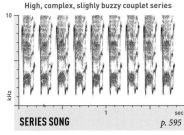

SERIES SONG *p. 595*

At least Apr.–July, in interactions, possibly by males in close courtship. May be a truncated version of typical Song; more study needed.

High, downslurred | Variably buzzy

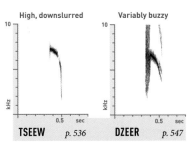

TSEEW *p. 536* | **DZEER** *p. 547*

Possibly all year; function poorly understood. Some versions given in chases. Plastic; the two versions shown grade into one another.

High sharp Chips and Tseets

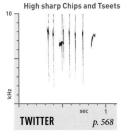

TWITTER *p. 568*

All year, in chases. Highly plastic; Tseets often at end, or in repeated rhythms.

Harsh but high

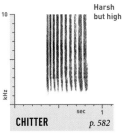

CHITTER *p. 582*

All year, in agitation. Grades smoothly into Chittits and single Chits.

Often a doubled Chittit (p. 582)

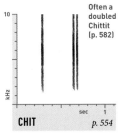

CHIT *p. 554*

All year; most common call. Distinctive; no hummingbird in our area sounds similar.

VIOLET-CROWNED HUMMINGBIRD

Amazilia violiceps

Rare and local in riparian woodlands in canyons, where it nests in sycamores. A medium-large hummingbird; note plain white underparts and red bill.

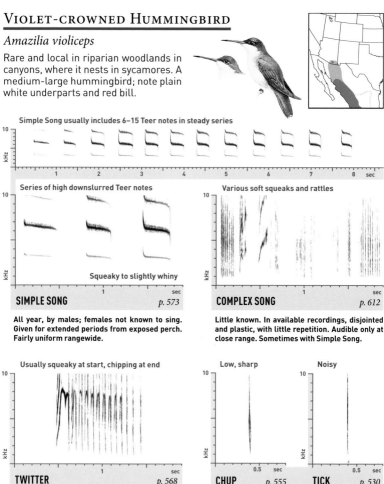

Simple Song usually includes 6–15 Teer notes in steady series

Series of high downslurred Teer notes

Various soft squeaks and rattles

Squeaky to slightly whiny

SIMPLE SONG   *p. 573*

COMPLEX SONG   *p. 612*

All year, by males; females not known to sing. Given for extended periods from exposed perch. Fairly uniform rangewide.

Little known. In available recordings, disjointed and plastic, with little repetition. Audible only at close range. Sometimes with Simple Song.

Usually squeaky at start, chipping at end

Low, sharp

Noisy

TWITTER   *p. 568*

CHUP   *p. 555*

TICK   *p. 530*

All year, in interactions. Highly plastic; some versions lack squeaky notes at start. Other versions almost entirely squeaky.

All year. Loud Chup given in excitement, often in series; softer Tick given singly while foraging. Both grade into one another and into Twitter.

SWIFTS (Order Apodiformes, Family Apodidae)

Swifts are highly aerial birds that pursue flying insects on the wing. They resemble swallows, but tend to fly faster and higher, on stiffer, narrower wings. Though they differ greatly from hummingbirds in shape and habits, anatomical details show them to be a related lineage that developed from a common ancestor, and they share the order Apodiformes.

Swifts nest in rock crevices, in hollow trees, inside chimneys, or on cliffs, rarely perching on anything other than a vertical surface. In some species, some individuals may stay aloft for months at a time, even sleeping on the wing. Vocalizations tend to be rather simple and highly plastic; they are apparently innate.

White-throated Swift

Aeronautes saxatalis

Nests on cliffs and canyon walls in arid country, as well as on sea cliffs and occasionally on man-made structures. Often highly vocal on the wing.

Speed, pitch, and tone quality constantly changing, slowly or abruptly

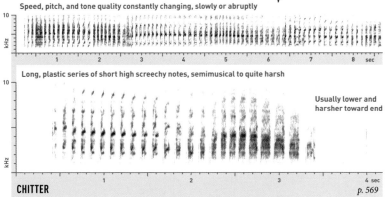

Long, plastic series of short high screechy notes, semimusical to quite harsh

Usually lower and harsher toward end

CHITTER

p. 569

All year, by both sexes, on the wing or from inside nest crevice. Extremely plastic but uniform rangewide. Juveniles beg with similar calls. Multiple individuals often join in screechy chorus. In the nest, also gives single soft high Grates (p. 559) like single notes from end of Chitter.

Black Swift

Cypseloides niger

Nests in the spray zone of waterfalls and on sea cliffs. Away from nest, flies high; easiest to see and hear at dawn and dusk, or in inclement weather.

Groups give Twitter of Chips and Pips with occasional Pseeps, lower and slower than Chimney Swift's

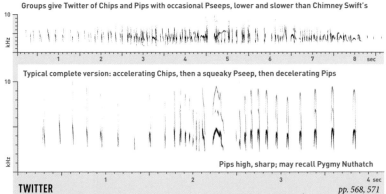

Typical complete version: accelerating Chips, then a squeaky Pseep, then decelerating Pips

Pips high, sharp; may recall Pygmy Nuthatch

TWITTER

pp. 568, 571

Likely all year, but mostly heard near nest, June–Aug. Extremely plastic but uniform rangewide. Also gives single Pip (p. 534). Complete version of Twitter call, shown above, quite distinctive; decelerating series of Pips may recall Western Kingbird.

CHIMNEY SWIFT

Chaetura pelagica

Often seen over towns. Note dark color, cigar-shaped body, and stiff wingbeats. Once nested in hollow trees, but now primarily in chimneys.

Short Twitters often vaguely downslurred; notes sometimes accelerate into brief buzzes

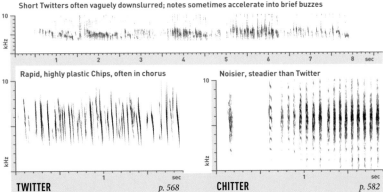

Rapid, highly plastic Chips, often in chorus

Noisier, steadier than Twitter

TWITTER   *p. 568*   **CHITTER**   *p. 582*

Twitter given all year, by adults, generally in flight. Mostly associated with social behavior, but function unstudied. Extremely plastic; variations in speed and tone may correspond to different behaviors. Chitter given in chimneys by young birds; by 5 weeks of age, Chitter starts becoming Twitter.

VAUX'S SWIFT

Chaetura vauxi

Very similar to Chimney Swift, but does not overlap in typical U.S. range. Nests in hollow trees, less often in chimneys; roosts in chimneys in fall migration.

Twitter significantly slower, more musical, and more varied than Chimney Swift's, and slightly higher

Often gives short Songlike phrases

Listen for brief high burry notes

Can recall Brown Creeper Song, but more chipping and plastic

TWITTER   *p. 568*

Twitter given all year, by adults, generally in flight. Mostly associated with social behavior. Extremely plastic, but certain Songlike patterns have a tendency to recur; more study needed. Nestlings give Chitter (p. 582) like Chimney Swift's. Also reportedly makes a booming sound with wings inside nest cavity.

RAILS (Family Rallidae)

Rails are related to cranes and to the Limpkin of the East; the three families share the order Gruiformes. The rails are medium-sized to small, brownish or blackish, rather chickenlike birds that live in dense marsh vegetation. Most species are much more often heard than seen, though the coots and gallinules are highly visible as they swim in areas of open water.

In many species, sounds tend to consist of rather simple harsh, grunting, or screechy notes, but vocal repertoires are complex, and duets are common. Sounds are apparently innate.

CRANES (Family Gruidae)

The long-necked, long-legged cranes have played symbolic roles in many cultures worldwide owing to their graceful beauty, their elaborate mating dances, and their loud trumpeting vocalizations. They resemble herons, but fly with necks outstretched, nest on the ground, and sport a "bustle" of drooping tertial feathers that usually cover the tail. Vocal duetting is well developed in the North American species. As far as is known, all sounds are innate.

YELLOW RAIL

Coturnicops noveboracensis

A tiny rail, highly local in summer in sedge marshes and in winter in salt marshes. Very difficult to see. Prefers to run from danger rather than fly.

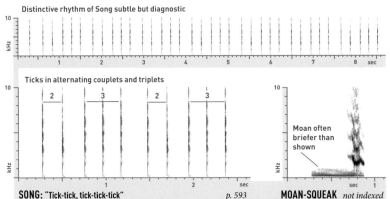

Distinctive rhythm of Song subtle but diagnostic

Ticks in alternating couplets and triplets

SONG: "Tick-tick, tick-tick-tick" *p. 593*

MOAN-SQUEAK *not indexed*

Moan often briefer than shown

Song given May–Aug., by males, mostly after dark; rarely by day early in season. Northern Cricket Frog has similar song, but with different rhythm. Moan-Squeak given in high alarm; Moan and Squeak sometimes given separately. Also a descending Cackle rarely in migration; no recordings.

BLACK RAIL

Laterallus jamaicensis

Very local in salt marshes and certain inland freshwater marshes. Tiny and rarely seen. Note that juveniles of other rail species are small and black.

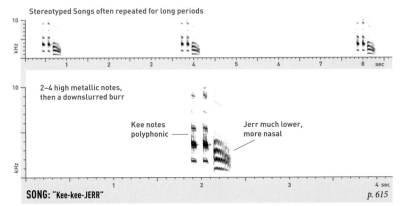

Stereotyped Songs often repeated for long periods

2–4 high metallic notes, then a downslurred burr

Kee notes polyphonic

Jerr much lower, more nasal

SONG: "Kee-kee-JERR"
p. 615

Mostly Mar.–June, but given all year in some regions. Fairly uniform and stereotyped. Up to 4 Kee notes, but usually 2; up to 3 Jerr notes, but almost always just 1. Given mostly at night, especially in East, but occasionally during day.

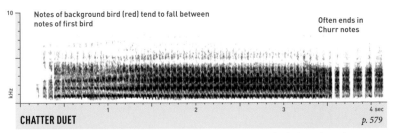

Notes of background bird (red) tend to fall between notes of first bird

Often ends in Churr notes

CHATTER DUET
p. 579

Mostly during breeding season; function little known. Often given by pairs in duet, alternating notes so rapidly that they sound like a single bird. Sometimes mixed with Churrs; one bird may give Churrs in response to partner's Chatter. Strikingly similar to call given by duetting Least Grebes.

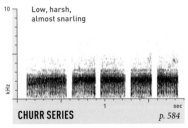

Low, harsh, almost snarling

CHURR SERIES
p. 584

All year, but especially in breeding season, in agitation or alarm. Notes fairly constant in pitch, but series plastic in speed.

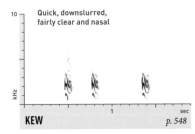

Quick, downslurred, fairly clear and nasal

KEW
p. 548

All year, by both sexes. Possibly a contact call. Higher versions in faster series may be associated with alarm.

Virginia Rail

Rallus limicola

Common but secretive. Breeds in dense cattails or bulrushes, rarely in salt marshes. Only two-thirds the size of Ridgway's Rail.

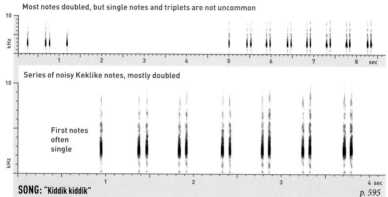

Most notes doubled, but single notes and triplets are not uncommon

Series of noisy Keklike notes, mostly doubled

First notes often single

SONG: "Kiddik kiddik" *p. 595*

Mostly Apr.–June, possibly only from males; apparently functions primarily in attracting a mate, though also given after eggs hatch. Consistent doubling of notes diagnostic, but occasional birds give 10 or more single notes before doubling. Higher than calls of Ridgway's Rail, lacking the bass register.

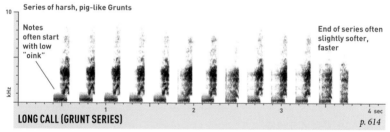

Series of harsh, pig-like Grunts

Notes often start with low "oink"

End of series often slightly softer, faster

LONG CALL (GRUNT SERIES) *p. 614*

All year, often by mates in unsynchronized duets, apparently to maintain pair bond; also in territorial disputes. Pair duets tend to be longer and less accelerating than solo versions. Averages shorter than Grunt Series of Ridgway's Rail, with lower, more growling, more disyllabic notes.

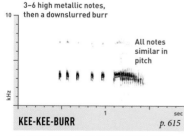

3–6 high metallic notes, then a downslurred burr

All notes similar in pitch

KEE-KEE-BURR *p. 615*

In early spring, perhaps by females. Introductory notes and final burry note may be given separately.

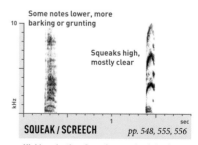

Some notes lower, more barking or grunting

Squeaks high, mostly clear

SQUEAK / SCREECH *pp. 548, 555, 556*

Highly plastic; Squeaks grade into longer Screeches and sometimes lower Barks. Given singly in alarm by both sexes.

RIDGWAY'S RAIL

Rallus obsoletus

The largest rail in our area, a rare and local resident of coastal saltmarshes and Colorado River wetlands. Recently split from the Clapper Rail of the East.

Often gives irregular, slow Keks at length before accelerating to top speed and then slowing

Series of abrupt noisy notes, reaching 2.5–3.5 notes per second

SONG (KEK SERIES) *p. 581*

Mostly Mar.–June, apparently by unmated males advertising for a mate. Usually delivered in irregular bouts lasting 4–10 seconds, accelerating and then decelerating, but may be given continuously. Speed of series increases with level of agitation.

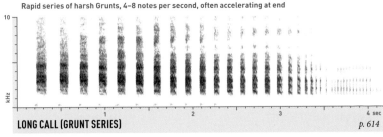

Rapid series of harsh Grunts, 4–8 notes per second, often accelerating at end

LONG CALL (GRUNT SERIES) *p. 614*

All year, often by mates in unsynchronized duets, in courtship and to maintain pair bond; also used in territorial disputes. Pair duets tend to be longer and less accelerating than solo versions. Fastest versions have shortest, most Keklike notes, but still maintain some grunting quality.

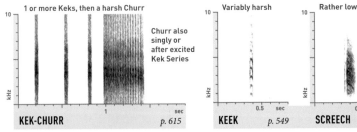

1 or more Keks, then a harsh Churr

Churr also singly or after excited Kek Series

KEK-CHURR *p. 615*

Variably harsh

KEEK *p. 549*

Rather low

SCREECH *p. 556*

Given by unmated females to attract a male. Churr without Kek given by both sexes, perhaps as alarm or contact call near nest.

Given singly but repeatedly in agitation. Also gives Chuk (p. 554) like single Kek from Song. Some versions recall Screech of Virginia Rail.

SORA

Porzana carolina

Breeds mostly in freshwater marshes; winters in brackish and salt marshes also. Rather bold for a rail, but still more often heard than seen.

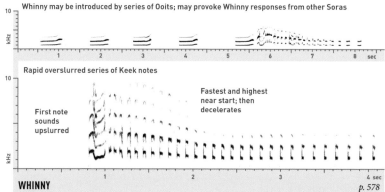

Whinny may be introduced by series of Ooits; may provoke Whinny responses from other Soras

Rapid overslurred series of Keek notes

First note sounds upslurred

Fastest and highest near start; then decelerates

WHINNY   *p. 578*

Mostly in spring migration and breeding season, apparently in territorial defense and pair contact; often associated with chases. May be provoked by sounds of other Soras, Virginia Rails, or any loud noise. Females often respond to mate's Whinny, often starting halfway through his call (see top).

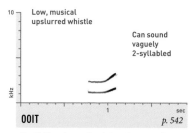

Low, musical upslurred whistle

Can sound vaguely 2-syllabled

OOIT   *p. 542*

Mostly May–June, often in long bouts, perhaps to attract mate. Similar calls reportedly given singly by night migrants.

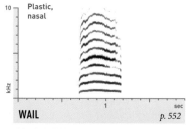

Plastic, nasal

WAIL   *p. 552*

Mostly in fall, especially by juveniles; possibly also by night migrants. Length and pitch pattern highly plastic; grades into Ooit.

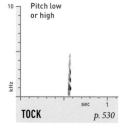

Pitch low or high

TOCK   *p. 530*

Mostly in contact between family members near nest. Like coot calls, but softer.

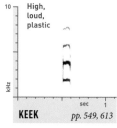

High, loud, plastic

KEEK   *pp. 549, 613*

Mostly during breeding, perhaps in alarm; sometimes in long regular bouts.

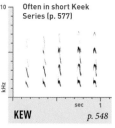

Often in short Keek Series (p. 577)

KEW   *p. 548*

Variable, plastic; grades into Keek. May be related to female version of Whinny.

Common Gallinule

Gallinula galeata

Common in southern marshes; local farther north. Not particularly shy, but often inconspicuous. Formerly lumped with the Common Moorhen of Eurasia.

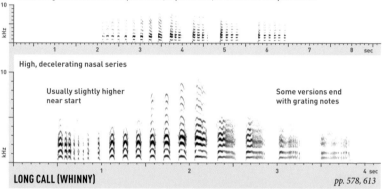

Some Long Calls break into couplet series, triplet series, or even more complex series

High, decelerating nasal series

Usually slightly higher near start

Some versions end with grating notes

LONG CALL (WHINNY)

pp. 578, 613

All year, but mostly Apr.–June. Believed to be used by males to advertise territory; perhaps also in mate contact. Often given in response to mates or neighbors. Versions that end in grates or complex series fairly distinctive, but some overlap with pattern of Purple Gallinule Long Call.

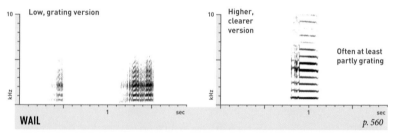

Low, grating version

Higher, clearer version

Often at least partly grating

WAIL

p. 560

Exceedingly variable and plastic category of sounds; often loud. Varies in pitch and tone, but strained quality rather distinctive when present, as though bird is trying and failing to keep voice clear of grating sounds. Function and behavioral context poorly understood.

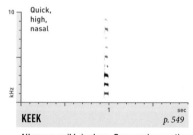

Quick, high, nasal

KEEK

p. 549

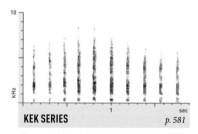

KEK SERIES

p. 581

All year, possibly in alarm. Rare versions grating or clucking Tocks (p. 530), or low Barks (p. 550). Averages higher than Purple Gallinule Keek.

Not well known. Kek Series variable, plastic; speed can be irregular. Note type can change during series, but does not alternate.

AMERICAN COOT

Fulica americana

Abundant on vegetated ponds as well as larger lakes. In migration and winter, occasionally found on salt water. Swims and dives more than other rails.

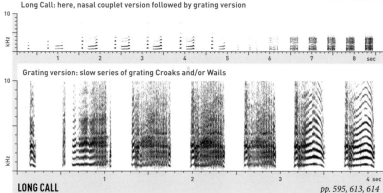

Long Call: here, nasal couplet version followed by grating version

Grating version: slow series of grating Croaks and/or Wails

LONG CALL

pp. 595, 613, 614

Highly plastic and variable. Often introduced by single-note calls. Two main types, possibly corresponding to different sexes or purposes. More common version consists mostly of long Grates. Another version includes nasal couplet series, with second note of each couplet often long, rising.

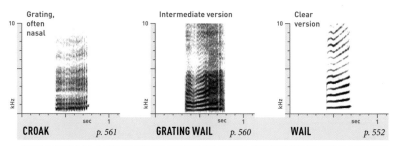

Grating, often nasal

Intermediate version

Clear version

CROAK *p. 561*

GRATING WAIL *p. 560*

WAIL *p. 552*

Exceedingly variable and plastic category of intergrading sounds; often loud. Most versions at least partly croaking or grating, but some are clear and nasal. Function and behavioral context need more study. Begging juveniles give whistled Screech (p. 556) or Pseep Series (p. 570).

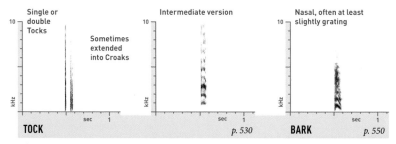

Single or double Tocks

Sometimes extended into Croaks

Intermediate version

Nasal, often at least slightly grating

TOCK

p. 530

BARK *p. 550*

Gives a wide variety of highly variable and plastic single-note sounds. Most versions at least partly grating. Function and behavioral context need more study.

SANDHILL CRANE

Antigone canadensis

Breeds in bogs and grassy wetlands. Locally common; most populations stable or increasing. In migration, often stages for weeks in enormous flocks.

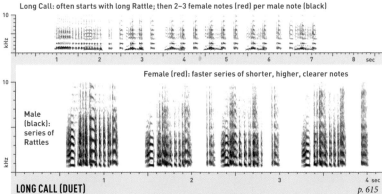

Long Call: often starts with long Rattle; then 2–3 female notes (red) per male note (black)

Female (red): faster series of shorter, higher, clearer notes

Male (black): series of Rattles

LONG CALL (DUET)   *p. 615*

All year, by mated pairs, sometimes joined by their older offspring. Variable, but often highly synchronized, given with display postures. Used in aggressive encounters with other pairs or family groups; also apparently to reinforce pair bond. Birds separated from mates may give their half of the duet alone.

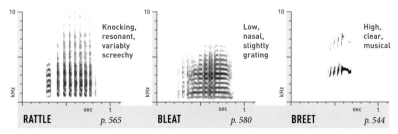

Knocking, resonant, variably screechy

RATTLE   *p. 565*

Low, nasal, slightly grating

BLEAT   *p. 580*

High, clear, musical

BREET   *p. 544*

Rattle given all year, in mild alarm, in aggressive encounters, and in contact, especially on the wing. Variable and plastic, but distinctive, frequent, and loud. Calls of pairs and family groups often overlap. Bleats infrequent; function unclear. Juveniles give Breet, grading into Rattle around the age of 9 months.

ELONGATED TRACHEAS

The distinctively resonant vocalizations of some birds may be connected to the unique anatomy of their tracheas, which are extremely long, looping around inside the body and sometimes penetrating the sternum (breastbone) before entering the lungs. Some have argued that the elongation of the trachea serves to lower the pitch of the calls, while others believe its primary purpose is to make the calls louder by using the sternum as a resonator. More study is needed. The North American birds known to possess this anatomy include some of our loudest species: Limpkin, Sandhill and Whooping Cranes, Trumpeter and Tundra Swans, and Plain Chachalaca.

SHOREBIRDS

This diverse group includes several related families of birds. The majority of species are adapted to feeding along shorelines, where they search for invertebrate prey in various ways, from probing in mud to wading in water. Some species are regularly found far from water. Despite great variation in size and shape, many shorebirds share a distinctive basic body plan, often with long legs, a long neck, and a long bill. Many species have extensive vocal repertoires.

STILTS AND AVOCETS (Family Recurvirostridae)

This family includes two species in our area, both with long legs and striking plumage. Their vocalizations are loud and frequent, but much simpler than those of most other shorebirds. All sounds are presumed to be innate.

OYSTERCATCHERS (Family Haematopodidae)

The oystercatchers are large, strikingly patterned birds with distinctive long, bright red bills that are deep but quite thin side-to-side, which they use in foraging for bivalves, worms, and other marine invertebrates. Our species inhabits beaches and barrier islands, especially those made almost entirely of mollusk shells. All sounds are presumed to be innate.

BLACK-NECKED STILT

Himantopus mexicanus

Unmistakable, with bold plumage and ridiculously long legs. Breeds on shallow briny ponds; winters in flooded fields and salt marshes.

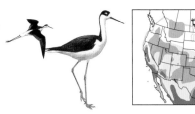

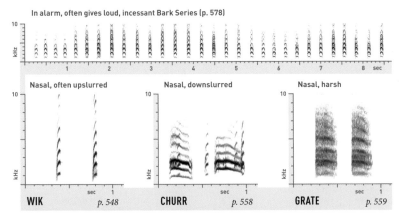

In alarm, often gives loud, incessant Bark Series (p. 578)

Nasal, often upslurred

Nasal, downslurred

Nasal, harsh

WIK   *p. 548*

CHURR   *p. 558*

GRATE   *p. 559*

Highly vocal at all times of year. Vocalizations plastic, grading into one another fairly smoothly according to level of agitation. Wik is most common call, given in alarm and in flight; plastic. Often noticeably upslurred, but some versions are Barks (p. 550). Grates given in high alarm; Churrs intermediate.

AMERICAN AVOCET

Recurvirostra americana

Beautiful and elegant. Rusty plumage turns white in winter. Often feeds by sweeping upturned bill back and forth just under the surface of the water.

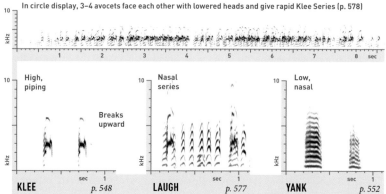

In circle display, 3–4 avocets face each other with lowered heads and give rapid Klee Series (p. 578)

High, piping	Nasal series	Low, nasal
Breaks upward		
KLEE *p. 548*	**LAUGH** *p. 577*	**YANK** *p. 552*

Klee is most common call, given all year in alarm and in flight, and in loud rapid chorus during circle display, apparently between 2 mated pairs or a mated pair and a single bird. Some versions are 2-syllabled Ka-lees. Laugh little known. Yank given during wing-raising distraction display.

BLACK OYSTERCATCHER

Haematopus bachmani

Often vocal and conspicuous on rocky shores and shellfish beds, but overall rather local and uncommon. True to its name, forages mostly on bivalves.

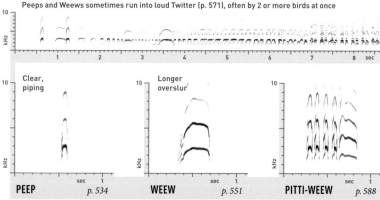

Peeps and Weews sometimes run into loud Twitter (p. 571), often by 2 or more birds at once

Clear, piping	Longer overslur	
PEEP *p. 534*	**WEEW** *p. 551*	**PITTI-WEEW** *p. 588*

All calls highly plastic, grading into one another. Peeps are most common, in contact and alarm. In courtship and territorial display, Peeps run into long Twitter with sudden changes in pitch and speed. Weew and combinations like Pitti-weew may function in pair bonding.

PLOVERS (Family Charadriidae)

The plovers are small to medium-sized shorebirds with short, stout bills and fairly long legs. Rather than probing in mud, they forage by alternately running and pausing, using their large eyes to spot prey. They are well known for their distraction displays, often feigning a broken wing or other injury in order to lead potential predators, including humans, away from nest or young. Sometimes these displays are accompanied by distinctive vocalizations.

All plover sounds are likely innate; most are whistles, with occasional trills or burrs. In several species, a flight display, with associated songlike sounds, forms an important part of courtship.

SNOWY PLOVER

Charadrius nivosus

Uncommon to rare; highly local. Nests on open sandy beaches and the edges of alkaline lakes. Populations fluctuate, but have declined in many areas.

Ter-weet and Peer-purr often given together, in plastic series (pp. 540, 611)

Rising whistle, usually sounding 2-syllabled

Ending loud, emphatic

TER-WEET   p. 587

Falling whistle, then a low burr

Here, with Burrts

PEER-PURR   p. 590

Mostly Jan.–Sept. Ter-weet given by males excavating nest scrapes in courtship, or in territorial confrontations, and by both sexes in response to potential predators. Peer-purr is given by males in courtship, possibly also in other contexts and by females. Both calls uniform but plastic.

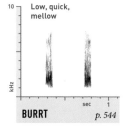

Low, quick, mellow

BURRT   p. 544

All year, in alarm. Plastic; in high alarm, can become longer, coarser.

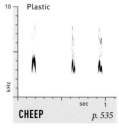

Plastic

CHEEP   p. 535

Mostly Aug.–Apr., in alarm, often in flight, including by flocks.

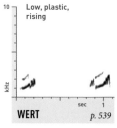

Low, plastic, rising

WERT   p. 539

Possibly all year. May simply be a short, plastic version of Ter-weet.

SEMIPALMATED PLOVER

Charadrius semipalmatus

Nests on patches of gravel or sand near water. In migration and winter, found on mudflats and beaches. Small, with one breast band, dark back.

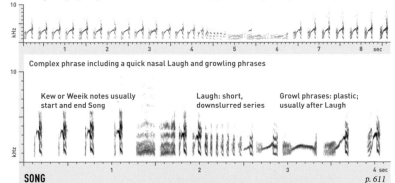

In flight display, long series of Weeiks are occasionally punctuated by a Laugh and 3–4 Growls

Complex phrase including a quick nasal Laugh and growling phrases

Kew or Weeik notes usually start and end Song

Laugh: short, downslurred series

Growl phrases: plastic; usually after Laugh

SONG   *p. 611*

All year. Highly stereotyped version given in flight display, mostly on breeding grounds but occasionally on spring migration. Softer, more plastic versions given frequently in migration and winter in aggressive interactions with other members of the same species, often with single rising Weeik notes.

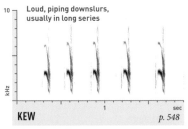

Loud, piping downslurs, usually in long series

KEW   *p. 548*

All year, in anxiety, possibly in aggression. Plastic; grades into Weeik notes of Song and likely also into Peeplike notes.

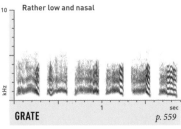

Rather low and nasal

GRATE   *p. 559*

Mostly June–Aug., in high-intensity distraction display near nest or young, while feigning injury.

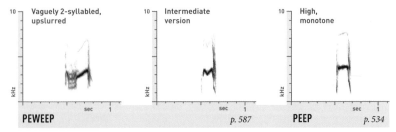

Vaguely 2-syllabled, upslurred

PEWEEP

Intermediate version

p. 587

High, monotone

PEEP   *p. 534*

All year. Most common calls, often given in flight. Also given in alarm, in low-intensity distraction display, and in conjunction with Kew calls. Highly plastic; 2-syllabled Peweep and 1-syllabled Peep are typical, but gives a range of intermediates.

PIPING PLOVER

Charadrius melodus

Uncommon and highly local; nests on open sand or gravel near water. Shares orange legs and bill with Semipalmated Plover, but much paler.

Song may continue for 10 seconds or more

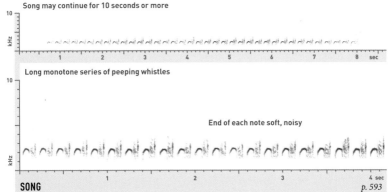

Long monotone series of peeping whistles

End of each note soft, noisy

SONG *p. 593*

Apr.–July, by males, usually in slow-flapping display flight over territory, or from ground after display flight ends. Uniform and stereotyped. Introduced by a long series of Weets. Female often responds to male Song with Peeps.

Fairly low, plastic, slightly noisy

Often accelerating

CHUCKLE *p. 583*

Mostly Apr.–July, by both sexes, in aggression, often accompanied by horizontal threat posture. Peeping version given from nest scrape.

Long, rising whistle

Consonant sound at start and end, audible at close range

WEET *p. 540*

Possibly all year, by both sexes, in alarm and aggression. Also associated with male Song, and sometimes given during flight display.

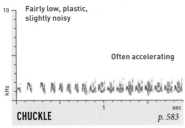

1-noted Peep

Here, given on 2 distinct pitches

PEEP *sec* 1 *p. 534*

2-noted Pee-lo

Breaks downward

PEE-LO *sec* 1 *p. 587*

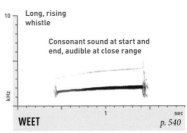

Sounds 2- or vaguely 3-syllabled

PEEA-LO *sec* 1 *p. 587*

A set of intergrading calls, given all year by both sexes in contexts ranging from contact to alarm; different versions have slightly different functions. 1-note Peeps run into rapid series by males excavating nest scrapes in courtship; some drawn-out versions given in broken-wing distraction display.

KILLDEER

Charadrius vociferus

Large, with two breast bands. Found in open fields and areas with short grass, not necessarily near water. Highly vocal, often calling nearly nonstop.

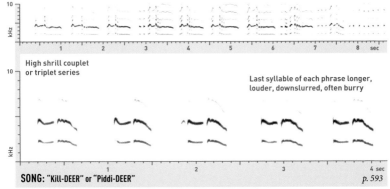

Excited versions of Song often burrier and more plastic, grading into Trill or other sounds

High shrill couplet or triplet series

Last syllable of each phrase longer, louder, downslurred, often burry

SONG: "Kill-DEER" or "Piddi-DEER"   *p. 593*

The species' namesake, given all year, especially in spring. Highly plastic, but geographically uniform; typically delivered on the wing, sometimes nonstop for many minutes. Some versions are quadruplet series. Associated with courtship, but also given in alarm and excitement.

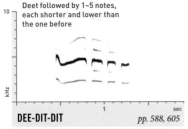

Deet followed by 1–5 notes, each shorter and lower than the one before

DEE-DIT-DIT   *pp. 588, 605*

Excited version of Deet, ranging from 2-noted Dee-dit (p. 587) to longer versions that grade into the Twitter.

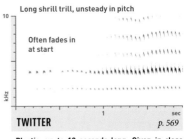

Long shrill trill, unsteady in pitch

Often fades in at start

TWITTER   *p. 569*

Plastic; up to 10 seconds long. Given in close courtship and in high alarm, including in broken-wing distraction displays.

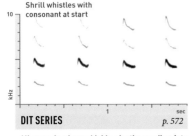

Shrill whistles with consonant at start

DIT SERIES   *p. 572*

All year, in alarm; highly plastic, grading into other sounds, especially Deet. Note length and shape highly variable.

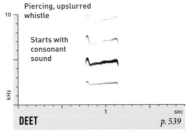

Piercing, upslurred whistle

Starts with consonant sound

DEET   *p. 539*

All year, in many situations. Highly plastic, especially in length; typical version shown. Grades into all other sounds.

MOUNTAIN PLOVER

Charadrius montanus

Breeds in well-grazed patches of arid shortgrass prairie, agricultural fields, and prairie dog towns. Population has apparently declined.

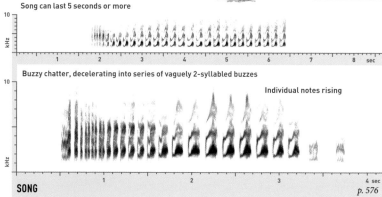

Song can last 5 seconds or more

Buzzy chatter, decelerating into series of vaguely 2-syllabled buzzes

Individual notes rising

SONG *p. 576*

Mostly Mar.–June, by both sexes, in territorial flight display, from the ground, or during aggressive interactions. Occasionally given in fall migration. Somewhat variable and slightly plastic; often truncated. Compare Dunlin Song.

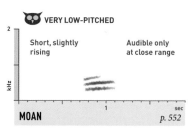

🦉 **VERY LOW-PITCHED**

Short, slightly rising

Audible only at close range

MOAN *p. 552*

Mostly Apr.–June, by courting male and sometimes female, at potential nest scrape, bowing head with fanned tail.

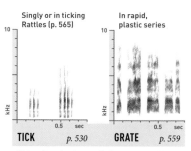

Singly or in ticking Rattles (p. 565)

In rapid, plastic series

TICK *p. 530*   **GRATE** *p. 559*

Ticks given in threat display to potential nest predators; Grates and soft, clear Wails given in injury-feigning distraction display.

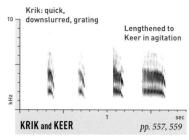

Krik: quick, downslurred, grating

Lengthened to Keer in agitation

KRIK and KEER *pp. 557, 559*

Apparently all year, by both sexes, in alarm and in interactions, often on the wing. Like Keer of Forster's Tern, but usually shorter.

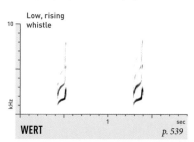

Low, rising whistle

WERT *p. 539*

Apparently all year, by both sexes, in mild alarm, often with Keer. Somewhat plastic.

AMERICAN GOLDEN-PLOVER

Pluvialis dominica

Breeds in Arctic; winters in southern South America. In migration, frequents fields, away from water. Lacks Black-bellied's white tail and wing stripe.

Away from breeding grounds, gives plastic, songlike phrases of short notes (p. 605)

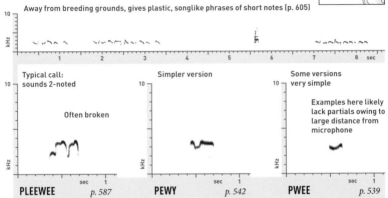

Typical call:
sounds 2-noted

Often broken

PLEEWEE *p. 587*

Simpler version

PEWY *p. 542*

Some versions
very simple

Examples here likely
lack partials owing to
large distance from
microphone

PWEE *p. 539*

Only sounds heard in migration and winter are shown. Pleewee and Pewy are typical of the variable, plastic calls given in flight and alarm; most versions are brief and complex, without much change in pitch. Plastic songlike phrases often quite soft; quality may recall Song of Willet.

PACIFIC GOLDEN-PLOVER

Pluvialis fulva

Breeds in Alaska; migrates in small numbers along Pacific Coast, and a few winter in California. Very rare inland. Very similar to American; voice differs.

Away from breeding grounds, gives plastic, songlike phrases of short notes (p. 605)

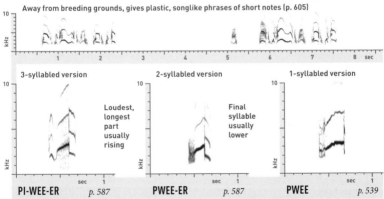

3-syllabled version

Loudest,
longest
part
usually
rising

PI-WEE-ER *p. 587*

2-syllabled version

Final
syllable
usually
lower

PWEE-ER *p. 587*

1-syllabled version

PWEE *p. 539*

Only sounds heard in migration and winter are shown. Pi-wee-er and Pwee-er are typical of the variable, plastic calls given in flight and alarm; most versions are 2- or vaguely 3-syllabled. Plastic songlike phrases often quite soft; may average higher than American's. Also gives high Weew in alarm (p. 551).

BLACK-BELLIED PLOVER

Pluvialis squatarola

Breeds in Arctic; winters mostly along coasts. Generally prefers mudflats to fields. In flight, shows large black patch at base of underwing.

Song (p. 605): Usually starts with trill, then 1–2 slurred, broken whistles, like Long-billed Curlew

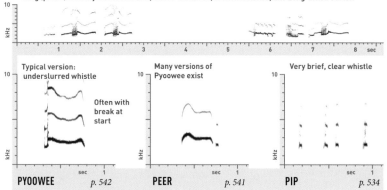

Typical version: underslurred whistle	Many versions of Pyoowee exist	Very brief, clear whistle
Often with break at start		
PYOOWEE p. 542	**PEER** p. 541	**PIP** p. 534

Only sounds heard in migration and winter are shown. Typical Pyoowee is highly distinctive, but variable and plastic; many kinds of whistles may be given. Pips soft, given during foraging, perhaps in contact. Songs given during interactions; nonbreeding versions are plastic and often truncated.

SANDPIPERS (Family Scolopacidae)   *next pages*

The sandpipers are a highly diverse family of shorebirds. Many species breed in the high Arctic, where they perform elaborate courtship displays, drawing on large, complex vocal repertoires. In many species, these breeding sounds are not heard south of the Arctic, and are therefore omitted from this book. All sandpiper sounds are presumed to be innate.

Most sandpipers frequent mudflats, beaches, and lakeshores, particularly outside the breeding season. Many species are gregarious, forming large flocks in good habitat. A group of species informally known as the "rockpipers" specializes in rocky coastlines. This group includes the turnstones, Surfbird, Rock Sandpiper, and Wandering Tattler, as well as the oystercatchers in the family Haematopodidae.

Snipe and woodcock are terrestrial sandpipers with short necks and legs and extremely long bills. They give aerial flight displays accompanied by mechanical sounds made by the tail (snipe) or the wings (woodcock).

The phalaropes swim more than other shorebirds, often spinning in tight circles to raise food to the surface. Females are brightly colored, while males incubate the eggs and raise the young. Unlike other sandpipers, phalaropes tend to have fairly simple vocal repertoires.

LONG-BILLED CURLEW

Numenius americanus

Large, with an amazing bill. Breeds on dry shortgrass prairie; in migration and winter, found on mudflats and in wet meadows. Populations have declined.

Song can be very long, sometimes including broken phrases like in Upland Sandpiper Song

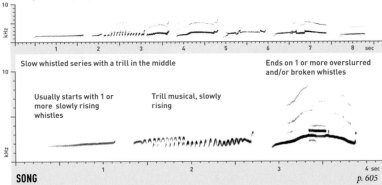

Slow whistled series with a trill in the middle

Ends on 1 or more overslurred and/or broken whistles

Usually starts with 1 or more slowly rising whistles

Trill musical, slowly rising

SONG

p. 605

All year, by both sexes, especially during the early breeding season. Highly variable and plastic. Often given from ground, especially after alighting; also given in circling flight and during territorial chases on breeding grounds.

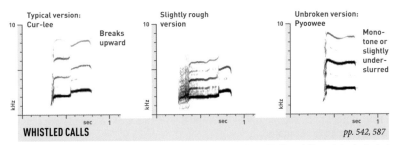

Typical version: Cur-lee

Breaks upward

Slightly rough version

Unbroken version: Pyoowee

Mono-tone or slightly under-slurred

WHISTLED CALLS

pp. 542, 587

All year, by both sexes; most common call, given in contact or alarm. Highly variable and plastic. Typical version is a clear, upward-breaking Cur-lee, but many variations exist. Given singly or in rapid series, from ground or in flight.

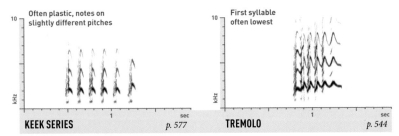

Often plastic, notes on slightly different pitches

First syllable often lowest

KEEK SERIES

p. 577

TREMOLO

p. 544

All year, especially during breeding season, by both sexes, in agitation, usually in flight. Sometimes used to mob predators, including humans. Highly plastic, grading into whistled calls. Some versions are Tremolos, some are Trills, and some are repeated screechy series of 2–3 notes.

WHIMBREL

Numenius phaeopus

Breeds on Arctic tundra. In migration and winter, found on tidal flats and beaches; a few remain on wintering range year-round.

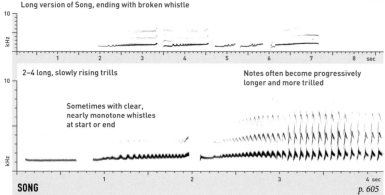

Long version of Song, ending with broken whistle

2–4 long, slowly rising trills

Notes often become progressively longer and more trilled

Sometimes with clear, nearly monotone whistles at start or end

SONG   *p. 605*

All year, but especially in aerial display on breeding grounds. Often given in flight. Variable and plastic; versions given in migration and winter are often truncated, sometimes with shorter and clearer notes.

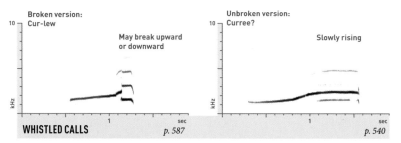

Broken version: Cur-lew

May break upward or downward

Unbroken version: Curree?

Slowly rising

WHISTLED CALLS   *p. 587*   *p. 540*

Probably all year, but apparently quite infrequent outside breeding season, when most instances may be better described as truncated songs. More study needed. Variable and plastic, but apparently averages longer than similar calls of Long-billed Curlew.

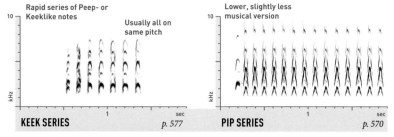

Rapid series of Peep- or Keeklike notes

Usually all on same pitch

Lower, slightly less musical version

KEEK SERIES   *p. 577*   **PIP SERIES**   *p. 570*

All year, in alarm. Most common call outside breeding season, given whenever birds are disturbed, often in flight. Somewhat variable in tone quality and plastic in length.

UPLAND SANDPIPER

Bartramia longicauda

A bird of tallgrass prairie. Related to curlews; note similar vocalizations and plumage. Flies on stiff wings, held high for a moment after landing.

Song sometimes interspersed with low, soft Groans

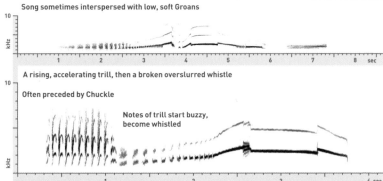

A rising, accelerating trill, then a broken overslurred whistle

Often preceded by Chuckle

Notes of trill start buzzy, become whistled

SONG *p. 605*

All year, by both sexes, but mostly Apr.–Aug., by males, often in gliding flight display. Sometimes given at night. Variable and somewhat plastic; frequently truncated, especially in late summer. Much like some versions of Long-billed Curlew Song, but averages more complex.

Clear, piping, usually 3-noted

Middle note often highest

QUIDDYQUIT *p. 570*

All year, including by night migrants. Plastic, especially in length.

Semimusical

CHUCKLE *p. 583*

All year, in alarm. Plastic; grades into Quiddyquit. Rarely shortened to 1 Cluck.

Nasal, variably burry

YANK *p. 552*

Mostly June–July, in high alarm, by parents with chicks.

BAR-TAILED GODWIT
Limosa lapponica

Rare visitor to Pacific Coast in winter and migration. Vocal repertoire like those of other godwits. Song speed variable.

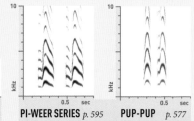

Rapid nasal couplet series (Pitty Pitty)

SONG *pp. 613, 590*

PI-WEER SERIES *p. 595*

PUP-PUP *p. 577*

Marbled Godwit

Limosa fedoa

Nests in well-grazed grasslands, often far from water; in migration and winter, frequents mudflats. Resembles Long-billed Curlew, but bill shape differs.

Song: rapid nasal triplet series (Gur-i-dik Gur-i-dik) that often starts and ends with simpler notes (p. 613)

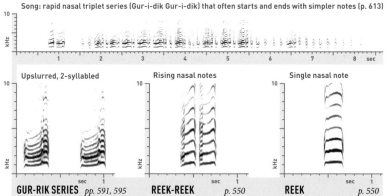

GUR-RIK SERIES *pp. 591, 595* **REEK-REEK** *p. 550* **REEK** *p. 550*

Upslurred, 2-syllabled   Rising nasal notes   Single nasal note

All vocalizations plastic, grading into one another, and given all year. Most distinctive sound is Song, a rapid series of Gur-i-dik phrases, given during interactions. Reek and Reek-reek given in alarm, often in flight. Gur-rik Series given repeatedly by males in circling courtship flights; also in other contexts.

Hudsonian Godwit

Limosa haemastica

Smaller than Marbled Godwit. Migrates mostly along East Coast in fall, through Great Plains in spring; winters in southern South America.

Song: rapid nasal triplet series (Poowitty Poowitty) that often starts and ends with simpler notes (p. 613)

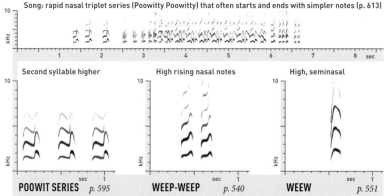

POOWIT SERIES *p. 595* **WEEP-WEEP** *p. 540* **WEEW** *p. 551*

Second syllable higher   High rising nasal notes   High, seminasal

Vocal repertoire much like that of Marbled Godwit, but all vocalizations higher and faster. Weew and Weep-weep given in alarm and in flight; often shortened to quick soft Wip. Poowit Series and Song possibly not often heard away from breeding grounds. Also gives soft Growls (not indexed).

Ruddy Turnstone

Arenaria interpres

Plumage and short, upturned bill are unique. Found mostly on rocky or sandy shores, occasionally mudflats. Uncommon to rare inland in migration.

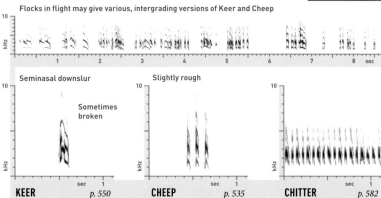

Flocks in flight may give various, intergrading versions of Keer and Cheep

Seminasal downslur

Sometimes broken

KEER *p. 550*

Slightly rough

CHEEP *p. 535*

CHITTER *p. 582*

Vocal all year; most sounds are plastic variants of those shown. Keer usually given singly, in alarm. Grades into Cheeps, given in short quick series in flight and in flocks. Chitter given in aggression, often with tail raised. Tone often varies within a Chitter; some versions are rattling.

Black Turnstone

Arenaria melanocephala

Breeds in Alaska. Winters along rocky Pacific shorelines, often with Surfbirds; note smaller size, white belly, darker legs and plumage. Very rare inland.

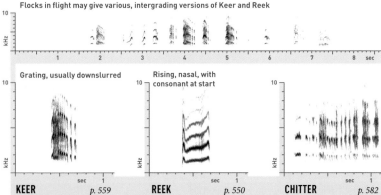

Flocks in flight may give various, intergrading versions of Keer and Reek

Grating, usually downslurred

KEER *p. 559*

Rising, nasal, with consonant at start

REEK *p. 550*

CHITTER *p. 582*

Vocal all year; most sounds are plastic variants of those shown. Keer given in agitation; highly plastic but usually high and grating. Reek given in alarm, often in short series. Chitter is most common call, given in interactions. Tone often varies within a Chitter; some versions are rattling.

Red Knot

Calidris canutus

Larger than other *Calidris* sandpipers. Winters on sandy beaches; uncommon inland in migration. North American population has declined rapidly.

Pips sometimes run into Twitter by agitated flocks

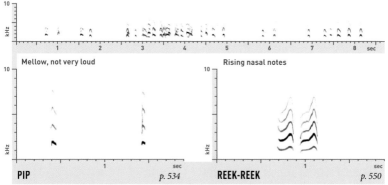

Mellow, not very loud

Rising nasal notes

PIP *p. 534*

REEK-REEK *p. 550*

Not very vocal away from breeding grounds. Pip is most common call in migration and winter; may sometimes run into a rapid mellow Twitter. Reek-reek given all year in alarm; sometimes single or in longer series. Complex flight song of slow whistles not known away from breeding grounds.

Surfbird

Calidris virgata

Breeds on alpine tundra. Winters along rocky Pacific shores, often with Black Turnstones; note larger size, spotted belly, yellow legs. Very rare inland.

In winter, sometimes gives low, plastic, burry songlike phrases (not indexed)

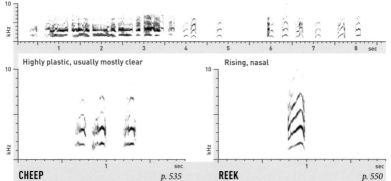

Highly plastic, usually mostly clear

Rising, nasal

CHEEP *p. 535*

REEK *p. 550*

Less vocal away from breeding grounds than Black Turnstone. On breeding grounds, repertoire is large, including complex songs given during flight display. In migration and winter, Cheeplike calls are most common vocalization. Cheep grades into Reek, given in alarm, mostly singly.

Stilt Sandpiper

Calidris himantopus

In migration and winter, found mostly on freshwater ponds, where it forages by wading up to its belly. Note slightly downcurved bill, white back and rump.

Agitated flocks may give a variety of Cheeps and Pewlike notes

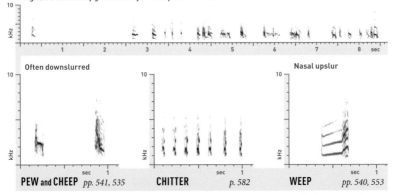

Often downslurred

Nasal upslur

PEW and CHEEP *pp. 541, 535* **CHITTER** *p. 582* **WEEP** *pp. 540, 553*

Cheep is most common call in migration and winter, given in alarm or upon taking flight. Some versions are clear single Pews recalling Lesser Yellowlegs. Screechy Chitter may be given in interactions; never very fast. Weep given in alarm, singly or in short series. All calls plastic.

Sanderling

Calidris alba

Winters on sandy beaches, habitually chasing the outgoing waves and fleeing the incoming ones. Uncommon inland in migration.

Agitated flocks may give a variety of plastic Peeplike notes

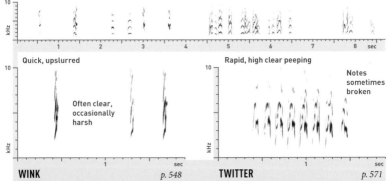

Quick, upslurred

Rapid, high clear peeping

Often clear, occasionally harsh

Notes sometimes broken

WINK *p. 548* **TWITTER** *p. 571*

Wink is most common call in migration and winter, given in mild alarm or upon taking flight. Twitter apparently given during interactions. Both calls highly plastic, grading into various Peeplike notes.

Dunlin

Calidris alpina

Winters on mudflats, mostly near coast, often in flocks of thousands. Forages with sewing-machine motion in shallow water. Bill long, slightly downcurved.

Song: series of semimusical burrs, higher at start, notes often becoming vaguely 2-syllabled (p. 576)

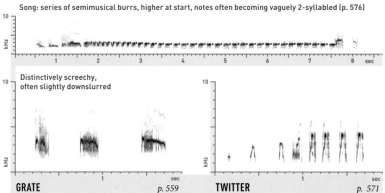

Distinctively screechy, often slightly downslurred

GRATE *p. 559*

TWITTER *p. 571*

Grate given in flight or alarm; classic version is quite distinctive, long and scratchy. Twitter may be given in interactions. All calls are exceedingly plastic; excited flocks make many sounds, from single Peeps to short trilled or noisy notes. Song sometimes given in migration.

Rock Sandpiper

Calidris ptilocnemis

Several distinct subspecies breed in Alaska, wintering on rocky shores and jetties exposed to the surf. Uncommon south of Canada; very rare inland.

Cheep highly plastic; occasionally runs into Chitter (p. 582)

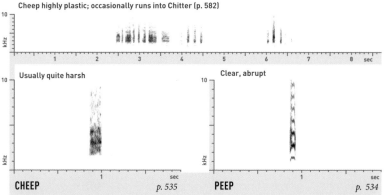

Usually quite harsh

Clear, abrupt

CHEEP *p. 535*

PEEP *p. 534*

Not very vocal away from the breeding grounds. Most sounds in migration and winter are variations on the Cheep, given when flushed; Peep is apparently quite rare. In flight, or when birds gather in flocks, Cheeps can run into long harsh Chitter.

Pectoral Sandpiper

Calidris melanotos

Fairly common migrant on mudflats. Resembles Least Sandpiper, but larger. Breeding male develops an inflatable throat sac used in hooting displays.

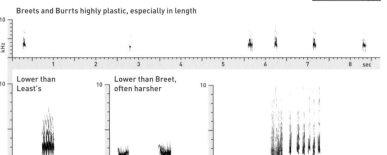

Breets and Burrts highly plastic, especially in length

BREET	BURRT	CHITTER
Lower than Least's	Lower than Breet, often harsher	
p. 544	p. 544	p. 582

Breet and Burrt given all year, in alarm and in flight, Breet possibly by females and Burrt by males. At least on breeding grounds, males give low, harsh Snarls (p. 560) without burriness. Chitter given infrequently all year, during aggressive interactions, sometimes with high soft squeaking notes.

Buff-breasted Sandpiper

Calidris subruficollis

Our only sandpiper in which males display on leks (p. 75). Once abundant, now uncommon. Migrants use mown fields and dry, sandy edges of ponds.

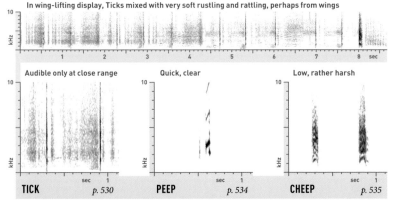

In wing-lifting display, Ticks mixed with very soft rustling and rattling, perhaps from wings

TICK	PEEP	CHEEP
Audible only at close range	Quick, clear	Low, rather harsh
p. 530	p. 534	p. 535

Very quiet for a sandpiper. Soft Ticks given only by male in display, lifting one wing at a time; mostly on breeding grounds but occasionally in spring migration. Peep known from a single recording, from male in display. Cheep given all year in alarm or in flight; plastic, but low for a sandpiper.

Baird's Sandpiper

Calidris bairdii

Abundant migrant on the Great Plains; rare on both coasts. Larger and longer-winged than Semipalmated. Prefers slightly drier ground than other peeps.

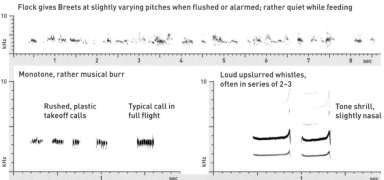

Flock gives Breets at slightly varying pitches when flushed or alarmed; rather quiet while feeding

Monotone, rather musical burr

Rushed, plastic takeoff calls | Typical call in full flight

BREET *p. 544*

Loud upslurred whistles, often in series of 2–3

Tone shrill, slightly nasal

WEEP and WEEP SERIES *pp. 540, 572*

Breet given all year, in alarm and in flight. Lower and more monotone than Least Sandpiper's; slightly higher and more musical than Pectoral's. Weep and Weep Series given in alarm, mostly on breeding grounds, but sometimes on migration and in winter; distinctive. Also gives peeping Twitter (p. 571).

White-rumped Sandpiper

Calidris fuscicollis

Size and shape like Baird's. Migrates mostly through Great Plains in spring, down East Coast in fall. Vocalizations distinctive, standing out in mixed flocks.

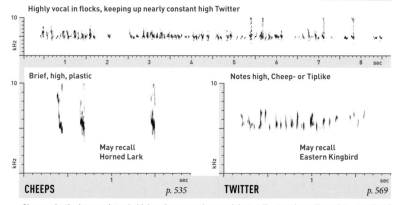

Highly vocal in flocks, keeping up nearly constant high Twitter

Brief, high, plastic

May recall Horned Lark

CHEEPS *p. 535*

Notes high, Cheep- or Tiplike

May recall Eastern Kingbird

TWITTER *p. 569*

Cheeps plastic, but consistently higher than any other sandpiper call; sometimes likened to squeaks of mice or bats. Each note very short. Twitter highly plastic; the end of a continuum of increasingly rapid Cheeps as excitement level rises.

LEAST SANDPIPER

Calidris minutilla

The smallest sandpiper in the world. Note yellow legs. Migrants found on mudflats, often feeding on the periphery of mixed shorebird flocks.

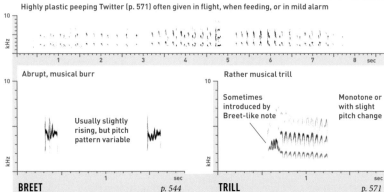

Highly plastic peeping Twitter (p. 571) often given in flight, when feeding, or in mild alarm

Abrupt, musical burr

Usually slightly rising, but pitch pattern variable

BREET   p. 544

Rather musical trill

Sometimes introduced by Breet-like note

Monotone or with slight pitch change

TRILL   p. 571

Breet given all year, by both sexes, usually in full flight. Plastic but fairly distinctive; higher than Breet of Baird's Sandpiper. Trill given in high alarm near nest or young; rare in migration and winter. Plastic and variable; averages lower than Western Sandpiper Trill, but much overlap.

PEEPING TWITTERS OF SANDPIPERS

Least, Semipalmated, and Western Sandpipers give highly plastic peeping notes when in groups, and sometimes while feeding or while flying; large flocks can generate a high-pitched muddle of constant peeping. Similar peeping Twitters are given by both dowitchers, Sanderling, Dunlin, and Baird's Sandpiper. White-rumped Sandpiper gives a Twitter that is distinctively high in pitch.

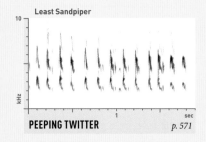

Least Sandpiper

PEEPING TWITTER   p. 571

TAKEOFF CALLS OF SANDPIPERS

Many sandpipers give a slightly modified version of the typical call upon taking flight in alarm. These "takeoff calls" tend to be shorter, harsher, somewhat lower, and repeated in brief rapid series. They often last only a few seconds before the bird reverts to more typical calls.

The takeoff calls of Long-billed Dowitcher can recall the Pew Series of Short-billed Dowitcher. The takeoff calls of Least and, especially, Western Sandpipers can resemble the Cheep of Semipalmated. When listening to sandpipers flushing, pay particular attention to the calls made after the initial burst, once the birds are completely aloft.

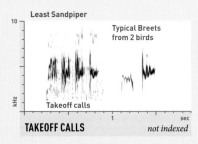

Least Sandpiper

Typical Breets from 2 birds

Takeoff calls

TAKEOFF CALLS   not indexed

Semipalmated Sandpiper

Calidris pusilla

Common in migration on the Great Plains; uncommon to rare on the West Coast. Can be difficult to separate from Western Sandpiper; note voice.

Highly plastic peeping Twitter (p. 571) often given in flight, when feeding, or in mild alarm

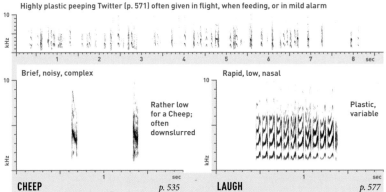

Brief, noisy, complex

Rather low for a Cheep; often downslurred

Rapid, low, nasal

Plastic, variable

CHEEP   p. 535   **LAUGH**   p. 577

Cheep given all year, by both sexes, in flight or in alarm. Western Sandpiper calls average higher, but can match Semipalmated. Laugh given in alarm near nest, in interactions on migration. Lower, more nasal, and far more often heard than Trills of Least and Western.

Western Sandpiper

Calidris mauri

Bill averages longer, thinner, and more downcurved than Semipalmated's, but with much overlap. Much more likely than Semipalmated to occur in winter.

Highly plastic peeping Twitter (p. 571) often given in flight, when feeding, or in mild alarm

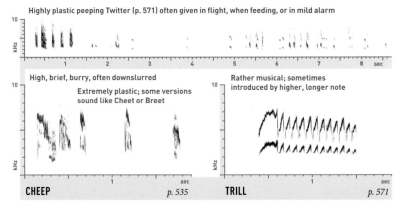

High, brief, burry, often downslurred

Extremely plastic; some versions sound like Cheet or Breet

Rather musical; sometimes introduced by higher, longer note

CHEEP   p. 535   **TRILL**   p. 571

Cheep given all year, mostly in full flight. Classic version notably higher than Semipalmated Sandpiper Cheep, but low, noisy versions can match Semipalmated. Trill given in high alarm near nest or young; rare in migration and winter. Plastic and variable; some versions match Trill of Least.

Long-billed Dowitcher

Limnodromus scolopaceus

Feeds in flocks, wading up to its belly and probing mud with a rapid sewing-machine motion of its bill. Prefers fresh water, often with emergent vegetation.

Song: Rattles, then 3 or more complex burry phrases in series (p. 611)

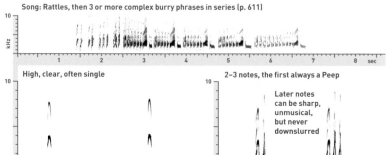

High, clear, often single

2–3 notes, the first always a Peep

Later notes can be sharp, unmusical, but never downslurred

PEEP *p. 534*

RATTLE *p. 570*

Peep given all year, even when calm, as well as in flight. Varies little in pitch and quality. Excited birds run Peeps together into long Twitters (p. 571). Peeping Rattle given by alarmed birds, on the ground or upon flushing. Faster than Short-billed calls; consistently different in quality. Song sometimes given in migration.

Short-billed Dowitcher

Limnodromus griseus

Very similar to Long-billed Dowitcher, and sometimes flocks with it. Best identified by voice. Generally prefers coastlines and mudflats.

Song: Pews, then rattling phrases, each ending with 2–4 upslurred burrs (p. 611)

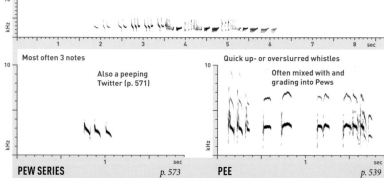

Most often 3 notes

Also a peeping Twitter (p. 571)

Quick up- or overslurred whistles

Often mixed with and grading into Pews

PEW SERIES *p. 573*

PEE *p. 539*

Pews somewhat plastic; like calls of yellowlegs, but slightly faster and higher, more likely to sound like Tew. Can become slightly noisy in alarm. Pee calls given by excited birds on the ground. Highly plastic, but generally longer and more musical than Long-billed Peeps. Song sometimes given in migration.

WILSON'S SNIPE

Gallinago delicata

Common but rather secretive in dense wetlands. Occasionally sits up on low perches such as fenceposts, often when giving Bark Series or Kik-a Series.

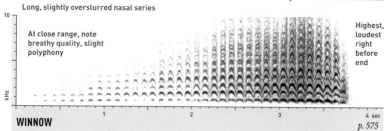

Long, slightly overslurred nasal series

At close range, note breathy quality, slight polyphony

Highest, loudest right before end

WINNOW — p. 575

Given at bottom of dives during display flight; sound created by air rushing through 2 outermost tail feathers. Very like Boreal Owl Song, especially at a distance. At close range, note slight polyphony owing to 2 tail feathers vibrating in near-unison. Heard day or night, but especially at dawn and dusk.

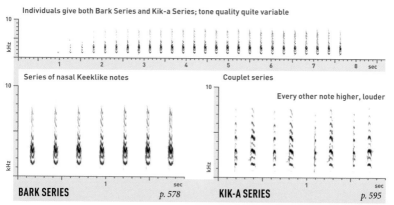

Individuals give both Bark Series and Kik-a Series; tone quality quite variable

Series of nasal Keeklike notes

BARK SERIES — p. 578

Couplet series

Every other note higher, louder

KIK-A SERIES — p. 595

Given by both sexes, usually from ground or low perch, but sometimes in flight. Function not entirely clear; couplet version reportedly used more often when mate present. Both versions used to communicate with chicks, especially in alarm. Low, noisy versions sometimes heard in alarm.

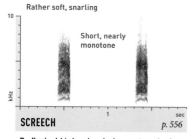

Rather soft, snarling

Short, nearly monotone

SCREECH — p. 556

By flushed birds; also during nocturnal migration and courtship pursuit flights. Only sound likely to be heard in migration and winter.

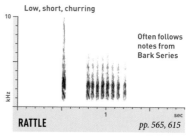

Low, short, churring

Often follows notes from Bark Series

RATTLE — pp. 565, 615

Little known. Possibly an alarm call to chicks; function needs more study.

SPOTTED SANDPIPER

Actitis macularius

Widespread along riverbanks and pond edges. Bobs tail often. Flies on stiff, shivering wings. Females larger; males tend nests and raise young.

Songs 2–10 seconds long, often in flight display; sometimes preceded by Pee notes

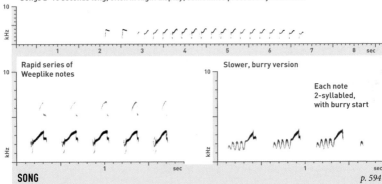

Rapid series of Weeplike notes

Slower, burry version

Each note 2-syllabled, with burry start

SONG   *p. 594*

Two main song patterns. Faster, simpler song (left) given by both sexes on ground or in flight, in territorial advertisement. Slower, more complex song (right) given on ground, mostly by female, in courtship. Intermediate songs also used in courtship.

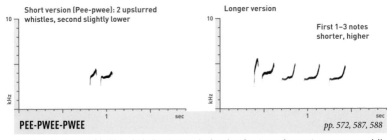

Short version (Pee-pwee): 2 upslurred whistles, second slightly lower

Longer version

First 1–3 notes shorter, higher

PEE-PWEE-PWEE   *pp. 572, 587, 588*

All year, in alarm and in flight, including nocturnal migration; 2-note version most common, especially from ground. Longer versions sometimes very long, sounding like (and perhaps grading into) Song. Lower and slower than Solitary Sandpiper Pwee-pwee-pwee, with series almost always downslurred.

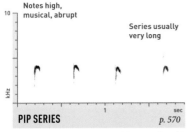

Notes high, musical, abrupt

Series usually very long

PIP SERIES   *p. 570*

Given in alarm by breeding adults, usually with chicks present. Loud, often incessant.

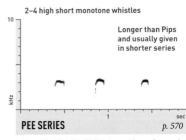

2–4 high short monotone whistles

Longer than Pips and usually given in shorter series

PEE SERIES   *p. 570*

Given by birds in flight at any time of year, including during sustained migration (day or night).

Solitary Sandpiper

Tringa solitaria

Aptly named. Breeds in the abandoned nests of other birds, in trees around muskeg bogs. At other seasons, favors wooded ponds or dense wetlands.

Songs usually embedded in plastic series of Klee notes

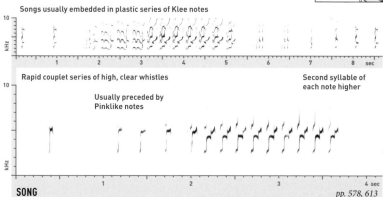

Rapid couplet series of high, clear whistles

Second syllable of each note higher

Usually preceded by Pinklike notes

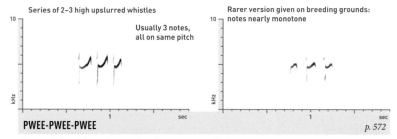

SONG

pp. 578, 613

On breeding grounds, often in display flight, but also from ground or exposed perches. Variable; sometimes consists of 2 consecutive couplet series (top). Level and significance of variation needs more study. Associated with series of loud Klee notes, like Pink but longer, often broken.

Series of 2–3 high upslurred whistles

Usually 3 notes, all on same pitch

Rarer version given on breeding grounds: notes nearly monotone

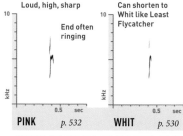

PWEE-PWEE-PWEE

p. 572

Commonly heard from migrants. Given in alarm or upon flushing; also during nocturnal flight. Almost never more than 3 notes. Infrequently heard version on breeding grounds (right) associated with Song; may represent intergrade with Pinklike notes.

Loud, high, sharp

End often ringing

Can shorten to Whit like Least Flycatcher

High, musical, monotone

Notes almost slow enough to count

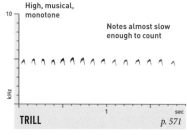

PINK *p. 532* **WHIT** *p. 530* **TRILL** *p. 571*

Variable, but calls from one bird usually identical, at least during migration. More plastic on breeding grounds.

Little known; apparently given by migrants in aggressive altercations. Sounds higher than most other vocalizations of the species.

GREATER YELLOWLEGS

Tringa melanoleuca

Common on mudflats in migration and winter; breeds in boreal bogs. Larger than Lesser Yellowlegs, with stouter, longer, slightly upturned bill.

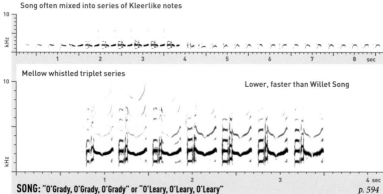

Song often mixed into series of Kleerlike notes

Mellow whistled triplet series

Lower, faster than Willet Song

SONG: "O'Grady, O'Grady, O'Grady" or "O'Leary, O'Leary, O'Leary"
p. 594

Mostly Apr.–June, by breeding males, but also occasionally by northbound migrants. Somewhat variable, but fairly stereotyped. Often given in flight display. Triplet pattern distinctive; tone quality may recall certain car alarms.

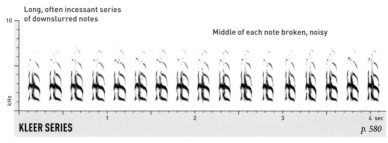

Long, often incessant series of downslurred notes

Middle of each note broken, noisy

KLEER SERIES
p. 580

All year, in high alarm. Variable and slightly plastic. Given on breeding grounds in defense of nest or chicks, but also by excited birds in a variety of situations in migration and winter. Grades into Klee Series.

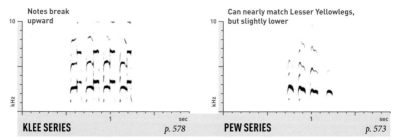

Notes break upward

KLEE SERIES
p. 578

Can nearly match Lesser Yellowlegs, but slightly lower

PEW SERIES
p. 573

All year; most common calls. Given in alarm, often when taking wing. Variable and plastic; most versions are series of at least 3 notes, but occasionally gives 1 or 2 notes alone. When notes are broken, higher second portion gives entire call the impression of high pitch.

LESSER YELLOWLEGS

Tringa flavipes

Common on mudflats in migration and winter; breeds in open boreal forest. Like Greater Yellowlegs, shows white rump and tail in flight.

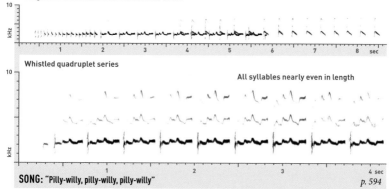

Song often mixed into series of Peerlike notes

Whistled quadruplet series

All syllables nearly even in length

SONG: "Pilly-willy, pilly-willy, pilly-willy"   *p. 594*

Mostly Apr.–June, by breeding males, but also occasionally by northbound migrants. Females occasionally sing. Somewhat variable, but fairly stereotyped. Often given in flight display. Quadruplet pattern distinctive.

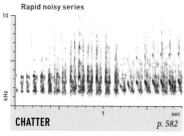

Rapid noisy series

CHATTER   *p. 582*

All year, in altercations. Highly plastic; grades into other calls. Similar notes not known from Greater Yellowlegs.

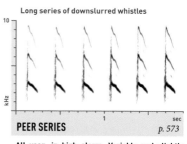

Long series of downslurred whistles

PEER SERIES   *p. 573*

All year, in high alarm. Variable and slightly plastic, but never with harsh or grating quality of Greater Yellowlegs' Kleer Series.

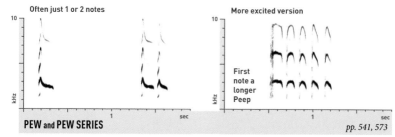

Often just 1 or 2 notes

PEW and PEW SERIES   sec

More excited version

First note a longer Peep

pp. 541, 573

All year; most common call. Given in alarm, often when taking wing. Variable and plastic; as excitement level rises, notes increase in number and are more likely to break upward, increasing potential for confusion with Greater Yellowlegs.

WILLET (WESTERN)

Tringa semipalmata inornata

Large; wing pattern striking. Breeds in freshwater marshes. Eastern subspecies is subtly smaller and shorter-billed; song differs slightly.

Slightly slower than Eastern Willet Song (8–12 phrases in 10 seconds)

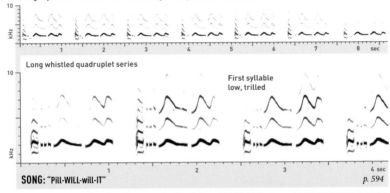

Long whistled quadruplet series

First syllable low, trilled

SONG: "Pill-WILL-will-IT"   *p. 594*

By both sexes, mostly on breeding grounds; often in long, high flight displays, sometimes at night. Females may join courting males in Song and on the wing. Female Song slightly lower than male's, with slightly less change in pitch. Eastern birds mostly unresponsive to playback of Western Song.

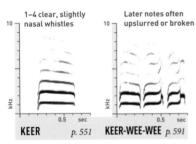

1–4 clear, slightly nasal whistles

Later notes often upslurred or broken

KEER *p. 551*   **KEER-WEE-WEE** *p. 591*

All year, in flight or in flock contact. Highly plastic; number of notes varies, and voice may break in any note.

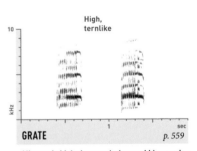

High, ternlike

GRATE   *p. 559*

All year, in high alarm and when mobbing predators. Highly plastic; grades into both Keer and Klip. Often in series.

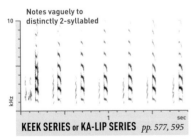

Notes vaguely to distinctly 2-syllabled

KEEK SERIES or KA-LIP SERIES *pp. 577, 595*

Given in long, rapid series, mostly during breeding season, in alarm and after Song. Variable, plastic, especially in speed.

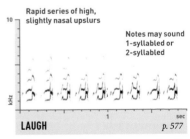

Rapid series of high, slightly nasal upslurs

Notes may sound 1-syllabled or 2-syllabled

LAUGH   *p. 577*

All year, mostly in aggressive encounters between Willets; also sometimes upon landing. Rather soft and plastic.

WANDERING TATTLER

Tringa incana

Uncommon and solitary visitor to rocky seacoasts, late summer through spring. Rare inland. Note plain gray plumage, yellow legs, tail-bobbing behavior.

Song (pp. 572, 593): fast whistled couplet series, second note in each couplet barely lower in pitch

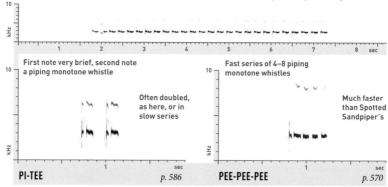

First note very brief, second note a piping monotone whistle

Often doubled, as here, or in slow series

PI-TEE　　　　　　　　　*p. 586*

Fast series of 4–8 piping monotone whistles

Much faster than Spotted Sandpiper's

PEE-PEE-PEE　　　　　　*p. 570*

Pi-tee given mostly May–July, but sometimes at other seasons, in contact and in alarm; usually in series. Pee-pee-pee given all year, in alarm, in agitation, and in flight; most common call. Song, occasionally heard in winter, slightly variable and plastic; some versions are simple series of single monotone notes.

WILSON'S PHALAROPE

Phalaropus tricolor

Our largest phalarope. Breeds in wet meadows; never found offshore. Non-breeding adults like Lesser Yellowlegs, but paler, especially on face.

Bark Series usually given for short periods; sometimes switches to Reeks

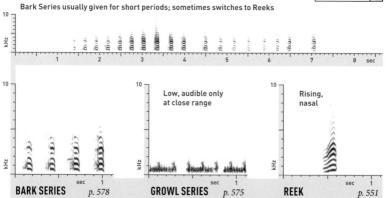

Low, audible only at close range

Rising, nasal

BARK SERIES　　*p. 578*　　　　**GROWL SERIES**　　*p. 575*　　　　**REEK**　　*p. 551*

Bark Series apparently given mostly Apr.–July; may function in courtship. Growl Series given in close courtship. Reek appears to be most common call, given all year, by male in alarm, possibly by female in display. Barks and Reeks may recall short notes of Red-breasted Nuthatch.

Red-necked Phalarope

Phalaropus lobatus

Our smallest phalarope; bill short, very thin. Spends most of the year at sea, but some migrate through the center of the continent.

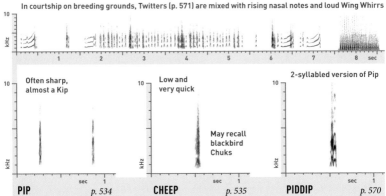

In courtship on breeding grounds, Twitters (p. 571) are mixed with rising nasal notes and loud Wing Whirrs

| PIP *p. 534* | CHEEP *p. 535* | PIDDIP *p. 570* |

Often sharp, almost a Kip

Low and very quick

2-syllabled version of Pip

May recall blackbird Chuks

Pip, Cheep, and Piddip are all variations on the most common call, given all year in contact and alarm. Highly plastic; all versions frequently heard in the constant pipping Twitter or Chitter of flocks. Other call types and Wing Whirrs not known away from breeding grounds.

Red Phalarope

Phalaropus fulicarius

Slightly larger than Red-necked; note thicker bill. Spends most of the year at sea; very rare inland. Fall and winter birds average paler than Red-necked.

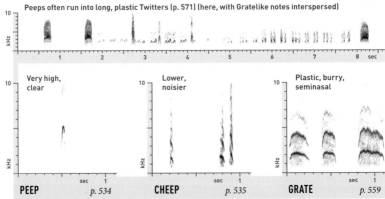

Peeps often run into long, plastic Twitters (p. 571) (here, with Gratelike notes interspersed)

| PEEP *p. 534* | CHEEP *p. 535* | GRATE *p. 559* |

Very high, clear

Lower, noisier

Plastic, burry, seminasal

Peep given all year; most common call in migration and winter. Plastic. At least on breeding grounds, also gives lower, hoarser Cheeps more like Red-necked. Grates given all year, in altercations. In courtship on breeding grounds, gives Wing Whirrs like Red-necked plus brief, noisy rising notes.

AUKS, PUFFINS, AND RELATIVES (Family Alcidae)

Known as "alcids," these mostly black-and-white seabirds are almost exclusively pelagic, coming ashore primarily for breeding. They dive in pursuit of fish and krill, propelling themselves underwater with their wings. All species in our area nest in seaside colonies, either on cliff ledges or in burrows, except for the Marbled Murrelet, which nests high in trees. One North Atlantic species, the flightless, penguinlike Great Auk, was hunted to extinction by the middle of the nineteenth century.

The vocalizations of most alcid species consist of low grunts and groans, which accompany various types of visual displays to comprise a fairly complex system of communication. Vocal repertoire is most complex in the *Cepphus* guillemots, which give long series and phrases of high-pitched whistles, unlike other alcids.

PIGEON GUILLEMOT

Cepphus columba

Nests along rocky coasts, under rocks or in crevices. Unlike other alcids, stays mostly close to shore. Bright red mouth lining conspicuous when vocalizing.

Seet Series (p. 571): Series of monotone whistles alternated with series of Pseeps or Songlike trills

Song: High overslurred accelerating trill

Usually starts with accelerating, rising series of Pseeps or Srees

SONG   p. 603

Song and Seet Series given mostly May–July, near nest, by both sexes. Both sounds plastic, grading into one another. Song may be given to defend a mate or territory after pairing. Pairs also give Song in unsynchronized duets. Males may give Seet Series to attract a mate, and different versions in aggression.

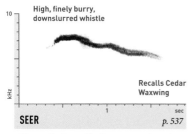

High, finely burry, downslurred whistle

Recalls Cedar Waxwing

SEER   p. 537

All year, in alarm. Plastic; some versions clear. Begging juveniles likely give coarsely trilled versions, July–Oct., like calls of young gulls.

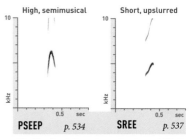

High, semimusical

PSEEP   p. 534

Short, upslurred

SREE   p. 537

Plastic; often in irregular series, in courtship, in flight, and in other contexts. These forms grade into one another and into Seer.

ANCIENT MURRELET *Synthliboramphus antiquus*

Breeds on Alaskan islands; winters mostly offshore, south to N. California. Vocal repertoire large; most common call shown.

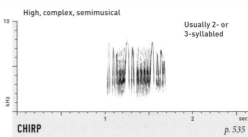

High, complex, semimusical

Usually 2- or 3-syllabled

CHIRP   *p. 535*

SCRIPPS'S MURRELET *Synthliboramphus scrippsi*

Breeds on California's Channel Islands; disperses offshore in late summer and fall north to Washington. Most common call shown.

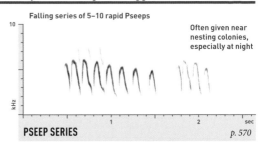

Falling series of 5–10 rapid Pseeps

Often given near nesting colonies, especially at night

PSEEP SERIES   *p. 570*

GUADALUPE MURRELET *Synthliboramphus hypoleucus*

Breeds on islands off Baja California; disperses offshore in late summer and fall north to central California. Voice not well known.

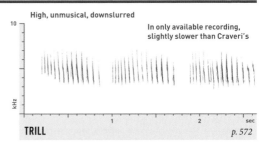

High, unmusical, downslurred

In only available recording, slightly slower than Craveri's

TRILL   *p. 572*

CRAVERI'S MURRELET *Synthliboramphus craveri*

Breeds on islands off Baja California; rare offshore in late summer and fall north to central California. Voice not well known.

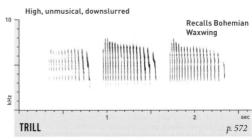

High, unmusical, downslurred

Recalls Bohemian Waxwing

TRILL   *p. 572*

Marbled Murrelet

Brachyramphus marmoratus

Nests mostly in trees in old-growth forests up to 50 miles inland; commutes to ocean after dusk, back to nest at dawn. Often vocal near nest and in flight.

All calls plastic; some versions of Keer shorter, more overslurred, and/or clearer

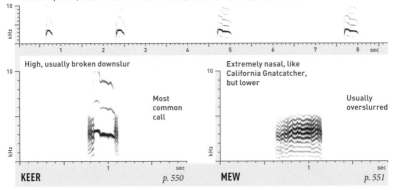

High, usually broken downslur

Most common call

KEER　　　　　　　　p. 550

Extremely nasal, like California Gnatcatcher, but lower

Usually overslurred

MEW　　　　　　　　p. 551

Keer and Mew given all year by both sexes, in flight above treetops, especially just after dusk and before dawn, and at sea. Keer is easy to confuse with begging call of juvenile gull. Wings can make distinctive rushing sound like helicopter blades (p. 557); in display dive near nest, also a loud jetlike Whoosh (p. 567).

Rhinoceros Auklet　*Cerorhinca monocerata*

Local breeder on offshore rocks, Alaska to California. Vocalizes mostly from inside nest burrow; also occasionally in groups at sea.

6–10 low, overslurred Moans; may recall mooing of cattle

Soft clicks sometimes audible between Moans; single Moans and Grunts also reported

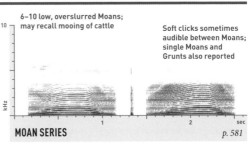

MOAN SERIES　　　　p. 581

Tufted Puffin　*Fratercula cirrhata*

Local breeder on offshore rocks, Alaska to California. Vocalizes mostly from inside nest burrow; also occasionally in groups at sea.

Long, extremely low, overslurred

Plastic, grading into various croaking notes

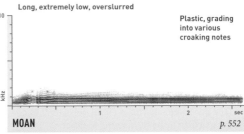

MOAN　　　　　　　　p. 552

COMMON MURRE

Uria aalge

Breeds in colonies on rocks and sea cliffs. Winters at sea, sometimes near shore. Related Thick-billed Murre, rare in our area, has thicker bill, lower voice.

Bleat (p. 580) is long, burry, overslurred Bark; Long Call (p. 613) is plastic pattern of Barks in series

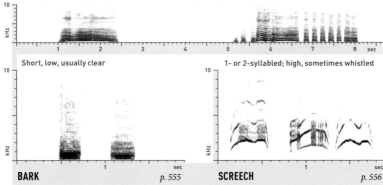

Short, low, usually clear

1- or 2-syllabled; high, sometimes whistled

BARK *p. 555*

SCREECH *p. 556*

Most vocal at nest, but adults sometimes give Barks and Bleats at sea. At nest, Barks given in alarm, often with bowing of head. Bleats and Long Call intergrade; both given in aggression as well as family contact. Also gives low Groan. Young birds beg with loud whistling Screeches (p. 587) in nest and at sea, June–Oct.

JAEGERS (Genus *Stercorarius*)

These gull-like Arctic breeders migrate mostly offshore and winter at sea. They have large vocal repertoires, but are nearly silent in migration and winter.

Pomarine Jaeger *Stercorarius pomarinus*

Mostly silent at sea, but reportedly gives sharp Witch-yew and high Week when interacting with other Pomarine Jaegers at a food source or when attacking other species.

Parasitic Jaeger *Stercorarius parasiticus*

At sea, reportedly gives high Weet calls when interacting with other Parasitic Jaegers at a food source; less vocal than Pomarine and silent when pursuing other birds.

Long-tailed Jaeger *Stercorarius longicaudus*

No vocalizations reported at sea.

GULLS (Family Laridae, Subfamily Larinae)

Enduring symbols of the seashore, gulls are not restricted to it. All behaviors and sounds are likely innate. Sounds tend to be highly variable and plastic, but the general quality of the voice differs noticeably between species.

The types of sounds described on the following pages are typical of large gull species. Display postures tend to vary subtly between species, especially Long Call postures, but the postures may be perfunctory or absent in any given performance.

The vocal repertoires of small gulls tend to be less complex than those of large gulls, and display postures tend to differ.

Long Calls: Used in pair-bonding displays and aggressive encounters, often by multiple birds at once, Long Calls are given all year but are most frequent during the breeding season. Most have three distinct parts:

Introductory notes: Short, sometimes accelerating series of yelps or wails, often omitted; given in fairly neutral posture

High notes: The 1-3 highest and loudest notes, occasionally omitted; usually given with a deep bow of the head

Terminal series: A series of 1- or 2-syllabled notes, the least variable part of the call. Head sometimes tossed back at start; most notes given with head at 45 degree angle. Some species toss head with each note.

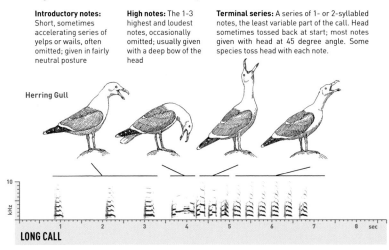

Wails ("Mew" display): Important in courtship and pair-bonding, Wails are also given all year by rivals in territorial encounters, typically in slow series, with the neck extended and the bill pointing downward. Two birds often Wail while walking side by side, sometimes with grass or other objects in the bill, before or after Long Calls.

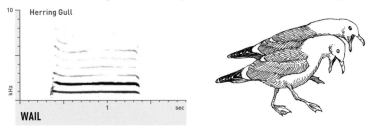

Begging calls (head-toss display): These calls are repeated at regular intervals to solicit the regurgitation of food. They are given by adults as well as juveniles, especially females begging from their mates during pair-bonding displays, with a characteristic quick toss of the head to a nearly vertical position.

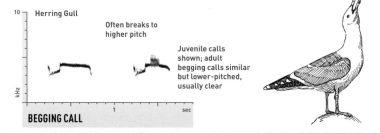

Other Gull Sounds and Displays

Chuckles: Not strongly associated with any given posture, Chuckles are given in aggression, during fights, and in alarm.

Grunts ("choking" display): In courtship or high-intensity aggressive encounters, one or both birds may point the head downward and lower the tongue bone to change the facial expression, usually while giving a series of soft, low Grunts. Especially in courtship, a bird may actually settle on the ground as though preparing a nest.

Copulation calls: Intermediate between the Grunt and the Chuckle, this sound is made only by the male during copulation.

Single-note calls: Individual gulls give different versions of calls that more or less resemble single notes from the Long Call. Distinct versions may be given in nest defense, in contact, or in aggression. These calls are usually highly plastic, but at least some versions likely serve to identify individual birds.

Notes on the Voices of Young Gulls

Throughout much of their first year of life, most gulls have very high-pitched whistling or keening voices that can sound quite different from those of adults. In most species, not much is known about the transition between juvenile calls and adult calls, but it is apparently quite gradual. More study needed.

BLACK-LEGGED KITTIWAKE

Rissa tridactyla

Breeds in colonies on sea cliffs, often on offshore islands; occasionally on vertical manmade structures. Winters mostly at sea; rare inland.

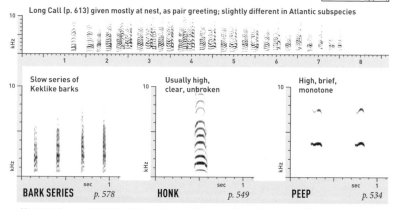

Long Call (p. 613) given mostly at nest, as pair greeting; slightly different in Atlantic subspecies

Slow series of Keklike barks	Usually high, clear, unbroken	High, brief, monotone
BARK SERIES *p. 578*	**HONK** *p. 549*	**PEEP** *p. 534*

Most sounds apparently given all year, but especially at breeding colony, where kittiwakes can drown out all other species. Bark Series given in alarm; distinctively slow. Honks also often given in slow series. Peeps given by begging juveniles and adults, including at sea. Also gives long nasal Wails (p. 552).

BONAPARTE'S GULL

Chroicocephalus philadelphia

A small gull, rather like a tern in shape, habits, and voice. The only North American gull that regularly nests in trees. Breeds in boreal forests.

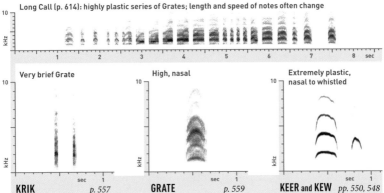

Long Call (p. 614): highly plastic series of Grates; length and speed of notes often change

Very brief Grate	High, nasal	Extremely plastic, nasal to whistled
KRIK *p. 557*	**GRATE** *p. 559*	**KEER** and **KEW** *pp. 550, 548*

Krik given in agitation, especially near nest; almost never extended into a Chuckle. Most adult calls are versions of highly plastic Grate. Keers, Kews, and other whistled calls are most common sounds of first-winter birds; perhaps also given by feeding adults.

SABINE'S GULL

Xema sabini

An elegant species with a striking wing pattern. Breeds in the Arctic; migrates and winters mostly well offshore, but a few are found inland each fall.

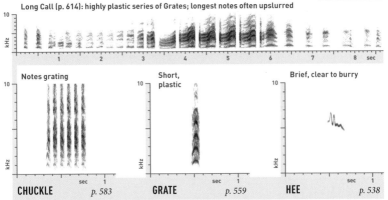

Long Call (p. 614): highly plastic series of Grates; longest notes often upslurred

Notes grating	Short, plastic	Brief, clear to burry
CHUCKLE *p. 583*	**GRATE** *p. 559*	**HEE** *p. 538*

Chuckle given in alarm, usually on the wing near nest. Grate is most common call of adults all year; highly plastic, sometimes given in rapid series. Hee is most common call of young birds, at least through fall; also apparently by adults at sea.

FRANKLIN'S GULL

Leucophaeus pipixcan

Found mostly inland. Breeds on marshy ponds. Migrants often gather to hunt insects in plowed fields. Winters in southern South America.

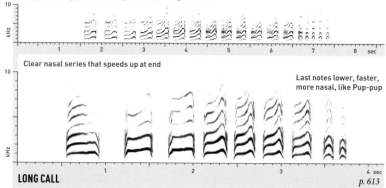

Long Call: long notes clear, upslurred, usually unbroken

Clear nasal series that speeds up at end

Last notes lower, faster, more nasal, like Pup-pup

LONG CALL
p. 613

Variable and plastic, but distinctive: starts with slow upslurs (2–3 notes/second), ends faster (4 notes/second). Often given on the wing. Does not bow during call, but extends head nearly horizontally; sometimes tosses head back on 1–2 final notes.

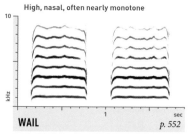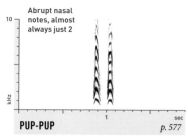

High, nasal, often nearly monotone

WAIL
p. 552

Quite plastic. Usually in series. Given in courtship, at nest, and in contact with chicks.

Abrupt nasal notes, almost always just 2

PUP-PUP
p. 577

All year, in alarm or aggression. Single notes sometimes heard. Long series of similar notes given during copulation and nest defense.

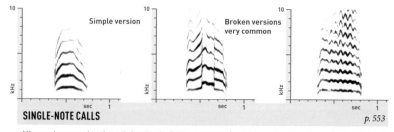

Simple version

Broken versions very common

SINGLE-NOTE CALLS
p. 553

All year, in many situations. Quite plastic. Broken versions (center) may be given in courtship with head-toss display. Tremolo versions (right) given in high agitation.

Heermann's Gull

Larus heermanni

Distinctive. Breeds only in Mexico, but disperses north along coast to British Columbia May–Nov.; nonbreeders remain in California all year. Rare inland.

Long Call: often truncated to just a decelerating series of nasal Barks (p. 578)

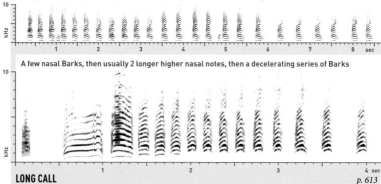

A few nasal Barks, then usually 2 longer higher nasal notes, then a decelerating series of Barks

LONG CALL *p. 613*

Variable and plastic, but distinctive, lower and more nasal than Long Call of any other gull in the West. During longest and highest notes, holds neck high but bends head down; then raises head to 45 degree angle at start of terminal series, gradually lowering head as series continues.

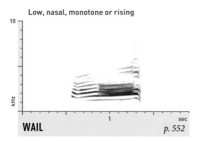

Low, nasal, monotone or rising

WAIL *p. 552*

Not often heard. Usually in series. Not well known; likely given in similar contexts to Wails of other gull species.

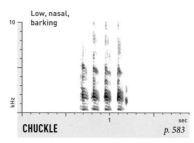

Low, nasal, barking

CHUCKLE *p. 583*

All year, in alarm or aggression. Speed somewhat plastic. Most versions clear.

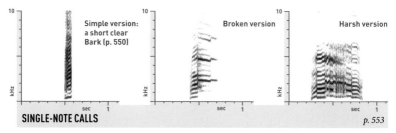

Simple version: a short clear Bark (p. 550)

Broken version

Harsh version

SINGLE-NOTE CALLS *p. 553*

All year, in many situations. Extremely varied and plastic. Short clear Barks are most common version, but longer versions frequent. Broken and harsh versions mostly in aggression or agitation. Young birds beg with very high keening notes, usually clear and monotone.

Ring-billed Gull

Larus delawarensis

The most common gull in much of North America, especially inland. Much smaller than Herring Gull. Acquires adult plumage in three years.

Long Call: 1–5 long Squeals; then a series of Yelps or Squawks, often 2-syllabled

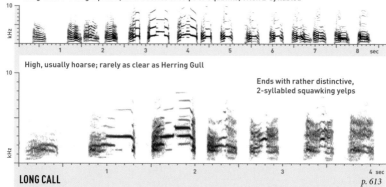

High, usually hoarse; rarely as clear as Herring Gull

Ends with rather distinctive, 2-syllabled squawking yelps

LONG CALL

p. 613

Variable and plastic. Final series relatively slow, usually 2–3 notes/second. Distinctive display postures shared with Mew Gull: long Squeals given with head down, sometimes nearly touching feet; most or all later notes delivered with bill vertical, head pumped up and down on each note.

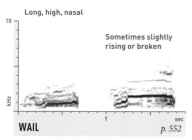

Long, high, nasal

Sometimes slightly rising or broken

WAIL

p. 552

All year, in slow series. Apparently variable and plastic, but few recordings. Example above may be intergrade with Long Call.

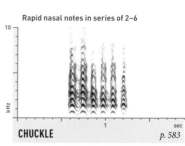

Rapid nasal notes in series of 2–6

CHUCKLE

p. 583

All year, in alarm or aggression. Fairly stereo-typed but variable. Averages faster and longer than Chuckles of larger gulls.

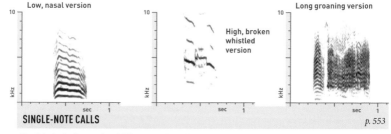

Low, nasal version

High, broken whistled version

Long groaning version

SINGLE-NOTE CALLS

p. 553

Bewilderingly diverse. In addition to versions shown, also gives short Barks, high whistled Trills, and all manner of intermediates. High whistled versions heard more often from subadults; trilled and groaning versions more often in fights over food; other generalizations difficult.

Mew Gull

Larus canus

Breeds in fresh- or saltwater wetlands; winters mostly along the Pacific Coast. Resembles Ring-billed Gull, but slightly smaller, with subtly different plumage.

Long Call: 1–5 long Squeals, then a series of high, clear Yelps or Yaps

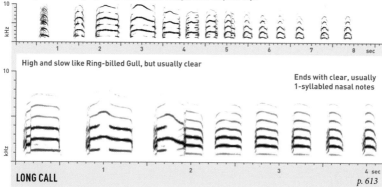

High and slow like Ring-billed Gull, but usually clear

Ends with clear, usually 1-syllabled nasal notes

LONG CALL
p. 613

Variable and plastic. Final series relatively slow, usually 2–3 notes/second. Some low-pitched versions quite nasal, recalling Franklin's Gull or even Heermann's Gull. Display postures apparently similar to those of Ring-billed Gull.

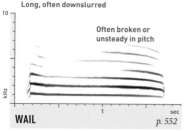

Long, often downslurred

Often broken or unsteady in pitch

WAIL
p. 552

All year, often in series. Many versions shorter than shown. Can be confused with calls of a raptor, especially when broken and given in series.

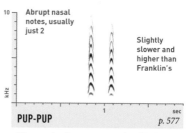

Abrupt nasal notes, usually just 2

Slightly slower and higher than Franklin's

PUP-PUP
p. 577

All year, in alarm or aggression. Single notes frequently heard. Long, laughing series of similar notes sometimes given; context unknown.

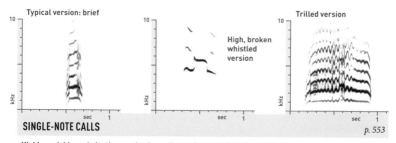

Typical version: brief

High, broken whistled version

Trilled version

SINGLE-NOTE CALLS
p. 553

Highly variable and plastic sounds given all year in many situations. Generally similar to similar calls of Ring-billed Gull, but pitch averages higher, and typical versions are briefer. May grade into Wail.

WESTERN GULL

Larus occidentalis

The common resident gull of the Pacific
Coast from Washington State south, but
very rare even a few miles from the sea.
Northern birds slightly paler.

Long Call: a rapid series of Barks or Yelps, often lacking 3-part structure of other gull Long Calls

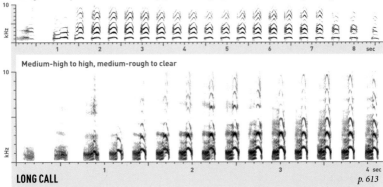

Medium-high to high, medium-rough to clear

LONG CALL   p. 613

All year. Highly variable and plastic, but distinctively simple: often omits introductory notes, and almost
never includes high notes before the terminal series. Often slows slightly at end. Averages faster than
Glaucous-winged (3–4 notes/second) and lower than Herring, but overlaps both in these characters.

Long, often slightly rising

WAIL   p. 552

All year, in slow series. Variable and plastic, but
most versions apparently fairly high and clear,
without voice breaks.

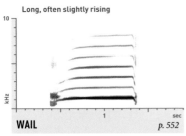

Nasal, barking
version

Low, grunting
version

CHUCKLE   p. 583

All year, in alarm or aggression. Fairly stereotyped but highly variable; some versions low and
rather clicking.

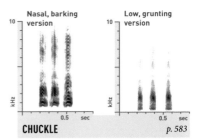

High, clear
version

Barking
version

Low, rather
grunting
version

SINGLE-NOTE CALLS   p. 553

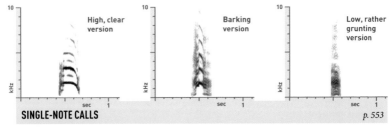

Highly variable and plastic sounds given all year in many situations. Averages slightly lower and shorter
than corresponding calls of Herring Gull, but much overlap. As in other large gulls, juveniles gives higher-
pitched versions (not shown).

GLAUCOUS-WINGED GULL

Larus glaucescens

Common along the north Pacific Coast; interbreeds freely with Western Gull in Washington and Oregon. Hybrids likely sound intermediate. Rare inland.

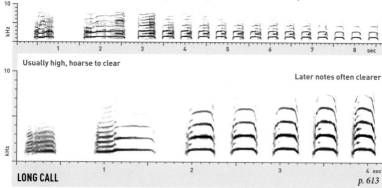

Long Call: 1–2 short Barks and/or 1–2 long Wails or Squeals; then a series of Yelps

Usually high, hoarse to clear

Later notes often clearer

LONG CALL — p. 613

All year. Highly variable and plastic; generally intermediate between Long Calls of Herring and Western. Often slows slightly at end. Averages slower than Western (2–2.5 notes/second) and often higher and clearer, with more complex beginning, but overlaps in all characters.

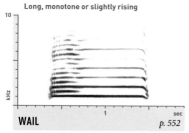

Long, monotone or slightly rising

WAIL — p. 552

All year, in slow series. Variable and plastic; usually high, clear, and unbroken, but a few versions are hoarse or squealing.

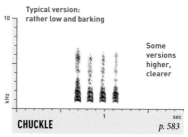

Typical version: rather low and barking

Some versions higher, clearer

CHUCKLE — p. 583

All year, in alarm or aggression. Fairly stereotyped but variable.

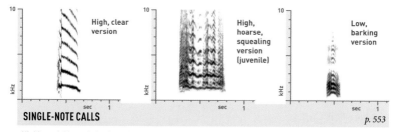

High, clear version

High, hoarse, squealing version (juvenile)

Low, barking version

SINGLE-NOTE CALLS — p. 553

Highly variable and plastic sounds given all year in many situations. Typical versions are rather high and clear; low barking versions are less common.

HERRING GULL

Larus argentatus smithsonianus

The most common large gull in many areas, especially in winter. Like other large gulls, acquires adult plumage in 4 years; young birds highly variable.

Long Call: 1–2 short Barks and/or 1–2 long Wails or Squeals; then a series of Yelps

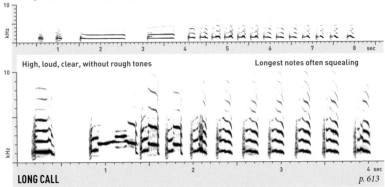

High, loud, clear, without rough tones

Longest notes often squealing

LONG CALL
p. 613

A quintessential sound of the seashore, given all year. Highly variable and plastic; may be as short as a single broken note followed by 4–5 Yelps. Final series may be 1- or 2-noted; speed 3–4 notes/second. Sometimes ends with 1–2 more widely spaced notes, or occasionally with Wails.

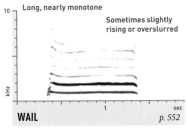

Long, nearly monotone

Sometimes slightly rising or overslurred

WAIL
p. 552

All year, in slow series. Variable and plastic; some versions low and strongly nasal; some versions with 1–2 voice breaks.

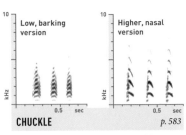

Low, barking version

Higher, nasal version

CHUCKLE
p. 583

All year, in alarm or aggression. Fairly stereotyped but highly variable; some versions rougher than shown.

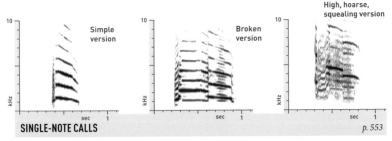

Simple version

Broken version

High, hoarse, squealing version

SINGLE-NOTE CALLS
p. 553

A catch-all category for a wide variety of highly variable and plastic sounds ranging from Barks to Yelps to Squeals. Most common call; given all year in many situations. Usually high and clear, though low and rough versions exist. High trilled version (not shown) used against predators or in fights over food.

ICELAND GULL

Larus glaucoides

Western form, "Thayer's Gull," formerly considered a separate species. Smaller than Herring, with more rounded head, smaller bill, and less black in wingtip.

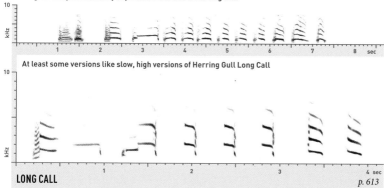

Long Call may not be safely separable from that of Herring Gull

At least some versions like slow, high versions of Herring Gull Long Call

LONG CALL   *p. 613*

Voice poorly known. Long Call seems to resemble the higher-pitched Long Calls of Herring Gull. In the few available recordings, final series is of high, clear 1-syllabled notes; speed 2–2.5 notes/second. Other calls likely resemble those of Herring Gull; more study needed.

YELLOW-FOOTED GULL

Larus livens

Breeds only in Mexico; uncommon at the Salton Sea, mostly in summer and fall. Closely related to Western Gull and once considered conspecific with it.

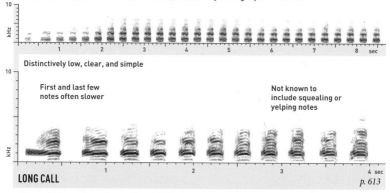

Long Call: a rapid series of low, clear Barks, consistently lacking 3-part structure

Distinctively low, clear, and simple

First and last few notes often slower

Not known to include squealing or yelping notes

LONG CALL   *p. 613*

All year. Rather uniform for a gull, and distinctive: even simpler than Western Gull Long Call, containing only one note type, and consistently clearer and lower in pitch. Speed like Western (3–4 notes/second). Other calls generally low and clear as well. Juveniles beg with high, keening whistles.

CALIFORNIA GULL

Larus californicus

Breeds on fresh water and saline lakes; winters mostly along the coast. Intermediate in size between Ring-billed and Herring Gulls.

Long Call: higher versions like hoarse Herring Gull; 1–2 Wails or Squeals, then a series of Yelps

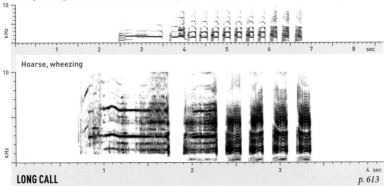

Hoarse, wheezing

LONG CALL *p. 613*

All year. Highly variable and plastic. Often distinctively low, nasal, and hoarse, but many versions nearly as high and clear as Herring Gull Long Call. Final series almost always of single notes, not couplets; speed 3–5 notes/second. May slow slightly at end. Last few notes occasionally hoarser.

Long, low, overslurred

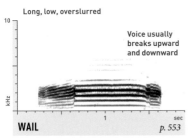

Voice usually breaks upward and downward

WAIL *p. 553*

All year, in slow series. Variable and plastic; low, groaning versions distinctive, but some are higher, more like Herring Gull.

Low nasal notes, usually in series of 2–3

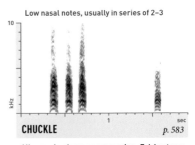

CHUCKLE *p. 583*

All year, in alarm or aggression. Fairly stereotyped but variable. Generally lower and more nasal than in most other gulls.

Hoarse, broken Yelps

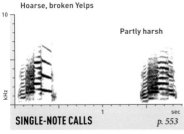

Partly harsh

SINGLE-NOTE CALLS *p. 553*

Distinctive, low, braying

BLEAT *p. 580*

Highly variable and plastic sounds given all year in many situations. Most are Yelps and Squeals like those of Herring Gull, but noticeably hoarse. Distinctive Bleat given by birds of all ages in many situations. Bleating calls of other gull species are higher and rarer, restricted mostly to altercations.

LESSER BLACK-BACKED GULL

Larus fuscus

A European species, uncommon but increasing in North America. Most are of the subspecies *graellsii*, from Western Europe; darker *fuscus* is very rare.

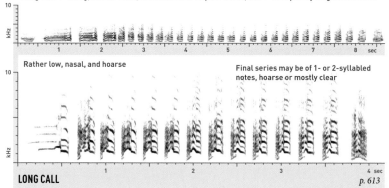

Long Call: 1–4 long, hoarse Wails; then a series of Yelps or Barks, last 1–2 frequently longer and slower

Rather low, nasal, and hoarse

Final series may be of 1- or 2-syllabled notes, hoarse or mostly clear

LONG CALL   *p. 613*

All year. Highly variable and plastic. Usually distinctly lower, hoarser, and more nasal than Herring Gull Long Call, but some versions are difficult to separate. Final series rapid, 4–5 notes/second. Longest notes sometimes squealing. Call sometimes ends with Wails.

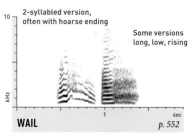

2-syllabled version, often with hoarse ending

Some versions long, low, rising

WAIL   *p. 552*

All year, in slow series. Variable and plastic. Relatively high, 2-parted versions may be from females.

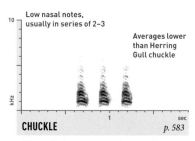

Low nasal notes, usually in series of 2–3

Averages lower than Herring Gull chuckle

CHUCKLE   *p. 583*

All year, in alarm or aggression. Variable.

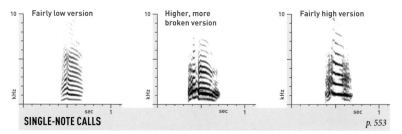

Fairly low version

Higher, more broken version

Fairly high version

SINGLE-NOTE CALLS   *p. 553*

Highly variable and plastic sounds given all year in many situations. On average, lower, more nasal, and hoarser than corresponding sounds of Herring Gull, but much overlap.

Glaucous Gull

Larus hyperboreus

One of our largest gulls. Rare but regular south of Canada in winter. The much smaller Iceland is the only other gull that often shows pure white wingtips.

Long Call: 1–2 short Barks and/or 1–2 Wails or Squeals; then a series of Yelps

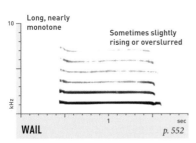

High, clear, and often squealing

Terminal series averages slower than Herring Gull

LONG CALL
p. 613

All year. Highly variable and plastic. Terminal series usually of 1-syllabled notes, around 2–2.5 notes/second. Sometimes ends with 1–2 more widely spaced notes, or occasionally with Wails.

Long, nearly monotone

Sometimes slightly rising or overslurred

WAIL
p. 552

All year, in slow series. Variable and plastic; some versions with 1–2 voice breaks.

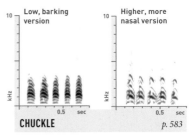

Low, barking version

Higher, more nasal version

CHUCKLE
p. 583

All year, in alarm or aggression. Fairly stereotyped but highly variable; some versions harsher than shown.

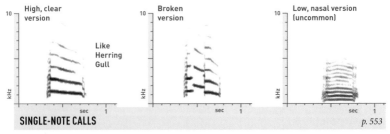

High, clear version

Like Herring Gull

Broken version

Low, nasal version (uncommon)

SINGLE-NOTE CALLS
p. 553

Most common call; given all year in many situations. Usually high and clear, though low and rough versions exist. High trilled version (not shown) used against predators or in fights over food. Not known to be separable from similar calls of Herring Gull.

SKIMMERS (Family Laridae, Subfamily Rynchopidae)

Represented by three species worldwide, the skimmers are highly distinctive ternlike birds that feed by flying along the surface of the water and dragging their elongated, razor-thin lower mandibles through it, snapping their bills shut upon contact with a fish. Sounds are likely innate, and vocal repertoires are quite simple.

TERNS (Family Laridae, Subfamily Sterninae)

Closely related to gulls, terns average smaller and more slender, often with longer tails. More closely tied to water than gulls, they feed primarily by plunge-diving, or by picking food from the water's surface in flight. Most species nest on the ground in dense, noisy colonies. Sounds are likely innate, and vocal repertoires are often highly complex. Many sounds accompany elaborate visual displays rather like those of gulls.

LOONS (Family Gaviidae)

Superficially similar to grebes, but not closely related; feet are webbed. Vocalizations are likely innate. Vocal repertoires are moderately complex; some species engage in unsynchronized duets and choruses, including at night.

BLACK SKIMMER

Rynchops niger

Unique in North America. Flight style distinctive even when not skimming surface of water for fish: wingbeats shallow and graceful, head held low.

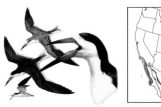

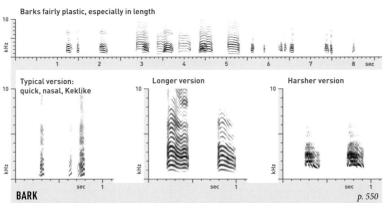

Barks fairly plastic, especially in length

Typical version: quick, nasal, Keklike

Longer version

Harsher version

BARK

p. 550

Quite vocal in groups at any time of year. All vocalizations are variations on a nasal Bark. Short, low versions given in contact and mild alarm, becoming longer and somewhat higher in high alarm. Short harsh and burry versions sometimes heard during courtship and interactions near nest.

LEAST TERN

Sternula antillarum

Our smallest tern. Yellow bill and white forehead distinctive. Nests on beaches and river sandbars. Some populations endangered.

Possible Long Call (p. 614): Screeches and Ki-deeks mixed in complex pattern; context little known

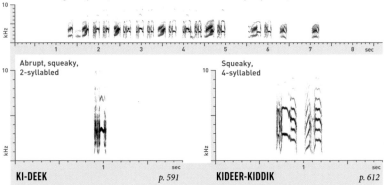

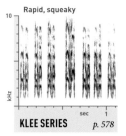

Abrupt, squeaky, 2-syllabled

KI-DEEK   *p. 591*

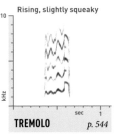

Squeaky, 4-syllabled

KIDEER-KIDDIK   *p. 612*

Most common and distinctive call. Variable, but fairly stereotyped within individuals; the two versions shown are the most typical. 4-syllabled version reportedly given by males carrying fish in courtship display flights; both versions given by both sexes when returning to nest, and in other contexts.

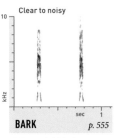

Rapid, squeaky

KLEE SERIES   *p. 578*

Mostly Apr.–May, by males in close courtship, often in duet with female Tremolo.

Rising, slightly squeaky

TREMOLO   *p. 544*

Mostly Apr.–May, by female in close courtship, often in duet with male Klee Series.

Clear to noisy

BARK   *p. 555*

All year, in high alarm, usually in series, often with Screech.

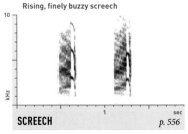

Rising, finely buzzy screech

SCREECH   *p. 556*

All year, in alarm, often in flight. Barks often mixed in as alarm level rises.

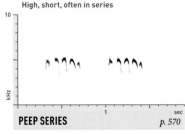

High, short, often in series

PEEP SERIES   *p. 570*

By begging juveniles, at least well into fall. Highly variable and plastic; may resemble flight calls of some sandpipers.

Caspian Tern

Hydroprogne caspia

Our largest tern, as large as a Ring-billed Gull. Bill dark red-orange and quite thick; wingtips usually show more black below than above.

At nest colony, gives a variety of screechy Snarls and grunting notes

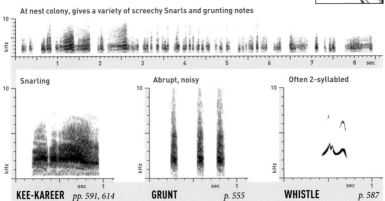

| **KEE-KAREER** *pp. 591, 614* | **GRUNT** *p. 555* | **WHISTLE** *p. 587* |

Snarling — Abrupt, noisy — Often 2-syllabled

Voice distinctive. Kee-kareer mostly Mar.–Aug., by adults, in contact and flight; noisy, but never grating. 1-syllabled versions given all year. Grunt given mostly near nest. Whistle by juveniles to at least Mar.; varies from very short to long, broken Squeal like that of young Red-tailed Hawk.

Gull-billed Tern

Gelochelidon nilotica

Medium-sized, with a very stout black bill. Found mostly along coasts; local inland. Nests on beaches, on barrier islands, and in marshes.

In high alarm, Yap Series sometimes extended into long, rapid Chuckle (p. 583)

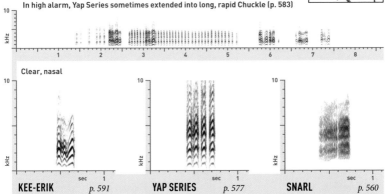

| **KEE-ERIK** *p. 591* | **YAP SERIES** *p. 577* | **SNARL** *p. 560* |

Clear, nasal

Kee-erik is most common call of adults, given in family contact and in flight. Variable; often a 2-syllabled Kee-rik. Yap Series is given in alarm; usually 3- or 4-noted, but becomes longer, faster, and noisier as alarm increases. Snarl grades into short Wail (not shown), which may be associated with courtship.

ROYAL TERN

Thalasseus maximus

Found along sea coasts, foraging close to shore. Smaller than Caspian Tern, with orange bill; wingtips usually show more black above than below.

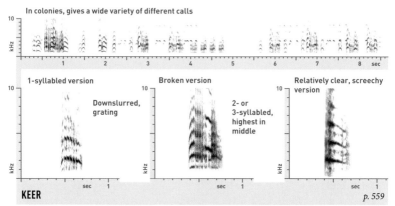

In colonies, gives a wide variety of different calls

1-syllabled version
Downslurred, grating

Broken version
2- or 3-syllabled, highest in middle

Relatively clear, screechy version

KEER
p. 559

All year, by adults in flight or in alarm; most common call. Typically downslurred and grating, but variable and plastic; some versions monotone. Different versions may serve different functions.

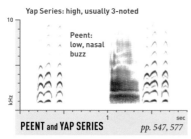

Yap Series: high, usually 3-noted
Peent: low, nasal buzz

PEENT and YAP SERIES
pp. 547, 577

Apr.–June, in courtship, Yap Series given with bill pointed down, likely by females; Peent with wings drooped and crest erect, likely by males.

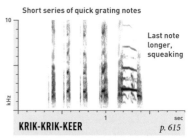

Short series of quick grating notes
Last note longer, squeaking

KRIK-KRIK-KEER
p. 615

At least Apr.–June, by males in aggressive encounters; possibly also in other contexts. Rather variable and plastic.

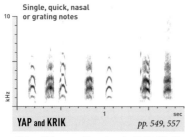

Single, quick, nasal or grating notes

YAP and KRIK
pp. 549, 557

Not well known; often given in flight. Plastic; clear Yaps and grating Kriks may intergrade.

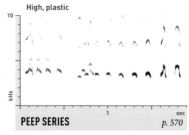

High, plastic

PEEP SERIES
p. 570

Mostly July–Oct., by juveniles, often incessantly. Highly plastic; gradually grades into high, clear version of Grate as birds mature.

ELEGANT TERN

Thalasseus elegans

Much like Royal Tern; note slightly longer, thinner bill. In late summer and fall, disperses north along the Pacific Coast; winters mostly in South America.

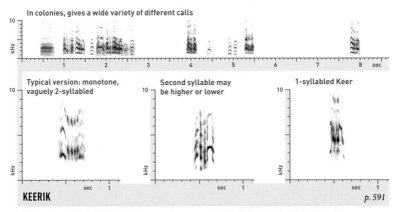

In colonies, gives a wide variety of different calls

Typical version: monotone, vaguely 2-syllabled

Second syllable may be higher or lower

1-syllabled Keer

KEERIK *p. 591*

All year, by adults in flight or in alarm; most common call. Much like Keer (p. 559) of Royal Tern; averages slightly higher and shorter and more often 2-syllabled, but some versions are difficult to distinguish.

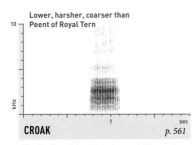

Lower, harsher, coarser than Peent of Royal Tern

CROAK *p. 561*

Mostly Apr.–July, likely in courtship, perhaps also in flight. Infrequent. Females may court with Peep Series.

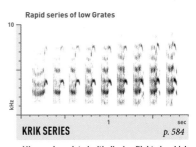

Rapid series of low Grates

KRIK SERIES *p. 584*

All year. Associated with display flights in which pair flies in tandem, one above the other, both birds periodically gliding with wings in a V.

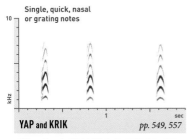

Single, quick, nasal or grating notes

YAP and KRIK *pp. 549, 557*

Often given in flight, likely in contact or alarm. Plastic; most versions clear and nasal, but some are grating Kriks (not shown).

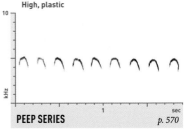

High, plastic

PEEP SERIES *p. 570*

July–Oct., by juveniles, often incessantly; also perhaps by courting females. Highly plastic; grades into Grate as birds mature.

COMMON TERN

Sterna hirundo

Nests on rocky islands, in salt marshes, and on sandy barrier beaches. Similar to other *Sterna* terns; plumage of all species varies with age and season.

Attack sounds include rapid ticking Rattle followed by Snarl (p. 615), much like Arctic Tern

Series of high, 2- or 3-syllabled Grates

3-syllabled notes underslurred: Kee-er-rip

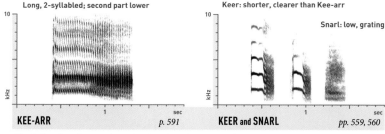

LONG CALL
p. 614

Mostly May–Aug. Somewhat variable and plastic; distinct versions likely given in courtship and aggression. Often given in flight chases. When given on ground, bill and head often lifted 45 degrees above horizontal, sometimes followed by head bows.

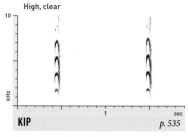

Long, 2-syllabled; second part lower

KEE-ARR
p. 591

Keer: shorter, clearer than Kee-arr

Snarl: low, grating

KEER and SNARL
pp. 559, 560

Mostly May–Aug. Kee-arr given when foraging, when approaching nest site with food, and in family contact. Typical version long, distinctly 2-syllabled, but grades into shorter, clearer Keers, given in high alarm and nest defense, often with long or short Snarls.

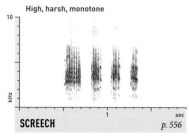

High, clear

KIP
p. 535

In contact or alarm. Shorter, lower than Arctic's; sounds monotone. Some versions very brief.

High, harsh, monotone

SCREECH
p. 556

Possibly July–Mar. Given by begging juveniles, in series or singly.

ARCTIC TERN

Sterna paradisaea

Nests in the Arctic and on offshore islands; at other seasons, found almost exclusively at sea. Rare inland. Bill and legs distinctively short.

In flight chases, Long Calls often mixed with Kews and Peeps

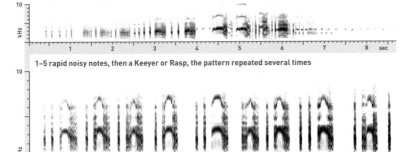

1–5 rapid noisy notes, then a Keeyer or Rasp, the pattern repeated several times

LONG CALL   *pp. 614, 615*

Mostly May–Aug. Distinct versions likely given in courtship and aggression. Often given in flight chases. When given on ground, bill and head often lifted 45 degrees above horizontal, sometimes followed by head bows. In attack, gives rapid ticking Rattle followed by Snarl (p. 615).

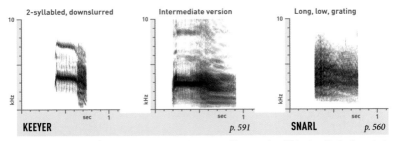

2-syllabled, downslurred   Intermediate version   Long, low, grating

KEEYER   *p. 591*   **SNARL**   *p. 560*

Mostly May–Aug. Keeyer given when foraging, when approaching nest site with food, and in family contact. Typical version distinctly 2-syllabled. Grades into Snarls, given in high alarm and nest defense. Intermediate versions can resemble Common Tern Kee-arr, but usually start higher.

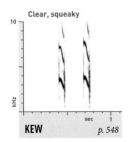

Clear, squeaky

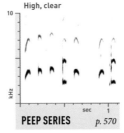

High, clear

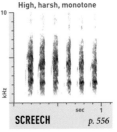

High, harsh, monotone

KEW   *p. 548*

PEEP SERIES   *p. 570*

SCREECH   *p. 556*

In contact or alarm. Higher, longer than Common's; monotone or downslurred.

Given in flight chases; possibly at other times. Usually in series.

Possibly July–Mar. Given by begging juveniles, in series or singly.

FORSTER'S TERN

Sterna forsteri

Breeds in freshwater and saltwater marshes; winters primarily along the coast. Winter head pattern distinctive: white with black patch behind eye.

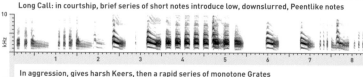

Long Call: in courtship, brief series of short notes introduce low, downslurred, Peentlike notes

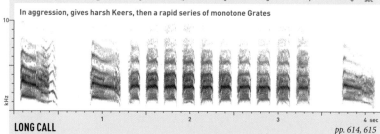

In aggression, gives harsh Keers, then a rapid series of monotone Grates

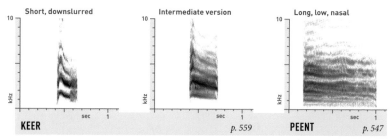

LONG CALL

pp. 614, 615

Mostly May–Aug. Distinct versions given in courtship and aggression, but these are highly plastic and intergrade. Often given during flight chases. When given on ground, courtship version often accompanied by bowed head, aggressive version by head thrown back to nearly vertical.

Short, downslurred	Intermediate version	Long, low, nasal

KEER

KEER *p. 559*

PEENT *p. 547*

Mostly May–Aug. Keer given when approaching nest site with food, in courtship flights with fish in bill, and in family contact. Typical version is distinctly brief and monosyllabic, but grades into Peents, given in high alarm and nest defense, often with Ticks and Kews. Peent lower, more nasal than Snarls of other *Sterna*.

Extremely brief noisy notes	Clear, downslurred	High, harsh, monotone

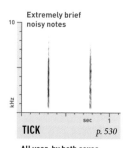

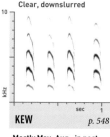

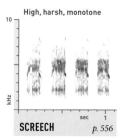

TICK *p. 530*

KEW *p. 548*

SCREECH *p. 556*

All year, by both sexes, singly in contact, and in rapid series in aggression.

Mostly May–Aug., in nest defense, usually in series, often with Peents and Ticks.

July–Mar., by begging juveniles, in series or singly.

BLACK TERN

Chlidonias niger

Quite small and short-tailed. Breeds in freshwater marshes; in winter, found largely offshore. Young birds and winter adults lack black body plumage.

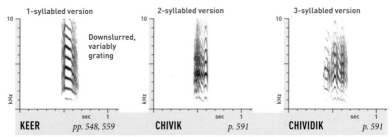

1-syllabled version — Downslurred, variably grating

KEER *pp. 548, 559*

2-syllabled version

CHIVIK *p. 591*

3-syllabled version

CHIVIDIK *p. 591*

Mostly May–Aug., by both sexes, in display flights and when approaching nest. Highly variable and fairly plastic; 1- and 2-syllabled versions are most common. Usually at least slightly grating. Averages somewhat lower than similar calls of most other terns.

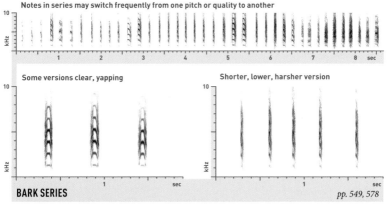

Notes in series may switch frequently from one pitch or quality to another

Some versions clear, yapping

Shorter, lower, harsher version

BARK SERIES *pp. 549, 578*

All year. A group of plastic, intergrading calls given in contact and alarm. In alarm, usually given in long series. Some versions can be quite similar to Bark Series of Wilson's Snipe or Keek Series of Willet, but usually more plastic in speed and quality.

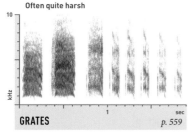

Often quite harsh

GRATES *p. 559*

Possibly all year, likely in aggression. Highly plastic; most versions are single notes. Multinote patterns like the one shown are rare.

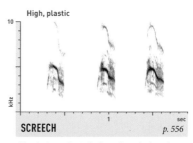

High, plastic

SCREECH *p. 556*

Mostly June–Sept., by juveniles, singly or in series. Develops gradually into Keerlike calls.

Common Loon

Gavia immer

A symbol of the north woods, its haunting calls nearly synonymous with wilderness. Breeds on wooded lakes; winters on lakes and along coast.

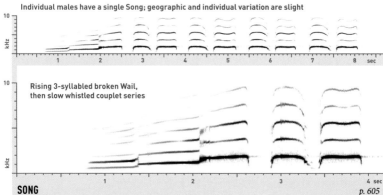

Individual males have a single Song; geographic and individual variation are slight

Rising 3-syllabled broken Wail, then slow whistled couplet series

SONG

p. 605

Mostly on breeding grounds, by males in aggressive conflicts or during nocturnal choruses. Can carry for miles. Individuals apparently recognize the Songs of neighbors and mates. Often given in "vulture" posture with body partway out of water, hunched back, extended neck, and partially spread wings.

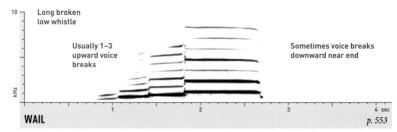

Long broken low whistle

Usually 1–3 upward voice breaks

Sometimes voice breaks downward near end

WAIL

p. 553

All year, but mostly heard on breeding grounds. An evocative sound, reminiscent of the howl of a wolf, often given in nocturnal choruses. Number of voice breaks varies with motivation; simplest version, an unbroken whistle, used in mate contact. More breaks indicates higher level of excitement or alarm.

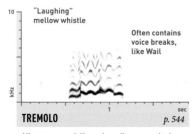

"Laughing" mellow whistle

Often contains voice breaks, like Wail

TREMOLO

p. 544

All year, especially on breeding grounds; in response to danger or territorial threat, in nocturnal choruses, and sometimes in flight.

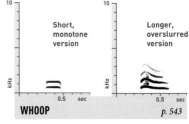

Short, monotone version

Longer, overslurred version

WHOOP

p. 543

All year, grading into one another and into Wail. Most common calls in migration and winter, mostly between mates.

Yellow-billed Loon

Gavia adamsii

Larger than Common Loon, with pale, slightly upturned bill. Rare breeder in far northern Arctic, and even rarer in winter along West Coast and inland.

Individual males likely have a single Song; geographic and individual variation are slight

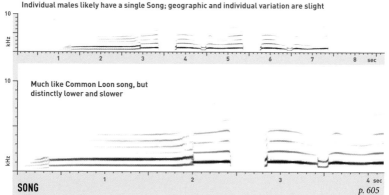

Much like Common Loon song, but distinctly lower and slower

SONG *p. 605*

Mostly on breeding grounds, by males in aggressive conflicts or during choruses; reportedly also by mated pairs. Courtship displays and calling postures seem generally similar to those of Common Loon. First note of Song often given by itself (p. 553), resembling low, slow Wail of Common Loon.

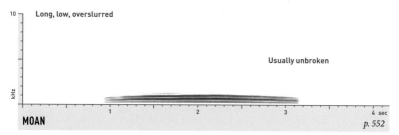

Long, low, overslurred

Usually unbroken

MOAN *p. 552*

All year, but mostly heard on breeding grounds, during territorial conflict. Most versions quite different from Wail of Common Loon, but some versions break upward to a higher pitch that is still quite low.

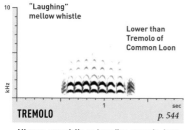

"Laughing" mellow whistle

Lower than Tremolo of Common Loon

TREMOLO *p. 544*

All year, especially on breeding grounds; in response to danger or territorial threat, in nocturnal choruses, and sometimes in flight.

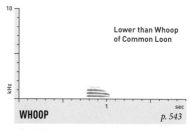

Lower than Whoop of Common Loon

WHOOP *p. 543*

Possibly all year; soft and rarely heard, but to be listened for in migration and winter.

Pacific Loon

Gavia pacifica

Smaller than Common Loon. Breeds in Arctic and winters on salt water; rare inland. In winter, note abrupt, unbroken border to dark brown cap and nape.

Some versions of Song are a series of broken Wails

Rising 3-syllabled broken Wail, then whistled couplet series

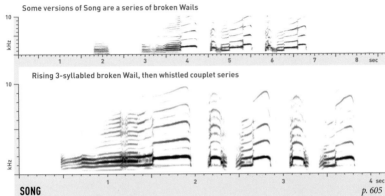

SONG   *p. 605*

Mostly on breeding grounds, by males, likely to defend territory and attract mate. Two song patterns illustrated above; difference in function, if any, not known. Pattern ending in couplet series recalls Song of Common Loon, but significantly lower and faster.

Unbroken version: rather low, moaning

Broken version: long Groan, breaking upward once or twice at end

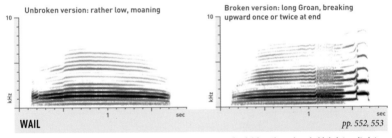

WAIL   *pp. 552, 553*

Mostly during breeding season, by both sexes. May serve territorial function; given in high-intensity interactions and in response to other Pacific Loons flying over the territory. Sometimes given by several individuals together in a chorus. Much lower and often rougher than Common Loon Wail.

Very low, downslurred

Lower than Whoop of Common Loon

Loud, screechy

With loud splash made by feet

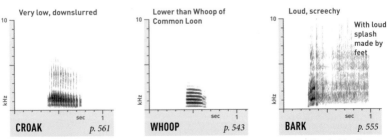

CROAK   *p. 561*

WHOOP   *p. 543*

BARK   *p. 555*

Often given by both members of a pair, in response to disturbance.

All year, in close contact. Plastic. Does not carry far.

Given with splashing dive when threatened by humans, other loons, etc.

Red-throated Loon

Gavia stellata

Our smallest loon, with a thin, upturned bill. Breeds in Arctic and winters on salt water; rare inland. Voice quite different from those of other loons.

Song: often a synchronized duet, with individuals alternating and overlapping phrases

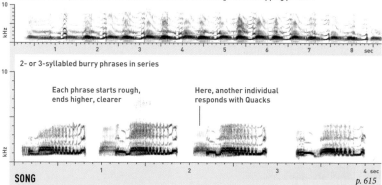

2- or 3-syllabled burry phrases in series

Each phrase starts rough, ends higher, clearer

Here, another individual responds with Quacks

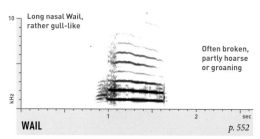

SONG   *p. 615*

Mostly May–Aug., usually by pairs in duet or small groups in chorus, often with body raised 45 degrees out of water, neck extended, and head pointed downward. Often introduced by 1–2 Wails. Tone quality growling and slightly screeching. Male and female versions apparently differ slightly.

Long nasal Wail, rather gull-like

Often broken, partly hoarse or groaning

WAIL   *p. 552*

All year, but mostly during breeding season. Plastic. By both sexes in territorial defense and in response to potential predators. Sometimes given in pair duet, usually unsynchronized.

Loud, screechy

With loud splash made by feet

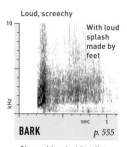

BARK   *p. 555*

Given with splashing dive when threatened by humans, other loons, etc.

Single long ducklike Quack

Varies from croaking to rather nasal

QUACK   *p. 559*

Short Quacklike notes accelerate, then end in longer note

KUK-KUK-QUACK   *p. 615*

All year; the most common calls in winter. Variable and somewhat plastic; Quacks often in series, but also singly. Given in contact, in alarm, and frequently in flight. Mostly directed at other loons, but sometimes given in response to eagles or other predators.

PELICANS (Family Pelecanidae)

Well known for the large pouches below their bills which they use to catch fish, the pelicans are among our largest birds. Adults are nearly silent, giving only rare grunting sounds; juveniles beg loudly at the nest. Sounds are likely innate.

CORMORANTS (Family Phalacrocoracidae)

These long-necked, dark-plumaged birds catch fish by diving underwater. Their bills are long and hooked, and after swimming they often sit with wings spread to dry their plumage.

These grunting and croaking vocalizations of cormorants are given mostly near nests or roosts. Sounds are likely innate.

AMERICAN WHITE PELICAN

Pelecanus erythrorhynchos

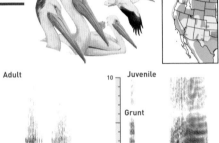

Very large and quite distinctive. Often soars. Flocks fly high in lines or V-formations. Feeds while swimming by dipping head below surface of water, often in cooperative groups.

Adults are usually silent, but do give brief, soft Croaks, mostly at nest site, in courtship or aggression. Begging juveniles give a range of plastic sounds, from Grunts (p. 555) to Groans.

CROAK *p. 561*

GROAN *p. 543*

BROWN PELICAN

Pelecanus occidentalis

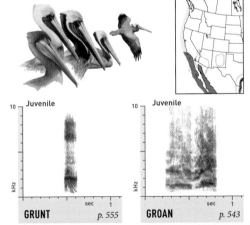

Smaller than American White Pelican. Almost exclusively coastal. Feeds by plunge-diving into the water from the air. Groups often glide low over the water in single file.

Adults are usually silent, except for soft Grunt or Groan given during courtship display; no recordings available. Begging juveniles give a range of plastic sounds, from Grunts to Groans to higher Screeches.

GRUNT *p. 555*

GROAN *p. 543*

DOUBLE-CRESTED CORMORANT

Phalacrocorax auritus

Our most widespread cormorant, occurring both inland and coastally. The largest species in the West. In flight, note kinked neck.

Colonies make a varied cacophony of Croaks, Grunts, and similar calls

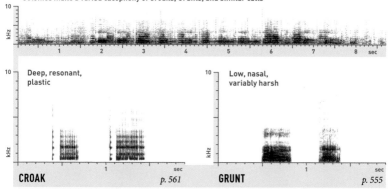

CROAK	GRUNT
Deep, resonant, plastic	Low, nasal, variably harsh
p. 561	*p. 555*

Calls often at colonies, rarely elsewhere, but may give Croaks when disturbed or in flight. Also gives Croaks to greet mate, with neck outstretched. In courtship, gives series of calls with head and tail vertical, wings lifted on each note; no recordings available. Also vocalizes during other displays. Young shriek in nest.

NEOTROPIC CORMORANT

Phalacrocorax brasilianus

Obviously smaller than Double-crested; tail proportionately longer. In the West, found primarily in fresh water habitats. In flight, note kinked neck.

Colonies give near-constant Croaks, with the occasional Grunt

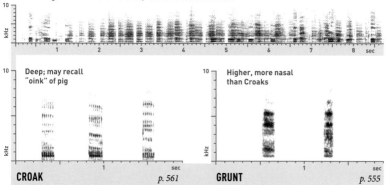

CROAK	GRUNT
Deep; may recall "oink" of pig	Higher, more nasal than Croaks
p. 561	*p. 555*

Calls often at colonies, rarely elsewhere, but may give Croaks when disturbed or in flight. Also gives Croaks to greet mate, with neck outstretched. In courtship, gives series of calls with head and tail vertical, wings lifted on each note; no recordings available. Also vocalizes during other displays.

Brandt's Cormorant

Phalacrocorax penicillatus

Almost exclusively coastal, nesting in colonies on rocky islands or sea cliffs. Note dark face, pale throat patch. In flight, neck often slightly kinked.

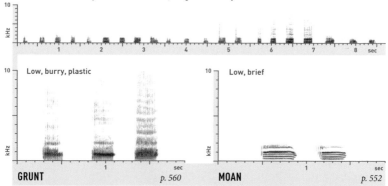

Not as vocal at colony as other cormorants, but gives a variety of sounds

Low, burry, plastic

Low, brief

GRUNT   *p. 560*

MOAN   *p. 552*

Calls fairly often at colonies, rarely elsewhere, but may give Grunt when disturbed or in flight. Some versions of Grunt are coarser, more Croaklike than shown. Moan is infrequent; function unknown. Displays are generally similar to those of other cormorants, and likely vocalizations are used in similar ways.

Pelagic Cormorant

Phalacrocorax pelagicus

Almost exclusively coastal, nesting on sea cliffs, singly or in loose colonies. Rather small and slender. In flight, note straight, thin neck and small head.

At nest, gives Honks and Moans, especially upon arrival and departure

Clear to slightly bleating

High, metallic, broken or unbroken

Soft, ticking

MOAN   *p. 552*

HONK   *p. 549*

RATTLE   *p. 565*

Calls fairly often at nest, rarely elsewhere. Voice rather different from other cormorants in our area. Moan and Honk may be given by different sexes; more study needed. Ticking Rattle, audible at close range, given with bill wide open; similar sounds may form components of other displays.

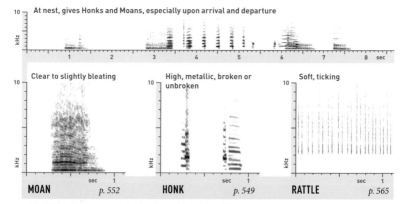

BITTERNS, HERONS, AND EGRETS (Family Ardeidae)

IBIS AND SPOONBILLS (Family Threskiornithidae)

These two families include medium-to-large wading birds with long legs, long necks, and long bills. Most species have limited vocal repertoires of grunting or croaking sounds, but bitterns give repeated songlike vocalizations.

Many species have a large number of visual displays performed in courtship or aggression, some accompanied by sounds. Common displays include an aggressive forward display with feathers/plumes raised and neck coiled as if to strike; a courtship neck-stretch display with the bill pointed upward, silently or with a call; and a greeting display between pairs when one arrives at the nest.

AMERICAN BITTERN

Botaurus lentiginosus

Stays well hidden in dense freshwater marshes. Like Least Bittern, sometimes responds to threats by freezing in place with bill and neck vertical.

Song preceded by slow, accelerating series of soft clicks, as bird appears to take gulps of air

VERY LOW-PITCHED

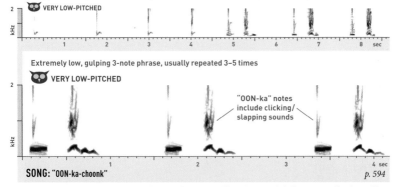

Extremely low, gulping 3-note phrase, usually repeated 3–5 times

VERY LOW-PITCHED

"OON-ka" notes include clicking/slapping sounds

SONG: "OON-ka-choonk" p. 594

Song is bizarre, unmistakable, and far-carrying; sounds like a huge rock being repeatedly plunged into water in a repeating 3-note pattern. Given mostly Mar.–July, especially from dusk to dawn. Not known whether both sexes sing. Sound appears to be made by expelling air from the inflated esophagus.

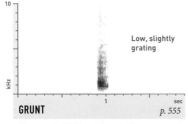

Low, slightly grating

GRUNT p. 555

All year, in flight and in agitation, apparently only during the day. Infrequent and not very loud.

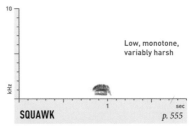

Low, monotone, variably harsh

SQUAWK p. 555

All year, in nocturnal migration, usually singly. Has not been recorded during the day. Lower than Black-crowned Night-Heron Squawk.

LEAST BITTERN

Ixobrychus exilis

Our smallest heron, breeding in dense freshwater or brackish marshes. Often skulking and difficult to see, but can be quite confiding when in the open.

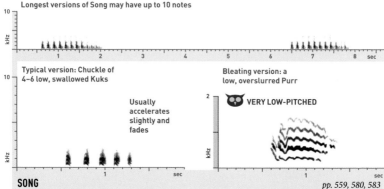

Longest versions of Song may have up to 10 notes

Typical version: Chuckle of 4–6 low, swallowed Kuks

Usually accelerates slightly and fades

Bleating version: a low, overslurred Purr

🦉 VERY LOW-PITCHED

SONG

pp. 559, 580, 583

Mostly Mar.–Oct., reportedly only by males. Often given from deep cover. Quality varies from nasal to grunting. Some versions recall Song of Black-billed Cuckoo, but lower, less musical. Bleating version little known; may serve a slightly different function, or may represent individual variation.

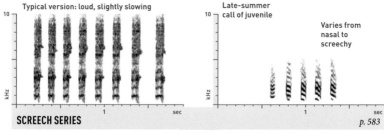

Typical version: loud, slightly slowing

Late-summer call of juvenile

Varies from nasal to screechy

SCREECH SERIES

p. 583

All year, apparently by both sexes, possibly in alarm or contact. Usually given infrequently, at long intervals, but variable, plastic versions may be given frequently in late summer, possibly by both adults and juveniles. Can be as short as 3 notes. Often mistaken for the call of a rail.

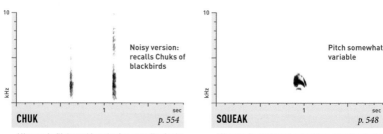

Noisy version: recalls Chuks of blackbirds

CHUK

p. 554

All year, in flight and in agitation, usually singly. Like Keks of large rails, and sometimes given in response to them.

Pitch somewhat variable

SQUEAK

p. 548

All year, in nocturnal migration, usually singly. Has not been recorded during the day. Generally lower and less plastic than Virginia Rail Squeak.

GREAT BLUE HERON

Ardea herodias

Our largest and most familiar heron, found in wetlands and along rivers and lakeshores. Breeds in colonies, building large stick nests high in trees.

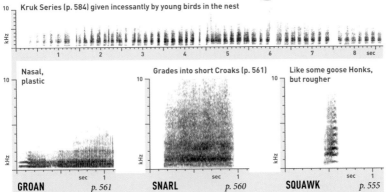

Kruk Series (p. 584) given incessantly by young birds in the nest

GROAN — Nasal, plastic — *p. 561*

SNARL — Grades into short Croaks (p. 561) — *p. 560*

SQUAWK — Like some goose Honks, but rougher — *p. 555*

All calls plastic. Croaking Groans given in early spring by males in neck-stretch display. Snarls, Croaks, and Squawks intergrade; all given in alarm or in flight. Kruk Series varies with age of young, but typically harsher than calls of young Great Egrets.

GREAT EGRET

Ardea alba

Locally common in wetland habitats. Much larger than Snowy Egret, with yellow bill and black legs. Forages deliberately, usually in an upright posture.

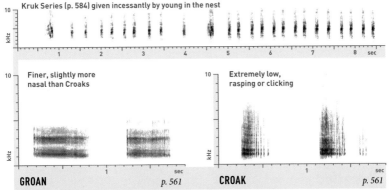

Kruk Series (p. 584) given incessantly by young in the nest

GROAN — Finer, slightly more nasal than Croaks — *p. 561*

CROAK — Extremely low, rasping or clicking — *p. 561*

Nearly silent in courtship; most displays are purely visual. All adult vocalizations seem to be plastic, on the continuum from Groans to Croaks, ranging from long to short; given in alarm and in interactions. Kruk Series of young in nest variable, but usually more nasal than Great Blue Heron's.

SNOWY EGRET

Egretta thula

Widespread in wetlands. Much smaller than Great Egret, with black bill and legs, yellow feet. Has recovered from ferocious hunting trade of the 1800s.

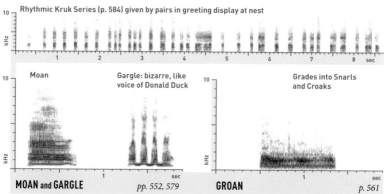

Rhythmic Kruk Series (p. 584) given by pairs in greeting display at nest

Moan

Gargle: bizarre, like voice of Donald Duck

Grades into Snarls and Croaks

MOAN and GARGLE *pp. 552, 579*

GROAN *p. 561*

Moan and Gargle given mostly Apr.–June, in courtship at nesting sites; Moan likely by both sexes, highly plastic Gargle reportedly only by males. Several other courtship vocalizations not shown. Groan, Snarls, and Croaks (p. 561) given all year, in alarm, in flight, and in aggressive interactions.

LITTLE BLUE HERON

Egretta caerulea

Found in freshwater and saltwater wetlands. First-year birds very similar to Snowy Egret; birds molting into adult plumage are variably patchy.

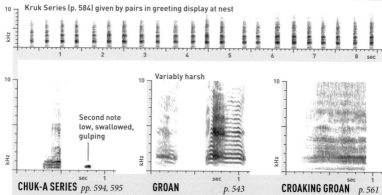

Kruk Series (p. 584) given by pairs in greeting display at nest

Second note low, swallowed, gulping

Variably harsh

CHUK-A SERIES *pp. 594, 595*

GROAN *p. 543*

CROAKING GROAN *p. 561*

Chuk-a given in series by courting male during neck-stretch display; descriptions vary, suggesting that some display sounds may be 1-noted. Highly plastic Groans not well known; may be given mostly at nest. Plastic Croaking Groans and Croaks (p. 561) given all year, in alarm and aggressive encounters.

CATTLE EGRET

Bubulcus ibis

Originally an Old World species. Began colonizing North America in the 1950s; now widespread. Forages in fields, often in close association with cattle.

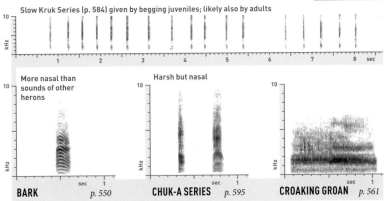

Slow Kruk Series (p. 584) given by begging juveniles; likely also by adults

More nasal than sounds of other herons

Harsh but nasal

BARK *p. 550*

CHUK-A SERIES *p. 595*

CROAKING GROAN *p. 561*

All sounds plastic. Barks given singly, grading into shorter, noisier Kruks in series, or into Chuk-a Series. All of these given at colony and possibly in alarm; single Chuk-a sounds sometimes given in flight. Croaking Groan rather harsh, almost snarling; given at nest, during altercations.

GREEN HERON

Butorides virescens

A small, squat, solitary heron of densely vegetated swamps, ponds, and wetlands. Often perches low above water, remaining motionless for long periods.

Rapid Kek Series given by adults in alarm; similar notes given by begging juveniles

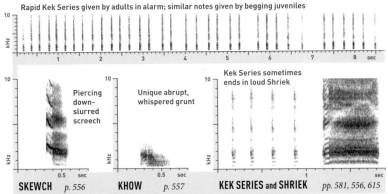

Piercing down-slurred screech

Unique abrupt, whispered grunt

Kek Series sometimes ends in loud Shriek

SKEWCH *p. 556*

KHOW *p. 557*

KEK SERIES and SHRIEK *pp. 581, 556, 615*

Skewch is most common call, given in alarm and flight; highly variable, grading into Keks. Khow often given in early breeding season from high in tree, possibly by males; rather quiet and difficult to locate. Kek Series given in alarm, grading into one or more plastic Shrieks in high alarm.

Glossy Ibis

Plegadis falcinellus

Main range is along the East Coast, but wanders westward, and a few breed among White-faced Ibis in the West. Legs and facial skin bluish; eyes dark.

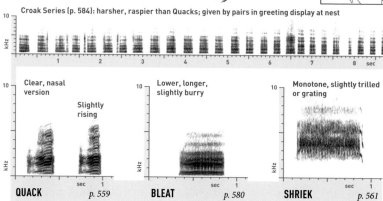

Croak Series (p. 584): harsher, raspier than Quacks; given by pairs in greeting display at nest

Clear, nasal version

Slightly rising

Lower, longer, slightly burry

Monotone, slightly trilled or grating

QUACK *p. 559*   **BLEAT** *p. 580*   **SHRIEK** *p. 561*

Quack given all year, in alarm, but most often near nest; most common call. Plastic; ranges from 1- to vaguely 2-syllabled, monotone to rising, clear and nasal to grunting or grating. Bleat given near nest, at least in spring; may function in courtship. Shrieks given by begging juveniles, including in flight.

White-faced Ibis

Plegadis chihi

Found in freshwater marshes. Very similar to Glossy Ibis, but legs, facial skin, and eye reddish. Only breeding adults have white facial border.

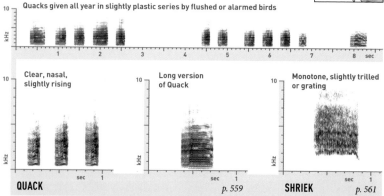

Quacks given all year in slightly plastic series by flushed or alarmed birds

Clear, nasal, slightly rising

Long version of Quack

Monotone, slightly trilled or grating

QUACK   *p. 559*   **SHRIEK** *p. 561*

Quack similar to Quack of Glossy Ibis, and not usually separable, but averages shorter and clearer, with roughest versions heard mostly at nest. Likely gives a Bleat (p. 580) near nest and a Croak Series (p. 584) in greeting like Glossy Ibis; no recordings available. Shrieks given by begging juveniles, including in flight.

BLACK-CROWNED NIGHT-HERON

Nycticorax nycticorax

Found nearly worldwide in various wetland habitats. Active mostly between late evening and early morning; often overlooked during the day.

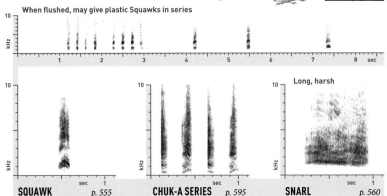

When flushed, may give plastic Squawks in series

SQUAWK *p. 555*

CHUK-A SERIES *p. 595*

SNARL *p. 560* — Long, harsh

Squawk is most common call, often given in flight; plastic but distinctive. Chuk-a Series given by young birds, which also give series of Clucks while still in the nest. Snarl given infrequently, during altercations. In display, gives a soft Pwut (p. 554), audible only at close range.

RAPTOR COPULATION SOUNDS

Most hawks and eagles have an excited vocalization, a Squeal Series (p. 545), that is given only during copulation. Often the female is most vocal, but both birds may participate.

These sounds are given infrequently and only during the early breeding season, but they tend to be loud, far-carrying, and often characteristic of the species.

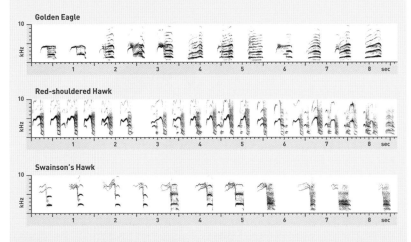

Golden Eagle

Red-shouldered Hawk

Swainson's Hawk

HAWKS, EAGLES, KITES, AND OSPREY
(Order Accipitriformes)

The sound of a raptor, in many people's minds, is the scream of a Red-tailed Hawk. But many raptors actually sound much like gulls, both in tone quality and in pattern. All sounds in this group are innate, as far as we know, but highly plastic.

VULTURES (Order Cathartiformes)

Adapted for scavenging carrion, American vultures appear to lack the muscles that attach to the syrinx in most birds. Thus, they give only rough grunts and hisses, when they vocalize at all. Most sounds do not carry far.

BLACK VULTURE

Coragyps atratus

Locates carrion primarily by sight. Often follows Turkey Vulture to carcasses and then displaces it. Soars on horizontal wings; wingbeats distinctively quick and snappy.

Distinctive short Grunts given by adults in interactions with other vultures at roosts or carcasses. Hiss given when birds feel threatened; juvenile version shown. Wings make a thrumming sound (p. 567) audible at close range.

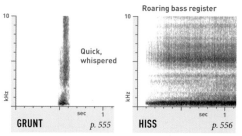

Roaring bass register

Quick, whispered

GRUNT *p. 555*

HISS *p. 556*

TURKEY VULTURE

Cathartes aura

Our most familiar vulture. Locates prey primarily by scent. Soars with wings in a shallow V, erratically tipping from side to side; wingbeats slow and labored. Juveniles have dark gray heads.

Heard even less often than Black Vulture. Juveniles and adults may give Hisses when threatened or in aggressive interactions near carcasses. Adults may also occasionally give shorter Grunts.

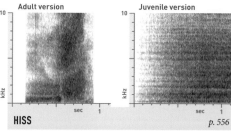

Adult version

Juvenile version

HISS *p. 556*

California Condor

Gymnogyps californianus

Enormous; wingspan nearly twice that of a Turkey Vulture. Note white underwing linings. Critically endangered and once extinct in the wild, but now reintroduced and breeding in parts of California, Arizona, and Utah. Voice like that of other vultures, but even less likely to be heard in the wild. Swish of air through wings is sometimes audible at a distance.

Northern Harrier

Circus hudsonius

Distinctively shaped, with long wings and tail. White rump often conspicuous in flight. Hunts low over marshes, with wings held in a shallow V.

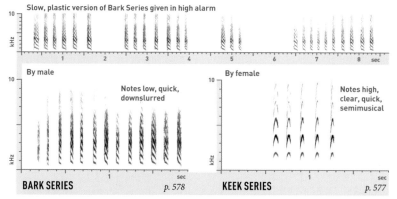

Slow, plastic version of Bark Series given in high alarm

By male — Notes low, quick, downslurred

By female — Notes high, clear, quick, semimusical

BARK SERIES *p. 578*

KEEK SERIES *p. 577*

All year, from alarmed or agitated birds in response to predators near the nest, harassment by blackbirds, etc. Male Bark Series and female Keek Series quite different in quality. In mild alarm, series are short; in high alarm, they become slower, longer, and more plastic.

Nasal but harsh, slightly grating

2–3 syllables run together

GRUNT *p. 560*

Known only from males near the nest, but may occur in other situations; more study needed.

High, clear, overslurred

Typically in series of 2–4

WHEEW *p. 542*

Mostly Mar.–July, during acrobatic courtship flight, apparently by male; also reportedly in aggression and by female on the nest.

WHITE-TAILED KITE

Elanus leucurus

Hunts on the wing over wetlands and savannas, hovering frequently, then dropping into the grass after small prey. Roosts communally in winter.

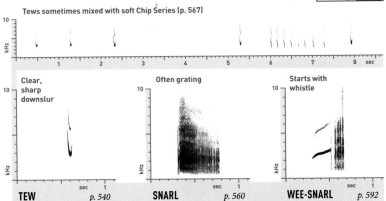

Tews sometimes mixed with soft Chip Series (p. 567)

TEW	*p. 540*

Clear, sharp downslur

SNARL	*p. 560*

Often grating

WEE-SNARL	*p. 592*

Starts with whistle

All sounds given all year; all plastic. Tew given in pair contact and in territorial defense; most common call. Snarl given in territorial defense. Wee-snarl given in a variety of situations including courtship and alarm; whistle usually upslurred and snarl usually brief, but can be longer than 1 second.

MISSISSIPPI KITE

Ictinia mississippiensis

Locally common in forests and savannas, sometimes nesting in parks and old neighborhoods. Catches insects in graceful soaring flight.

Keek-tew-tew-tew usually given in series beginning or ending with Keek-keer

High, 2-syllabled, second note downslurred

KEEK-KEER	*p. 586*

Like Keek-keer, but extending into short series

KEEK-TEW-TEW-TEW	*p. 606*

Keek-keer is most common call, given all year, but mostly near nest or in response to predator. Grades into Keek-tew-tew-tew, given mostly by excited birds near nest, in pair interactions or in reponse to nestlings. High, broken Squeal (p. 553) sometimes given, likely a variant of Keek-keer.

SHARP-SHINNED HAWK

Accipiter striatus

Our smallest accipiter, a fierce hunter of small birds. Smaller than Cooper's Hawk, with sharper tail corners, snappier wingbeats, different voice.

Keet Series fairly uniform, but plastic in number of notes

Series of high, sharp, seminasal notes

First note often lower

Higher, clearer, shriller than other accipiters

KEEK SERIES
p. 577

Mostly Apr.–Sept., by adults near nest, especially in response to intruders; also frequently heard from fledglings in late summer. Series sometimes followed by a few Tews.

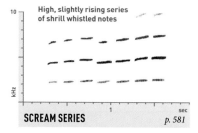

High, slightly rising series of shrill whistled notes

SCREAM SERIES
p. 581

Likely Mar.–May, during copulation, possibly by female, but poorly known. Plastic. Notes and series both tend to be barely upslurred; tone clear.

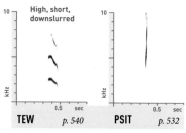

High, short, downslurred

TEW
p. 540

PSIT
p. 532

All year. Tews given near nest and on fall migration; grade into excited, sometimes twittering Psit. Also gives a clear Chip (p. 531; not shown).

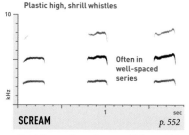

Plastic high, shrill whistles

Often in well-spaced series

SCREAM
p. 552

By adults near the nest, apparently in courtship; possibly in other contexts also. Like Killdeer Deet, but higher.

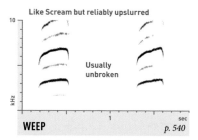

Like Scream but reliably upslurred

Usually unbroken

WEEP
p. 540

July–Oct., by begging juveniles. Juvenile version of Scream, often given with Keet Series. Plastic.

COOPER'S HAWK

Accipiter cooperii

Intermediate in size between Sharp-shinned Hawk and Northern Goshawk. Found in a variety of habitats; in some areas, nests readily in towns.

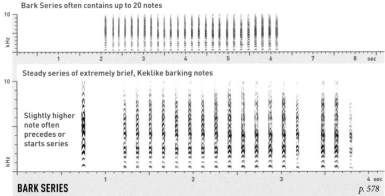

Bark Series often contains up to 20 notes

Steady series of extremely brief, Keklike barking notes

Slightly higher note often precedes or starts series

BARK SERIES
p. 578

Mostly Mar.–May, especially from females, and especially in response to intruders near the nest. Also heard from agitated birds in fall. Recently fledged juveniles in late summer give a higher-pitched version. Much lower and more nasal than series of other accipiters or Northern Flicker.

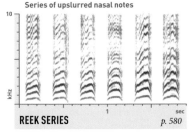

Series of upslurred nasal notes

REEK SERIES
p. 580

Mar.–May, during copulation, mostly by females; males sometimes join in near the end. Fairly plastic.

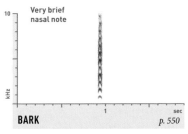

Very brief nasal note

BARK
p. 550

Mostly Mar.–May, especially by males, in contact and near nest. Most common call. Given regularly at dawn in breeding season.

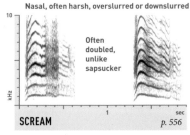

Nasal, often harsh, overslurred or downslurred

Often doubled, unlike sapsucker

SCREAM
p. 556

Mostly Jan.–June, by breeding females; males give lower, harsher versions. Plastic. Some versions are clear Mews (p. 551), like a sapsucker's.

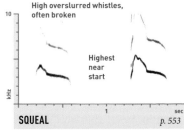

High overslurred whistles, often broken

Highest near start

SQUEAL
p. 553

Mostly July–Oct. Juvenile version of the Scream, given by begging fledglings and later by independent birds. Plastic and variable.

Northern Goshawk

Accipiter gentilis

Our largest accipiter. Breeds in mature forests, hunting squirrels, rabbits, and birds. Voice generally more like Sharp-shinned's than like Cooper's.

Here, Keklike version of Screech Series introduced by Tew

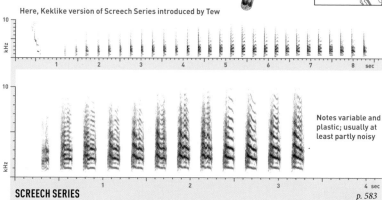

Notes variable and plastic; usually at least partly noisy

SCREECH SERIES
p. 583

Mostly Mar.–June. Given in aggressive defense of nest or territory against all threats, especially by females. Some less excited versions have shorter notes, more like Cooper's Hawk Bark Series, but always higher-pitched and slower. Notes in series sometimes broken.

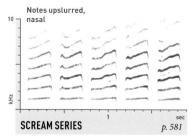

Notes upslurred, nasal

SCREAM SERIES
p. 581

Mar.–May, by females during copulation, with males sometimes joining in near the end. Lower than Sharp-shinned Hawk Scream Series.

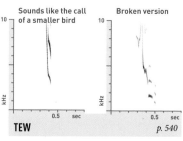

Sounds like the call of a smaller bird

Broken version

TEW
p. 540

Apparently by both sexes, all year, likely in mild alarm. Variable and plastic, not very loud. Broken versions can recall Northern Flicker Kleer.

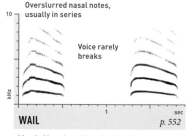

Overslurred nasal notes, usually in series

Voice rarely breaks

WAIL
p. 552

Mostly Mar.–Aug. Given frequently by pairs near the nest; some Wails in each series may have a slight tremolo.

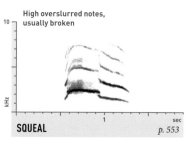

High overslurred notes, usually broken

SQUEAL
p. 553

Plastic, variable. Juvenile version of the Wail, easily confused with Cooper's Hawk Squeal, but more often broken.

Harris's Hawk

Parabuteo unicinctus

Prefers mesquite and saguaro deserts; often breeds in small colonies, hunts in groups. Popular in falconry. Vocal repertoire may be larger than shown.

Squeal Series slow and steady, usually 3–6 notes

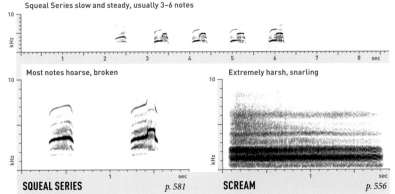

Most notes hoarse, broken

Extremely harsh, snarling

SQUEAL SERIES *p. 581*

SCREAM *p. 556*

Both sounds given all year. Squeal Series apparently given as a signal to mates or hunting partners; occasionally shortened to a single note. Scream given in several situations; most common call. Distinctively harsh, but fading at end. Juveniles give a clearer version.

Common Black Hawk

Buteogallus anthracinus

Widespread in Central America, but rare and local north of Mexico in mature deciduous riparian woods. Note very short tail with single white band.

Keek Series (p. 577): fast versions recall Sharp-shinned Hawk, but notes slightly hoarser and less abrupt

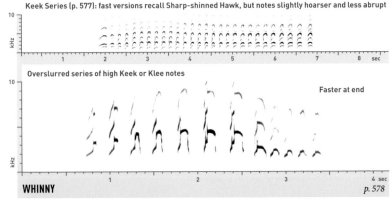

Overslurred series of high Keek or Klee notes

Faster at end

WHINNY *p. 578*

Whinny is most common call, given all year in various situations. Plastic but distinctive; like Bald Eagle Keek Series, but slower and longer. Common Black Hawk Keek Series given in courtship, to solicit food, and near nest; rate varies, 1–4 notes/second, depending on level of hunger or perhaps excitement.

RED-TAILED HAWK

Buteo jamaicensis

Our most common hawk, found almost everywhere. Variable in plumage, but vocalizations are the same throughout range and among subspecies.

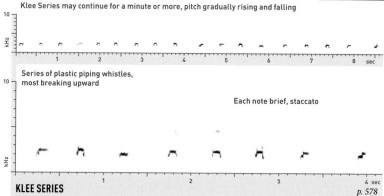

Klee Series may continue for a minute or more, pitch gradually rising and falling

Series of plastic piping whistles, most breaking upward

Each note brief, staccato

KLEE SERIES   p. 578

Mostly Mar.–June, by both sexes. Given in swooping, high-altitude courtship flight, sometimes by both members of a pair. Often sounds very faint because of great distance from the ground. Distinctive; Swainson's Klee Series is similar but averages slower, with longer notes. Beware jay imitations (p. 307).

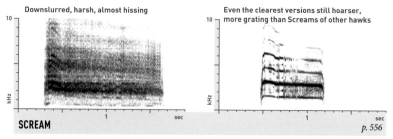

Downslurred, harsh, almost hissing

Even the clearest versions still hoarser, more grating than Screams of other hawks

SCREAM   p. 556

All year, from both sexes. An iconic sound, heard in hundreds of movies and commercials, often from the mouths of other species. Hissing, screechy quality distinctive; sounds louder than it actually is. Sexes may differ in pitch and tone quality; more study needed.

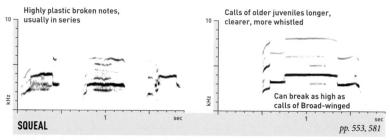

Highly plastic broken notes, usually in series

Calls of older juveniles longer, clearer, more whistled

Can break as high as calls of Broad-winged

SQUEAL   pp. 553, 581

Mostly July–Sept. Given by immature birds in their first summer and fall, including apparently independent birds. May transition gradually into adult Scream over course of first fall and winter; more study needed. Far more likely to cause identification confusion than adult Scream.

Gray Hawk

Buteo plagiatus

A small buteo of riparian areas and adjacent thorn scrub or savanna. Once rare in the U.S., but population seems to have increased in recent years.

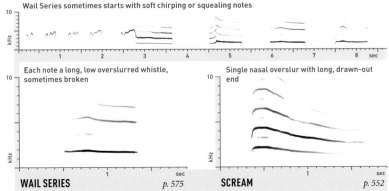

Wail Series sometimes starts with soft chirping or squealing notes

Each note a long, low overslurred whistle, sometimes broken

Single nasal overslur with long, drawn-out end

WAIL SERIES   *p. 575*

SCREAM   *p. 552*

Wail Series given mostly Mar.–Aug. Plastic; usually 3–5 notes, most smoothly overslurred, but some may break or sound 2-syllabled. Scream given all year; most common call. Sometimes noisy; juveniles beg with a higher, hoarser, often broken version. Version with Squeals given during copulation.

Zone-tailed Hawk

Buteo albonotatus

Uncommon in western canyons. Almost exactly matches Turkey Vulture in flight style, shape, and coloration, perhaps to fool prey into allowing close approach.

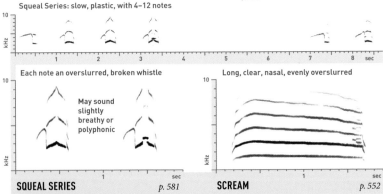

Squeal Series: slow, plastic, with 4–12 notes

Each note an overslurred, broken whistle

May sound slightly breathy or polyphonic

Long, clear, nasal, evenly overslurred

SQUEAL SERIES   *p. 581*

SCREAM   *p. 552*

Squeal Series given mostly Mar.–Aug., by females near nest, by begging juveniles, and by adults in high, swooping courtship flight. Scream given all year, in alarm and also in courtship; most common call. Longer and more monotone than Gray Hawk Scream.

Red-shouldered Hawk

Buteo lineatus

A small buteo of mature forests; hunts small prey mostly from perches. Highly vocal in spring, calling loudly in flight display just above treetops.

In courtship flight with dangling legs, gives Squeal Series (p. 581) of broken Kleerlike notes

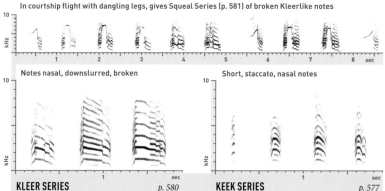

Notes nasal, downslurred, broken

Short, staccato, nasal notes

KLEER SERIES *p. 580*

KEEK SERIES *p. 577*

All sounds most common Feb.–Apr. Kleer Series is most common sound, often given in flight. Sometimes continues for long periods. Keek Series given in alarm or excitement; intergrades with Kleer Series. Occasional Keek given singly.

Broad-winged Hawk

Buteo platypterus

Our smallest buteo, found in deciduous or mixed forests, where it hunts mostly from perches. Migrates in large flocks called kettles. Voice distinctively high.

Pee-tee Series typically 4–6 seconds long

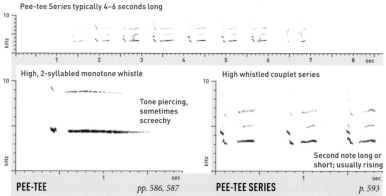

High, 2-syllabled monotone whistle

Tone piercing, sometimes screechy

High whistled couplet series

Second note long or short; usually rising

PEE-TEE *pp. 586, 587*

PEE-TEE SERIES *p. 593*

Pee-tee given all year in many situations; most common call. Rarely, second note breaks or is downslurred, or the 2 notes run together. Pee-tee Series not well known; highly plastic. Apparently used in territorial defense, possibly also in courtship.

Short-tailed Hawk

Buteo brachyurus

Widespread in the tropics. Rare in our area; has nested in southeast Arizona. Nests in mature, often wet forests; soars over other habitats.

Screams and Squeals can intergrade over the course of a short Squeal Series (p. 581)

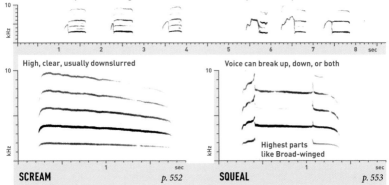

High, clear, usually downslurred

SCREAM *p. 552*

Voice can break up, down, or both

Highest parts like Broad-winged

SQUEAL *p. 553*

Scream and Squeal may be just variants of a single plastic call; both given all year. Distinctive in range, often nasal but never noisy; always clearer and higher than adult Red-tailed. Squeals sometimes given in slow series; function of series not known.

Swainson's Hawk

Buteo swainsonii

Common and widespread in the open country of the West. Migrates to South America in large flocks called kettles. Soars with wings held in a shallow V.

Klee Series much like Red-tailed's

Notes high, brief, often broken

KLEE SERIES *p. 578*

Nasal, downslurred, often quite noisy

SCREAM *p. 556*

Klee Series given mostly Mar.–June in swooping, high-altitude courtship flight, sometimes with Screams interspersed. Scream given all year; most common call. Usually clearer, more nasal than Red-tailed Scream. Female Scream is reportedly shorter and lower than male's; may also tend to be clearer.

Ferruginous Hawk

Buteo regalis

Our largest buteo, uncommon in short-
to mid-grass prairies and sagebrush
steppes. Soars with long, pointed wings
held in a shallow V.

Variety of Screamlike sounds given in nest defense

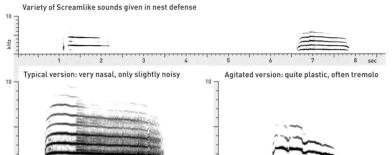

Typical version: very nasal, only slightly noisy

Agitated version: quite plastic, often tremolo

SCREAM

p. 552

Scream given all year, but infrequent except near nest. Scream lower than in most other hawks; some versions perhaps best called Wails (p. 552). Likely has a series call given in courtship, like other buteos, but no
recordings available.

Rough-legged Hawk

Buteo lagopus

Breeds on Arctic cliffs; in winter, found
in open areas. Plumage ranges from
light to dark. Frequently hovers in place
while hunting.

Variety of Screamlike sounds given in nest defense, including Squeals (p. 553)

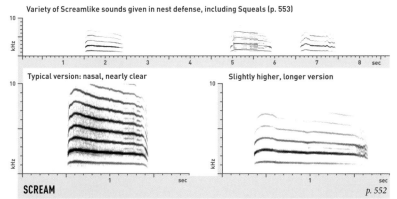

Typical version: nasal, nearly clear

Slightly higher, longer version

SCREAM

p. 552

Scream given all year, but infrequent except near nest. Scream lower than in most other hawks; some versions perhaps best called Wails (p. 552). Likely has a series call given in courtship, like other buteos, but no
recordings available.

BALD EAGLE

Haliaeetus leucocephalus

Found mostly along rivers, lakeshores, and coasts; dives into water for live fish or scavenges on shore. Populations are rebounding after sharp declines.

Screams usually given in series of 2–4

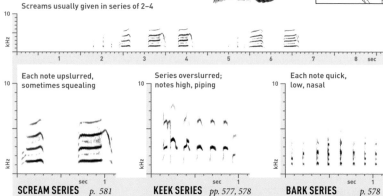

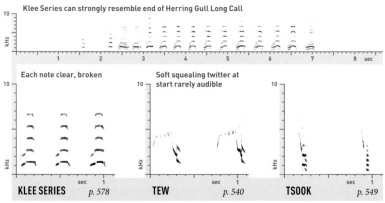

Each note upslurred, sometimes squealing	Series overslurred; notes high, piping	Each note quick, low, nasal
SCREAM SERIES *p. 581*	**KEEK SERIES** *pp. 577, 578*	**BARK SERIES** *p. 578*

All sounds given all year, and all quite plastic; function of most is poorly known. Scream Series plastic; some versions resemble Reek of female Wood Duck. Keek Series is most common call, often introduced by 4–5 well-spaced single Keeks. Bark Series given mostly near nest, likely in alarm.

GOLDEN EAGLE

Aquila chrysaetos

A large, powerful raptor of open areas and mountains, rare in the East. Not closely related to Bald Eagle. Hunts mostly rabbits and ground squirrels.

Klee Series can strongly resemble end of Herring Gull Long Call

Each note clear, broken	Soft squealing twitter at start rarely audible	
KLEE SERIES *p. 578*	**TEW** *p. 540*	**TSOOK** *p. 549*

Not very vocal. Klee Series given at least Jan.–July; may function in courtship. Tew and Tsook may be variants of one another; given by adults near the nest, and similar calls given by older begging juveniles. Tew lower than Osprey Tew; Tsook sometimes run into short rapid series.

OSPREY

Pandion haliaetus

A unique raptor; catches live fish by plunging feet-first into the water. Nests readily on artificial platforms, sometimes even in towns.

Screams plastic; series may continue for long periods

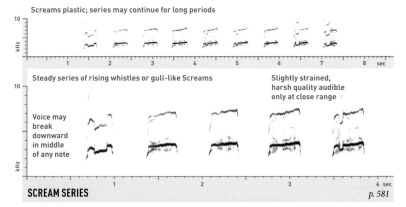

Steady series of rising whistles or gull-like Screams

Slightly strained, harsh quality audible only at close range

Voice may break downward in middle of any note

SCREAM SERIES

p. 581

Mostly during breeding season: Nov.–Mar. in Florida, Mar.–June in New England. Given by both sexes, usually on the wing. Given continuously during male's fish flight display in courtship, along a shallowly undulating path high above the nest, legs typically dropped, sometimes carrying a fish.

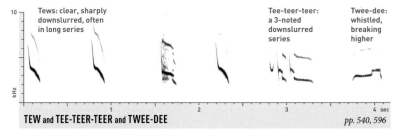

Tews: clear, sharply downslurred, often in long series

Tee-teer-teer: a 3-noted downslurred series

Twee-dee: whistled, breaking higher

TEW and TEE-TEER-TEER and TWEE-DEE

pp. 540, 596

All year. A few Tews often given alone in series, but long series typically ends in distinctive pattern: broken Tew, normal Tew, Tee-teer-teer, Twee-dee. Exact order varies, but typical performance is quite uniform rangewide. Function not well known; may generally signal alertness or excitement.

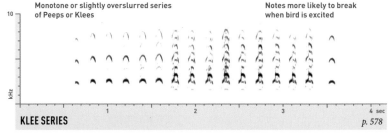

Monotone or slightly overslurred series of Peeps or Klees

Notes more likely to break when bird is excited

KLEE SERIES

p. 578

All year, by both sexes, but especially by females near nest. Highly plastic. Similar calls also used by females to beg for food from males in courtship.

OWLS (Order Strigiformes)

With large, forward-facing eyes adapted to hunting in low light, these birds of prey are primarily nocturnal, although some are also active in daylight. Owls can rotate their heads up to 270 degrees, and specially modified wing feathers make their wingbeats nearly inaudible. Most North American species are classified in the family Strigidae, except for the Barn Owl, which is in the family Tytonidae.

The songs of large owl species are low-pitched and hooting; smaller species give whistled or nasal songs. Although females are larger than males in body size, males tend to have lower-pitched voices. Many species give a variety of barking, wailing, or shrieking calls. All vocalizations are thought to be innate, and most are uniform throughout a species' range.

BARN OWL

Tyto alba

Widespread in open country, sometimes near human habitation. Nests in enclosed spaces ranging from barns to riverbank cavities. Highly nocturnal.

When cornered in nest or roost, gives long screeching Hiss in defensive posture (p. 556)

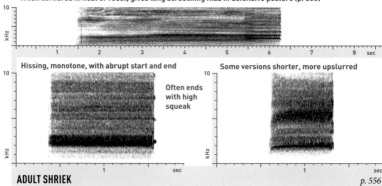

Hissing, monotone, with abrupt start and end

Often ends with high squeak

Some versions shorter, more upslurred

ADULT SHRIEK — p. 556

All year, especially during breeding season; most common vocalization. Often given repeatedly by adult males on the wing, sometimes with soft Wing Claps (p. 566). Shrieks of most other owl species usually given singly in flight or repeatedly from perch. Female version reportedly rarer, more broken.

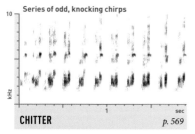

Series of odd, knocking chirps

CHITTER — p. 569

Variable and plastic. All year, mostly near nest, in food exchanges; also by young, by pairs in courtship, and occasionally in flight.

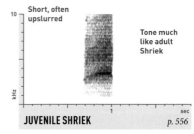

Short, often upslurred

Tone much like adult Shriek

JUVENILE SHRIEK — p. 556

By begging juveniles, usually every few seconds for long periods, often from a perch. Also by adult females. Variable and plastic.

GREAT HORNED OWL

Bubo virginianus

Our most familiar and widespread owl. Primarily nocturnal, but often active and vocal before full dark. Note large size and prominent ear tufts.

Male (black) and female (red) often sing in unsynchronized duets

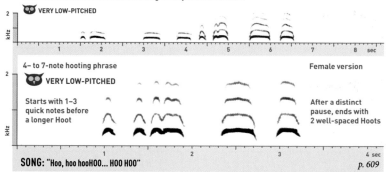

🦉 VERY LOW-PITCHED

4- to 7-note hooting phrase · Female version

🦉 VERY LOW-PITCHED

Starts with 1–3 quick notes before a longer Hoot

After a distinct pause, ends with 2 well-spaced Hoots

SONG: "Hoo, hoo hooHOO... HOO HOO" · *p. 609*

All year, especially Dec.–Mar., by both sexes. Fairly uniform and stereotyped. Females average slightly higher-pitched, with more syllables per phrase. Number of syllables varies somewhat with agitation level, but individuals are often identifiable by subtle differences in song.

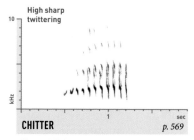

High sharp twittering

CHITTER · *p. 569*

Given infrequently, during copulation and around the nest. Quite plastic; some versions less staccato and more squealing.

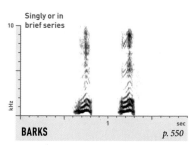

Singly or in brief series

BARKS · *p. 550*

Highly variable and plastic; given in agitation and close courtship. Sometimes Barks may introduce Song, or replace first few notes of Song.

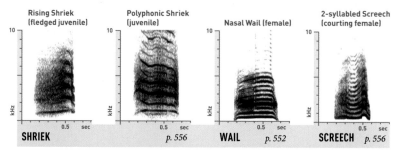

Rising Shriek (fledged juvenile)

Polyphonic Shriek (juvenile)

Nasal Wail (female)

2-syllabled Screech (courting female)

SHRIEK · *p. 556*

WAIL · *p. 552*

SCREECH · *p. 556*

A catch-all category of variable, plastic sounds heard all year. Food-begging Shrieks of juveniles begin in nest and continue for months after fledging; usually very noisy and upslurred. Adults, especially females, give Screeches and Wails that average more nasal, in close courtship and in agitation.

Snowy Owl

Bubo scandiacus

Breeds in Arctic; winters on prairies, beaches, and other open areas. Active both day and night. In some winters, irrupts far south of normal range.

Hoots sometimes in steady series of up to 8; here, introduced by bill snaps and chuckles

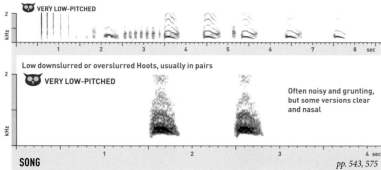

👁 **VERY LOW-PITCHED**

Low downslurred or overslurred Hoots, usually in pairs

🦉 **VERY LOW-PITCHED**

Often noisy and grunting, but some versions clear and nasal

SONG   *pp. 543, 575*

Mostly on breeding grounds, by males; females sing infrequently. Males sing with head lowered and tail raised, bowing with each note. Song given in territorial advertisement, often by multiple neighboring birds at once, as well as in response to threats.

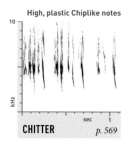

High, plastic Chiplike notes

CHITTER   *p. 569*

Mostly near nest, by juveniles and likely also adults.

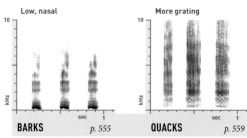

Low, nasal

BARKS   *p. 555*

More grating

QUACKS   *p. 559*

A varied category of intergrading calls. Barks and Quacks given all year as a threat, usually in slow series. Faster, softer Chuckles (p. 583) given in close courtship, among other situations.

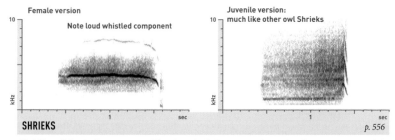

Female version

Note loud whistled component

Juvenile version: much like other owl Shrieks

SHRIEKS

Variable and plastic. Most common call in winter, in clashes over territory; also given near nest, mostly by female. Some versions less noisy, almost a pure whistle. Shrieks given by juveniles for some time after fledging, often with Chitters.

NORTHERN HAWK OWL

Surnia ulula

A large owl of boreal forests, active mostly during the day. Shape and behavior recall *Accipiter* hawks. In some winters, irrupts south of normal range.

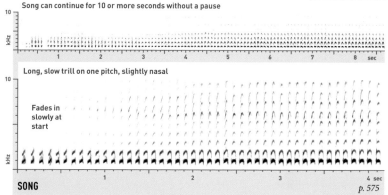

Song can continue for 10 or more seconds without a pause

Long, slow trill on one pitch, slightly nasal

Fades in slowly at start

SONG *p. 575*

Mostly Feb.–Apr., by both sexes. Uniform. Male sings from perch or in circular display flight above territory that includes gliding and Wing Claps. Female song may average hoarser, shorter, and more unsteady in rhythm and pitch. Like Boreal Owl Long Song, but higher, faster, and more nasal.

Like soft, slow snippets of Song

KEEK SERIES *p. 577*

Like flicker Keek Series, but short, high, and plastic. All year, in high alarm.

Occasionally gives single Screeches

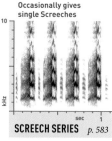

SCREECH SERIES *p. 583*

In alarm, often in flight, perhaps mostly near nest.

High, squeaky, chipping

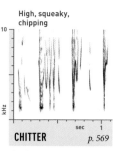

CHITTER *p. 569*

In alarm, perhaps mostly near nest. Extremely plastic.

Rapid nasal series or slow trills, variably screechy

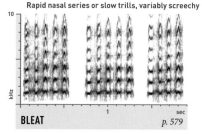

BLEAT *p. 579*

All year, by adults and juveniles in high alarm. Lower-intensity versions of this call reported from some observers in winter.

Long, slightly rising, ending in high Squeak

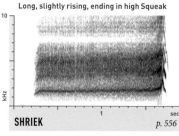

SHRIEK *p. 556*

By juveniles begging for food and by adults in pair interactions, in food exchanges, and in alarm near the nest.

Northern Pygmy-Owl (Southern Rockies)

Glaucidium gnoma pinicola

A tiny but fierce predator of small birds and mammals, generally uncommon to rare in open pine and mixed-conifer forests. Sometimes active during the day.

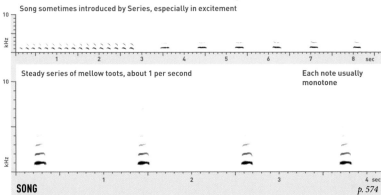

Song sometimes introduced by Series, especially in excitement

Steady series of mellow toots, about 1 per second

Each note usually monotone

SONG *p. 574*

All year by both sexes, but especially Feb.–June, mostly at dawn and dusk; not often heard during the day or in full darkness. Easily imitated by human whistling. Female version averages slightly higher-pitched than male's; female may respond to male Song with Song of her own, or with other calls.

Quick repeating rhythm

Notes often more Piplike and seminasal than shown

Recalls Pygmy Nuthatch

CHITTER *p. 569* **TRILL (MUSICAL)** *p. 571* **TWITTER** *p. 571*

Chitter given in close courtship and near the nest, possibly by both sexes. Musical Trill given in courtship, near the nest, and in contact with fledglings, at least by females. Audible mostly at close range. Higher Twitter of loud sharp Peeps given only during copulation, usually in one brief burst.

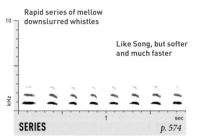

Rapid series of mellow downslurred whistles

Like Song, but softer and much faster

SERIES *p. 574*

By both sexes in close courtship, often in unsynchronized duet. Often precedes copulation. Non-Song calls apparently similar in all populations.

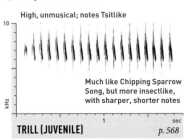

High, unmusical; notes Tsitlike

Much like Chipping Sparrow Song, but more insectlike, with sharper, shorter notes

TRILL (JUVENILE) *p. 568*

Mostly May–Aug., by begging juveniles, sometimes at regular intervals for long periods. Very easily overlooked.

Northern Pygmy-Owl (California)

Glaucidium gnoma californicum

Almost identical to southern Rockies group, but differs in Song; may represent a different species. Like other pygmy-owls, nests in cavities.

Song sometimes introduced by Series, especially in excitement

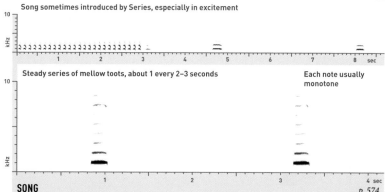

Steady series of mellow toots, about 1 every 2–3 seconds

Each note usually monotone

SONG

p. 574

All year by both sexes, but especially Feb.–June, mostly at dawn and dusk; not often heard during the day or in full darkness. Similar to Song of southern Rockies birds, but less than half as fast. In all three U.S. populations, song appears to be fairly uniform and stereotyped, though rate is somewhat plastic.

Northern Pygmy-Owl (Mountain)

Glaucidium gnoma gnoma

Almost identical to southern Rockies group, but differs in Song; may represent a different species. Widespread in Mexico; fairly common in limited U.S. range.

Song sometimes introduced by Series, especially in excitement

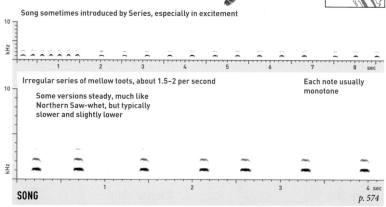

Irregular series of mellow toots, about 1.5–2 per second

Some versions steady, much like Northern Saw-whet, but typically slower and slightly lower

Each note usually monotone

SONG

p. 574

All year by both sexes, but especially Feb.–June, mostly at dawn and dusk; more likely to be heard during day than songs of other U.S. pygmy-owls. Birds in the Santa Catalina and Pinaleño Mountains northeast of Tucson may sound intermediate between this population and southern Rockies birds; more study needed.

FERRUGINOUS PYGMY-OWL

Glaucidium brasilianum

Widespread in the tropics. In Arizona, rare and local in saguaro cactus. Most active at dawn and dusk, but sometimes during the day.

Song sometimes introduced by Chweeklike notes that gradually morph into whistles

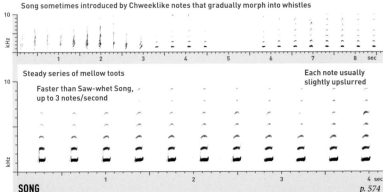

Steady series of mellow toots

Faster than Saw-whet Song, up to 3 notes/second

Each note usually slightly upslurred

SONG

p. 574

All year by both sexes, but especially Feb.–June, mostly at dawn and dusk; not often heard during the day or in full darkness. Easily imitated by human whistling. Female version averages slightly higher-pitched than male's; female may respond to male Song with Song of her own, or with Trill.

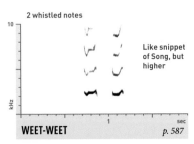

Male Chitter

Note quick repeating rhythm

CHITTER *p. 569*

Female Trill

Musical

TRILL *pp. 571, 572*

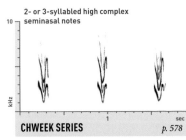

Juvenile begging Trill

High, cricketlike

Chitter given by male in close courtship. Female Trill given in courtship and during prey exchanges; also in response to male Song. Higher, more insectlike Trill of juvenile given for a month or so after fledging.

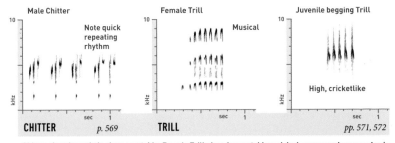

2 whistled notes

Like snippet of Song, but higher

WEET-WEET *p. 587*

Reportedly given by female near nest in response to potential predators, including humans.

2- or 3-syllabled high complex seminasal notes

CHWEEK SERIES *p. 578*

By both sexes, especially females, reportedly in aggression. Variable and plastic. Sometimes given by pairs in synchronized duet.

GREAT GRAY OWL

Strix nebulosa

Our largest owl, huge and distinctive. Rare in spruce bogs and boreal forest; active mostly at night. In some winters, irrupts south of normal range.

Song: 8–12 low hooting notes, slightly softer at start and end

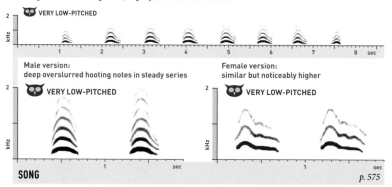

👁 VERY LOW-PITCHED

Male version: deep overslurred hooting notes in steady series

👁 VERY LOW-PITCHED

Female version: similar but noticeably higher

👁 VERY LOW-PITCHED

SONG *p. 575*

Mostly Mar.–June, by both sexes, often near the nest. Usually given in full darkness. Slow, steady rhythm and low pitch are distinctive; notes are usually long enough to be distinctly overslurred.

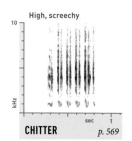

High, screechy

CHITTER *p. 569*

Highly plastic. Given at nest during feedings of mate or young.

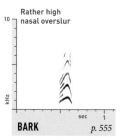

Rather high nasal overslur

BARK *p. 555*

Little known; this example given at nest after male fed incubating female.

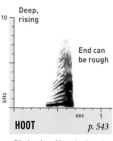

Deep, rising

End can be rough

HOOT *p. 543*

Distinctive. Given by female asking to be fed by male; also in family contact.

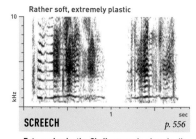

Rather soft, extremely plastic

SCREECH *p. 556*

Extremely plastic. Similar sounds given in distraction display, near nest, and possibly also during copulation.

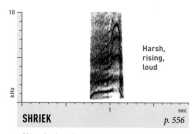

Harsh, rising, loud

SHRIEK *p. 556*

Given by begging young, up to several months after fledging. Quite variable; can overlap with Shrieks of other owl species.

BARRED OWL

Strix varia

A spooky voice of deep forests. Mostly nocturnal, but occasionally active and vocal by day. Range has expanded in the West at the expense of Spotted Owl.

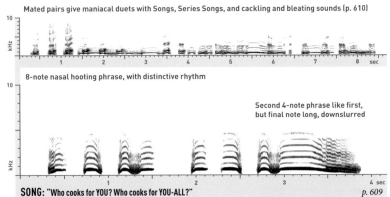

Mated pairs give maniacal duets with Songs, Series Songs, and cackling and bleating sounds (p. 610)

8-note nasal hooting phrase, with distinctive rhythm

Second 4-note phrase like first, but final note long, downslurred

SONG: "Who cooks for YOU? Who cooks for YOU-ALL?"   *p. 609*

All year, by both sexes, but especially Feb.–Apr. Uniform and fairly stereotyped, though occasionally truncated. "Who," "cooks," and "you" notes all on same pitch. Female voice averages higher, with burrier final note. Higher and louder than Great Horned Owl Song, with different rhythm.

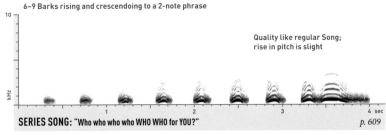

6–9 Barks rising and crescendoing to a 2-note phrase

Quality like regular Song; rise in pitch is slight

SERIES SONG: "Who who who who WHO WHO for YOU?"   *p. 609*

All year, by both sexes. Reportedly given during confrontations between mated pairs; also given within pairs and by solo birds, often in association with Song. A key component of caterwauling duets.

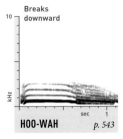

Breaks downward

HOO-WAH   *p. 543*

Like last note of Song. All year by both sexes, perhaps in pair contact.

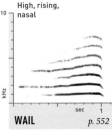

High, rising, nasal

WAIL   *p. 552*

By breeding females, to beg food from mate, and in contact.

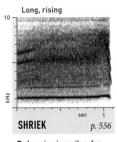

Long, rising

SHRIEK   *p. 556*

By begging juveniles, for months after fledging. Note length (1.5–2 seconds).

SPOTTED OWL

Strix occidentalis

Related to Barred Owl, but less common and more retiring. Inhabits old-growth coniferous forests; in Rockies and Mexico, usually near cliff faces.

In this pair duet, female (red) gives Wail and Song; male gives Series Song and then Song

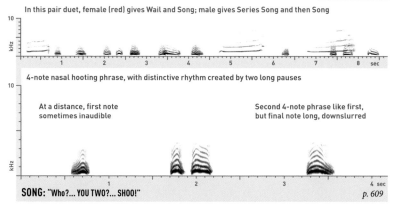

4-note nasal hooting phrase, with distinctive rhythm created by two long pauses

At a distance, first note sometimes inaudible

Second 4-note phrase like first, but final note long, downslurred

SONG: "Who?... YOU TWO?... SHOO!"   *p. 609*

All year, by both sexes, but especially Feb.–Apr., in territorial advertisement and disputes. Uniform and fairly stereotyped, though occasionally truncated; extra notes occasionally added at beginning or end. Female voice averages higher.

6–9 barking Hoots, the series slightly rising and slowing

In excitement, can end with part or all of typical Song

SERIES SONG: "Who who who WHO... WHO... WHO?"   *pp. 575, 609*

All year, by both sexes, in agitation, especially during territorial altercations. Male version shown; female version higher, often with faster, steadier rhythm. Pair duets do not include cackling or bleating sounds, unlike Barred Owl. Occasionally gives high Chitter (p. 569) like other owls.

In agitation, series runs long

BARK SERIES   *p. 578*

All year, mostly by females, in territorial disputes. Barks grade into Wails.

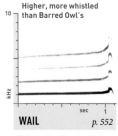

Higher, more whistled than Barred Owl's

WAIL   *p. 552*

By breeding females, to beg food from mate, and in contact. Less often by males.

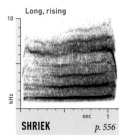

Long, rising

SHRIEK   *p. 556*

By begging juveniles, for months after fledging, gradually becoming Wail.

LONG-EARED OWL

Asio otus

Medium-sized. Nests and roosts in forests, conifers, or dense thickets; hunts in adjacent meadows, marshes, and other open areas. Highly nocturnal.

Song often given in flight display, along with well-spaced single Wing Claps (p. 566)

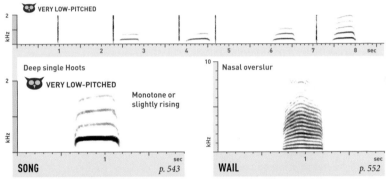

🦉 VERY LOW-PITCHED

Deep single Hoots

🦉 VERY LOW-PITCHED

Monotone or slightly rising

Nasal overslur

SONG *p. 543*

WAIL *p. 552*

All year, by males, but mostly Feb.–June. Hoots given singly, usually 2–4 seconds apart, from perch, but also in zigzag display flight.

Mostly Feb.–Apr., by females, especially near nest. May be female version of Song. Can end with soft, low second syllable.

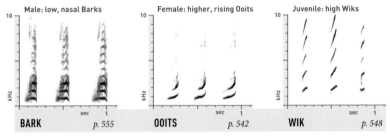

Male: low, nasal Barks

Female: higher, rising Ooits

Juvenile: high Wiks

BARK *p. 555*

OOITS *p. 542*

WIK *p. 548*

Highly variable, especially by age and sex, and somewhat plastic. Given all year in alarm or in response to disturbance, especially Mar.–Aug., near nest or dependent young. Grades into Wails.

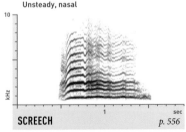

Unsteady, nasal

High, piercing, slightly metallic

1- or 2-syllabled, monotone or downslurred

SCREECH *p. 556*

KEER *p. 550*

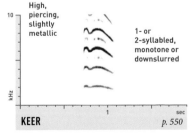

Extremely plastic calls given in high alarm near nest, including during injury-feigning distraction display. Some are clear, wailing.

Mostly May–Aug., by begging juveniles; often repeated every few seconds for long periods. Unlike any other owl sound.

SHORT-EARED OWL

Asio flammeus

A medium-sized owl of open prairies and other open areas. Nests and usually roosts on the ground. Often active before sunset or after sunrise.

Song often given in flight display, along with rapid rattling series of Wing Claps (p. 566)

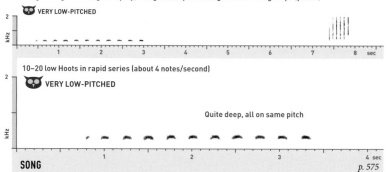

🦉 VERY LOW-PITCHED

10–20 low Hoots in rapid series (about 4 notes/second)

🦉 VERY LOW-PITCHED

Quite deep, all on same pitch

SONG   p. 575

Mostly Feb.–Apr., by courting males during acrobatic "sky dancing" display flight. Sometimes also delivered from ground or prominent perch. Females apparently do not sing, but may respond to male Song with barking or screeching calls.

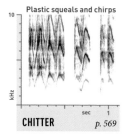

Plastic squeals and chirps

CHITTER   p. 569

By adults in high alarm near the nest; similar calls given by juveniles.

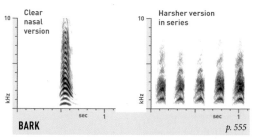

Clear nasal version

Harsher version in series

BARK   p. 555

All year; given in alarm and during interactions, often in flight. Variable and somewhat plastic, ranging from clear and nasal to harsh and noisy; may be given singly or in series.

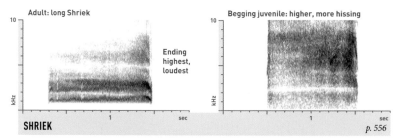

Adult: long Shriek

Ending highest, loudest

Begging juvenile: higher, more hissing

SHRIEK   p. 556

Variable and plastic. Adults give Shrieks all year, in aggressive interactions. Juveniles give a variety of calls, including Shrieks, chittering Rattles, and more rarely a high whistled call rather like the Keer of Long-eared Owl, but generally noisier.

BOREAL OWL

Aegolius funereus

A little-seen bird of coniferous forests, where it nests in tree cavities. Highly nocturnal. In some winters, irrupts south of normal range.

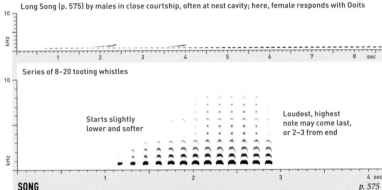

Long Song (p. 575) by males in close courtship, often at nest cavity; here, female responds with Ooits

Series of 8–20 tooting whistles

Starts slightly lower and softer

Loudest, highest note may come last, or 2–3 from end

SONG *p. 575*

All year, but mostly Feb.–June, by males. Female not known to sing. Recalls winnowing of Wilson's Snipe, but shorter, more whistled, and almost never given before full dark. Long Song, in presence of female, can continue for more than 10 seconds. A few soft Songlike notes announce food deliveries at the nest.

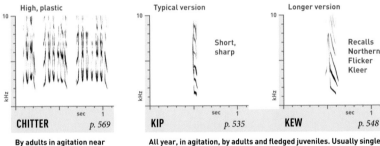

High, plastic

CHITTER *p. 569*

By adults in agitation near nest; similar calls also by fledglings.

Typical version

Short, sharp

KIP *p. 535*

Longer version

Recalls Northern Flicker Kleer

KEW *p. 548*

All year, in agitation, by adults and fledged juveniles. Usually single, but occasionally in short series. Usually higher and sharper than Northern Saw-whet Kews.

Often up- or overslurred

WAIL *p. 552*

High Peeps interspersed

Peep

WAIL-PEEP *p. 592*

Soft low upslur

OOIT *p. 542*

A set of intergrading, often plastic sounds, usually lower and shorter than Northern Saw-whet's. Wails and Wail-peeps have been recorded in early spring from agitated birds in response to playback of Song. Ooits are softer and appear to be a common call given in close-range contact.

Northern Saw-whet Owl

Aegolius acadicus

Breeds in coniferous and mixed forests, and occasionally in more open areas. In winter, roosts in dense vegetation, especially conifers. Highly nocturnal.

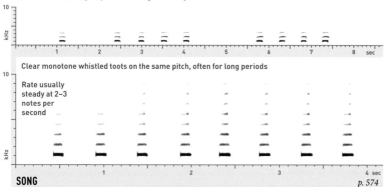

In excitement, Song may become irregular in rhythm

Clear monotone whistled toots on the same pitch, often for long periods

Rate usually steady at 2–3 notes per second

SONG

p. 574

Mostly Jan.–May, by males. Female Song infrequent, given only in courtship; usually softer, with less steady rhythm than male's. Songs of individual males vary slightly in pitch. In excitement, may speed up to nearly 5 notes/second for short periods. Song is easily imitated by human whistling.

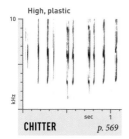

High, plastic

CHITTER *p. 569*

By adults in agitation; similar calls also by fledglings.

High, downslurred, barking

KEW *p. 548*

Often in brief, rapid series

KEEK SERIES *p. 577*

All year, in agitation, by adults and fledged juveniles; also in response to playback of Northern Saw-whet or Boreal Owl Song. Highly variable and plastic.

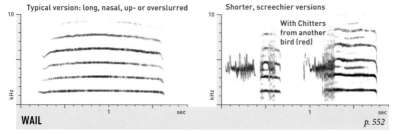

Typical version: long, nasal, up- or overslurred

WAIL

Shorter, screechier versions

With Chitters from another bird (red)

p. 552

All year, from agitated birds, often males, during interactions and in response to playback. Sometimes repeated every few seconds for extended periods. Variable and plastic, but most versions higher and longer than corresponding calls of Boreal Owl.

WESTERN SCREECH-OWL

Megascops kennicottii

Found in wooded habitats, especially deciduous riparian woods. Voice very different from Eastern Screech-Owl's; the two species rarely hybridize.

Accelerating series of 5–15 mellow monotone whistles

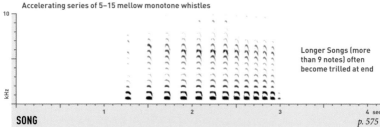

Longer Songs (more than 9 notes) often become trilled at end

SONG   *p. 575*

All year, by both sexes, but especially Dec.–Mar. by males. Heard mostly after dusk and before dawn; very rare during the day. Female version averages higher-pitched. Songs average longer in northern part of range, with faster ending, but length and speed vary in any given locale.

Bark Series (p. 578): given by singing males when female approaches, or in response to playback

Mellow whistled 2-part trill, each part barely downslurred

First part 3–6 notes   Second part 10–20 notes

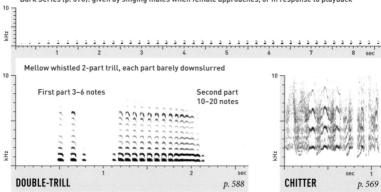

DOUBLE-TRILL   *p. 588*

CHITTER   *p. 569*

All year, by both sexes, apparently in contact and to strengthen pair bond. Second part sometimes given alone. Pairs sometimes duet with one bird giving Song and the other Double-trill.

In close interactions. Soft. Extremely plastic, ranging from rattling to squealing.

Kew-du-du: common 3-noted variant   Kew: downslurred, usually nasal, barking

Some long, monotone versions easily confused with Eastern Screech-Owl Song

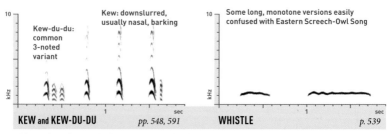

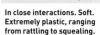

KEW and KEW-DU-DU   *pp. 548, 591*

WHISTLE   *p. 539*

All year, by both sexes, in alarm and contact. All of these sounds are highly variable, plastic, and intergrading. In agitation, Barks tend to get higher and louder; they are often 2- or 3-noted, the first note highest. A whistled version of the Kew-du-du is a common female contact call. Also snaps bill in agitation.

EASTERN SCREECH-OWL

Megascops asio

Breeds in wooded habitats, sometimes including residential areas. Almost exclusively nocturnal. Most are gray, but some are bright rufous in color.

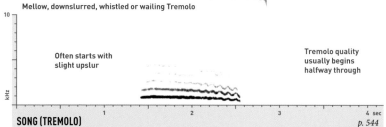

Mellow, downslurred, whistled or wailing Tremolo

Often starts with slight upslur

Tremolo quality usually begins halfway through

SONG (TREMOLO) *p. 544*

All year, by both sexes, but especially Mar.–June, mostly after dusk and before dawn; very rare during the day. Female version averages higher-pitched. More common than Trill later in breeding season; sometimes given along with Trill.

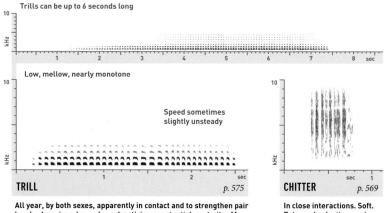

Trills can be up to 6 seconds long

Low, mellow, nearly monotone

Speed sometimes slightly unsteady

TRILL *p. 575*

All year, by both sexes, apparently in contact and to strengthen pair bond; also given by males advertising a potential nest site. More common than Song in early breeding season. Rare during day.

CHITTER *p. 569*

In close interactions. Soft. Extremely plastic, ranging from rattling to squealing.

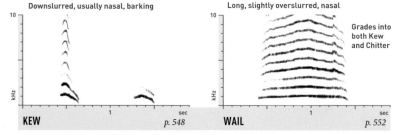

Downslurred, usually nasal, barking

KEW *p. 548*

Long, slightly overslurred, nasal

Grades into both Kew and Chitter

WAIL *p. 552*

All year, by both sexes, in alarm and contact. These sounds are highly variable, plastic, and intergrading. Most versions average longer and more nasal than corresponding sounds of Western Screech-Owl, but some versions are quite similar. Also snaps bill in agitation.

Whiskered Screech-Owl

Megascops trichopsis

Found in riparian woods and adjacent pine-oak forest, mostly at higher elevations than Western Screech-Owl, but the two overlap in many areas.

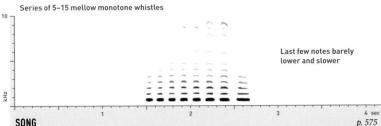

Series of 5–15 mellow monotone whistles

Last few notes barely lower and slower

SONG *p. 575*

Mostly Feb.–June, by both sexes, but especially males, to advertise a nest cavity, defend territory, and maintain pair contact. Heard mostly after dusk and before dawn; very rare during the day. Female version averages higher-pitched. Speed varies regionally; version shown is typical.

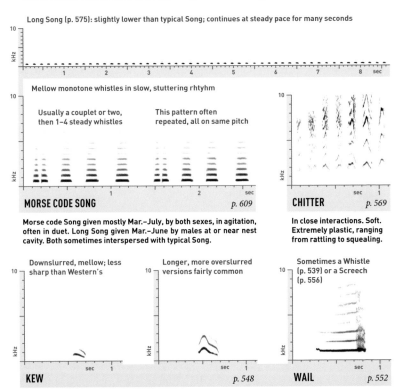

Long Song (p. 575): slightly lower than typical Song; continues at steady pace for many seconds

Mellow monotone whistles in slow, stuttering rhtyhm

Usually a couplet or two, then 1–4 steady whistles

This pattern often repeated, all on same pitch

MORSE CODE SONG *p. 609*

Morse code Song given mostly Mar.–July, by both sexes, in agitation, often in duet. Long Song given Mar.–June by males at or near nest cavity. Both sometimes interspersed with typical Song.

CHITTER *p. 569*

In close interactions. Soft. Extremely plastic, ranging from rattling to squealing.

Downslurred, mellow; less sharp than Western's

KEW

Longer, more overslurred versions fairly common

p. 548

Sometimes a Whistle (p. 539) or a Screech (p. 556)

WAIL *p. 552*

All year, by both sexes, in alarm and contact. These sounds are highly variable, plastic, and intergrading. Kews sometimes run together into decelerating series of 5–8 notes; several such series can be run together, creating a Morse code–like rhythm. Wail infrequent, given only in high alarm.

Flammulated Owl

Psiloscops flammeolus

An uncommon and elusive summer resident of open yellow pine forests and aspen groves. Eats mostly moths and beetles. Migratory. Note dark eyes.

Complex Song (p. 609): often a 4-syllabled hooting phrase (Hoot-it HOO-doo)

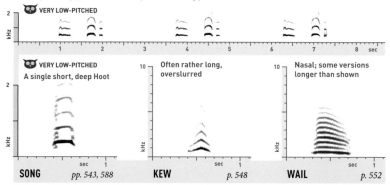

👀 VERY LOW-PITCHED

👀 VERY LOW-PITCHED
A single short, deep Hoot

Often rather long, overslurred

Nasal; some versions longer than shown

SONG pp. 543, 588   **KEW** p. 548   **WAIL** p. 552

Song mostly Apr.–July, by males. Single-note version much like Long-eared Owl Song, but Hoots much briefer. Multinote Complex Song given in territorial disputes. Kew grades into Wail; Kewlike notes given in contact, Wails in alarm. Most sounds given by both sexes, though female sounds average higher and longer.

Elf Owl

Micrathene whitneyi

The world's smallest owl, smaller than a starling. Breeds in old woodpecker holes in saguaros, thorn scrub, and riparian woods. Normally nocturnal.

Long Song can last a minute or longer; given by males in close courtship and in excitement

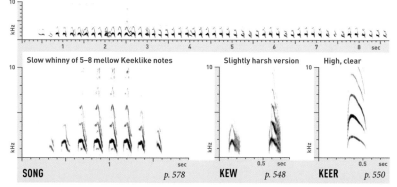

Slow whinny of 5–8 mellow Keeklike notes

Slightly harsh version

High, clear

SONG p. 578   **KEW** p. 548   **KEER** p. 550

Mostly Mar.–June, by males, mostly shortly after dusk and before dawn. Females not known to sing. In excitement, middle section of Song may rise in pitch. Kew and Keer are most common calls, often in series; variable and plastic. Soft Kews given in close contact, high shrill Keers in alarm. Also gives Chitter (p. 569).

BURROWING OWL

Athene cunicularia

Active both day and night. Most nest in prairie dog or ground squirrel burrows, but Florida birds can dig their own burrows. Population has declined.

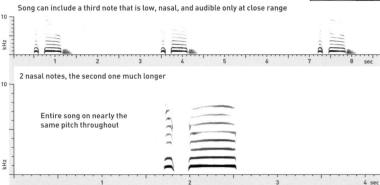

Song can include a third note that is low, nasal, and audible only at close range

2 nasal notes, the second one much longer

Entire song on nearly the same pitch throughout

SONG: "Bip-BEEEP"

p. 591

Mostly Mar.–May, by males, night or day. Uniform and fairly stereotyped, though first note is occasionally omitted, and soft, variable versions may be given in close courtship. Female may respond to male Song with "eep" notes similar to second note of Song, but softer, grading into Snarl.

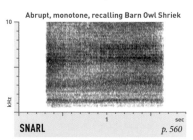

Abrupt, monotone, recalling Barn Owl Shriek

SNARL

p. 560

By juveniles and adults, begging for food and in high alarm. May startle predators by resembling rattlesnake rattles.

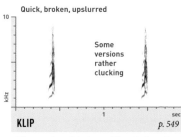

Quick, broken, upslurred

Some versions rather clucking

KLIP

p. 549

All year, in mild alarm. Like single note of Chuckle, which it grades into.

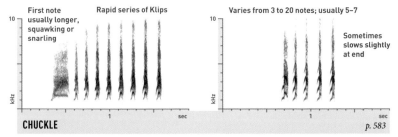

First note usually longer, squawking or snarling

Rapid series of Klips

Varies from 3 to 20 notes; usually 5–7

Sometimes slows slightly at end

CHUCKLE

p. 583

All year, in alarm; most common call. Slightly variable and quite plastic, becoming longer as agitation level rises. Notes usually slow enough to count. In high alarm, initial snarling note may be given by itself or repeated later in Chuckle.

TROGONS (Family Trogonidae)

Trogons are colorful tropical cavity-nesting birds; they use their broad, serrated bills to eat fruit and insects. Sounds are simple and innate.

KINGFISHERS (Family Alcedinidae)

Our two species of kingfishers are rarely found far from water. Medium to large birds, with distinctively long, stout bills and ragged crests, they specialize in eating fish, which they capture by plunge-diving headfirst into slow-moving water. Nests are in cavities, usually in holes excavated in earthen banks. Vocalizations are simple and apparently innate, primarily unmusical chatters, clicks, and clucks.

Elegant Trogon

Trogon elegans

A remarkable and much sought-after Central American species, highly local in the U.S. in sycamore forests along streams in pine-oak canyons.

Song: usually 5–10 Kweeah notes; consecutive songs tend to be similar

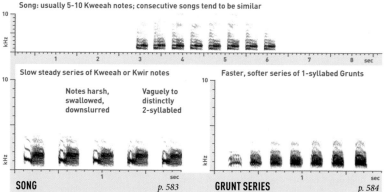

Slow steady series of Kweeah or Kwir notes

Notes harsh, swallowed, downslurred

Vaguely to distinctly 2-syllabled

SONG   *p. 583*

Faster, softer series of 1-syllabed Grunts

GRUNT SERIES   *p. 584*

Mostly Mar.–Aug., by both sexes, sometimes in unsynchronized duet. Song given loudly in advertisement. Grunt Series usually softer, given near nest and in response to playback; may be a version of Song, or a call with distinct function. Some versions of Song are clearer, more Yelping, recalling Wild Turkey.

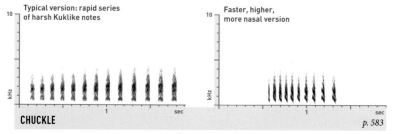

Typical version: rapid series of harsh Kuklike notes

Faster, higher, more nasal version

CHUCKLE   *p. 583*

All year, by both sexes, in alarm and agitation. Rather infrequent. Faster, higher versions may indicate higher alarm; slower, lower versions may grade into Grunt Series. Young birds beg with screechy Peeps (not indexed).

BELTED KINGFISHER

Megaceryle alcyon

Found near water, almost anywhere suitable nest sites are available. Solitary, but often conspicuous and vocal, especially during defense of territory.

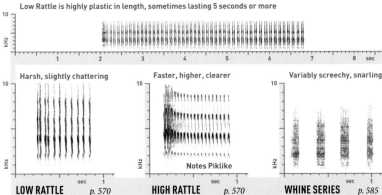

Low Rattle is highly plastic in length, sometimes lasting 5 seconds or more

Harsh, slightly chattering | Faster, higher, clearer | Variably screechy, snarling

Notes Piklike

LOW RATTLE *p. 570* | **HIGH RATTLE** *p. 570* | **WHINE SERIES** *p. 585*

All sounds variable and plastic. Most common call is Low Rattle, given all year in alarm and aggression. Semimusical High Rattle given by both sexes during and after confrontations, and as a pair greeting. Whine Series given mostly by male early in breeding season; short, often run into plastic Chitter.

GREEN KINGFISHER

Chloroceryle americana

Our smallest kingfisher, about the size of a starling; rather quiet and inconspicuous. Found near clear freshwater streams and ponds.

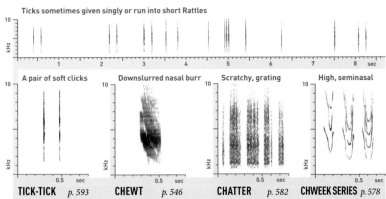

Ticks sometimes given singly or run into short Rattles

A pair of soft clicks | Downslurred nasal burr | Scratchy, grating | High, seminasal

TICK-TICK *p. 593* | **CHEWT** *p. 546* | **CHATTER** *p. 582* | **CHWEEK SERIES** *p. 578*

Tick-tick given all year; most common call. Uniform but somewhat plastic, especially in rhythm. Does not carry far. Chewt not well known; given during interactions, and in flight, possibly in alarm. Chatter and Chweek Series given by excited birds during interactions, often together.

WOODPECKERS (Family Picidae)

The woodpeckers are medium-sized to large birds well known for perching on the vertical surfaces of tree trunks and using their chisel-like bills to extract insect larvae from the wood. They are equally well known for drumming rapidly on resonant surfaces—dead limbs, telephone poles, houses, and even metal objects like chimney covers and transformer boxes—in order to communicate over long distances.

As far as is known, all sounds of woodpeckers are innate. Most vocal sounds are quite simple, with little individual or regional variation.

Identifying Woodpecker Species by Drum

Adult males and females of all North American woodpecker species drum, either in territorial advertisement, to strengthen the pair bond, or both. Sapsuckers can be instantly recognized by their distinctive rhythm. Williamson's is unique; Red-naped and Red-breasted sound similar but can often be identified by range. Among other woodpeckers, the most important feature to listen to is the speed of the Drum.

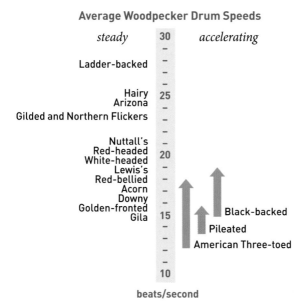

Average Woodpecker Drum Speeds

steady   30   *accelerating*

Ladder-backed

Hairy — 25
Arizona
Gilded and Northern Flickers

Nuttall's
Red-headed — 20
White-headed
Lewis's
Red-bellied
Acorn
Downy
Golden-fronted
Gila — 15

Black-backed
Pileated
American Three-toed

10

beats/second

Drum speed is variable, and there is much overlap between species that are close to one another on the table. However, it is often possible to separate the Drums of species that are far apart on the table, such as those of Hairy and Downy Woodpeckers.

Drumming vs. Tapping

In addition to drumming, most woodpeckers frequently engage in tapping, which is similar to drumming but generally softer, slower, and much more plastic. Tapping appears to be used mostly in short-range communication between members of a pair. The Red-cockaded Woodpecker of the East occasionally drums its tongue on the surface of trees in mild excitement, and other species may do this as well; the resulting sound is said to resemble the rattle of a rattlesnake's tail. A woodpecker excavating a nest site or in search of food makes a more methodical tapping sound.

LEWIS'S WOODPECKER

Melanerpes lewis

Uncommon in open brushy forests and
burn areas. Quieter than other woodpeckers; mostly silent away from nest.
Frequently catches insects on the wing.

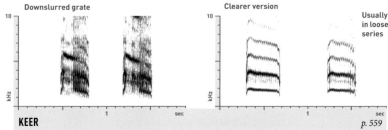

Downslurred grate   Clearer version

Usually
in loose
series

KEER *p. 559*

Mostly Apr.–June, by male, apparently to defend territory and attract mate. Variable; some versions
harsher than shown, some squeakier, but almost always grating. Generally resembles Kweeah of Red-
headed Woodpecker, but always downslurred, and usually in series.

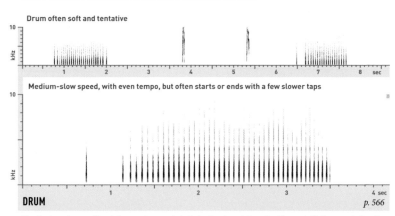

Drum often soft and tentative

Medium-slow speed, with even tempo, but often starts or ends with a few slower taps

DRUM *p. 566*

Mostly Apr.–June; rather infrequent, repeated at short or long intervals. Given mostly by unmated males,
so may serve an advertising function. Often associated with Keer or Pseep. May grade into tapping more
frequently than in other woodpeckers, creating plastic, unsteady rhythm.

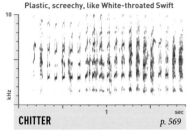 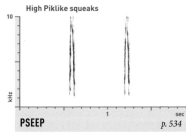

Plastic, screechy, like White-throated Swift   High Piklike squeaks

CHITTER *p. 569*

PSEEP *p. 534*

All year, but infrequent away from nest. Mostly
by male, probably in aggression. Sometimes in
flight, including slow-flapping display flight.

All year, in alarm, but infrequent away from
nest. Form of call reportedly differs slightly between sexes.

RED-HEADED WOODPECKER

Melanerpes erythrocephalus

Locally common in forests and in open areas with a few mature trees. Often sits on snags or telephone poles, sallying out to catch insects on the wing.

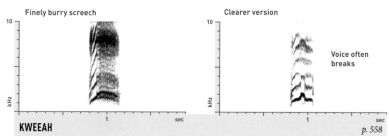

Finely burry screech

Clearer version

Voice often breaks

KWEEAH *p. 558*

Mostly Feb.–July; often alternated with Drum. Variable and plastic. Higher and finer than Red-bellied Kwirr; often harsher.

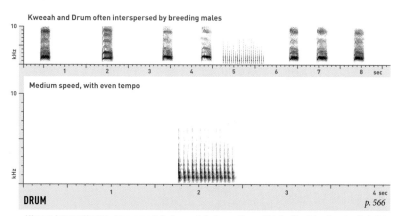

Kweeah and Drum often interspersed by breeding males

Medium speed, with even tempo

DRUM *p. 566*

All year, but mostly Apr.–June; especially frequent during and immediately after intrusions by rivals into territory. Generally drums for shorter periods than other woodpeckers. Sometimes drums on manmade surfaces, including metal and ceramics.

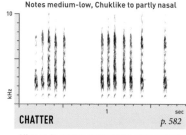

Notes medium-low, Chuklike to partly nasal

CHATTER *p. 582*

All year, in contact and possibly other contexts. Some frogs can sound very similar, especially Wood Frog and Gray Tree Frog.

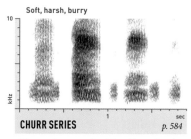

Soft, harsh, burry

CHURR SERIES *p. 584*

All year, in aggressive interactions of various kinds.

ACORN WOODPECKER

Melanerpes formicivorus

Conspicuous and vocally social. Closely tied to oak trees; famous for storing large numbers of acorns in "granary trees." Breeds cooperatively.

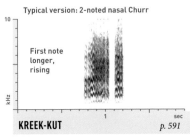

Typical version: 2-noted nasal Churr

First note longer, rising

KREEK-KUT — p. 591

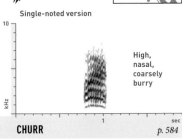

Single-noted version

High, nasal, coarsely burry

CHURR — p. 584

All year; most common and distinctive calls, given in territorial interactions and in response to predators. Kreek-kut was evidently the inspiration for the laugh of the Woody Woodpecker cartoon character. Plastic; some versions have 3 or more syllables; one-syllabled Churrs are common, especially in alarm.

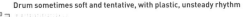

Drum sometimes soft and tentative, with plastic, unsteady rhythm

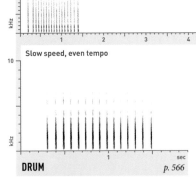

Slow speed, even tempo

DRUM — p. 566

All year, especially Feb.–July, often with Kreek-kut. Slow, averaging about 17 notes/second. Rather infrequent.

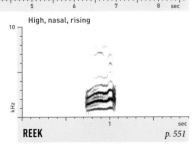

High, nasal, rising

REEK — p. 551

Mostly in spring, in territorial conflicts, often with Kreek-kut and Drum. Plastic; a similar but much lower call is given during chases.

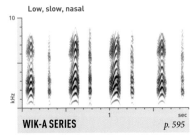

Low, slow, nasal

WIK-A SERIES — p. 595

All year, in interactions. Loud and frequent; distinctively slow, staccato, and nasal. Often given with wings spread.

Coarse, harsh, nasal

CHURR SERIES — p. 558

Mostly in summer, during high-intensity interactions. Quality rather like Churr call, but usually harsher.

GILA WOODPECKER

Melanerpes uropygialis

Loud, conspicuous resident of riparian areas, residential areas, and deserts, especially those with saguaro cactus. No similar woodpeckers in its range.

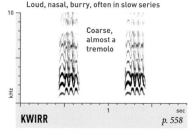

Loud, nasal, burry, often in slow series

Coarse, almost a tremolo

KWIRR *p. 558*

All year, in territorial interactions and in pair contact. Apparently not reliably distinguishable from Kwirr of Red-bellied; no range overlap.

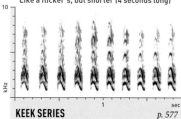

Like a flicker's, but shorter (4 seconds long)

KEEK SERIES *p. 577*

Little known; perhaps just a variant of the Yap Series, but almost twice as fast (5 notes/second) and given in discrete 4-second bursts.

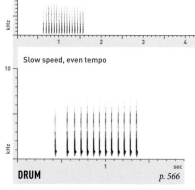

Slow speed, even tempo

DRUM *p. 566*

All year, especially Feb.–July. One of the slowest western woodpecker Drums, averaging about 15 notes/second. Rather infrequent.

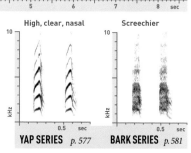

High, clear, nasal

YAP SERIES *p. 577*

Screechier

BARK SERIES *p. 581*

All year, in alarm or interactions. Yap Series is loud, frequent, distinctive; Bark Series a rare variant. Single Yaps and Barks are very rare.

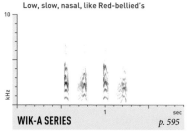

Low, slow, nasal, like Red-bellied's

WIK-A SERIES *p. 595*

All year; in interactions. Infrequent. Much like Wik-a Series of Red-bellied Woodpecker.

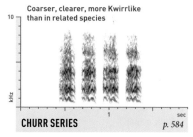

Coarser, clearer, more Kwirrlike than in related species

CHURR SERIES *p. 584*

All year; especially Feb.–May. Given in close contact, often by mated pairs.

Golden-fronted Woodpecker

Melanerpes aurifrons

Common and vocally conspicuous in mesquite scrub and open dry woodlands. Occasionally hybridizes with related Red-bellied
Woodpecker where ranges overlap.

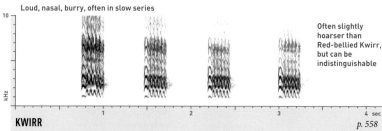

Loud, nasal, burry, often in slow series

Often slightly
hoarser than
Red-bellied Kwirr,
but can be
indistinguishable

KWIRR *p. 558*

All year, especially Feb.–May. Most common vocalization; often alternated with Drum as a territorial call.
No equivalent of Red-bellied's Chuckle is known from Golden-fronted.

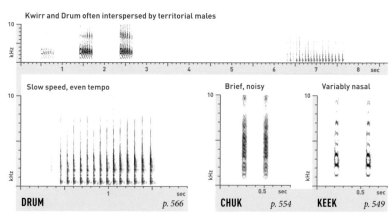

Kwirr and Drum often interspersed by territorial males

Slow speed, even tempo

DRUM *p. 566*

All year, especially Feb.–July. One of the slowest
western woodpecker Drums, averaging about 15
notes/second.

Brief, noisy

CHUK *p. 554*

Variably nasal

KEEK *p. 549*

All year, in alarm or pair contact, often in long
series. Chuk and Keek intergrade; each note
usually shorter than Red-bellied Bark.

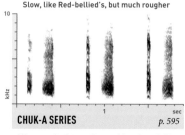

Slow, like Red-bellied's, but much rougher

CHUK-A SERIES *p. 595*

All year; in interactions with other Golden-
fronteds. Function not well understood.

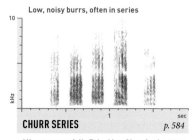

Low, noisy burrs, often in series

CHURR SERIES *p. 584*

All year; especially Feb.–May. Given in close contact, often by mated pairs. Averages finer than
Red-bellied Woodpecker Churr.

Red-bellied Woodpecker

Melanerpes carolinus

Common in forests and in residential neighborhoods with large trees. Calls loudly and frequently.

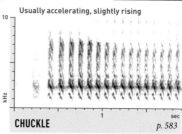

Loud, nasal, burry, often in slow series

Coarse, almost a tremolo

KWIRR   *p. 559*

All year, especially Feb.–July. Most common vocalization; often alternated with Drum as a territorial call.

Usually accelerating, slightly rising

CHUCKLE   *p. 583*

All year, especially Feb.–July, in interactions. Infrequent. Faster, harsher, shorter than Northern Flicker Keek Series (2–3 seconds total).

Kwirr and Drum often interspersed by territorial males

Medium speed, even tempo

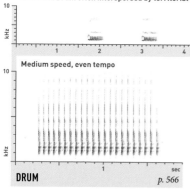

DRUM   *p. 566*

All year, especially Feb.–July. Averages 18 notes/second. Often given in conjunction with Kwirr.

Loud, downslurred, abrupt

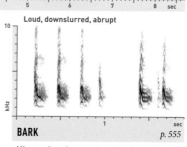

BARK   *p. 555*

All year, in pair contact and in alarm. Variable, but generally sharply downslurred, unlike Chuk of Golden-fronted, and rather screechy.

Low, slow, nasal, like Gila's

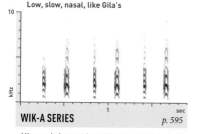

WIK-A SERIES   *p. 595*

All year, in interactions with other Red-bellieds. Function not well understood.

Rather coarse, almost chattering

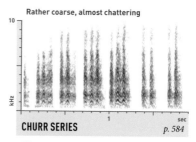

CHURR SERIES   *p. 584*

All year; especially Feb.–May, in close contact, often by mated pairs.

LADDER-BACKED WOODPECKER

Dryobates scalaris

Common in deserts, thorn scrub, riparian woodlands, and areas dominated by cactus. Generally prefers drier habitats than Downy or Nuttall's Woodpeckers.

Rapid series of Piklike notes, becoming lower, slower, and rougher

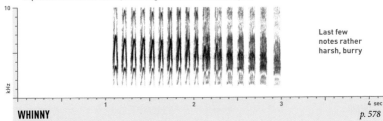

Last few notes rather harsh, burry

WHINNY   *p. 578*

All year; shorter version often given Mar.–May. Often given with Pik calls and Drums. Like Downy Woodpecker Whinny, but usually becomes distinctively rough in quality at end. Quite different from Nuttall's Woodpecker Rattle.

Drum fast, often repeated at medium to long intervals (5 seconds or more)

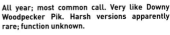

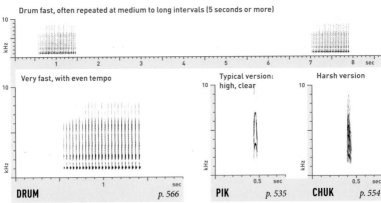

Very fast, with even tempo

DRUM   *p. 566*

All year, especially winter and spring. Our fastest woodpecker Drum on average, overlapping in speed with Hairy, Arizona, and flickers.

Typical version: high, clear

Harsh version

PIK   *p. 535*   **CHUK**   *p. 554*

All year; most common call. Very like Downy Woodpecker Pik. Harsh versions apparently rare; function unknown.

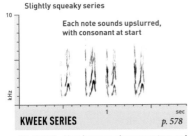

Slightly squeaky series

Each note sounds upslurred, with consonant at start

KWEEK SERIES   *p. 578*

Mostly Feb.–May, in aggressive encounters and in fluttering flight displays. Usually in short series; also occasionally single notes.

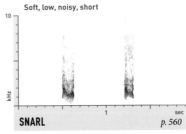

Soft, low, noisy, short

SNARL   *p. 560*

Mostly Feb.–May, in similar situations as Kweek, and often in conjunction with it. In some variants, notes are reportedly 2-syllabled.

NUTTALL'S WOODPECKER

Dryobates nuttallii

Found almost exclusively in California, mostly in oaks and in other deciduous forests. Occasionally hybridizes with Downy and Ladder-backed.

Rapid monotone series of Piklike notes, high-pitched and harsh

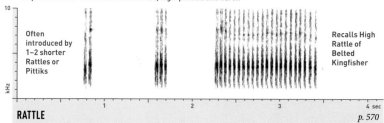

Often introduced by 1–2 shorter Rattles or Pittiks

Recalls High Rattle of Belted Kingfisher

RATTLE   p. 570

All year, but especially Oct.–May, in contact and in order to establish territories. Uniform, but highly plastic in length, grading into Pittik; some versions are quite long. In agitation, gives repeated short Rattles over and over.

Drum fast, often repeated at medium to long intervals (5 seconds or more)

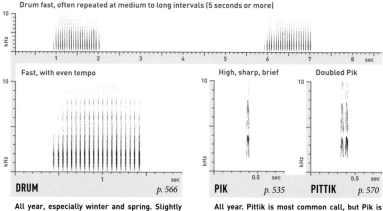

Fast, with even tempo

DRUM   p. 566

All year, especially winter and spring. Slightly faster than Downy's and slower than Ladder-backed's, but difficult to distinguish from both.

High, sharp, brief   Doubled Pik

PIK   p. 535   **PITTIK**   p. 570

All year. Pittik is most common call, but Pik is also common, as are longer versions. Pik briefer, less musical than Downy's.

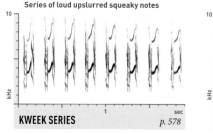

Series of loud upslurred squeaky notes

KWEEK SERIES   p. 578

All year, by both sexes, in aggression and in pair contact. Much more common than Kweek Series of other woodpeckers.

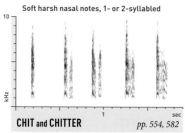

Soft harsh nasal notes, 1- or 2-syllabled

CHIT and CHITTER   pp. 554, 582

Mostly Feb.–May, in close interactions; not very loud. Highly plastic. Like Chit of Downy Woodpecker, but squeakier.

DOWNY WOODPECKER

Dryobates pubescens

Widespread in woodlands and towns. Smaller and shorter-billed than Hairy Woodpecker; note differences in voice. Coastal populations are grayer below.

Rapid series of Pik notes, usually slightly downslurred

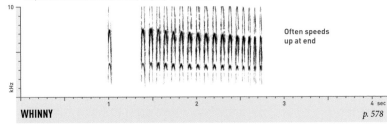

Often speeds up at end

WHINNY *p. 578*

All year. Often follows Pik calls. Clearer than Hairy Woodpecker Rattle, and usually downslurred and accelerating, becoming slightly less musical at end. Beginning of Whinny can vary in speed; some versions begin with Kweek Series.

Drum relatively slow, but often repeated at short intervals (4 seconds or less)

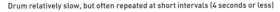

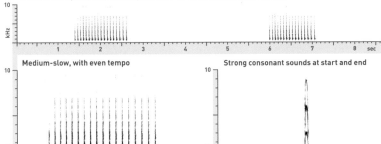

Medium-slow, with even tempo

Strong consonant sounds at start and end

DRUM *p. 566*

All year, but especially winter and spring. Averages 16 notes/second; slow speed alone may be sufficient for identification in some regions.

PIK *p. 535*

All year; most common call. Somewhat variable in pitch, but fairly stereotyped. Averages lower than Hairy Peek, but sometimes similar.

Notes like Pik, but longer, often upslurred

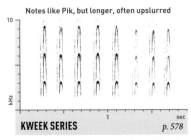

Soft and noisy

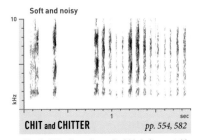

KWEEK SERIES *p. 578*

All year, by excited birds, often in slow-flapping "butterfly" display flight. Varies from clear to noisy; can end in Whinny.

CHIT and CHITTER *pp. 554, 582*

Mostly Feb.–June, by excited birds in courtship displays or disputes. Plastic, but fairly distinctive.

HAIRY WOODPECKER

Dryobates villosus

Widespread, with a greater preference for dense forest and coniferous trees than Downy Woodpecker. Coastal populations are grayer below.

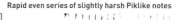
Rapid even series of slightly harsh Piklike notes

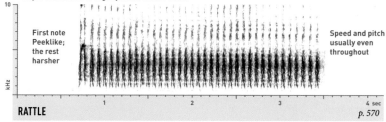

First note Peeklike; the rest harsher

Speed and pitch usually even throughout

RATTLE   p. 570

All year. Often follows Peek calls. Noisier, more monotone, and often longer than similar Downy Woodpecker Whinny. Agitated birds give a less noisy extended version, up to 10 seconds long, that may rise and fall in pitch, interspersed with fast series of Peeks.

Drum fast, often repeated at fairly long intervals (5 seconds or more)

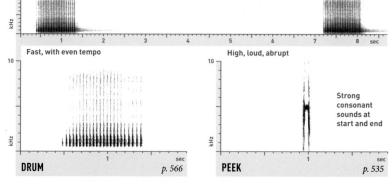

Fast, with even tempo

High, loud, abrupt

Strong consonant sounds at start and end

DRUM   p. 566

All year, but especially winter and spring. One of the fastest woodpecker Drums, averaging 25 notes/second.

PEEK   p. 535

All year; most common call. Variable, but usually distinctly higher than Downy Woodpecker Pik.

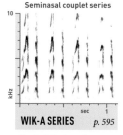
Seminasal couplet series

WIK-A SERIES   p. 595

Mostly Feb.–June, by excited birds in courtship displays or disputes.

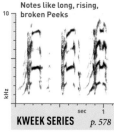
Notes like long, rising, broken Peeks

KWEEK SERIES   p. 578

All year, from excited birds, often in slow-flapping display flight.

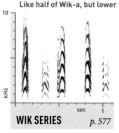
Like half of Wik-a, but lower

WIK SERIES   p. 577

All year. Variable; in pair greetings and low-grade disputes.

ARIZONA WOODPECKER

Dryobates arizonae

A primarily Mexican species, found in mountain pine-oak forests as well as pure oak woodlands and riparian areas with sycamores.

Rapid even series of harsh burry notes

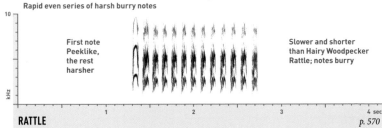

First note Peeklike, the rest harsher

Slower and shorter than Hairy Woodpecker Rattle; notes burry

RATTLE   *p. 570*

All year. Often follows Peek calls. Distinctive; usually monotone, but some versions are overslurred. Some versions begin with Kweek notes. Generally infrequent, but can be given repeatedly during altercations and in agitation. Sometimes interspersed with Drum by territorial birds.

Drum fast, repeated at short or long intervals

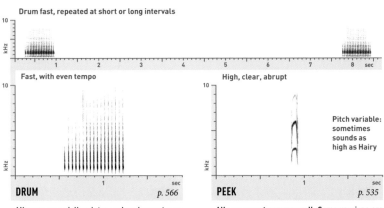

Fast, with even tempo

High, clear, abrupt

Pitch variable: sometimes sounds as high as Hairy

DRUM   *p. 566*

All year, especially winter and spring; not very frequent. Not reliably distinguishable from drums of Hairy Woodpecker or Northern Flicker.

PEEK   *p. 535*

All year; most common call. Some versions are broken. Much like Hairy Woodpecker Peek; subtly but consistently longer.

Notes soft, slightly harsh

Sometimes single Kweeks

Squeakier than Downy's

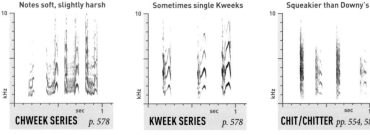

CHWEEK SERIES   *p. 578*

Infrequent, mostly during aggressive encounters. Grades into Kweek Series.

KWEEK SERIES   *p. 578*

Infrequent, in aggressive encounters, reportedly mostly by females.

CHIT/CHITTER   *pp. 554, 582*

By excited birds in interactions, often with other calls. Highly plastic.

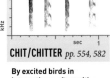

WHITE-HEADED WOODPECKER

Dryobates albolarvatus

Closely associated with pine, especially ponderosa pine. Frequently drills into pine cones for seeds, much less often into tree trunks for insects.

Rapid even series of slightly harsh Piklike notes

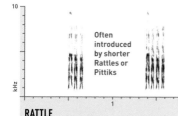

Often introduced by shorter Rattles or Pittiks

Slightly lower than Hairy Woodpecker, with first note more similar to the rest

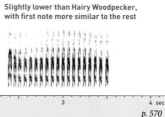

RATTLE   *p. 570*

All year, in interactions, sometimes in flight. Uniform, but highly plastic in length, grading smoothly into Pittik. Quality much like Hairy Woodpecker Rattle, but often shorter; agitated Hairy can sound quite similar, but usually alternates Rattles with many single Peek notes, unlike White-headed.

Drum medium fast, repeated at long or short intervals

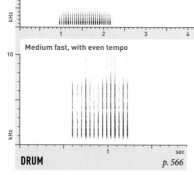

Medium fast, with even tempo

2 or more Piklike notes run together

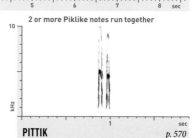

DRUM   *p. 566*

All year, but especially Mar.–June. Usually relatively short; very similar to Drum of Nuttall's Woodpecker.

PITTIK   *p. 570*

All year; most common call. Very similar to Pittik of Nuttall's Woodpecker, but usually contains 2–3 Piklike notes; single Piks are rare.

Slow, high, a little screechy

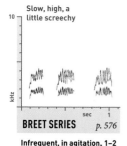

Notes upslurred

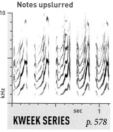

Highly plastic

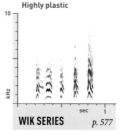

BREET SERIES   *p. 576*

Infrequent, in agitation. 1–2 Pittiks sometimes followed by 1–2 Breets.

KWEEK SERIES   *p. 578*

Mostly in spring and summer, in interactions. Plastic.

WIK SERIES   *p. 577*

By excited birds in interactions. Also gives a harsh Chip (p. 531).

AMERICAN THREE-TOED WOODPECKER

Picoides dorsalis

Breeds in northern spruce forests; often forages for bark beetles by flaking off bark
rather than drilling holes. Shares lack of
fourth toe with Black-backed Woodpecker.

Rapid, irregular series of Piklike notes

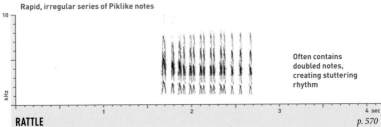

Often contains
doubled notes,
creating stuttering
rhythm

RATTLE   *p. 570*

All year, in threat and territorial displays; rather infrequent. Averages briefer and more plastic than
Whinny and Rattle calls of other woodpeckers; stuttering rhythm distinctive when present. Last few notes
may be lower, more churring.

Drums rather frequently, at long or short intervals

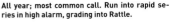

Fast version: speeds up, fades slightly at end

Start less sharp, more squeaky than Hairy

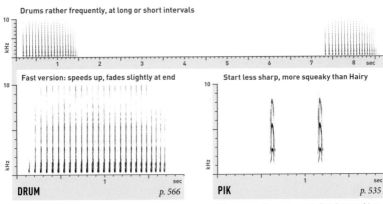

DRUM   *p. 566*

Our most noticeably accelerating woodpecker
Drum. Birds apparently give both fast and slow
versions; some slow versions do not speed up.

PIK   *p. 535*

All year; most common call. Run into rapid series in high alarm, grading into Rattle.

Fast series of nasal underslurs

First few notes
can be 2-parted

Notes like Pik but lower, more nasal

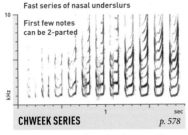
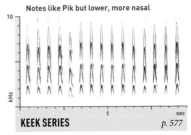

CHWEEK SERIES   *p. 578*

Infrequent, mostly Feb.–May, in aggressive encounters with same or other species.

KEEK SERIES   *p. 577*

Possibly all year, by excited birds in a variety of
situations, often between members of a mated
pair.

BLACK-BACKED WOODPECKER

Picoides arcticus

Breeds in northern coniferous forests. Like Three-toed, often irrupts into areas where fire or beetle outbreaks have killed many trees.

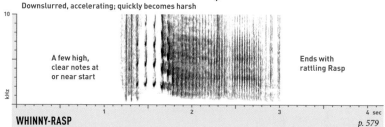

Downslurred, accelerating; quickly becomes harsh

A few high, clear notes at or near start

Ends with rattling Rasp

WHINNY-RASP *p. 579*

All year, by both sexes, mostly in aggressive encounters with same or other species; possibly also in establishing territories in spring. Plastic, but always distinctive. Abbreviated versions may include just the opening Whinny notes, or just the ending Rasp.

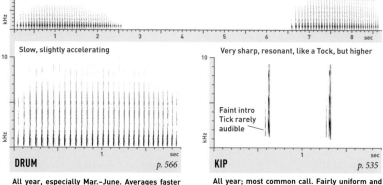

Drums rather frequently, at long or short intervals

Slow, slightly accelerating

DRUM *p. 566*

All year, especially Mar.–June. Averages faster and more uniform than Three-toed Drum, with less acceleration, but difficult to separate.

Very sharp, resonant, like a Tock, but higher

Faint intro Tick rarely audible

KIP *p. 535*

All year; most common call. Fairly uniform and stereotyped; unique among woodpeckers.

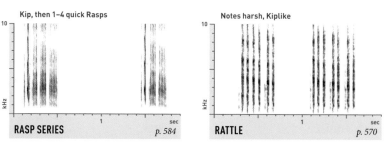

Kip, then 1–4 quick Rasps

RASP SERIES *p. 584*

Mostly Apr.–June, apparently as a summons to mate. Distinctive, but compare Three-toed Rattle.

Notes harsh, Kiplike

RATTLE *p. 570*

Apparently given in high alarm; sole recording is of captive bird. Rhythm irregular.

NORTHERN FLICKER

Colaptes auratus

Common and familiar. "Yellow-shafted" form of the East, "Red-shafted" form of the West mix freely in the Great Plains. All populations sound similar.

Long steady series on same pitch; starts slightly softer, lower, slower

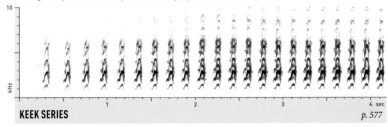

KEEK SERIES *p. 577*

All year, especially Mar.–June, mostly by unmated males, but also by females. Often long, up to 20 seconds. Usually given from a prominent perch along with Drum. Compare Pileated Woodpecker Keek Series.

Territorial males often alternate Keek Series with Drum, sometimes without pauses

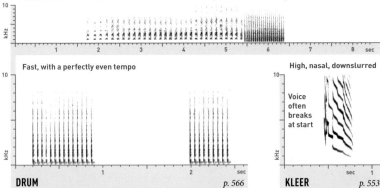

Fast, with a perfectly even tempo

DRUM *p. 566*

Mar.–Aug. Can be indistinguishable from other fast-drumming species; listen for Keek Series or other calls to confirm identification. Often given on metal to increase volume.

High, nasal, downslurred

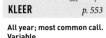

Voice often breaks at start

KLEER *p. 553*

All year; most common call. Variable.

High nasal couplet series: Wik louder, upslurred; second note low, brief

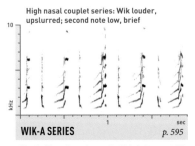

WIK-A SERIES *p. 595*

Mostly Mar.–June, by excited birds in courtship or in disputes. Variable; some versions omit second note, becoming Wik Series (p. 577).

Soft, low, somewhat musical

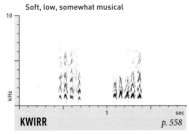

KWIRR *p. 558*

All year. Recalls Kwirr of Gila and related woodpeckers, but usually given in flight and only audible at close range.

GILDED FLICKER

Colaptes chrysoides

Found in saguaro cactus and riparian areas. Like "Red-shafted" Northern Flicker, but brown of forehead extends to nape, and underwings are golden.

Long steady series on same pitch; starts slightly softer, lower, slower

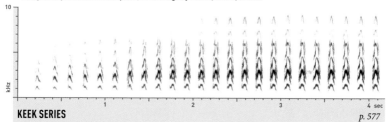

KEEK SERIES

p. 577

All year, especially Mar.–June, mostly by unmated males, but also by females. Often long, up to 20 seconds. Usually given from a prominent perch along with Drum. Not known to differ from Northern Flicker's Keek Series. The two species occasionally hybridize.

Territorial males often alternate Keek Series with Drum, sometimes without pauses

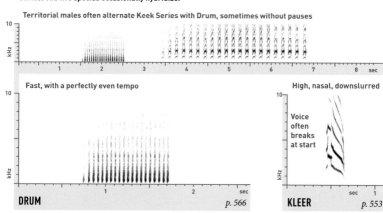

Fast, with a perfectly even tempo

DRUM

p. 566

Mar.–Aug. Can be indistinguishable from other fast-drumming species; listen for Keek Series or other calls to confirm identification. Not known to differ from Northern Flicker Drum.

High, nasal, downslurred

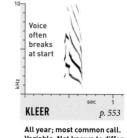

Voice often breaks at start

KLEER

p. 553

All year; most common call. Variable. Not known to differ from Northern's Kleer.

High nasal couplet series: Wik louder, upslurred; second note low, brief

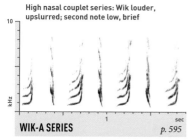

WIK-A SERIES

p. 595

Mostly Mar.–June, by excited birds in courtship or in disputes. Some versions are Wik Series (p. 542). Not known to differ from Northern's.

Soft, low, somewhat musical

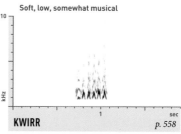

KWIRR

p. 558

All year, usually given in flight and only audible at close range. Not known to differ from Northern Flicker's Kwirr.

Red-naped Sapsucker

Sphyrapicus nuchalis

Uncommon in coniferous, deciduous, and mixed forests. Sometimes hybridizes with Red-breasted Sapsucker and with the Yellow-bellied Sapsucker of the East.

Harsh, broken notes in slow series

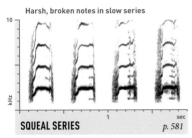

SQUEAL SERIES *p. 581*

By both sexes, but especially males, Apr.–June; often with Drum. Like Red-headed Woodpecker Kweeah, but not burry; almost always in series.

Like Red-Bellied Woodpecker Chuckle, but lower, harsher, faster

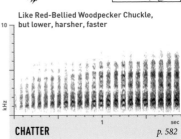

CHATTER *p. 582*

Mostly early in breeding season, between members of a pair. Also in fluttering flight display and in response to intruders.

Drum often very long, later taps in couplets or Morse code–like pattern

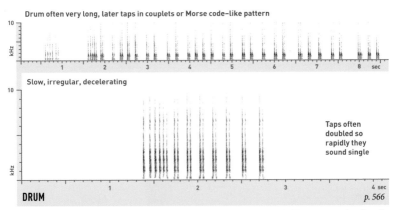

Slow, irregular, decelerating

Taps often doubled so rapidly they sound single

DRUM *p. 566*

By both sexes, but especially by unpaired males in spring. Instantly recognizable as a sapsucker drum by its extremely slow, decelerating rhythm, but length and exact pattern are highly plastic. Not known to be separable from Red-breasted Sapsucker Drum, but consistently different from Williamson's.

Harsh Screeches in simple or couplet series

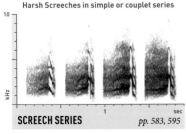

SCREECH SERIES *pp. 583, 595*

All year, especially in breeding season, in contact between mates and territorial squabbles. Highly plastic, but usually harsh; soft or loud.

Nasal downslur or overslur

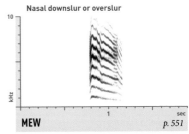

MEW *p. 551*

All year, in mild alarm; most common call. Quite plastic; often noisy and/or burry, occasionally broken. Often in short series.

Red-breasted Sapsucker

Sphyrapicus ruber

Uncommon in coniferous, deciduous, and mixed forests. Southern subspecies has some black and white on face, more like Red-naped Sapsucker.

Harsh, broken notes in slow series

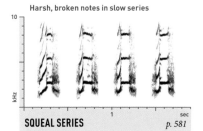

SQUEAL SERIES *p. 581*

By both sexes, but especially males, Apr.–June; often with Drum. Not reliably separable from Squeal Series of Red-naped Sapsucker.

Similar to Red-naped Sapsucker Chatter

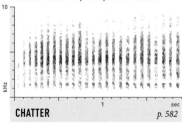

CHATTER *p. 582*

Mostly early in breeding season, between members of a pair. Also in fluttering flight display and in response to intruders.

Drum often very long, later taps in couplets or Morse code–like pattern

Slow, irregular, decelerating

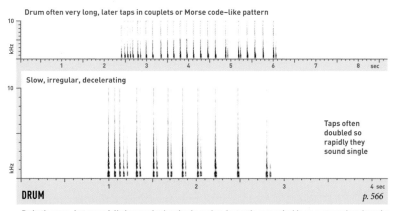

Taps often doubled so rapidly they sound single

DRUM *p. 566*

By both sexes, but especially by unpaired males in spring. Instantly recognizable as a sapsucker drum by its extremely slow, decelerating rhythm, but length and exact pattern are highly plastic. Not known to be separable from Red-naped Sapsucker Drum, but consistently different from Williamson's.

Harsh Screeches in simple or couplet series

SCREECH SERIES *pp. 583, 595*

All year, especially in breeding season, in contact between mates and territorial squabbles. Highly plastic, but usually harsh; soft or loud.

Nasal downslur or overslur

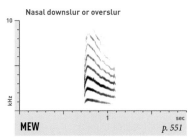

MEW *p. 551*

All year, in mild alarm; most common call. Quite plastic; often burry or partly harsh. Often in short series. Like Mew of Red-naped.

WILLIAMSON'S SAPSUCKER

Sphyrapicus thyroideus

Uncommon in coniferous forests, especially those of ponderosa pine, Douglas-fir, or western larch. Sexes differ greatly in plumage, unlike most woodpeckers.

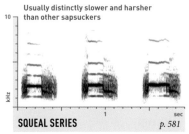

Usually distinctly slower and harsher than other sapsuckers

SQUEAL SERIES *p. 581*

By both sexes, but mostly males, Apr.–June; often with Drum. Plastic; many versions are harsh Screech Series (p. 583); rare versions are clear.

Like other sapsucker Chatters

CHATTER *p. 582*

Mostly early in breeding season, in fluttering flight display at end of encounters; possibly also between members of a pair.

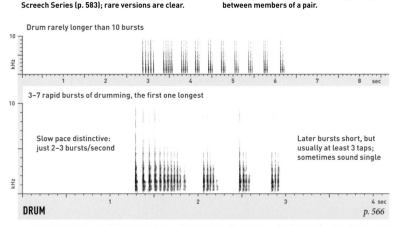

Drum rarely longer than 10 bursts

3–7 rapid bursts of drumming, the first one longest

Slow pace distinctive: just 2–3 bursts/second

Later bursts short, but usually at least 3 taps; sometimes sound single

DRUM *p. 566*

By both sexes, but especially by unpaired males in spring. The most distinctive woodpecker Drum in the West, almost always diagnostic; compare other sapsuckers. Bursts of drumming usually rather evenly spaced, but because they become successively shorter, overall rhythm may seem to accelerate.

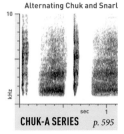

Alternating Chuk and Snarl

CHUK-A SERIES *p. 595*

In high aggression, often with Chuckle. Distinctively slow and harsh.

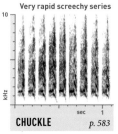

Very rapid screechy series

CHUCKLE *p. 583*

Given in pair contact and in aggression. Slower, slightly less harsh than Chatter.

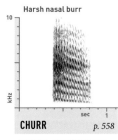

Harsh nasal burr

CHURR *p. 558*

In alarm, especially near nest. Usually slightly downslurred.

PILEATED WOODPECKER

Dryocopus pileatus

The largest surviving woodpecker in North America, the size of a crow, with distinctive plumage. Requires forests with large-diameter trees.

Keek Series can last 60 seconds or more

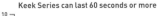

Notes nasal, sometimes almost clucking

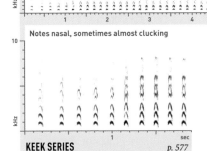

KEEK SERIES   p. 577

All year, by both sexes, in alarm and in interactions, often in flight. Slower, often lower, and more plastic than Northern Flicker Keek Series.

Higher, faster, shorter than Keek Series

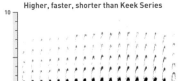

WHINNY   p. 578

Loud; rarely longer than 2 seconds. Last note typically lower. In spring, given repeatedly from high perch, suggesting territorial function.

Rather slow; slightly faster and softer at end

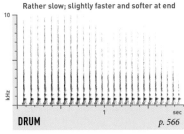

DRUM   p. 566

All year, but especially in early spring. Typically loud and far-carrying, acceleration subtle. Length variable, up to 3 full seconds.

Slow series of up- or underslurred nasal notes

Soft to medium-loud

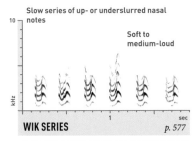

WIK SERIES   p. 577

Equivalent to Wik-a Series of other woodpeckers. Given in close pair interactions; louder versions may be given in disputes or in flight.

FALCONS AND CARACARAS (Family Falconidae)   *next pages*

Long thought to be closely related to hawks and eagles, falcons and caracaras are now recognized as a very distinct family on the basis of genetic evidence. As far as is known, all vocalizations are innate.

Falcons (genus *Falco*) are built for aerial hunting. Their long, pointed wings and long tails give them great speed and agility, perfect for chasing down birds in flight, though they will also take a variety of other prey. For centuries, falconers have taken advantage of the hunting skills of captive birds. The North American species are all closely related and have very similar vocal repertoires.

Caracaras are more terrestrial than falcons, with longer legs and more rounded wings. Our sole species is a slow flier that feeds mostly on carrion.

AMERICAN KESTREL

Falco sparverius

Our smallest, most common, and most colorful falcon, found in various open habitats; often seen on roadside telephone wires. Nests in tree cavities.

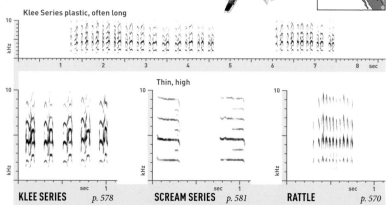

Klee Series plastic, often long

Thin, high

KLEE SERIES *p. 578* **SCREAM SERIES** *p. 581* **RATTLE** *p. 570*

Klee Series given all year, in alarm and in aerial displays; most common call by far. Notes can sound 1- or 2-syllabled. Scream Series given in courtship, in high alarm, and by begging fledglings; can be upslurred or monotone. Rattles heard mostly during pair interactions. All calls intergrade.

CRESTED CARACARA

Caracara cheriway

A unique raptor, local in open country. Eats carrion as well as live prey, which it sometimes steals from other birds. Soars infrequently.

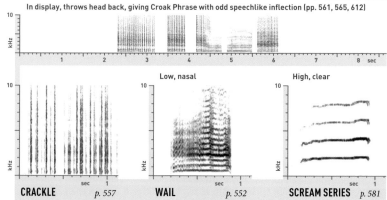

In display, throws head back, giving Croak Phrase with odd speechlike inflection (pp. 561, 565, 612)

Low, nasal

High, clear

CRACKLE *p. 557* **WAIL** *p. 552* **SCREAM SERIES** *p. 581*

Mostly silent except during breeding season, or when groups gather to feed. Crackle given in agitation, Croak Phrase (p. 612) in aggressive displays, with head thrown back. Low, harsh Wails given rarely by adults in high alarm. High, keening Scream Series given frequently by young birds.

MERLIN

Falco columbarius

Slightly larger than American Kestrel, with white bands in tail. Breeds in open forests; in winter, found along woodland edges and in open areas.

Whinny usually slightly overslurred; plastic in length

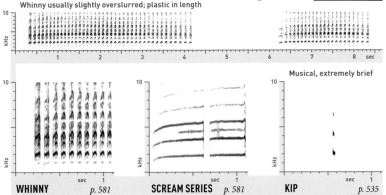

WHINNY *p. 581*	**SCREAM SERIES** *p. 581*	Musical, extremely brief **KIP** *p. 535*

Whinny given all year by both sexes in courtship and aggression; most common call. Can resemble Pileated Woodpecker Whinny, but generally screechier. Scream Series given by juveniles and females begging for food. Kip given in courtship and pair contact; male version slightly higher. Not very vocal in winter.

APLOMADO FALCON

Falco femoralis

Once more widespread, but extirpated from the U.S. in the twentieth century. Has been reintroduced to New Mexico; a few may wander north from Mexico.

Keek Series often slightly overslurred; plastic in length

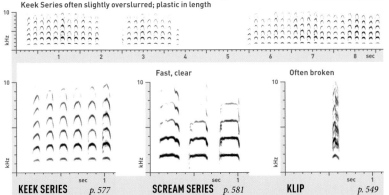

KEEK SERIES *p. 577*	Fast, clear **SCREAM SERIES** *p. 581*	Often broken **KLIP** *p. 549*

Keek Series given all year in aggression; distinctively clear, nasal. Only slightly faster than Scream Series, given by female near nest and by male in copulation. Klip given all year in pair contact and other interactions. Highly plastic; ranges from unbroken, nasal Kip to broken, squeaky Klee (p. 549).

PRAIRIE FALCON

Falco mexicanus

Breeds on mountain cliffs and prairie bluffs in the West; wanders farther east in winter. In flight, note dark feathers where the wing meets the body.

Screams often broken

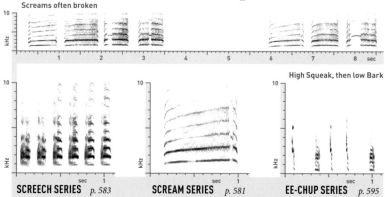

High Squeak, then low Bark

SCREECH SERIES *p. 583*   **SCREAM SERIES** *p. 581*   **EE-CHUP SERIES** *p. 595*

Screech Series and Scream Series given all year, but primarily Mar.–May, near nest. Screech Series given mostly in alarm, Scream Series in courtship, Ee-chup Series in courtship display from prospective nest site on cliff ledge. All vocalizations plastic and similar to sounds of Peregrine Falcon.

GYRFALCON

Falco rusticolus

The largest falcon. Rare in winter in our area. Plumage ranges from nearly pure white (mostly Greenland breeders) to very dark; most birds are gray.

Scream Series slower than in other falcons

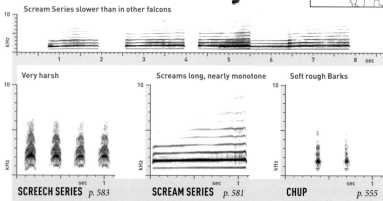

Very harsh   Screams long, nearly monotone   Soft rough Barks

SCREECH SERIES *p. 583*   **SCREAM SERIES** *p. 581*   **CHUP** *p. 555*

Vocalizes primarily on breeding grounds. Screech Series, given during aggressive encounters, is most likely call to be heard in winter. Scream Series and Chup associated mostly with courtship and nesting. Chup usually given in rapid plastic series, grading into louder, more regular Screech Series.

PEREGRINE FALCON

Falco peregrinus

Legendary for aerial speed. Once rare and endangered, but now recovering. Many populations descend in part from released birds.

Scream Series ranges from nearly clear to quite screechy

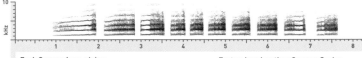

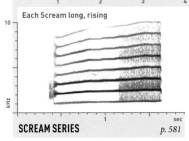

Each Scream long, rising

SCREAM SERIES
p. 581

All year, in a variety of contexts, but especially Mar.–June. Plastic; different forms may serve slightly different functions.

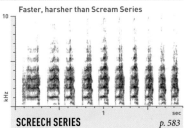

Faster, harsher than Scream Series

SCREECH SERIES
p. 583

All year, in alarm, but especially in nest defense. Rather plastic; may increase in pitch and speed in accordance with agitation level.

A Squeak and a Chip in couplet series

Triplet versions include low Tick

EE-CHUP SERIES
p. 595

Mostly Mar.–May, in courtship and nest defense. Quite plastic; in agitation, speeds up and syllables merge into Tsooklike notes (p. 549).

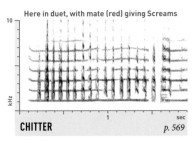

Here in duet, with mate (red) giving Screams

CHITTER
p. 569

Mostly Mar.–May, by males during copulation; also by either sex in high alarm. All our falcons give similar calls, but recordings are few.

PARROTS AND PARAKEETS (Family Psittacidae)   *next pages*

Our only native parrots are gone: the Carolina Parakeet became extinct in 1918, and the Thick-billed Parrot, extirpated from the United States in the 1930s, now survives only in Mexico. But flocks of escaped or released parrots again roam parts of the region, especially southern California, where dozens of species have been seen in the wild. Most nest in cavities, though Monk Parakeets build elaborate stick nests.

Parrots are among the most intelligent birds, with complex social interactions and vocal learning abilities. Although parrots are famous for imitation in captivity, wild parrots rarely imitate environmental sounds. Their natural sounds include Yelps, Shrieks, Squeals, and Chatters, often in chorus. Identification by ear often requires experience; only a fraction of each species' repertoire is shown here.

Rose-ringed Parakeet *Psittacula krameri*

Local in Los Angeles area and in Bakersfield, California. Native to India and Africa. Most calls are variations on the Kleer Series.

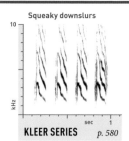

Squeaky downslurs

KLEER SERIES *p. 580*

Grating (as here) or clear

KEER *pp. 550, 559*

Rosy-faced Lovebird *Agapornis roseicollis*

Local in the Phoenix, Arizona, area. Native to southern Africa. Voice varied, high, musical. Also gives a seminasal Wik (p. 548).

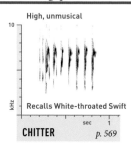

High, unmusical

Recalls White-throated Swift

CHITTER *p. 569*

High, clear, musical

CHIRP *p. 548*

Monk Parakeet *Myiopsitta monachus*

Rare in cities in the West. Native to S. America. Most sounds are versions of the Grate, but also gives Yelps and squealing Chirps.

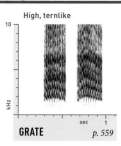

High, ternlike

GRATE *p. 559*

Rather low for a parakeet

CHIRP *p. 548*

Nanday Parakeet *Aratinga nenday*

Local in S. California. Native to S. America. Voice distinctive; all sounds are ternlike Grates, without squealing or chirping.

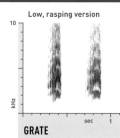

Low, rasping version

GRATE *p. 559*

High version, like Royal Tern

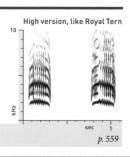

p. 559

MITRED PARAKEET *Psittacara mitratus*

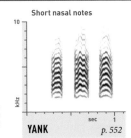

Short nasal notes

YANK   *p. 552*

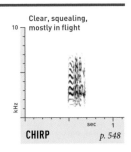

Clear, squealing, mostly in flight

CHIRP   *p. 548*

Local in California. Native to S. America. Most common sound is laughing series of Yanks, like loud Red-breasted Nuthatch.

RED-MASKED PARAKEET *Psittacara erythrogenys*

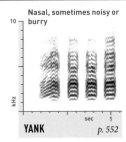

Nasal, sometimes noisy or burry

YANK   *p. 552*

Harsh, squealing

CHIRP   *p. 548*

Local in California. Native to Peru and Ecuador. All sounds like those of Mitred, but on average lower and slightly harsher.

BLUE-CROWNED PARAKEET *Thectocercus acuticaudatus*

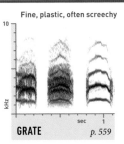

Fine, plastic, often screechy

GRATE   *p. 559*

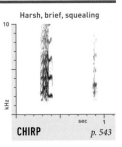

Harsh, brief, squealing

CHIRP   *p. 543*

Local in California. Native to S. America. Grates average long, high, and screechy. Chirps relatively rare.

YELLOW-CHEVRONED PARAKEET *Brotogeris chiriri*

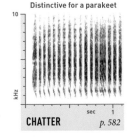

Distinctive for a parakeet

CHATTER   *p. 582*

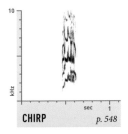

CHIRP   *p. 548*

Local in California. Native to S. America. Sounds from native range shown. In U.S., some may sound slightly different.

Red-crowned Parrot *Amazona viridigenalis*

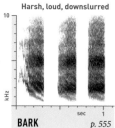

Harsh, loud, downslurred

BARK   *p. 555*

1- or 2-syllabled musical, downslurred Pee-loo

WHISTLE   *p. 587*

Locally common in S. California. Voice hugely varied, as in all *Amazona*. Most distinctive calls are shown.

Lilac-crowned Parrot *Amazona finschi*

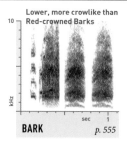

Lower, more crowlike than Red-crowned Barks

BARK   *p. 555*

Typical version distinctive: brief, rising, squeaky

CHIRP   *p. 548*

Local in S. California. Native to W. Mexico. Repertoire includes complex musical Chirps in flight, soft ravenlike Croaks on perch.

Red-lored Parrot *Amazona autumnalis*

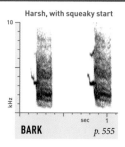

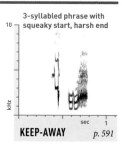

Harsh, with squeaky start

BARK   *p. 555*

3-syllabled phrase with squeaky start, harsh end

KEEP-AWAY   *p. 591*

Local in S. California. Native to Cen. and S. America. Voice high and squeaky, with metallic clinking; "Keep-away" call distinctive.

Yellow-headed Parrot *Amazona oratrix*

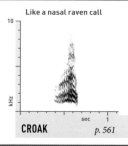

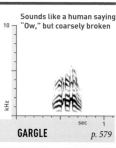

Like a nasal raven call

CROAK   *p. 561*

Sounds like a human saying "Ow," but coarsely broken

GARGLE   *p. 579*

Local in California. Native to Mexico and Cen. America. Voice low, clear, distinctively humanlike. Gargle is most common call.

TYRANT FLYCATCHERS (Order Passeriformes, Family Tyrannidae)

The tyrant flycatchers are small to medium-sized birds that sally out from perches to capture insects on the wing. "Tyrant" describes the highly aggressive behavior of certain species, especially kingbirds, which will readily defend territories against birds many times their size. In some species, even members of mated pairs may act aggressively toward one another. Because many flycatchers look very similar, vocalizations can be important in identifying them.

Flycatchers apparently do not learn any of their vocalizations; all sounds are innate. Sounds tend to be quite uniform and stereotyped, with little regional variation. Most species have extensive and complex vocal repertoires.

Dawn Songs and Day Songs

Flycatcher songs tend to be constructed of short phrases, often combined with one another according to strict syntactic rules that allow only certain combinations. For example, the Brown-crested Flycatcher's song phrases A, B, and C are never given alone, but only in combinations AC, ACC, BC, and BCC. Many species have a **Dawn Song** that males sing during the hour or so prior to sunrise during the breeding season. In some species the Dawn Song and the Day Song are different; in some species they are the same.

Flight Songs

Several flycatcher species—including the phoebes and Willow, Hammond's, Dusky, and Least Flycatchers—have an additional type of complex song given very infrequently by breeding males in display flights. This **Flight Song** tends to be a precise, stereotyped recombination of calls and song phrases that are also given in other contexts. It may be given more often at dusk, or possibly upon detection of aerial predators. In some flycatchers, such as the kingbirds, the song given during flight displays is much like the typical Day Song or Dawn Song.

One complete Flight Song performance of Say's Phoebe

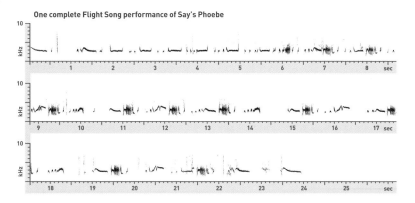

Other Sounds

Most flycatchers give several kinds of calls. Most kingbirds perform pair duets, albeit rather poorly synchronized ones; in some kingbirds, the male and the female duet calls are quite different. In some flycatchers, such as phoebes, males court females with unique call types given during "nest site showing" displays. All flycatchers snap their bills loudly when in pursuit of flying insects; many species also snap their bills in agitation.

Northern Beardless-Tyrannulet

Camptostoma imberbe

Nests in riparian and live oak woods, in clumps of Spanish moss and ball moss. Like a small *Empidonax*, but note dark eyeline and broken eye-ring.

Dawn Song (p. 598): "Pip... Peer... PEER-peeper"

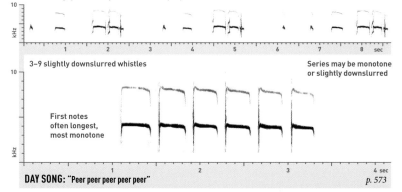

3–9 slightly downslurred whistles

Series may be monotone or slightly downslurred

First notes often longest, most monotone

DAY SONG: "Peer peer peer peer peer"

p. 573

Mostly Mar.–Aug. Rather uniform and stereotyped, but some birds give more plastic versions in which notes break. Day Song may have territorial function; sometimes given by females. Dawn Song reportedly given only by males prior to sunrise in breeding season.

Notes usually burry, sometimes in series

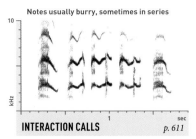

INTERACTION CALLS p. 611

Mostly May–Aug., in interactions. Highly plastic; all pitch patterns occur. Often mixed with various other calls by excited birds.

High semimusical trill, often downslurred

Here, with another bird giving Peers

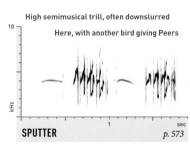

SPUTTER p. 573

Mostly May–Aug., by both sexes, apparently in contact and some courtship situations. Plastic; some versions longer, more rattling.

Like Peer, but 2-syllabled, breaking downward at end

PEE-UK p. 587

Simple downslurred whistle, like single note of Day Song

PEER p. 541

All year; most common calls. Pee-uk reportedly given only by female, while Peer is given by both sexes. Not known to what extent the 2 variants may differ in function. Both calls given at intervals while foraging; Peer also in response to playback of Dawn Song.

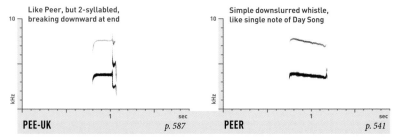

WESTERN WOOD-PEWEE

Contopus sordidulus

Common in western pine forests and riparian woodlands. Nearly identical to Eastern Wood-Pewee, a rare vagrant to the West, but voice differs.

Dawn Song (p. 597): A and B phrases alternate: "Dzeer... Pee-widdit... Dzeer... Pee-widdit"

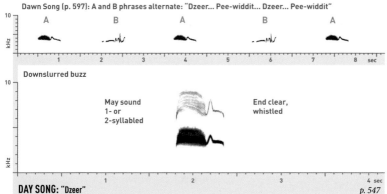

Downslurred buzz

May sound 1- or 2-syllabled

End clear, whistled

DAY SONG: "Dzeer" p. 547

All year, but mostly May–Aug. Variable but stereotyped and distinctive. Presumed to be given mostly by males, in territorial advertisement and communication with mate. Dawn Song consists of Dzeer alternated with clear, whistled Pee-widdit phrase. Only Dzeer given during day; Dawn Song sometimes at Dusk.

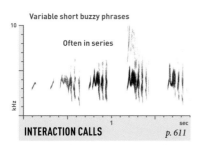

Variable short buzzy phrases

Often in series

INTERACTION CALLS p. 611

Mostly May–Sept. A catch-all category of plastic sounds given in interactions, sometimes in flight; some versions rhythmic, repeating.

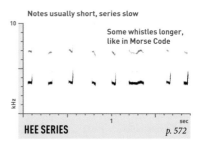

Notes usually short, series slow

Some whistles longer, like in Morse Code

HEE SERIES p. 572

Mostly May–Aug. Function unclear, but apparently given in agitation; associated with Interaction Calls and Day Song.

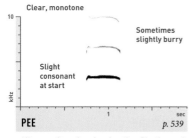

Clear, monotone

Sometimes slightly burry

Slight consonant at start

PEE p. 539

All year; often given in migration. Plastic, grading into Dzree; occasionally slightly upslurred.

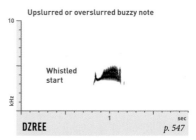

Upslurred or overslurred buzzy note

Whistled start

DZREE p. 547

All year, in mate contact and nest defense; most common call. Slightly plastic. Similar to Day Song and often confused with it.

GREATER PEWEE

Contopus pertinax

Breeds in mountain pine-oak forests. Larger than Western Wood-Pewee, with distinct wispy crest. Bill long, with pale lower mandible.

Dawn Song (p. 598): almost always in AAB pattern: "Where to eat? Where to eat? José María"

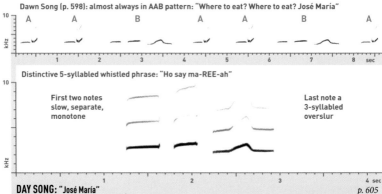

Distinctive 5-syllabled whistled phrase: "Ho say ma-REE-ah"

First two notes slow, separate, monotone

Last note a 3-syllabled overslur

DAY SONG: "José María"   *p. 605*

Dawn Song mostly Mar.–Aug., by male, before sunrise; female not known to sing. Complete Dawn Song can be given during the day, especially in spring, but more often birds give B phrase alone, usually omitting the last syllable: "Ho say ma-REE?" Day Song also given in winter. A complex Flight Song is given rarely.

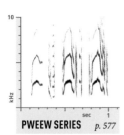

PWEEW SERIES   *p. 577*

Not well known; given in aggressive interactions, sometimes with snap of bill.

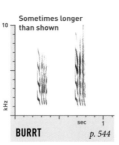

Sometimes longer than shown

BURRT   *p. 544*

Not well known; given in aggressive interactions, often with Twitter.

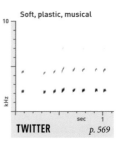

Soft, plastic, musical

TWITTER   *p. 569*

Soft call given in agitation, often with Burrt.

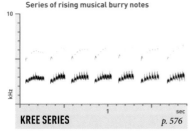

Series of rising musical burry notes

KREE SERIES   *p. 576*

Given rather frequently, sometimes for extended periods, sometimes in middle of Dawn Song. Function poorly understood.

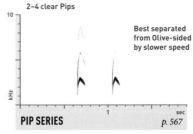

2–4 clear Pips

Best separated from Olive-sided by slower speed

PIP SERIES   *p. 567*

All year; most common call. Excited birds can give long series. Speed usually 3 notes/second or slower; Olive-sided, 4 notes/second or faster.

Olive-sided Flycatcher

Contopus cooperi

Larger than wood-pewees, with larger head and shorter tail. Breeds in coniferous forests; frequently perches at the very top of tall dead snags.

Possible Dawn Song from California: third note occasionally omitted

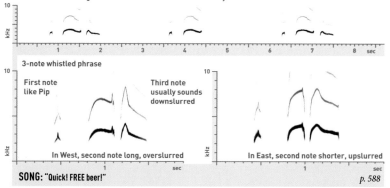

3-note whistled phrase

First note like Pip

Third note usually sounds downslurred

In West, second note long, overslurred

In East, second note shorter, upslurred

SONG: "Quick! FREE beer!" p. 588

Subtle but consistent geographic variation; songs from Rockies and Pacific Coast slightly lower than those from Alaska east. Given all day; apparently no difference between Day Song and Dawn Song in many individuals, but in one dawn recording, third note is occasionally omitted, creating two different phrase types.

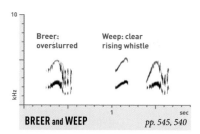

Breer: overslurred

Weep: clear rising whistle

BREER and WEEP pp. 545, 540

Mostly May–July, in aggressive interactions. Breers and Weeps rather plastic, often mixed together in variable order.

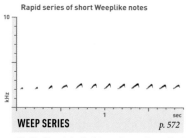

Rapid series of short Weeplike notes

WEEP SERIES p. 572

Mostly May–July, in interactions, often mixed with other calls. Plastic.

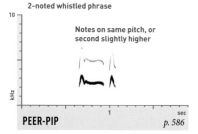

2-noted whistled phrase

Notes on same pitch, or second slightly higher

PEER-PIP p. 586

Mostly May–July, repeated regularly for long intervals or mixed in with Pip Series, which it grades into. Function unclear.

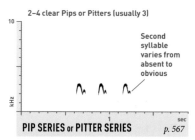

2–4 clear Pips or Pitters (usually 3)

Second syllable varies from absent to obvious

PIP SERIES or PITTER SERIES p. 567

All year; most common call, in pair contact. Excited birds can give long series. Pitch somewhat variable; Pips more common than Pitters.

ALDER FLYCATCHER

Empidonax alnorum

Nearly identical to Willow; the two were long considered one species, "Traill's Flycatcher." Breeds in northern willow and alder swamps and wet thickets.

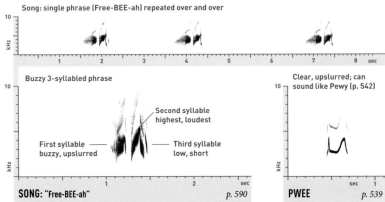

Song: single phrase (Free-BEE-ah) repeated over and over

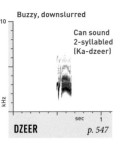

Buzzy 3-syllabled phrase

Second syllable highest, loudest

First syllable buzzy, upslurred

Third syllable low, short

SONG: "Free-BEE-ah" *p. 590*

Mostly Apr.–Aug., by males; females may sing rarely. Combination of buzzy, upslurred first syllable and 3-syllabled structure rule out all sounds of Willow Flycatcher, but final syllable can be faint.

Clear, upslurred; can sound like Pewy (p. 542)

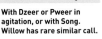

PWEE *p. 539*

With Dzeer or Pweer in agitation, or with Song. Willow has rare similar call.

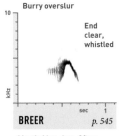

Burry overslur

End clear, whistled

BREER *p. 545*

Mostly May–Aug. Often repeated from a perch, like Song. Function unclear.

Buzzy, downslurred

Can sound 2-syllabled (Ka-dzeer)

DZEER *p. 547*

Mostly Apr.–Aug., in aggression, often with Twitter or Pweer. Plastic.

Soft, semimusical

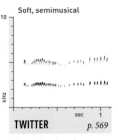

TWITTER *p. 569*

Mostly May–Aug., in aggression, often introducing Dzeer. Plastic.

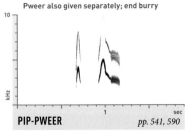

Pweer also given separately; end burry

PIP-PWEER *pp. 541, 590*

May–Aug. Pweer may be given without other calls, or with Dzeer or Twitter in agitation. Pip-pweer combination recalls Willow's Pete's-Beer.

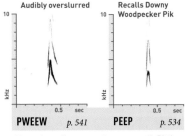

Audibly overslurred

PWEEW *p. 541*

Recalls Downy Woodpecker Pik

PEEP *p. 534*

Peep given all year; most common call. Distinctive, uniform, and stereotyped. Pweew a plastic, infrequent intergrade between Peep and Pweer.

WILLOW FLYCATCHER

Empidonax traillii

A brownish Empid with a faint eye-ring. Best separated from Alder Flycatcher by voice. Song of "Southwestern" subspecies differs slightly (see p. 291).

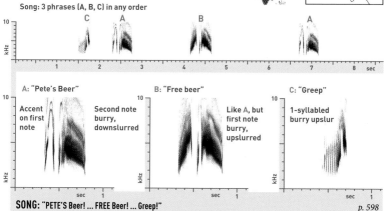

Song: 3 phrases (A, B, C) in any order

A: "Pete's Beer"
Accent on first note
Second note burry, downslurred

B: "Free beer"
Like A, but first note burry, upslurred

C: "Greep"
1-syllabled burry upslur

SONG: "PETE'S Beer! ... FREE Beer! ... Greep!" *p. 598*

Mosty Apr.–Aug., by males, but females occasionally sing. Three-part syntax and burry phrases distinctive; A, B, and C phrases rarely given outside context of full Song. Song begins at dawn, but given all day. "Pete's Beer" also transliterated as "Fitz-bew." Also has a complex Flight Song (p. 611).

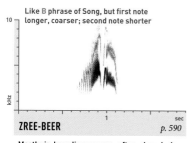

Like B phrase of Song, but first note longer, coarser; second note shorter

ZREE-BEER *p. 590*

Mostly in breeding season, often given in long bouts from exposed perch. Easy to mistake for Alder Flycatcher Song, but 2-syllabled.

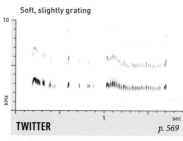

Soft, slightly grating

TWITTER *p. 569*

Mostly on breeding grounds. Plastic. Given in aggressive encounters, often during Song or with Zree-beer.

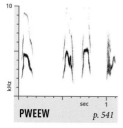

PWEEW *p. 541*

A catch-all category of plastic, sometimes burry notes given in excitement.

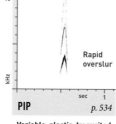

Rapid overslur

PIP *p. 534*

Variable, plastic; by excited birds. Very like some Alder Flycatcher calls.

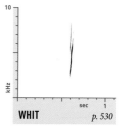

WHIT *p. 530*

All year; most common call. Never given by Alder Flycatcher.

HAMMOND'S FLYCATCHER

Empidonax hammondii

Uncommon in coniferous forests, often in dense stands. Very similar to Dusky; slightly longer wings, shorter tail, and shorter bill, but best identified by voice.

Song: 3 phrases (A, B, C) in any order

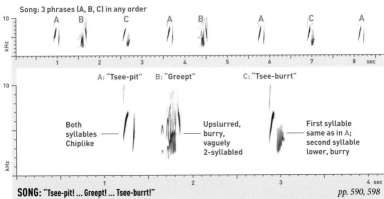

A: "Tsee-pit"   B: "Greept"   C: "Tsee-burrt"

Both syllables Chiplike

Upslurred, burry, vaguely 2-syllabled

First syllable same as in A; second syllable lower, burry

SONG: "Tsee-pit! ... Greept! ... Tsee-burrt!"   *pp. 590, 598*

Mostly May–July, by males; females not known to sing. Song begins at dawn, but given all day. Phrases more alike than in Dusky's song; A and C are particularly similar. In summer, males may repeat only phrase C (p. TKK), recalling Least Flycatcher song. Also gives a rare complex Flight Song (p. 574).

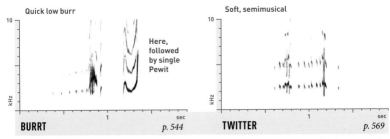

Quick low burr

Here, followed by single Pewit

Soft, semimusical

BURRT   *p. 544*   **TWITTER**   *p. 569*

A complex of rather plastic vocalizations, given mostly on breeding grounds, in interactions between males and females or adults and young. Also sometimes by adults upon landing after a brief flight. Rather soft; audible mostly at close range.

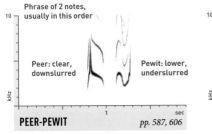

Phrase of 2 notes, usually in this order

Peer: clear, downslurred

Pewit: lower, underslurred

High, semimusical, recalling Pygmy Nuthatch

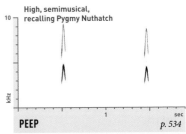

PEER-PEWIT   *pp. 587, 606*   **PEEP**   *p. 534*

Mostly May–July, by male, in short bouts in morning and evening. Peer often followed by several Pewits; both also given separately.

All year; most common call. Slightly higher than Peep of Alder Flycatcher, with weaker consonants; unlike any call of Dusky, Gray, or Least.

DUSKY FLYCATCHER

Empidonax oberholseri

Common in brushy habitats and open coniferous or mixed woods. Notoriously similar to Hammond's in appearance; the two often breed side by side.

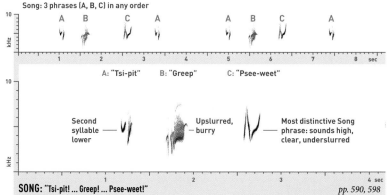

Song: 3 phrases (A, B, C) in any order

A: "Tsi-pit"   B: "Greep"   C: "Psee-weet"

Second syllable lower

Upslurred, burry

Most distinctive Song phrase: sounds high, clear, underslurred

SONG: "Tsi-pit! ... Greep! ... Psee-weet!"   *pp. 590, 598*

Mostly May–July, by males; females not known to sing. Song begins at dawn, but given all day. All three phrases quite different in quality; phrase C is most different from Hammond's. In summer, males may give only phrase A (p. TKK), often with Twitters or other calls. Also gives a rare complex Flight Song (p. 611).

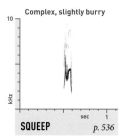

Complex, slightly burry

SQUEEP   *p. 536*

Mostly May–July, in high-intensity altercations, often with Pip or other calls.

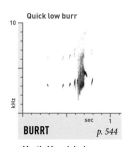

Quick low burr

BURRT   *p. 544*

Mostly May–July, in interactions, often with Twitter. Plastic.

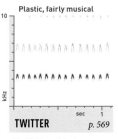

Plastic, fairly musical

TWITTER   *p. 569*

Mostly May–July, in courtship and other interactions. Not very loud.

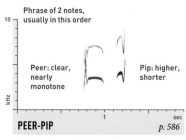

Phrase of 2 notes, usually in this order

Peer: clear, nearly monotone

Pip: higher, shorter

PEER-PIP   *p. 586*

All year, by male, especially in evening. Often several Peers precede Pip; both also given separately. Also transliterated as "Du-hic."

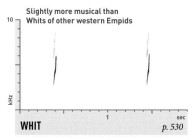

Slightly more musical than Whits of other western Empids

WHIT   *p. 530*

All year, in alarm and likely in contact, and sometimes without apparent cause; most common call.

GRAY FLYCATCHER

Empidonax wrightii

Breeds in pinyon-juniper woods; also locally in tall sagebrush and in open, brushy pine forests. Dips tail slowly down, then up, unlike other Empids.

Song: 2 phrases (A, B) in any order; A is much more common

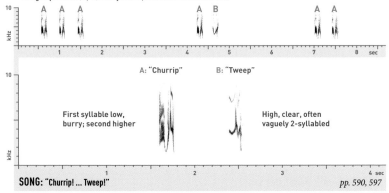

A: "Churrip" B: "Tweep"

First syllable low, burry; second higher

High, clear, often vaguely 2-syllabled

SONG: "Churrip! ... Tweep!"

pp. 590, 597

Mostly May–July, by males; females not known to sing. Song begins at dawn, but given all day. Phrases often organized into loose clusters; most common patterns are AA and AB. Sometimes gives A many times in a row before giving phrase B. Also gives a rare complex Flight Song; no recordings available.

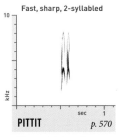

Fast, sharp, 2-syllabled

PITTIT p. 570

All year, in interactions. Sometimes repeated in songlike fashion.

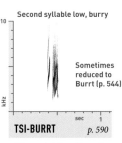

Second syllable low, burry

Sometimes reduced to Burrt (p. 544)

TSI-BURRT p. 590

All year, in interactions. Can be repeated in songlike fashion, even in winter.

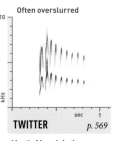

Often overslurred

TWITTER p. 569

Mostly May–July, in courtship and other interactions. Not very loud.

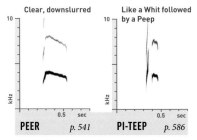

Clear, downslurred

PEER p. 541

Like a Whit followed by a Peep

PI-TEEP p. 586

Mostly May–July; function not clear. Peer often with Pittit or other calls. Versions include 2-syllabled Pee-uk (p. 587), brief Peep (p. 534).

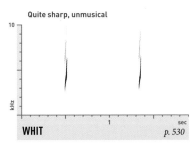

Quite sharp, unmusical

WHIT p. 530

All year, in alarm and likely in contact, and sometimes without apparent cause; most common call. Rather unmusical for an Empid Whit.

LEAST FLYCATCHER

Empidonax minimus

Local in the West in riparian woods. Like other members of the genus *Empidonax*, best identified by voice. A small, compact, rather dull-colored Empid.

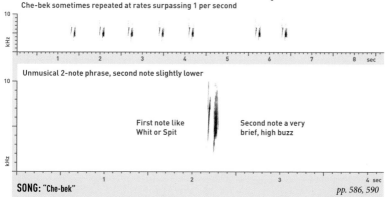

Che-bek sometimes repeated at rates surpassing 1 per second

Unmusical 2-note phrase, second note slightly lower

First note like Whit or Spit

Second note a very brief, high buzz

SONG: "Che-bek"   pp. 586, 590

Mostly Apr.–Aug., by males, but at least some singers are female. Highly uniform and stereotyped. Begins at dawn, but given all day. Higher and quicker than similar notes of other Empids, and often given at a far higher rate. Flight Song infrequent; includes rapid Che-beks with Pweew notes appended (p. 611).

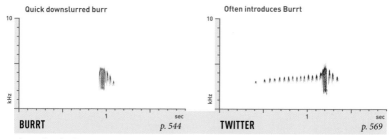

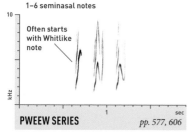

Quick downslurred burr

BURRT   p. 544

Often introduces Burrt

TWITTER   p. 569

A complex of rather plastic vocalizations, given mostly on breeding grounds, in interactions between males and females or adults and young. Also sometimes by adults upon landing after a brief flight. Rather soft; does not carry far.

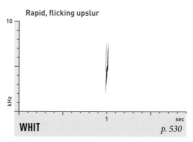

1–6 seminasal notes

Often starts with Whitlike note

PWEEW SERIES   pp. 577, 606

Mostly in breeding season, during aggressive defense of nest or territory, often with bill snaps; also in lead-up to Flight Song.

Rapid, flicking upslur

WHIT   p. 530

All year; most common call. In mild alarm and pair contact. Less frequent on fall migration.

PACIFIC-SLOPE FLYCATCHER

Empidonax difficilis

Breeds in humid, dense coniferous forest, especially along shaded streams. Formerly lumped with Cordilleran under the name "Western Flycatcher."

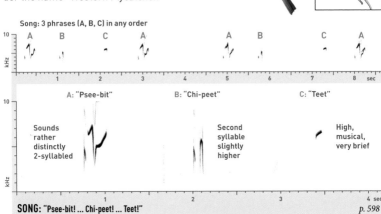

Song: 3 phrases (A, B, C) in any order

A: "Psee-bit"
Sounds rather distinctly 2-syllabled

B: "Chi-peet"
Second syllable slightly higher

C: "Teet"
High, musical, very brief

SONG: "Psee-bit! ... Chi-peet! ... Teet!"
p. 598

Mostly Apr.–July, by males, mostly before dawn, less often during day. Female not known to sing. Extremely similar to Cordilleran Song; both are higher, less burry than other Empid songs, with distinctive "Teet" phrase (C). Pacific-slope Song slightly higher overall, with subtly different A and B phrases.

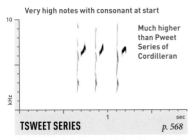

Very high notes with consonant at start

Much higher than Pweet Series of Cordilleran

TSWEET SERIES
p. 568

Mostly Apr.–July, in excited interactions. Audible mostly at close range. Also gives a soft sharp Pittit (p. 570).

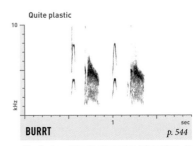

Quite plastic

BURRT
p. 544

Mostly Apr.–July, in aggressive interactions. In chases, often given in series. Also a pipping Twitter (p. 569), lower than in other Empids.

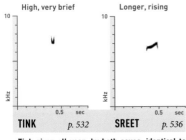

High, very brief

Longer, rising

TINK
p. 532

SREET
p. 536

Tink given all year, by both sexes; identical to Tink of Cordilleran. Sreet given by begging juveniles in summer; also likely by adults in alarm.

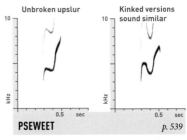

Unbroken upslur

Kinked versions sound similar

PSEWEET
p. 539

All year, by both sexes. Never broken into two notes. Almost always has a flat or downslurred central portion (kink) on the spectrogram.

CORDILLERAN FLYCATCHER

Empidonax occidentalis

Only separable from Pacific-slope Flycatcher by certain sounds. Prefers drier coniferous forests, but still seeks cool, shady areas, especially near water.

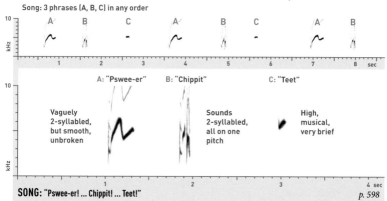

Song: 3 phrases (A, B, C) in any order

A | B | C | A | B | C | A | B

A: "Pswee-er"

Vaguely 2-syllabled, but smooth, unbroken

B: "Chippit"

Sounds 2-syllabled, all on one pitch

C: "Teet"

High, musical, very brief

SONG: "Pswee-er! ... Chippit! ... Teet!" p. 598

Mostly Apr.–July, by males, mostly before dawn, less often during day. Female not known to sing. Spectrographic analysis is helpful in identification. In a hybrid zone east of the Cascades from Oregon to British Columbia, birds are genetically mixed and songs and calls are intermediate.

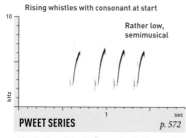

Rising whistles with consonant at start

Rather low, semimusical

PWEET SERIES p. 572

Mostly Apr.–July, in "nest-showing" courtship display and other excited interactions. Audible mostly at close range.

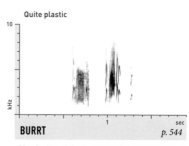

Quite plastic

BURRT p. 544

Mostly Apr.–July, in aggressive interactions. In chases, often given in series. Also a pipping Twitter (p. 569), lower than in other Empids.

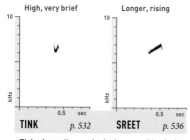

High, very brief

TINK p. 532

Longer, rising

SREET p. 536

Tink given all year, by both sexes; identical to Tink of Pacific-slope. Sreet given by begging juveniles; also likely by adults in alarm.

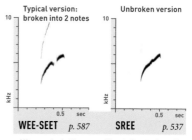

Typical version: broken into 2 notes

WEE-SEET p. 587

Unbroken version

SREE p. 537

All year, by both sexes. Classic 2-noted version easy to distinguish from Pacific-slope, but some birds also give unbroken versions (without kink).

BUFF-BREASTED FLYCATCHER

Empidonax fulvifrons

Our smallest and most distinctively colored Empid, local in open or patchy pine-oak forest with grassy understory, often near water.

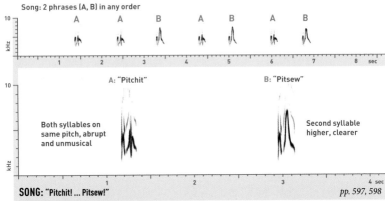

Song: 2 phrases (A, B) in any order

A: "Pitchit"
Both syllables on same pitch, abrupt and unmusical

B: "Pitsew"
Second syllable higher, clearer

SONG: "Pitchit! ... Pitsew!"   *pp. 597, 598*

Mostly May–Aug., by males; female has been observed to sing from nest. Song begins at dawn, but given all day. Burrt sometimes added in between phrases in the manner of a third song element; this three-part Song can recall Song of Hammond's.

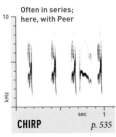

Often in series; here, with Peer

CHIRP   *p. 535*

Mostly May–Aug., in intense interactions. Some versions are 2-syllabled.

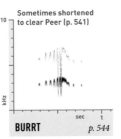

Sometimes shortened to clear Peer (p. 541)

BURRT   *p. 544*

Mostly May–Aug., in interactions, often in series connected by short Twitters.

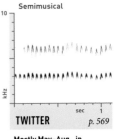

Semimusical

TWITTER   *p. 569*

Mostly May–Aug., in interactions, often with Burrt. Highly plastic.

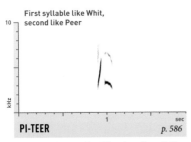

First syllable like Whit, second like Peer

PI-TEER   *p. 586*

Not well known. Possibly given in agitation; sometimes repeated for long periods or mixed with other calls.

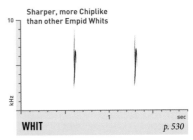

Sharper, more Chiplike than other Empid Whits

WHIT   *p. 530*

All year, in alarm and likely in contact; most common call. On average the sharpest, least musical Empid Whit.

SAY'S PHOEBE

Sayornis saya

Unlike other phoebes, prefers dry, open country, often far from water. Like other phoebes, readily nests under eaves of buildings, and migrates early in spring.

3 phrases (A, B, C); A by far most common; B and C almost never adjacent

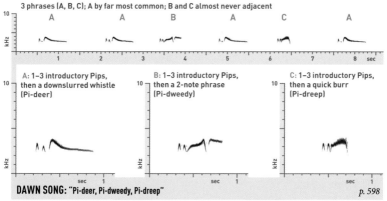

A: 1–3 introductory Pips, then a downslurred whistle (Pi-deer)

B: 1–3 introductory Pips, then a 2-note phrase (Pi-dweedy)

C: 1–3 introductory Pips, then a quick burr (Pi-dreep)

DAWN SONG: "Pi-deer, Pi-dweedy, Pi-dreep"   *p. 598*

Mostly Mar.–July. Phrase A predominates; B and C most frequent before dawn, becoming rarer throughout the morning; thus Dawn Song gradually transitions into Pi-deer calls, possibly a type of Day Song. One recording from Aug. includes a fourth phrase, a simple Pip; function unknown. See also Flight Song, p. 611.

Notes semimusical, slightly burry

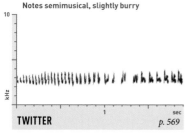

TWITTER   *p. 569*

Mostly Apr.–July, in family interactions near the nest, but apparently also sometimes in alarm. Highly plastic.

Rapid mix of burry phrases

Pyurr

Typical phrase: Preer-bit

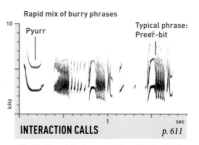

INTERACTION CALLS   *p. 611*

Mostly Apr.–July, in aggression. Highly plastic, except when repeated during Flight Song (p. 611).

1–3 introductory Pips, then a downslurred whistle

Sometimes shortened to downslurred Peer (p. 541)

Clear, musical, medium-low

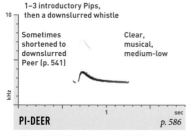

PI-DEER   *p. 586*

All year. Like Song phrase A. Most common call in summer; given in long strings during the day, near the nest, and in mild alarm.

Nearly monotone

Starts with consonant sound

Sometimes slightly tremolo

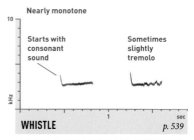

WHISTLE   *p. 539*

All year, in response to aerial or other predators; also sometimes from apparently calm birds. Grades into Pi-deer.

BLACK PHOEBE

Sayornis nigricans

Common near water, sometimes even beside backyard swimming pools. Dips tail frequently. Range expanding to the north and east.

2 phrases (A and B) usually alternated

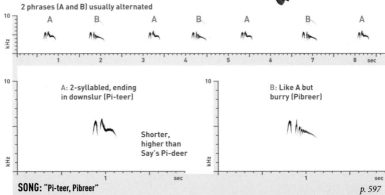

A: 2-syllabled, ending in downslur (Pi-teer)

Shorter, higher than Say's Pi-deer

B: Like A but burry (Pibreer)

SONG: "Pi-teer, Pibreer"   *p. 597*

Mostly Jan.–June; also on wintering grounds. Most frequent at dawn but given all day, especially by unmated males. Pattern like Eastern Phoebe song but higher-pitched, less burry, and often faster. Sometimes incorporates Teer like a third phrase. Rarely, if ever, repeats A or B phrases without the other.

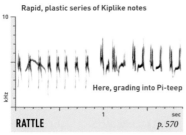

Rapid, plastic series of Kiplike notes

Here, grading into Pi-teep

RATTLE   *p. 570*

Highly plastic. Given near nest, when displaying potential nest site to mate, and in some aggressive situations.

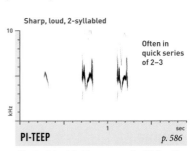

Sharp, loud, 2-syllabled

Often in quick series of 2–3

PI-TEEP   *p. 586*

Mostly Feb.–July. From aggressive birds; often used in flight display, along with Song phrases. Extremely similar to Eastern's Pi-teep.

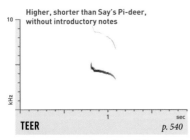

Higher, shorter than Say's Pi-deer, without introductory notes

TEER   *p. 540*

All year. Common at any time of day, often in long series, especially right before or after Song. Plastic; sometimes overslurred.

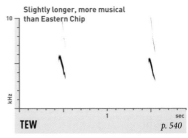

Slightly longer, more musical than Eastern Chip

TEW   *p. 540*

All year; most common call. Distinctively low and musical.

EASTERN PHOEBE

Sayornis phoebe

Found in open areas near water; nests on ledges under bridges or under eaves of buildings. Dips tail frequently. Range has expanded westward.

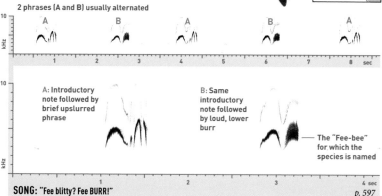

2 phrases (A and B) usually alternated

A: Introductory note followed by brief upslurred phrase

B: Same introductory note followed by loud, lower burr

— The "Fee-bee" for which the species is named

SONG: "Fee blitty? Fee BURR!"   *p. 597*

Mostly Mar.–July; Song is more plastic in fall. Most frequent at dawn but given all day, especially by unmated males, or in territorial altercations. Either phrase may be given repeatedly without the other: "Fee-burr . . . Fee-burr" (p. 590) or "Fee-blitty? Fee-blitty?" (p. 590). Also gives complex Flight Song (p. 611).

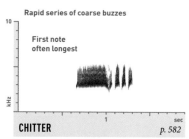

Rapid series of coarse buzzes

First note often longest

CHITTER   *p. 582*

Highly plastic. Given near nest, when displaying potential nest site to mate, and in some aggressive situations.

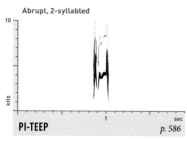

Abrupt, 2-syllabled

PI-TEEP   *p. 586*

Rather stereotyped. Given infrequently, in aggressive encounters and in run-up to Flight Song.

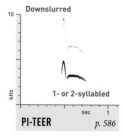

Downslurred

1- or 2-syllabled

PI-TEER   *p. 586*

Probably all year, but infrequent; during or after aggressive encounters.

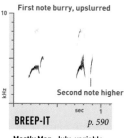

First note burry, upslurred

Second note higher

BREEP-IT   *p. 590*

Mostly Mar.–July; variable. In interactions, usually right after short flights.

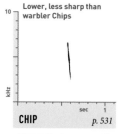

Lower, less sharp than warbler Chips

CHIP   *p. 531*

All year. Most common call; given while foraging, in alarm, or near the nest.

VERMILION FLYCATCHER

Pyrocephalus rubinus

Strikingly plumaged. One of the few flycatchers in which sexes differ. Found in
open areas with low bushes and trees,
often near water.

Single songtype repeated over and over

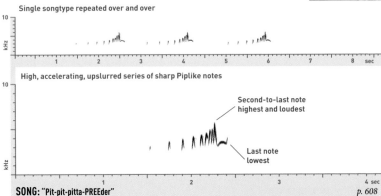

High, accelerating, upslurred series of sharp Piplike notes

Second-to-last note
highest and loudest

Last note
lowest

SONG: "Pit-pit-pitta-PREEder" *p. 608*

Mostly Feb.–July, by male. Quite uniform and stereotyped, but number of introductory Pips can vary. Given
at any time of day, even at night, but especially at dawn. Songs may be repeated almost without pause, or
with single bill snaps between Songs, especially during wing-fluttering flight display.

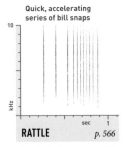

Quick, accelerating
series of bill snaps

RATTLE *p. 566*

By agitated birds. Also brief
wing Flutter (p. 566) in some
flight chases.

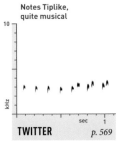

Notes Tiplike,
quite musical

TWITTER *p. 569*

Given in pair interactions,
perhaps courtship. Lower,
softer than Pseep; plastic.

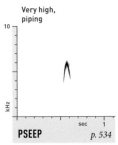

Very high,
piping

PSEEP *p. 534*

All year; most common call.
Somewhat plastic; clear to
slightly burry.

HYBRID PHOEBES

Range expansions by Black and
Eastern Phoebes have brought
the species into contact in
northern New Mexico and
southern Colorado, where they
now sometimes hybridize. First-
generation hybrids are intermediate between the parent species in both plumage and voice.
Some backcross hybrids may be
best detected by voice.

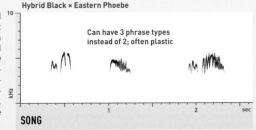

Hybrid Black × Eastern Phoebe

Can have 3 phrase types
instead of 2; often plastic

SONG

DUSKY-CAPPED FLYCATCHER

Myiarchus tuberculifer

Our smallest *Myiarchus* flycatcher. A bird of lower montane riparian woods and oak forest. Little rufous visible in tail, especially from below.

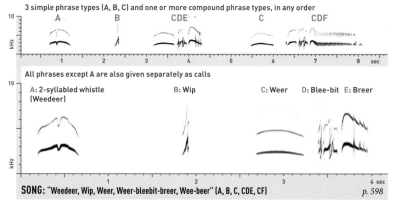

3 simple phrase types (A, B, C) and one or more compound phrase types, in any order

All phrases except A are also given separately as calls

A: 2-syllabled whistle (Weedeer) B: Wip C: Weer D: Blee-bit E: Breer

SONG: "Weedeer, Wip, Weer, Weer-bleebit-breer, Wee-beer" (A, B, C, CDE, CF)   p. 598

Mostly May–July. Given most vigorously at dawn, but sometimes during the day and often at dusk. Most phrases are simple (A, B, C). Compound phrases consist of C, D, E, and F (the "Beer" from Wee-beer), in order, typically in combinations CDE, CDF, or CF (Wee-beer), though D and E can be repeated.

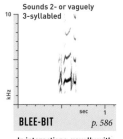

Sounds 2- or vaguely 3-syllabled

BLEE-BIT   p. 586

In interactions, usually with other calls. Complex but not burry.

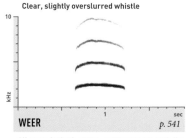

Burry, usually downslurred

Some versions much clearer, like Dawn Song phrase E

BREER   p. 545

In interactions, usually in loose series. Two versions intergrade somewhat.

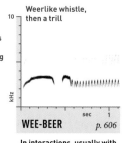

Weerlike whistle, then a trill

WEE-BEER   p. 606

In interactions, usually with other calls, especially Breer. Distinctive.

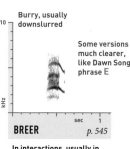

Clear, slightly overslurred whistle

WEER   p. 541

All year, by both sexes; most common and distinctive call, given even during the heat of the day. Unlikely to be confused with other species.

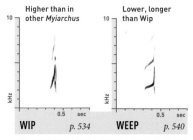

Higher than in other *Myiarchus*

WIP   p. 534

Lower, longer than Wip

WEEP   p. 540

Two versions intergrade somewhat; both are given near the nest and in other situations. Wip can be confused with Whit of Empids.

Ash-throated Flycatcher

Myiarchus cinerascens

Found in dry country, from deserts to pinyon-juniper woodlands to riparian corridors. Paler than other *Myiarchus*, but best identified by voice.

Two phrase types (A and B); A most common by far; barely audible burrs occasionally interspersed

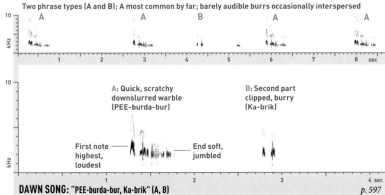

A: Quick, scratchy downslurred warble (PEE-burda-bur)

B: Second part clipped, burry (Ka-brik)

First note highest, loudest

End soft, jumbled

DAWN SONG: "PEE-burda-bur, Ka-brik" (A, B)   p. 597

Mostly Apr.–July, from predawn until just after sunrise; rarely if ever later in the day. Ending of phrase A sometimes doubled ("PEE-burda-burda-bur"). Rather similar to Brown-crested Flycatcher Dawn Song, but higher and scratchier, starting with a Piplike note rather than a whistle.

Like Dawn Song phrase B

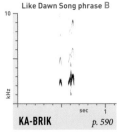

KA-BRIK   p. 590

Mostly Apr.–July. Distinctive. Second note barely longer and higher than first.

Coarse, musical burrs

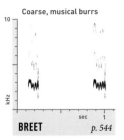

BREET   p. 544

Mostly Apr.–July. Short, monotone, and musical. Often in slow series.

Downslurred at start

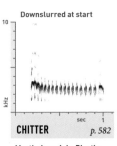

CHITTER   p. 582

Mostly Apr.–July. Plastic; slightly screechy. First and last notes often accented.

First note a Pip

Second note coarsely burry, downslurred or overslurred

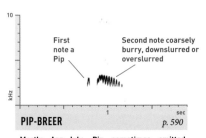

PIP-BREER   p. 590

Mostly Apr.–July. Pip sometimes omitted. Coarser than most similar flycatcher sounds.

Quick, fairly musical

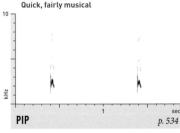

PIP   p. 534

All year. Slightly higher and more abrupt than Brown-crested or Great Crested Wips. Sometimes run into brief, rapid Pip Series (p. 570).

GREAT CRESTED FLYCATCHER

Myiarchus crinitus

A large treetop flycatcher of deciduous woodland edges. Nests in cavities, as do all our other *Myiarchus* species and Sulphur-bellied Flycatcher.

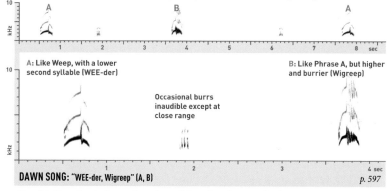

2 phrase types (A and B) usually alternating; barely audible burrs occasionally interspersed

A: Like Weep, with a lower second syllable (WEE-der)

B: Like Phrase A, but higher and burrier (Wigreep)

Occasional burrs inaudible except at close range

DAWN SONG: "WEE-der, Wigreep" (A, B)   p. 597

May–July. Dawn Song in above pattern given only by males prior to sunrise, but phrases A and B, and intergrades between them, also given separately during the day by both sexes, apparently for pair bonding.

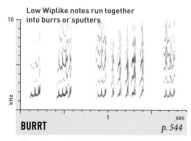

Low Wiplike notes run together into burrs or sputters

BURRT   p. 544

Given frequently May–July. Soft, highly plastic. Lower and shorter than Breet, given by calmer, often solo birds.

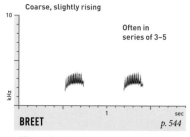

Coarse, slightly rising

Often in series of 3–5

BREET   p. 544

All year, in agitation or alarm. Loud and semi-musical. Can recall Evening Grosbeak Breet.

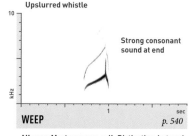

Upslurred whistle

Strong consonant sound at end

WEEP   p. 540

All year. Most common call. Distinctive, but variable; usually from solo birds. Some versions are burry, possibly given by female.

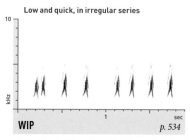

Low and quick, in irregular series

WIP   p. 534

May–July, in altercations. Plastic; can lack final downslur, sounding like Whit. Grades into Weep.

BROWN-CRESTED FLYCATCHER

Myiarchus tyrannulus

A large, vocally conspicuous treetop fly-catcher of riparian woods. Closely related to Great Crested, and very similar in plumage and habits, but voice differs.

Two compound phrases (AC, BC) in any order; endings often doubled (ACC, BCC)

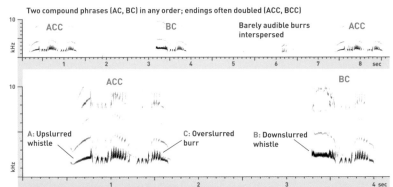

ACC   BC   Barely audible burrs interspersed   ACC

ACC   BC

A: Upslurred whistle   C: Overslurred burr   B: Downslurred whistle

DAWN SONG: "PLEASE put-it-HERE put-it-HERE... DEAR, put-it-HERE" (ACC, BC)   *p. 597*

Mostly Mar.–July, by males prior to sunrise. Uniform and stereotyped. Phrases combine only in patterns AC, ACC, BC, and BCC. Combinations starting with A are generally more common. Near nest, also gives long, soft Twitter (p. 569). In interactions, pairs and small groups chorus with a variety of calls.

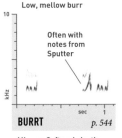

Low, mellow burr

Often with notes from Sputter

BURRT   *p. 544*

All year. Soft and plastic, from calm birds; grades into Sputter and Bree-burr.

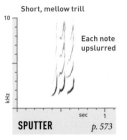

Short, mellow trill

Each note upslurred

SPUTTER   *p. 573*

All year, during interactions. Highly plastic; grades into Burrt and Bree-burr.

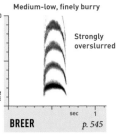

Medium-low, finely burry

Strongly overslurred

BREER   *p. 545*

All year, in agitation or alarm. Loud and frequent.

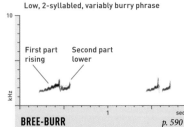

Low, 2-syllabled, variably burry phrase

First part rising   Second part lower

BREE-BURR   *p. 590*

All year, in excitement. Plastic; sometimes a single upslurred Weet, like Great Crested Weep but lower, with subtle burriness.

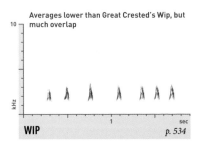

Averages lower than Great Crested's Wip, but much overlap

WIP   *p. 534*

Most common call. All year; from solo birds as well as in encounters. Some versions lack final downslur, sounding like Whit.

Sulphur-bellied Flycatcher

Myiodynastes luteiventris

Visually and vocally conspicuous in a variety of wooded habitats. Range has expanded northward. Diet more diverse than any other flycatcher's.

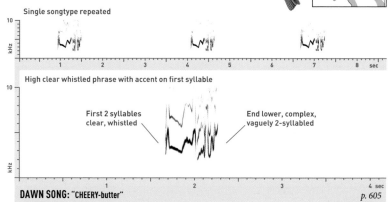

Single songtype repeated

High clear whistled phrase with accent on first syllable

First 2 syllables clear, whistled

End lower, complex, vaguely 2-syllabled

DAWN SONG: "CHEERY-butter" *p. 605*

Mostly May–Aug., reportedly by males, mostly prior to sunrise. Uniform and stereotyped. Lacks squeaky quality of the common daytime vocalizations, but distinctive. Second part ("butter" phrase) occasionally appended to Pip-William or Squeal or given alone during the day.

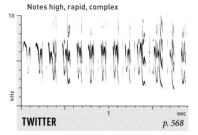

Notes high, rapid, complex

TWITTER *p. 568*

Mostly May–June, in close courtship, often accompanying copulation. Highly plastic; grades into Pseep Series.

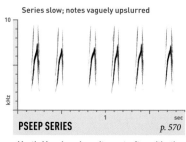

Series slow; notes vaguely upslurred

PSEEP SERIES *p. 570*

Mostly May–Aug., in excitement, often with other calls, especially Squeal. Not well known; function needs more study.

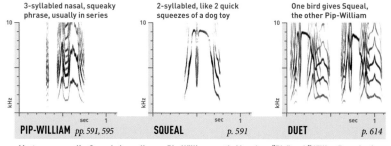

3-syllabled nasal, squeaky phrase, usually in series

PIP-WILLIAM *pp. 591, 595*

2-syllabled, like 2 quick squeezes of a dog toy

SQUEAL *p. 591*

One bird gives Squeal, the other Pip-William

DUET *p. 614*

Most common calls. Squeal given all year; Pip-William mostly May–Aug. "Pip" and "William" rarely given separately; Pip recalls Yap of Gila Woodpecker. Birds frequently give synchronized duets; sometimes several birds chorus. In some duets, both birds give Squeal. Unknown whether both sexes give both calls.

TROPICAL KINGBIRD

Tyrannus melancholicus

Widespread in the American tropics, but local in our area. Nearly identical to Couch's Kingbird, and best distinguished from it by voice.

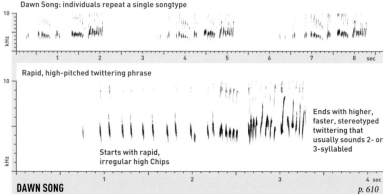

Dawn Song: individuals repeat a single songtype

Rapid, high-pitched twittering phrase

Ends with higher, faster, stereotyped twittering that usually sounds 2- or 3-syllabled

Starts with rapid, irregular high Chips

DAWN SONG   p. 610

Mostly Mar.–July, apparently only by males and only prior to dawn. Somewhat variable but generally stereotyped. Tone semimusical, chittering, like Chimney Swift but with many notes slightly longer and more grating. Differs from Eastern Kingbird Dawn Song in rhythm and lack of clear notes.

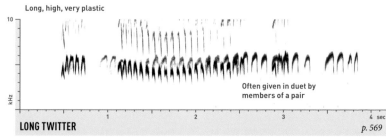

Long, high, very plastic

Often given in duet by members of a pair

LONG TWITTER   p. 569

Nearly all sounds of this species are high Twitters (p. 582). Except for the Dawn Song, all vocalizations tend to be highly plastic and poorly differentiated from one another. Long Twitter tends to be given during pair interactions, often accompanied by fluttering wings.

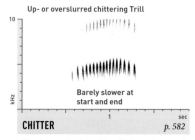

Up- or overslurred chittering Trill

Barely slower at start and end

CHITTER   p. 582

Plastic, grading into Short Twitter. Apparently given in mild alarm and sometimes during greetings or in between foraging flights.

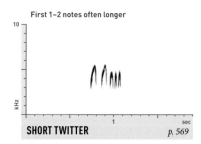

First 1–2 notes often longer

SHORT TWITTER   p. 569

Highly plastic, grading into Long Twitter. Given frequently in many contexts; some versions may function as a type of Day Song.

COUCH'S KINGBIRD

Tyrannus couchii

Vocally conspicuous in thorn scrub, riparian forest, overgrown fields, and other wooded areas. Highly local in the Big Bend region of Texas.

Dawn Song: burry A notes typically alternate with clusters of nasal B and BC phrases

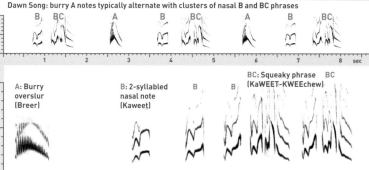

A: Burry overslur (Breer)

B: 2-syllabled nasal note (Kaweet)

BC: Squeaky phrase (KaWEET-KWEEchew)

DAWN SONG: "Breer... kaweet, kaWEET-KWEEchew" (A, B, BC)   *p. 598*

Mostly Apr.–July, reportedly by males, up to an hour after sunrise. Typical pattern: A phrase, then a rising and accelerating series of B phrases, then 1–2 BC phrases. After sunrise, BC phrases become gradually less common, and Pips and other calls are interspersed. BC phrases may be given all day in aggression.

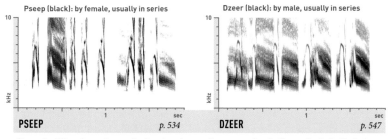

Pseep (black): by female, usually in series

Dzeer (black): by male, usually in series

PSEEP   *p. 534*   **DZEER**   *p. 547*

Two quite different sounds, usually given by pairs greeting one another in disorganized, plastic duets, accompanied by wing fluttering. Male gives downslurred buzzy Dzeer, female gives Pseep; both calls slightly plastic, each with a 2-syllabled version that starts with a harsh burr.

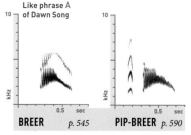

Like phrase A of Dawn Song

BREER   *p. 545*   **PIP-BREER**   *p. 590*

All year, especially by males guarding a territory, often with Pips. Breer finer, more overslurred than Ash-throated Flycatcher's.

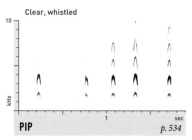

Clear, whistled

PIP   *p. 534*

All year, by both sexes, in interactions and in response to predators. Slightly plastic. No similar call known in Tropical Kingbird.

CASSIN'S KINGBIRD

Tyrannus vociferans

Breeds in open riparian forest, pinyon-juniper woodlands, and semideserts. Darker than Western Kingbird, with contrasting white chin, pale tail tip.

2 phrases (A, B), usually alternated, though A and its components are often repeated

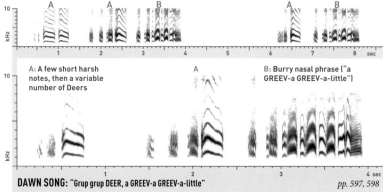

A: A few short harsh notes, then a variable number of Deers

A

B: Burry nasal phrase ("a GREEV-a GREEV-a-little")

DAWN SONG: "Grup grup DEER, a GREEV-a GREEV-a-little"   *pp. 597, 598*

Mostly Apr.–July, apparently only by males, mostly prior to dawn, but also sometimes given during the day or at night. In tumbling flight display, rises from perch to repeat abbreviated B phrase "a-GREEVurrr" phrase 3–6 times with loud Wing Whirr over last syllable of each phrase.

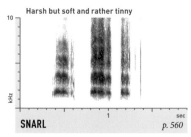

Harsh but soft and rather tinny

SNARL   *p. 560*

Mostly Apr.–July, in agitation and aggression. Highly plastic; grades into other calls.

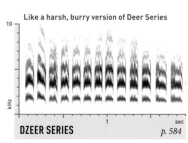

Like a harsh, burry version of Deer Series

DZEER SERIES   *p. 584*

Mostly Apr.–July, in agitation and in the lead-up to a flight display. Highly plastic.

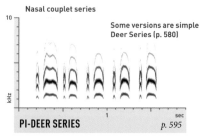

Nasal couplet series

Some versions are simple Deer Series (p. 580)

PI-DEER SERIES   *p. 595*

All year, in pair and territorial interactions. Males usually give Pi-deer Series, females Deer Series, often in unsynchronized duet. Plastic.

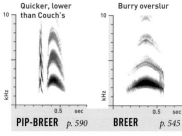

Quicker, lower than Couch's

Burry overslur

PIP-BREER *p. 590*   **BREER** *p. 545*

All year; most common and distinctive call, but compare similar sounds of other flycatchers, especially Couch's Kingbird.

THICK-BILLED KINGBIRD

Tyrannus crassirostris

A primarily Mexican species, local and uncommon in summer in riparian woods. Famously loud, with distinctive harsh, high-pitched voice.

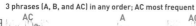
3 phrases (A, B, and AC) in any order; AC most frequent

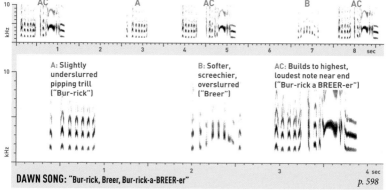

A: Slightly underslurred pipping trill ("Bur-rick")

B: Softer, screechier, overslurred ("Breer")

AC: Builds to highest, loudest note near end ("Bur-rick a BREER-er")

DAWN SONG: "Bur-rick, Breer, Bur-rick-a-BREER-er"   p. 598

Mostly Mar.–July, possibly only by males, almost exclusively prior to dawn. Not well known. Quite different from daytime vocalizations and easily misidentified; compare especially Dawn Song of Brown-crested Flycatcher.

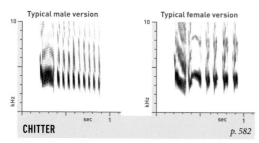

Typical male version

Typical female version

Notes downslurred, nasal

CHITTER   p. 582

DEER SERIES   pp. 580, 612

All year, in pair interactions, often in unsynchronized duet. Loud and frequent, with distinctive quality like arcing of high-voltage electricity. Quite plastic; female version typically shorter and slower.

Near nest, mostly by female. Extremely plastic, often incorporating Chitter notes.

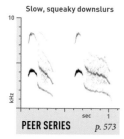
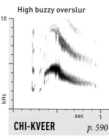
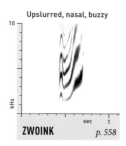

Slow, squeaky downslurs

High buzzy overslur

Upslurred, nasal, buzzy

PEER SERIES   p. 573

CHI-KVEER   p. 590

ZWOINK   p. 558

Poorly known; in the sole available example, Zwoinks regularly interspersed.

Mostly May–Aug. Kveer portion often appended to end of Zwoink.

All year, in contact and other situations; most common call. Highly distinctive.

WESTERN KINGBIRD

Tyrannus verticalis

Conspicuous, pugnacious, and vocal, common in a variety of open habitats. Range has expanded eastward. Note white outer tail feathers.

Single songtype repeated, but 1–3 Deer notes sometimes replace final phrase

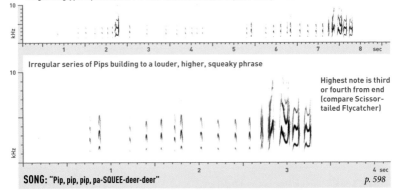

Irregular series of Pips building to a louder, higher, squeaky phrase

Highest note is third or fourth from end (compare Scissor-tailed Flycatcher)

SONG: "Pip, pip, pip, pa-SQUEE-deer-deer" *p. 598*

Mostly Apr.–July, reportedly only by males. Version shown above is given persistently prior to dawn, occasionally at other times during the day. In tumbling flight display, rises from perch to repeat "pa-SQUEE-deer-deer" phrase 3–6 times with 4–5 loud Wing Whirrs in between.

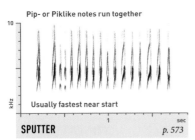

Pip- or Piklike notes run together

Usually fastest near start

SPUTTER *p. 573*

Apr.–Aug. Slightly plastic; used by both sexes in pair greetings and by males patrolling territory; female version often shorter.

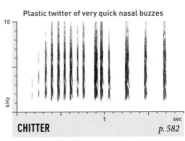

Plastic twitter of very quick nasal buzzes

CHITTER *p. 582*

All year, in frequent aggressive encounters. Highly plastic. Usually in conjunction with other calls.

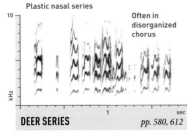

Plastic nasal series

Often in disorganized chorus

DEER SERIES *pp. 580, 612*

Mostly Apr.–Aug. Extremely plastic; from agitated birds in interactions. Often in conjunction with Chitter.

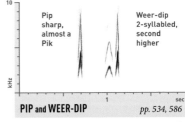

Pip sharp, almost a Pik

Weer-dip 2-syllabled, second higher

PIP and WEER-DIP *pp. 534, 586*

Apr.–Aug. These two calls also given separately, but often in above pattern. Pip is like individual notes from Sputter.

EASTERN KINGBIRD

Tyrannus tyrannus

Common and conspicuous in a variety of open and semi-open habitats. Well known for aggressively defending territory against all other birds.

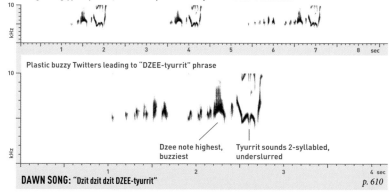

Single songtype repeated; introductory Twitters vary from absent to extensive

Plastic buzzy Twitters leading to "DZEE-tyurrit" phrase

Dzee note highest, buzziest

Tyurrit sounds 2-syllabled, underslurred

DAWN SONG: "Dzit dzit dzit DZEE-tyurrit" p. 610

Mostly May–July. Given persistently in hour before dawn; almost never heard during the day. Unlike other kingbird Dawn Songs, not used in tumbling flight display, though it may be given on the wing. Introductory Twitters become more prolonged as singing bout continues.

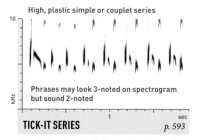

High, plastic simple or couplet series

Phrases may look 3-noted on spectrogram but sound 2-noted

TICK-IT SERIES sec p. 593

Mostly Apr.–Aug., by both sexes in pair greetings and by males patrolling territory; often with wing flutter. Female version often shorter.

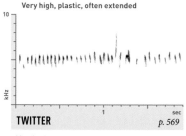

Very high, plastic, often extended

TWITTER sec p. 569

Mostly Apr.–Aug., by female during nest construction and by male patrolling territory. Male versions usually end in Dzeer.

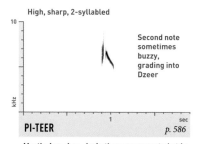

High, sharp, 2-syllabled

Second note sometimes buzzy, grading into Dzeer

PI-TEER sec p. 586

Mostly Apr.–Aug., by both sexes; repeated at intervals in agitation or aggression.

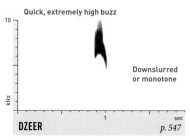

Quick, extremely high buzz

Downslurred or monotone

DZEER sec p. 547

Apr.–Aug., mostly by males. Most common call, often given in aggression. Plastic; grades smoothly into Pi-teer.

SCISSOR-TAILED FLYCATCHER

Tyrannus forficatus

Beautiful, elegant, and conspicuous in summer in grasslands and savannas with scattered shrubs or trees. Eats mostly grasshoppers and beetles.

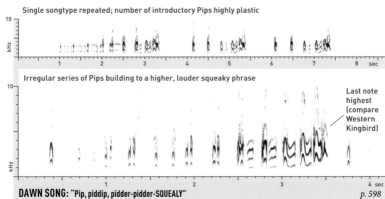

Single songtype repeated; number of introductory Pips highly plastic

Irregular series of Pips building to a higher, louder squeaky phrase

Last note highest (compare Western Kingbird)

DAWN SONG: "Pip, piddip, pidder-pidder-SQUEALY" *p. 598*

Mostly Apr.–July, reportedly only by males. Version shown above is given persistently prior to dawn, occasionally at other times during the day. In tumbling flight display, rises from perch to repeat Pidder-pidder-SQUEALY phrase 3–6 times with 4–5 loud Wing Whirrs in between.

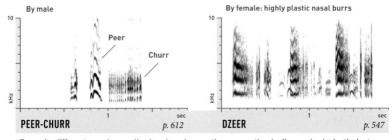

By male

Peer

Churr

By female: highly plastic nasal burrs

PEER-CHURR *p. 612*

DZEER *p. 547*

Two quite different sounds, usually given by pairs greeting one another in disorganized, plastic duets, accompanied by wing fluttering. Male gives Peer and Churr, together or separately; female gives Dzeer notes, each usually followed by a softer snarling burst. All calls quite plastic.

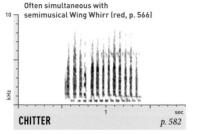

Often simultaneous with semimusical Wing Whirr (red, p. 566)

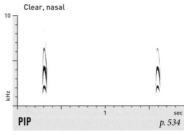

Clear, nasal

CHITTER *p. 582*

PIP *p. 534*

Mostly by males, often in Wing-Whirring flight display; females near the nest give slower, clearer versions, more like Pips in series.

Rather plastic. Frequent in many contexts; higher and sharper in alarm. Averages lower than in Western Kingbird, but much overlap.

WILLOW FLYCATCHER (SOUTHWESTERN)

Empidonax traillii extimus

This endangered subspecies of the Willow Flycatcher (p. 267) breeds in willows and deciduous woods along streams in the desert Southwest. Song shows slight but consistent differences from other populations.

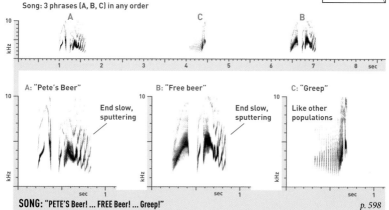

Song: 3 phrases (A, B, C) in any order

A

C

B

A: "Pete's Beer"

End slow, sputtering

B: "Free beer"

End slow, sputtering

C: "Greep"

Like other populations

SONG: "PETE'S Beer! ... FREE Beer! ... Greep!"

p. 598

Song like that of other Willow Flycatcher populations, except that final "Beer" note in song phrases A and B slows down audibly, ending in a coarse sputter with a rather distinctive quality. Song phrases also average slightly longer. No other vocal differences known.

SHRIKES (Family Laniidae)

next pages

The shrike family is primarily an Old World group. Only two species are found in North America. Both are almost entirely carnivorous, eating small birds and mammals as well as large insects such as grasshoppers. Shrikes are also known as "butcher birds" because of their habit of impaling prey on thorns or barbed wire fences, for reasons that are not entirely understood. Despite their hooked bills and carnivorous diet, shrikes are true songbirds, with learned and highly complex songs. In the two North American species, it is often difficult to distinguish between calls and songs; vocal communication in these birds needs more study.

VIREOS (Family Vireonidae)

pp. 294–302

Superficially similar to warblers, vireos have thicker, hook-tipped bills. They hunt caterpillars and other insects among leaves and branches with deliberate movements, pausing frequently. Most are subtly plumaged in shades of olive or gray.

The songs of vireos are complex and learned, usually consisting of short musical phrases with a burry or squeaky quality. Many species give longer, more complex phrases in close courtship. One eastern North American species, the White-eyed Vireo, regularly incorporates imitations of other bird species into its primary Song.

Call repertoires tend to be extensive, often including Whines, Chatters, Jits, and short musical Trills. All species snap their bills when highly agitated. Vireos are not known to call during nocturnal migration. However, many species sing regularly in fall, when song of other birds is infrequent.

LOGGERHEAD SHRIKE

Lanius ludovicianus

Widespread in open habitats, but many populations have declined. Not often found in the same region at the same season as Northern Shrike.

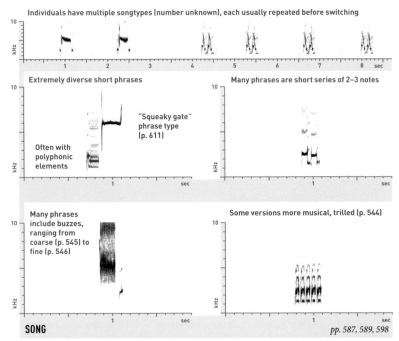

Individuals have multiple songtypes (number unknown), each usually repeated before switching

Extremely diverse short phrases

Many phrases are short series of 2–3 notes

Often with polyphonic elements

"Squeaky gate" phrase type (p. 611)

Many phrases include buzzes, ranging from coarse (p. 545) to fine (p. 546)

Some versions more musical, trilled (p. 544)

SONG   *pp. 587, 589, 598*

All year, by both sexes, but especially Feb.–June by male. Generally soft. Exceedingly variable, but usually stereotyped. Mostly consists of short, complex, rather musical phrases, often containing a brief trill or buzz, rarely totaling more than 2 syllables. Whines frequently mixed with Song; songs with more Whines reportedly serve a more territorial function. Immatures reported to give a continuous quiet Song, likely a form of subsong.

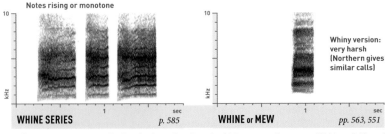

Notes rising or monotone

Whiny version: very harsh (Northern gives similar calls)

WHINE SERIES   *p. 585*   **WHINE or MEW**   *pp. 563, 551*

All year; most common call, given in alarm, but also mixed into song performances. Highly variable and plastic, ranging from metallic and buzzy to harsh and noisy. Different versions may serve different functions. Similar calls used in begging by both juveniles and adults.

NORTHERN SHRIKE

Lanius borealis

Slightly larger than Loggerhead Shrike, with thinner black face mask. Breeds in the Arctic, visiting our area only in winter. Uncommon in open habitats.

Individuals have multiple songtypes (number unknown), each usually repeated before switching

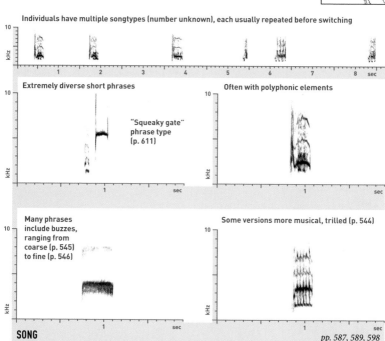

Extremely diverse short phrases

"Squeaky gate" phrase type (p. 611)

Often with polyphonic elements

Many phrases include buzzes, ranging from coarse (p. 545) to fine (p. 546)

Some versions more musical, trilled (p. 544)

SONG　　　　　　　　　　　　　　　　　*pp. 587, 589, 598*

All year, by both sexes, but especially Feb.–June by male. Generally soft. Exceedingly variable, but usually stereotyped. Mostly consists of short, complex, rather musical phrases, often containing a brief trill or buzz, rarely totaling more than 2 syllables. Whines frequently mixed with Song. More likely than Loggerhead to sing fast, complex Song in which consecutive phrases are different and separated by only short pauses, but most of the time, the two species are difficult to distinguish by voice.

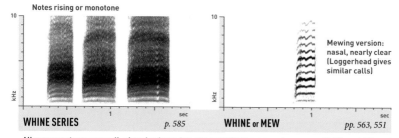

Notes rising or monotone

Mewing version: nasal, nearly clear (Loggerhead gives similar calls)

WHINE SERIES　　*p. 585*　　　　**WHINE or MEW**　　*pp. 563, 551*

All year; most common call, given in alarm, but also mixed into song performances. Highly variable and plastic, ranging from metallic and buzzy to harsh and noisy. Some versions recall Black-billed Magpie. Different versions may serve different functions. Similar calls used in begging by both juveniles and adults.

BLACK-CAPPED VIREO

Vireo atricapilla

Rare, local breeder in dense deciduous shrublands. Populations have seriously declined, primarily as a result of over-grazing and fire suppression.

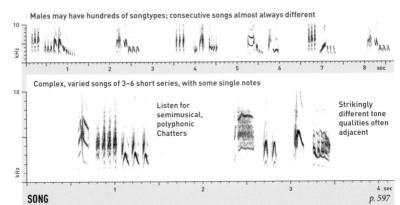

Males may have hundreds of songtypes; consecutive songs almost always different

Complex, varied songs of 3–6 short series, with some single notes

Listen for semimusical, polyphonic Chatters

Strikingly different tone qualities often adjacent

SONG *p. 597*

Distinctive. All year, but especially Mar.–July. Females not known to sing. At any given time, 3–10 different songtypes may be given in any order, including alternating 2 songtypes or repeating 1 for a while, before gradually switching to a different set of 3–10 songtypes. A few notes may be imitations.

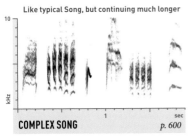

Like typical Song, but continuing much longer

COMPLEX SONG *p. 600*

Possibly all year. Reportedly only by male, in close courtship and territorial conflicts.

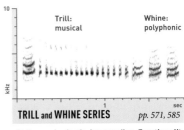

Trill: musical

Whine: polyphonic

TRILL and WHINE SERIES *pp. 571, 585*

Both sounds plastic, intergrading. Functions little known; apparently during pair interactions, often with Song.

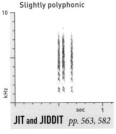

Slightly polyphonic

JIT and JIDDIT *pp. 563, 582*

All year. Plastic, but mostly 2-noted. Very like Ruby-crowned Kinglet Jiddit.

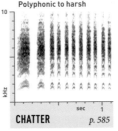

Polyphonic to harsh

CHATTER *p. 585*

All year, in alarm. Quite plastic; grades into Jiddit and Rasp.

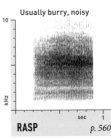

Usually burry, noisy

RASP *p. 560*

All year, in high alarm. Plastic, from polyphonic and burry to very harsh.

BELL'S VIREO

Vireo bellii

Local in dense shrubby habitats, often near water, from riparian woods in the Great Plains and California to mesquite bosques in southwestern deserts.

Males average 10 songtypes; 2 or 3 alternated many times before switching to another set

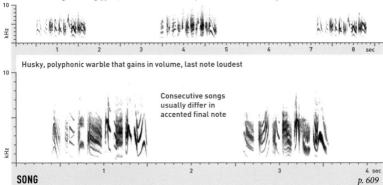

Husky, polyphonic warble that gains in volume, last note loudest

Consecutive songs usually differ in accented final note

SONG *p. 609*

All year, but especially Apr.–Aug. Females sing occasionally. Alternating pattern of 2 songtypes sometimes suggests a "question-and-answer" pattern, with one songtype ending on a rising note, the other on a falling note. Mated males often intersperse Chatters of Jitlike notes between Songs.

Rapid soft warble with many polyphonic notes and squeaks; no pauses or repetitions

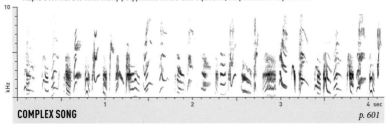

COMPLEX SONG *p. 601*

Reportedly only by males, mostly during breeding season, in a number of contexts, including close courtship and the aftermath of territorial disputes. Quality much like typical Song. Bell's apparently lacks an equivalent of the Trill call of other vireos.

Brief, usually polyphonic	Plastic, polyphonic	Usually in series
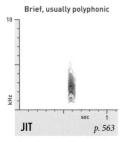	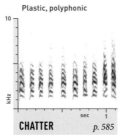	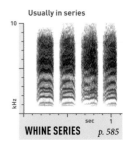
JIT *p. 563*	**CHATTER** *p. 585*	**WHINE SERIES** *p. 585*
In quiet contact between mated pairs, or with Song. Variable, plastic.	Infrequent, in alarm or agitation; grades into Whines.	All year, in alarm. Varies from polyphonic to rasping; usually in between.

GRAY VIREO

Vireo vicinior

A nondescript resident of very arid open woodlands and brushland, often in areas dominated by junipers. Compare Plumbeous Vireo.

Males have 20–50 songtypes; 4 or 5 given in the same order many times before switching to a new set

Short musical 2- to 3-syllabled phrases; consecutive phrases usually different

Most phrases clear, but some partly burry

Averages faster than similar vireo Songs, up to 1 phrase/second

SONG　　　　　　　　　　　　　　*p. 589*

All year, but especially Apr.–Aug. Females sing infrequently, generally only a few phrases at a time. Phrases generally shorter than in other vireos. Speed of Song varies with excitement level. Birds reportedly sometimes give the same songtype over and over for long periods, like Hutton's Vireo.

Faster than typical Song, with soft rising Squeaks between phrases

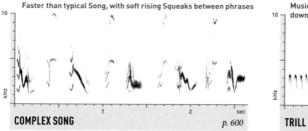

Musical, often downslurred

Very similar to trills of antelope ground squirrels

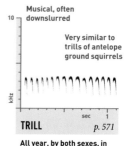

COMPLEX SONG　　　　　　*p. 600*

All year, reportedly only by male, in close courtship, after territorial conflicts, and in response to predators.

TRILL　　　　*p. 571*

All year, by both sexes, in many contexts involving pair contact. Quite plastic.

Soft, quick, noisy

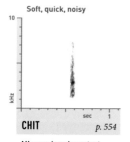

Noisy, polyphonic

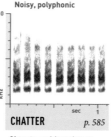

Long, harsh version

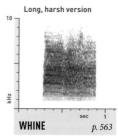

CHIT　　　*p. 554*

All year, in pair contact, especially near the nest. Does not carry far.

CHATTER　　　*p. 585*

Given to scold predators, often near nest. Variable, plastic; grades into Whine.

WHINE　　　*p. 563*

Infrequent, in alarm; varies from House Finch–like Zree to harsh rasp.

Hutton's Vireo

Vireo huttoni

Resident in mature mixed forests, usually near evergreen oaks. Beware the lookalike Ruby-crowned Kinglet, which is more active, with different voice.

Males typically sing 1 songtype for long periods before switching; repertoire size unknown

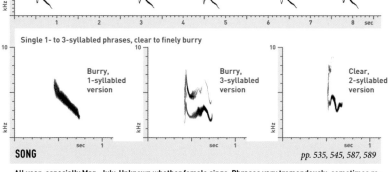

Single 1- to 3-syllabled phrases, clear to finely burry

Burry, 1-syllabled version

Burry, 3-syllabled version

Clear, 2-syllabled version

SONG   pp. 535, 545, 587, 589

All year, especially Mar.–July. Unknown whether female sings. Phrases vary tremendously, sometimes recalling House Sparrow Song or calls of finches; incessant repetition of the same sound can be an excellent field mark, but singing occasionally deviates from this pattern.

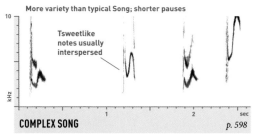

More variety than typical Song; shorter pauses

Tsweetlike notes usually interspersed

COMPLEX SONG   p. 598

In territorial encounters; possibly in close courtship. In some versions, males cycle through songtypes like other vireo species.

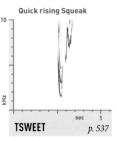

Quick rising Squeak

TSWEET   p. 537

Plastic and variable; perhaps given in aggression.

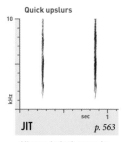

Quick upslurs

JIT   p. 563

All year, by both sexes, in contact. Variable; some versions rather noisy.

Notes rising, nasal, finely burry

LAUGH   p. 577

Distinctive; often slightly whiny. First note usually longest.

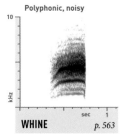

Polyphonic, noisy

WHINE   p. 563

All year, in high alarm. Plastic; varies from fairly clear to quite harsh.

Cassin's Vireo

Vireo cassinii

Nests mostly in dry deciduous and mixed forests. Formerly lumped with Plumbeous Vireo and the Blue-headed Vireo of the East as "Solitary Vireo."

Males have about 6–15 songtypes; consecutive songs usually different

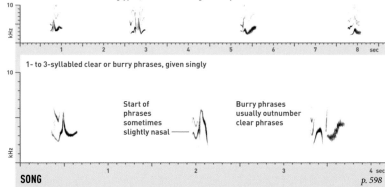

1- to 3-syllabled clear or bury phrases, given singly

Start of phrases sometimes slightly nasal —

Burry phrases usually outnumber clear phrases

SONG   *p. 598*

All year, especially May–July; some singing in fall. Female not known to sing. As in Plumbeous Vireo, percentage of phrases that are burry is highly geographically variable. Only consistent difference from Song of Plumbeous is that Cassin's averages slightly higher in pitch; the difference is subtle.

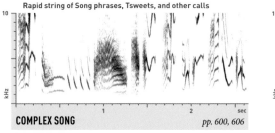

Rapid string of Song phrases, Tsweets, and other calls

COMPLEX SONG   *pp. 600, 606*

Reportedly given by males in close courtship, as well as mated males that have lost contact with mate. Rather soft.

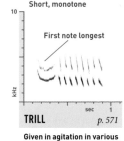

Short, monotone

First note longest

TRILL   *p. 571*

Given in agitation in various contexts; volume varies with excitement level.

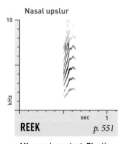

Nasal upslur

REEK   *p. 551*

All year, in contact. Plastic; rough versions exist.

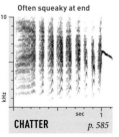

Often squeaky at end

CHATTER   *p. 585*

First note usually longest. Plastic; long harsh notes can be given singly.

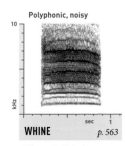

Polyphonic, noisy

WHINE   *p. 563*

All year, in high alarm. Plastic; varies from fairly clear to quite harsh.

PLUMBEOUS VIREO

Vireo plumbeus

Nests in pine, pinyon-juniper, and mixed forests. Plumage grayish; the dullest Cassin's are similar but still have greenish tinge in wings.

Males probably have about 10–20 songtypes; consecutive songs usually different

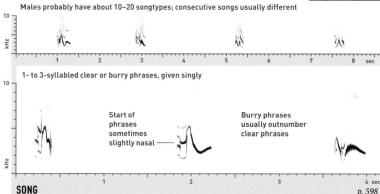

1- to 3-syllabled clear or burry phrases, given singly

Start of phrases sometimes slightly nasal —

Burry phrases usually outnumber clear phrases

SONG p. 598

All year, especially May–July; some singing in fall. Female not known to sing. Much like Gray Vireo Song, but usually with longer pauses between phrases (averaging 1 song phrase every 2.5 seconds); rate varies somewhat with excitement level.

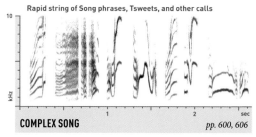

Rapid string of Song phrases, Tsweets, and other calls

COMPLEX SONG pp. 600, 606

Reportedly given by males in close courtship, as well as in response to cowbirds or predators near the nest. Rather soft.

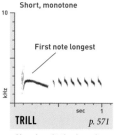

Short, monotone

First note longest

TRILL p. 571

Given in agitation in various contexts; volume varies with excitement level.

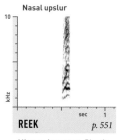

Nasal upslur

REEK p. 551

All year, in contact. Plastic; rough versions exist.

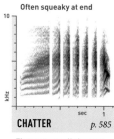

Often squeaky at end

CHATTER p. 585

First note usually longest. Plastic; long harsh notes can be given singly.

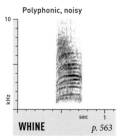

Polyphonic, noisy

WHINE p. 563

All year, in high alarm. Plastic; varies from fairly clear to quite harsh.

WARBLING VIREO (WESTERN)

Vireo gilvus [*swainsoni* group]

Breeds in deciduous woods, especially along streams. Nearly identical to the eastern subspecies, but differs somewhat in Song.

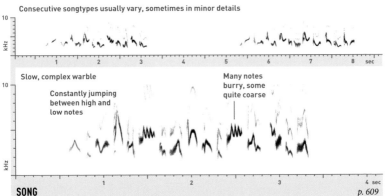

Consecutive songtypes usually vary, sometimes in minor details

Slow, complex warble

Constantly jumping between high and low notes

Many notes burry, some quite coarse

SONG *p. 609*

At least Mar.–Sept., by males; females also reported to sing. Song can become quite long (10 seconds or more) when male is excited, but most Songs are 2–4 seconds in length. Song less musical than Eastern, with more high squeaky notes throughout, like a high-pitched Blue Grosbeak. Repertoire size unknown.

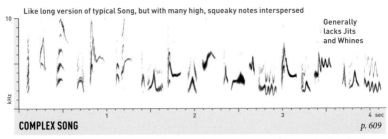

Like long version of typical Song, but with many high, squeaky notes interspersed

Generally lacks Jits and Whines

COMPLEX SONG *p. 609*

Given by males in close courtship and during aggressive encounters; more frequent than Complex Song of eastern subspecies. Both sexes occasionally give Jitlike and squeaky notes in agitation without Song. Females give rapid series of short whining notes in close courtship, somewhat like Trill calls of other vireos.

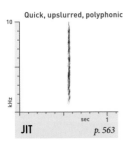

Quick, upslurred, polyphonic

JIT *p. 563*

By both sexes, in contact, in flight, and in mild alarm.

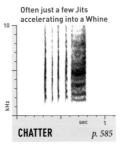

Often just a few Jits accelerating into a Whine

CHATTER *p. 585*

By both sexes in agitation. Plastic; some versions are longer, harsher.

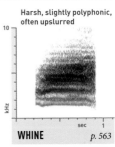

Harsh, slightly polyphonic, often upslurred

WHINE *p. 563*

All year, in alarm; most common call. Plastic, but typically quite harsh.

WARBLING VIREO (EASTERN)

Vireo gilvus [*gilvus* group]

Breeds in mature deciduous woods; also in parks and residential areas. Yellowish coloring below, when present, brightest on sides and flanks.

Consecutive songtypes usually vary, sometimes in minor details

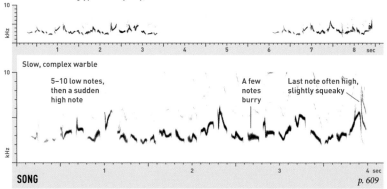

Slow, complex warble

5–10 low notes, then a sudden high note

A few notes burry

Last note often high, slightly squeaky

SONG *p. 609*

At least Mar.–Sept., by males; females also reported to sing. Song can become quite long (10 seconds or more) when male is excited, but most Songs are 2–4 seconds in length. Distinctive pitch pattern: pitch of entire Song rises and falls 1–2 times per second. Repertoire size unknown.

Numerous Jits, squeaky notes, and whiny notes before and during Song

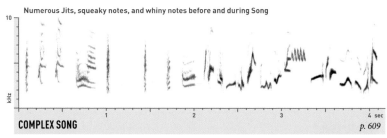

COMPLEX SONG *p. 609*

Given by males in close courtship and during aggressive encounters. Both sexes occasionally give Jitlike and squeaky notes in agitation without Song. Females give rapid series of short whining notes in close courtship, somewhat like Trill calls of other vireos.

Quick, upslurred, polyphonic

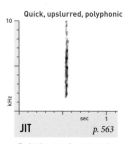

JIT *p. 563*

By both sexes, in contact, in flight, and in mild alarm.

Often just a few Jits accelerating into a Whine

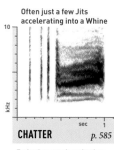

CHATTER *p. 585*

By both sexes in agitation. Plastic; some versions are longer, harsher.

Harsh, slightly polyphonic, often upslurred

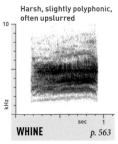

WHINE *p. 563*

All year, in alarm; most common call. Plastic, but typically quite harsh.

Red-eyed Vireo

Vireo olivaceus

One of the most abundant birds in eastern deciduous forests; less common in the West in deciduous riparian areas. Known for singing incessantly.

Individuals have 40 different songtypes on average; consecutive songs always differ

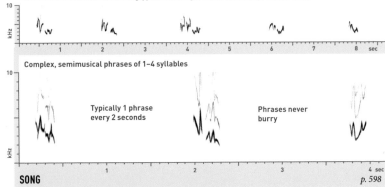

Complex, semimusical phrases of 1–4 syllables

Typically 1 phrase every 2 seconds

Phrases never burry

SONG   *p. 598*

Mostly Apr.–Aug., by males. Females not known to sing, and no song reported during fall migration. Song often given for long periods without a break, even on hot afternoons. Short intervals between song phrases and lack of burry quality distinctive in the West.

Like typical Song, but softer and faster, with high Psits and other notes interspersed

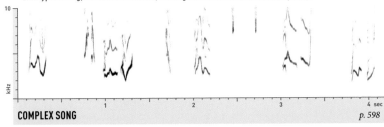

COMPLEX SONG   *p. 598*

Infrequent, mostly Apr.–Aug., likely by males in close courtship and during aggressive encounters. Not well known; even less frequent than other vireo Complex Songs. Does not carry far. Apparently grades into typical Song.

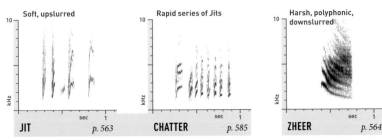

Soft, upslurred

Rapid series of Jits

Harsh, polyphonic, downslurred

JIT   *p. 563*   **CHATTER**   *p. 585*   **ZHEER**   *p. 564*

Mostly Apr.–June, in agitation and aggression. Highly plastic. Also gives a short, musical, overslurred Trill, perhaps a version of Chatter.

All year, in alarm. Most common call. Distinctive. Some versions monotone.

CROWS, RAVENS, JAYS, AND MAGPIES (Family Corvidae)

Members of this family, known as "corvids," are often considered to be among the most intelligent of all bird species, along with parrots. The crows and ravens are large and black; the jays and magpies are medium-sized, often with striking plumage. Some corvids cache enormous amounts of food, especially nuts and acorns, for retrieval during the colder months. Many species aggressively mob predators in noisy groups, sometimes making it easy to locate hawks and owls.

Although they belong to the oscine passerines, the group commonly known as "songbirds," the corvids rarely give "songs" in the traditional sense (though see p. 315). Many of their sounds appear harsh and simple to the human ear, but their vocal communication is apparently highly complex, as is their social behavior. Most corvid sounds likely have a learned component. At least in the Blue Jay, flockmates have been shown to match call types.

Rattle Calls of Female Corvids

All North American species of crows, ravens, and jays have a distinctive Rattle call that consists of a series of clicks, or sometimes two consecutive series of clicks, often delivered with a rapid up-and-down bobbing of the body. In the species that have been closely studied, these Rattle calls have been reported only from females. In ravens, they have been reported only from the dominant females in a group.

Rattles are apparently given in aggression toward other members of the species, in courtship, and sometimes in response to predators. They tend to be fairly quiet and are likely learned; in most species they vary geographically. No equivalent call has been reported in Black-billed or Yellow-billed Magpies.

BLACK-THROATED MAGPIE-JAY

Calocitta colliei

A spectacular and very loud species, native to west Mexico. A small feral population inhabits the Tijuana River valley near San Diego, California.

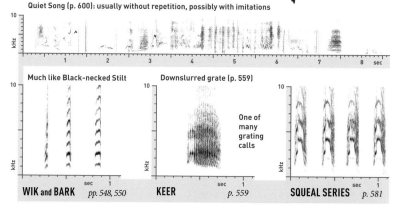

Quiet Song (p. 600): usually without repetition, possibly with imitations

Much like Black-necked Stilt

Downslurred grate (p. 559)

One of many grating calls

WIK and BARK   *pp. 548, 550*   KEER   *p. 559*   SQUEAL SERIES   *p. 581*

Vocal repertoire in native Mexico is exceedingly varied; most calls brief and loud, grating or squeaky. U.S. birds may give fewer calls, but Wiks, Barks, and Keers are frequent. Not known whether U.S. birds give Squeal Series; if so, beware confusion with lower, harsher California Scrub-Jay Zreek Series.

GRAY JAY

Perisoreus canadensis

Resident in spruce forests. Often bold and curious around people. Rather quiet, but vocal repertoire is very complex. Also called Canada Jay.

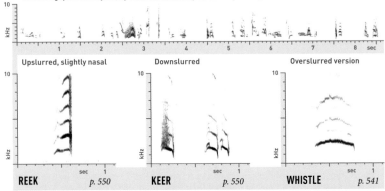

Quiet Song (p. 600): all year, by both sexes; usually without repetition, often with imitations

Upslurred, slightly nasal	Downslurred	Overslurred version
REEK p. 550	**KEER** p. 550	**WHISTLE** p. 541

A group of exceedingly variable sounds that seem to intergrade. Different versions likely serve different functions. Individuals may have several different Whistle types; one type usually given repeatedly, either singly or in series, before switching to another type. Some recall the Wail Series of Gray Hawk (p. 575).

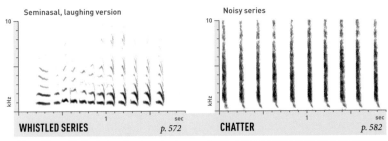

Seminasal, laughing version

Noisy series

WHISTLED SERIES p. 572 **CHATTER** p. 582

Many different kinds of notes given in series, in many situations; the 2 versions shown are regularly heard. Whistled Series may vary geographically. Chatters may be used in mobbing of predators, but are also given in other contexts.

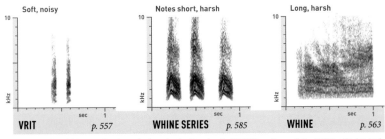

Soft, noisy	Notes short, harsh	Long, harsh
VRIT p. 557	**WHINE SERIES** p. 585	**WHINE** p. 563

A group of variable sounds, given singly or in short series. Soft Vritlike notes and short Whine Series typical in close contact. Long Whines typical of begging juveniles, but also given by adults in courtship, and occasionally directed at humans.

STELLER'S JAY

Cyanocitta stelleri

Vocally and visually conspicuous in coniferous and mixed forests. Vocal repertoire, like Blue Jay's, is enormous;
only most common sounds are shown.

Quiet Song (p. 600): all year, by both sexes; usually with little repetition, often with imitations

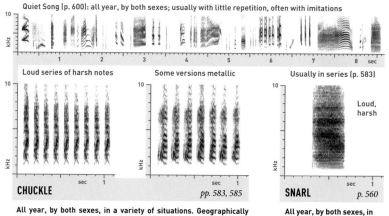

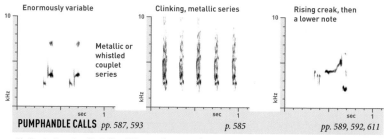

CHUCKLE   Loud series of harsh notes

Some versions metallic

Usually in series (p. 583)

Loud, harsh

SNARL   *p. 560*

All year, by both sexes, in a variety of situations. Geographically variable; West Coast versions may average more metallic and less harsh. Length usually 1–1.5 seconds.

All year, by both sexes, in many situations. Often upslurred or overslurred.

PUMPHANDLE CALLS   *pp. 587, 593*

Enormously variable

Metallic or whistled couplet series

Clinking, metallic series   *p. 585*

Rising creak, then a lower note   *pp. 589, 592, 611*

Similar to Pumphandle Calls of Blue Jay (next page). Repertoire size unknown; one version usually repeated multiple times before switching. Given in a wide variety of situations, including mild alarm. Most
versions are metallic simple or couplet series, or 2-syllabled phrases. Also imitates hawks (see p. 307).

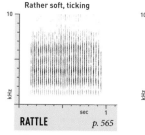

RATTLE   Rather soft, ticking   *p. 565*

All year, by females, in both
courtship and aggression.
Often 2-parted.

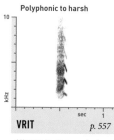

VRIT   Polyphonic to harsh   *p. 557*

All year, in close contact.
Variable, ranging from clear
and metallic to snarling.

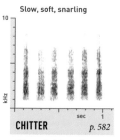

CHITTER   Slow, soft, snarling   *p. 582*

All year, in interactions.
Plastic; sometimes faster
than shown.

BLUE JAY

Cyanocitta cristata

Common and vocal in woodlands, forest edges, and residential areas. Readily visits feeders. Often moves in loose groups, especially in fall and winter.

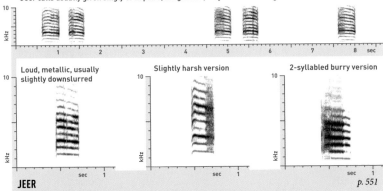

Jeer calls usually given singly or in pairs; in agitation, may be run into longer series

Loud, metallic, usually slightly downslurred

Slightly harsh version

2-syllabled burry version

JEER
p. 551

All year; most common call. Highly variable; individuals may have repertoires of multiple versions, serving slightly different functions. Flockmates apparently match call types. Jeers become harsher in alarm; 2-syllabled versions are common. Young birds beg with hoarse, highly plastic version.

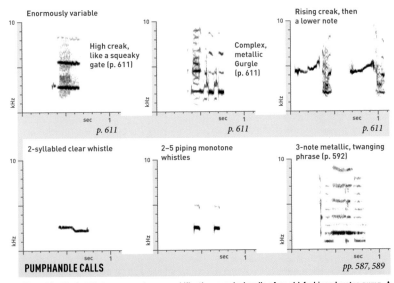

Enormously variable

High creak, like a squeaky gate (p. 611)

Complex, metallic Gurgle (p. 611)

Rising creak, then a lower note

p. 611

p. 611

p. 611

2-syllabled clear whistle

2–5 piping monotone whistles

3-note metallic, twanging phrase (p. 592)

PUMPHANDLE CALLS
pp. 587, 589

Named for the fact that some versions sound like the squeaky handle of an old-fashioned water pump. A hugely variable catch-all category of learned sounds. Individuals have multiple versions, but repertoire size unknown; 1 version usually repeated multiple times before switching. Given in a wide variety of situations, including mild alarm. Most versions are 2- to 3-syllabled, metallic, often slightly gurgling. May grade into the 2-syllable versions of the Jeer call.

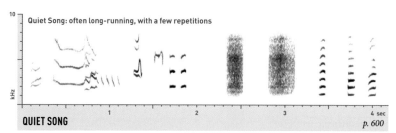

QUIET SONG   *p. 600*

Quiet Song: often long-running, with a few repetitions

Likely all year, by adults and older juveniles. May consist of whistled, whiny, nasal, and/or Jeerlike notes; sometimes contains a few imitations of other bird species. Usually given by solo birds. Does not carry far.

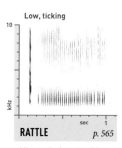

RATTLE   *p. 565*

Low, ticking

All year, by females. Often 2-parted, with 1 or more introductory clicks.

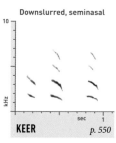

KEER   *p. 550*

Downslurred, seminasal

All year, sometimes after aggressive interactions. Not very loud.

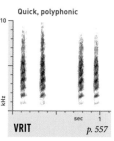

VRIT   *p. 557*

Quick, polyphonic

All year, in close contact. Often finely burry and upslurred. Not very loud.

JAY IMITATIONS

Blue Jays, Steller's Jays, and Gray Jays imitate the sounds of other birds. Blue and Steller's Jays mostly imitate hawks, for reasons that are poorly understood. Gray Jays imitate a wider variety of species, including shorebirds. In all three species, imitations range from poor to nearly perfect.

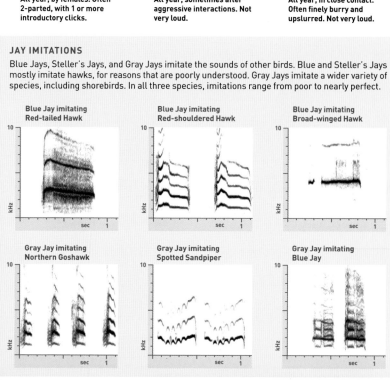

Blue Jay imitating Red-tailed Hawk

Blue Jay imitating Red-shouldered Hawk

Blue Jay imitating Broad-winged Hawk

Gray Jay imitating Northern Goshawk

Gray Jay imitating Spotted Sandpiper

Gray Jay imitating Blue Jay

Island Scrub-Jay

Aphelocoma insularis

Found exclusively on Santa Cruz Island in California. Larger than California Scrub-Jay, with larger bill and even more brightly colored plumage.

Quiet Song (p. 600): all year, by both sexes; usually without repetition, sometimes with imitations

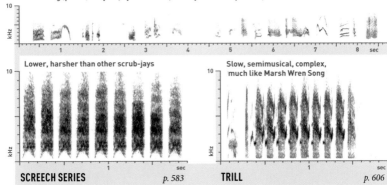

Lower, harsher than other scrub-jays

SCREECH SERIES   *p. 583*

Slow, semimusical, complex, much like Marsh Wren Song

TRILL   *p. 606*

Screech Series given all year, by both sexes, usually in undulating display flight during territorial interactions and possibly courtship. Males have two or more versions, each usually repeated before switching. Trill little known; does not resemble other scrub-jay sounds. Often given softly with Rattle.

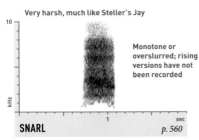

Very harsh, much like Steller's Jay

Monotone or overslurred; rising versions have not been recorded

SNARL   *p. 560*

All year, in a variety of situations, but seems to be less frequent than corresponding Zreek calls of other scrub-jays.

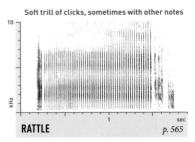

Soft trill of clicks, sometimes with other notes

RATTLE   *p. 565*

All year, by female, often while bobbing body, bill vertical. Usually in response to mate's Screech Series, often overlapping it. Variable.

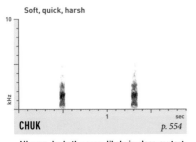

Soft, quick, harsh

CHUK   *p. 554*

All year, by both sexes, likely in close contact. Averages harsher, more monotone than corresponding Vrit calls of other scrub-jays.

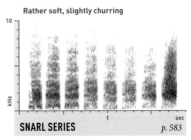

Rather soft, slightly churring

SNARL SERIES   *p. 583*

Not well known; possibly given all year in alarm. Also gives Whine call like other scrub-jays (p. 563).

CALIFORNIA SCRUB-JAY

Aphelocoma californica

Common and conspicuous in chapparal and residential areas. Much bolder than Woodhouse's Scrub-Jay, with slightly brighter plumage.

Quiet Song (p. 600): all year, by both sexes; usually without repetition, sometimes with imitations

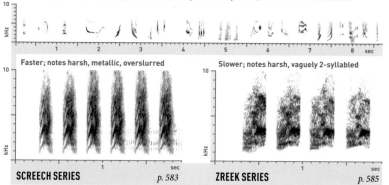

Faster; notes harsh, metallic, overslurred

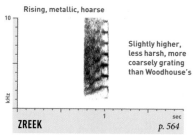

SCREECH SERIES *p. 583*

Slower; notes harsh, vaguely 2-syllabled

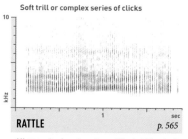

ZREEK SERIES *p. 585*

Likely two versions of one sound, given all year, by both sexes, usually in undulating display in territorial interactions and possibly courtship. Sometimes just a single note. Males have two or more versions, usually repeated before switching. Rivals match each other's Screech or Zreek Series types; mated pairs do not.

Rising, metallic, hoarse

Slightly higher, less harsh, more coarsely grating than Woodhouse's

ZREEK *p. 564*

All year, by both sexes, in a variety of situations; most common call. Variable, but distinctive.

Soft trill or complex series of clicks

RATTLE *p. 565*

All year, by female, often while bobbing body, bill vertical. Usually in response to mate's Screech or Zreek Series, often overlapping it.

Quick, soft, noisy, Chuklike

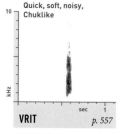

VRIT *p. 557*

All year, by both sexes, in close contact. Variable, plastic; sometimes whiny.

Harsh, noisy

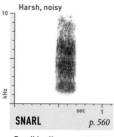

SNARL *p. 560*

Possibly all year; context not entirely clear. Not very loud.

Hoarse

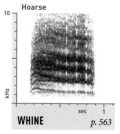

WHINE *p. 563*

Mostly June–Sept., by begging juveniles, but also by adult females. Plastic.

WOODHOUSE'S SCRUB-JAY

Aphelocoma woodhouseii

Found in arid scrub, especially among oaks or junipers. Formerly lumped with California Scrub-Jay under the name Western Scrub-Jay.

Quiet Song (p. 600): all year, by both sexes; usually without repetition, sometimes with imitations

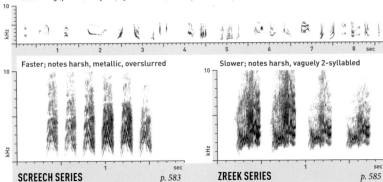

Faster; notes harsh, metallic, overslurred

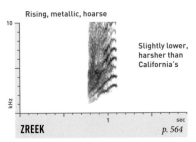

SCREECH SERIES   *p. 583*

Slower; notes harsh, vaguely 2-syllabled

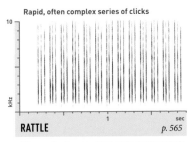

ZREEK SERIES   *p. 585*

Likely two versions of one sound, given all year, by both sexes, usually in undulating display in territorial interactions and possibly courtship. Sometimes just a single note. Males have two or more versions, usually repeated before switching. Rivals match each other's Screech or Zreek Series types; mated pairs do not.

Rising, metallic, hoarse

Slightly lower, harsher than California's

ZREEK   *p. 564*

All year, by both sexes, in a variety of situations; most common call. Variable, but distinctive.

Rapid, often complex series of clicks

RATTLE   *p. 565*

All year, by female, often while bobbing body, bill vertical. Usually in response to mate's Screech or Zreek Series, often overlapping it.

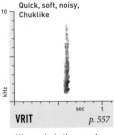

Quick, soft, noisy, Chuklike

VRIT   *p. 557*

All year, by both sexes, in close contact. Variable, plastic; sometimes whiny.

Harsh, noisy

SNARL   *p. 560*

Possibly all year; context not entirely clear. Not very loud.

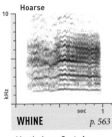

Hoarse

WHINE   *p. 563*

Mostly June–Sept., by begging juveniles, but also by adult females. Plastic.

MEXICAN JAY

ApJhelocoma wollweberi

Common in mountain pine-oak forests, moving in vocal flocks that breed cooperatively. Vocal repertoires of Texas and Arizona populations differ slightly.

Quiet Song (p. 600): all year, possibly by both sexes; usually without repetition, sometimes with imitations

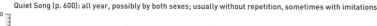

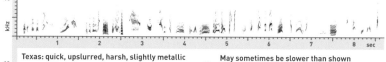

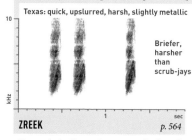

Texas: quick, upslurred, harsh, slightly metallic

Briefer, harsher than scrub-jays

ZREEK *p. 564*

All year; by far the most common call, given almost constantly by moving flocks in loose, plastic series. Geographically variable; see below.

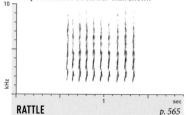

May sometimes be slower than shown

RATTLE *p. 565*

All year, by females, possibly in aggression toward other females. Known only from Texas and northeast Mexico; absent in Arizona birds.

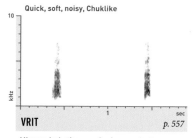

Quick, soft, noisy, Chuklike

VRIT *p. 557*

All year, by both sexes, in close contact. Variable, plastic; may grade into Snarl.

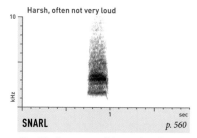

Harsh, often not very loud

SNARL *p. 560*

Possibly all year; not well known. Juveniles and adults occasionally beg for food with a Whine like scrub-jays (p. 563).

MEXICAN JAY (ARIZONA VS. TEXAS)

Arizona and Texas populations of Mexican Jay differ slightly in appearance, vocalizations, and other details. Arizona birds give Zreek calls that average slightly longer, higher, and less harsh; they also lack a Rattle call. The yellow bill of juveniles is retained for up to three years in Arizona, less than one year in Texas. Arizona birds are also slightly larger and paler.

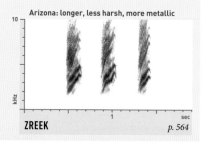

Arizona: longer, less harsh, more metallic

ZREEK *p. 564*

PINYON JAY

Gymnorhinus cyanocephalus

Uncommon in pinyon-juniper and some other arid habitats, mostly in large flocks that are highly vocal when on the move. Sometimes visits feeders.

Reer and Re-er-er may grade into each other within the repertoire of a single bird

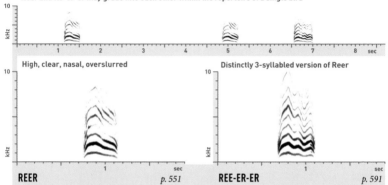

REER	REE-ER-ER
High, clear, nasal, overslurred	Distinctly 3-syllabled version of Reer
p. 551	*p. 591*

All year, in contact, especially between family members within a flock; birds recognize each other individually by these calls. The most common call types are shown, but different patterns sometimes occur, including multinote phrases. Quiet Song (p. 600) is also reported; no recordings available.

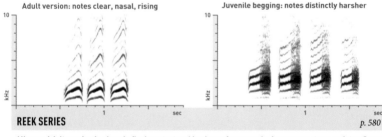

REEK SERIES — Adult version: notes clear, nasal, rising | Juvenile begging: notes distinctly harsher — *p. 580*

All year. Adult version is given in flock contact and in alarm; for example, in response to a predator. Sometimes shortened to a single note. Juvenile version is given loudly by fledged birds, usually quivering wings while begging from parents; at some times of year can be a dominant sound from moving flocks.

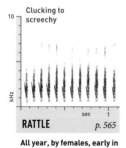 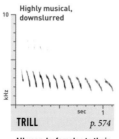 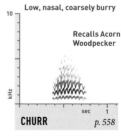

RATTLE — Clucking to screechy — *p. 565*

All year, by females, early in courtship and likely also in aggression. Variable.

TRILL — Highly musical, downslurred — *p. 574*

All year, by females to their mates, including to beg for food. Grades into Rattle.

CHURR — Low, nasal, coarsely burry. Recalls Acorn Woodpecker — *p. 558*

All year, in close contact. Does not carry far.

CLARK'S NUTCRACKER

Nucifraga columbiana

Local and social in coniferous forests. Caches huge numbers of pine seeds; some pines depend on this seed dispersal. Nomadic when cone crops fail.

Quiet Song (p. 600): may include all the calls in the repertoire, plus brief gurgling phrases

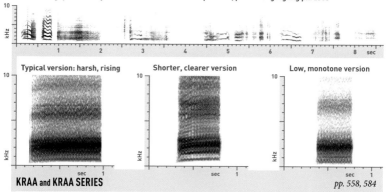

| Typical version: harsh, rising | Shorter, clearer version | Low, monotone version |

KRAA and KRAA SERIES   *pp. 558, 584*

All year; most common and distinctive call. Usually loud; apparently given in long-distance contact. Always extremely harsh, coarsely burry, and strongly nasal, usually slightly rising, sometimes monotone. Any version may be given in short series.

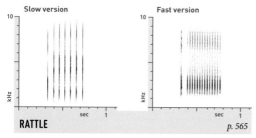
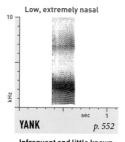

Slow version | Fast version | Low, extremely nasal

RATTLE   *p. 565*   **YANK**   *p. 552*

Rattles given at least by females; males have been reported to give similar sounds. Function and patterns of variation not well known.

Infrequent and little known; given in the Sierra Nevada, possibly elsewhere.

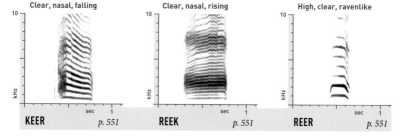

Clear, nasal, falling | Clear, nasal, rising | High, clear, ravenlike

KEER   *p. 551*   **REEK**   *p. 551*   **REER**   *p. 551*

Gives a variety of nasal, single-syllabled sounds. Keer given all year by both sexes, in contact or excitement. Version of Reek shown was given by begging juvenile; similar versions given in other contexts also. Reer little known. Some calls can recall those of Common Raven or Gray Jay.

Black-billed Magpie

Pica hudsonia

A social and conspicuous bird of dry, open country, woodland edges, and sometimes residential areas. Nest is a large ball of sticks.

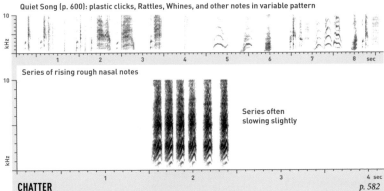

Quiet Song (p. 600): plastic clicks, Rattles, Whines, and other notes in variable pattern

Series of rising rough nasal notes

Series often slowing slightly

CHATTER　　p. 582

All year, by both sexes, in alarm. Uniform but plastic; length and speed increase with agitation level. Female version reportedly averages higher than male's. Quiet Song reportedly given mostly by aggressive young males, but also by females; most frequent fall through early spring.

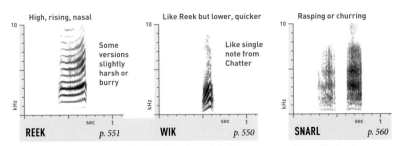

High, rising, nasal

Some versions slightly harsh or burry

REEK　　p. 551

Like Reek but lower, quicker

Like single note from Chatter

WIK　　p. 550

Rasping or churring

SNARL　　p. 560

These two sounds may intergrade. Both given all year by both sexes, Wik in contact, Reek in various situations. Reek repeated loudly by females near nest at start of breeding season.

Apparently given in response to predators near nest site.

QUIET SONGS OF CORVIDS

All North American species of jay and magpie occasionally give highly complex songs, almost always at very low volume. Reported to function in close courtship, they are also given often by lone birds, at almost any time of year. In at least some species, both sexes sing, though males may sing more frequently. Many Quiet Songs are built primarily of notes derived from the species' common calls, in addition to soft clicks, rattles, whines, and imitations of other bird species. Repetition tends to be fairly rare.

Crows and ravens apparently do not give complex Quiet Songs; in these species, the closest equivalent may be the Rattle, or in the case of American Crow, the Gurgle Phrase.

YELLOW-BILLED MAGPIE

Pica nuttalli

Almost identical to the closely related Black-billed Magpie; voice extremely similar. No range overlap; found only in California west of the Sierra Nevada.

Quiet Song (p. 600): plastic clicks, Rattles, Whines, and other notes in variable pattern

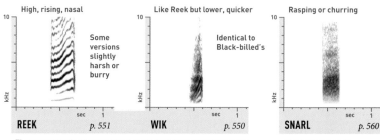

Series of rising rough nasal notes

Basically identical to Black-billed's

CHATTER   *p. 582*

All year, by both sexes, in alarm. Uniform but plastic; length and speed increase with agitation level. Female version reportedly averages higher than male's. Quiet Song reportedly given mostly by aggressive young males, but also by females; most frequent fall through early spring.

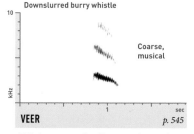

High, rising, nasal

Some versions slightly harsh or burry

REEK   *p. 551*

Like Reek but lower, quicker

Identical to Black-billed's

WIK   *p. 550*

Rasping or churring

SNARL   *p. 560*

These two sounds may intergrade. Both given all year by both sexes, Wik in contact, Reek in various situations. Reek repeated loudly by females near nest at start of breeding season.

Apparently given in response to predators near nest site.

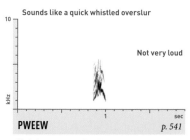

Downslurred burry whistle

Coarse, musical

VEER   *p. 545*

Sounds like a quick whistled overslur

Not very loud

PWEEW   *p. 541*

Little known; occasionally repeated at regular intervals. Similar sounds not known from Black-billed Magpie outside of Quiet Song.

Little known; occasionally repeated at regular intervals. Similar sounds not known from Black-billed Magpie outside of Quiet Song.

COMMON RAVEN

Corvus corax

Larger than crows; bill thicker, wings and tail longer. Found in mountains, forests, deserts, and some towns. Often soars; occasionally forms large flocks.

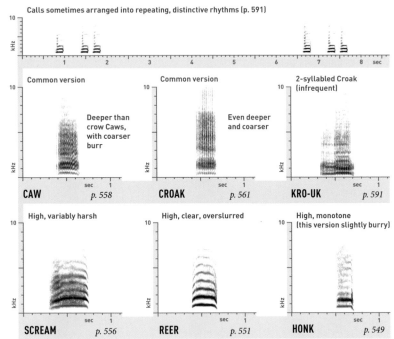

Calls sometimes arranged into repeating, distinctive rhythms (p. 591)

CAW — Deeper than crow Caws, with coarser burr — Common version — *p. 558*

CROAK — Even deeper and coarser — Common version — *p. 561*

KRO-UK — 2-syllabled Croak (infrequent) — *p. 591*

SCREAM — High, variably harsh — *p. 556*

REER — High, clear, overslurred — *p. 551*

HONK — High, monotone (this version slightly burry) — *p. 549*

All year. These calls play a key role in the complex social lives of ravens, but variation is poorly understood. Individuals give many different types of Croaklike calls, but consecutive calls from the same individual tend to be similar. Many versions briefer or longer than shown. Typical Caw resembles burrier versions of American Crow Caw, but deeper. Typical Croaks are distinctively deep, unlikely to be confused with any other bird sound. High Screams and Honks are quite diverse, ranging from coarsely burry to clear; a few versions approach Snow Goose Honks.

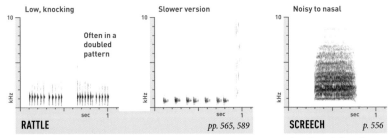

RATTLE — Low, knocking — Often in a doubled pattern — Slower version — *pp. 565, 589*

SCREECH — Noisy to nasal — *p. 556*

A highly variable set of sounds, given rather infrequently by females only, usually dominant females. Some versions more complex, gulping than shown. Often quite soft.

June–Sept., by juveniles. Variable, becoming clearer and lower with age.

Chihuahuan Raven

Corvus cryptoleucus

Smaller than Common Raven; slightly more crowlike in shape. Found in open arid country, and in some towns. Often congregates in large flocks.

Calls typically given in no particular pattern; repeating rhythms have not been documented

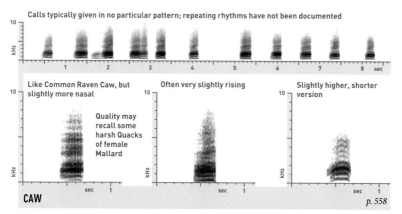

Like Common Raven Caw, but slightly more nasal

Quality may recall some harsh Quacks of female Mallard

Often very slightly rising

Slightly higher, shorter version

CAW *p. 558*

All year. Apparently less variable than Common Raven; most versions very like those shown, though some are higher-pitched. Common Raven Caw ranges from much lower than Chihuahuan's to much higher; can match Chihuahuan in pitch and pattern, but apparently rarely matches the nasal, slightly quacking quality.

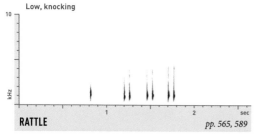

Low, knocking

RATTLE *pp. 565, 589*

Likely given in same situations as Common Raven Rattle. Range of variation poorly known; may have fast versions and gulping versions like Common Raven.

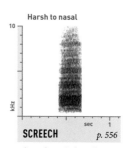

Harsh to nasal

SCREECH *p. 556*

June–Sept., by juveniles. Variable, becoming clearer and deeper with age.

IDENTIFICATION OF COMMON AND CHIHUAHUAN RAVENS

These two species present one of the most difficult identification problems among North American birds. Size differences are nearly impossible to judge except in the rare cases when the two species are seen side by side. Proportionally, Chihuahuan averages shorter-winged, shorter-tailed, and shorter-billed, with nasal bristles often extending more than halfway down the bill (usually less than halfway on Common). Some birds approach or match this description but sound like Common Ravens. It is presently unknown whether these are small Commons, vocally atypical Chihuahuans, birds of mixed ancestry, or some combination of these. Vocally typical Chihuahuans may be rare in portions of their mapped range, such as southeast Colorado. More study needed.

AMERICAN CROW

Corvus brachyrhynchos

Common and widespread in various habitats, including residential areas. Often gathers in large flocks, especially outside the breeding season.

When mobbing potential predators, gives long, harsh, snarling versions of Caw

High, relatively clear version

Burry version

CAW
p. 558

All year, by both sexes, in number of contexts; most common call. Often given in short series. Highly variable; consecutive calls from one bird usually the same, but individuals apparently have repertoires of different Caws. Along Pacific Coast, averages more nasal and groaning, the tendency increasing to the north.

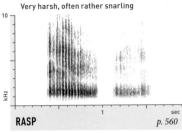

Very harsh, often rather snarling

RASP
p. 560

All year; highly plastic. Given when diving to attack potential predators, often with harsh Caws and rattling sounds.

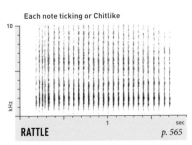

Each note ticking or Chitlike

RATTLE
p. 565

All year, especially July–Oct.; possibly by both sexes. Variable in quality, pitch, and speed; sometimes 2-parted. Function little known.

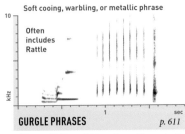

Soft cooing, warbling, or metallic phrase

Often includes Rattle

GURGLE PHRASES
p. 611

All year. Highly variable and sometimes plastic; most versions very soft, but occasionally louder. Function little known.

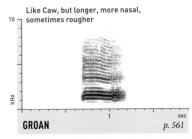

Like Caw, but longer, more nasal, sometimes rougher

GROAN
p. 561

Mostly June–Sept., by juveniles, but similar sounds given by females begging in courtship. Highly variable and plastic.

NORTHWESTERN CROW

Corvus caurinus

Possibly best considered a subspecies of American Crow. Nearly identical, but averages 10 percent smaller; voice also differs on average. Mostly coastal.

Flocks of crows often give several different types of Caws and Groans at once

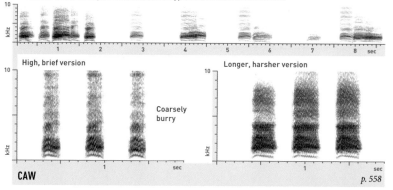

High, brief version

Coarsely burry

Longer, harsher version

CAW   *p. 558*

All year, by both sexes. Most common calls are long, harsh versions as illustrated at right; high clear Caws are rare north of the lower 48 and may represent vagrant American Crows or intergrades. Typical calls average harsher than American, with both higher noise content and coarser burr; also often more nasal.

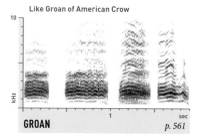

Like Groan of American Crow

GROAN   *p. 561*

At least June–Sept., by juveniles, but possibly given all year by apparent adults. Considered by some the "classic" sound of the species.

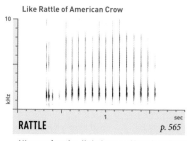

Like Rattle of American Crow

RATTLE   *p. 565*

All year; function little known. Also gives Rasp (p. 560) and Gurgle Phrases (p. 611) like those of American Crow.

IDENTIFICATION OF AMERICAN AND NORTHWESTERN CROWS

The taxonomic status of Northwestern Crow is controversial. Northwestern and American Crows differ slightly in measurements, but apparently intergrade smoothly across a broad area centered on northwest Washington state. Average differences in voice also appear to vary smoothly across locations, but not in the same geographic pattern as size: distinctly harsh, nasal Caws can be heard along the Pacific Coast at least to southern Oregon, considerably farther south than Northwestern-sized crows can be found. It is not at all clear that the taxon deserves species status, but the situation awaits a thorough investigation, including genetic data.

LARKS (Family Alaudidae)

This large family is represented in the New World by a single native species, plus one introduced species.

The Horned Lark is noteworthy for its apparently plastic calls and Song notes, rarely repeated in precisely the same way. It may be that the Horned Lark sings in a completely innovative fashion — each note a unique, one-time utterance — or it may be that the repertoire of notes from which it constructs songs is almost unfathomably vast. Some breeding birds give variable Teer calls that seem to repeat over time, almost like a form of song. More study needed.

HORNED LARK

Eremophila alpestris

Common in deserts, prairies, tundra, and other open areas with sparse vegetation. Plumage varies geographically, but voice varies little.

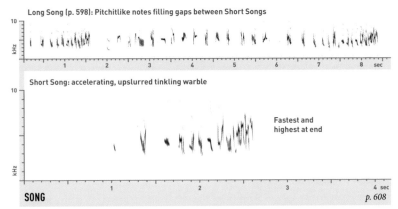

Long Song (p. 598): Pitchitlike notes filling gaps between Short Songs

Short Song: accelerating, upslurred tinkling warble

Fastest and highest at end

SONG   *p. 608*

Mostly Jan.–July. Short Song is more common. Long Song can last for several minutes and is more frequent later in the season and prior to dawn. Both may be given from the ground, from low perches, or in 300-foot-high flight display, though Short Songs predominate in flight.

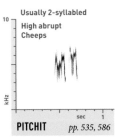

Usually 2-syllabled

High abrupt Cheeps

PITCHIT   *pp. 535, 586*

All year, given frequently; highly plastic. Function not well known.

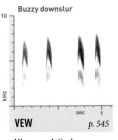

Buzzy downslur

VEW   *p. 545*

All year; relatively stereotyped. Given as birds flush, likely in alarm.

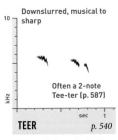

Downslurred, musical to sharp

Often a 2-note Tee-ter (p. 587)

TEER   *p. 540*

All year. Extremely plastic; can be monotone or upslurred.

EURASIAN SKYLARK

Alauda arvensis

Native to Europe and Asia. A small introduced population survives near Victoria, British Columbia. Also a very rare natural vagrant along the Pacific Coast.

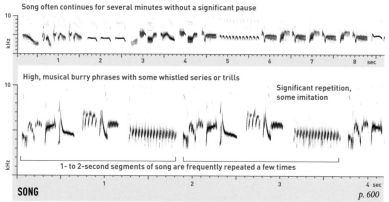

Song often continues for several minutes without a significant pause

High, musical burry phrases with some whistled series or trills

Significant repetition, some imitation

1- to 2-second segments of song are frequently repeated a few times

SONG
p. 600

An extraordinary sound, often given for many minutes without pause, usually in display flight over open field. Both sexes sing, though female Song is rarer, softer, and less varied. Birds also sing from ground, especially at dawn, often with more pauses. Sometimes repeats a single note from Song like a call.

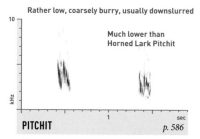

Rather low, coarsely burry, usually downslurred

Much lower than Horned Lark Pitchit

PITCHIT
p. 586

All year; most common call. Highly plastic; grades into Burrt and into single notes of Song. Some versions recall Purple Martin Veer.

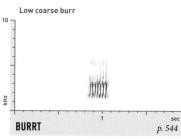

Low coarse burr

BURRT
p. 544

Rather infrequent. Plastic but usually semimusical; may recall Purple Martin Burrt.

SWALLOWS (Family Hirundinidae)

Adapted for the aerial pursuit of flying insects, swallows spend much of their time on the wing. They have small bills but wide mouths, and pointed wings for elegant, swooping flight. Several species are colonial nesters. Migrating swallows often gather in huge flocks, often including multiple species, perching on wires or swarming over bodies of water.

Unlike swifts, swallows are songbirds (oscine passerines), and most species give highly complex Songs that are likely learned. Some species, such as Purple Martin, Barn Swallow, and Tree Swallow, have distinctive Dawn Songs given mostly on the wing prior to sunrise. In some swallow species, certain calls may be learned; more study needed.

CAVE SWALLOW

Petrochelidon fulva

Nests in caves, in culverts, and under bridges. Nest is an open cup like Barn Swallow's. Range has expanded. Like Cliff Swallow, but note pale throat.

Typical Songs run 6 seconds or longer, but are often truncated

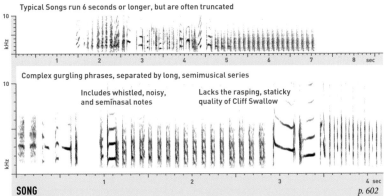

Complex gurgling phrases, separated by long, semimusical series

Includes whistled, noisy, and seminasal notes

Lacks the rasping, staticky quality of Cliff Swallow

SONG   p. 602

Mostly Mar.–Aug., presumably by male; not known whether female sings. Given often at nest site and on the wing nearby, including in response to disturbance; infrequent away from nest site. Variable and plastic; repertoire size unknown. Songs of Texas and Florida birds apparently similar.

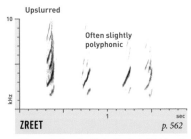

Upslurred

Often slightly polyphonic

ZREET   p. 562

All year, by foraging birds. Plastic but fairly uniform. Often higher, clearer than Barn Swallow Jit. Cliff Swallow lacks similar notes.

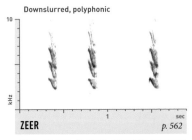

Downslurred, polyphonic

ZEER   p. 562

Likely all year, in alarm, especially near nest. Varible and plastic; much overlap with Cliff Swallow Zeer.

CLIFF SWALLOW

Petrochelidon pyrrhonota

Nests under bridges, in culverts, and on rock faces and canyon walls. Nest is an enclosed, gourd-shaped mud construction with a hole at the top.

Typical Songs run 6 seconds or longer, but can be truncated

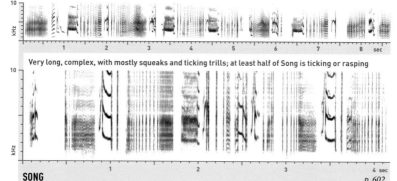

Very long, complex, with mostly squeaks and ticking trills; at least half of Song is ticking or rasping

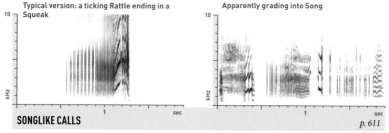

SONG
p. 602

Mostly Mar.–Aug., presumably by male; not known whether female sings. Given often at nest site and on the wing nearby, including in response to disturbance; infrequent away from nest site. Variable and plastic; repertoire size unknown. Usually not very loud.

Typical version: a ticking Rattle ending in a Squeak

Apparently grading into Song

SONGLIKE CALLS
p. 611

Mostly May–Aug., to recruit other members of the colony to a swarm of flying insects. Given mostly on chilly, overcast days when such swarms are infrequent. Variable and plastic; similar to brief snippets of Song. Unknown whether other swallow species have food recruitment calls; more study needed.

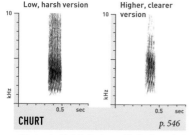

Low, harsh version

Higher, clearer version

CHURT
p. 546

All year, by birds in flocks, and in family contact near the nest. Slightly variable. Cave Swallow lacks similar notes.

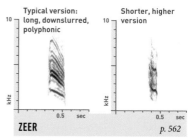

Typical version: long, downslurred, polyphonic

Shorter, higher version

ZEER
p. 562

Likely all year, in alarm, especially near nest. Variable, especially in pitch; rare versions are grating. Young beg with short, high Chirps.

BARN SWALLOW

Hirundo rustica

Common and widespread. Nests under bridges and the eaves of buildings, including houses in residential areas; forages over fields and wetlands.

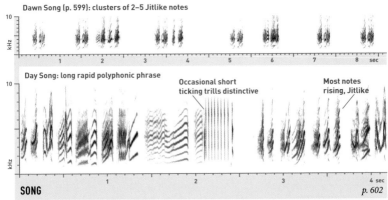

Dawn Song (p. 599): clusters of 2–5 Jitlike notes

Day Song: long rapid polyphonic phrase

Occasional short ticking trills distinctive

Most notes rising, Jitlike

SONG   *p. 602*

All year, especially Apr.–Aug., by males. Females also reported to sing. Often preceded by or mixed with Jits. Repertoire size unknown; consecutive songs sound similar but vary greatly in details and in length. Dawn Song given prior to sunrise, often on the wing; grades gradually into Day Song.

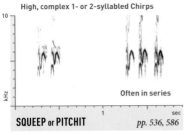

High, complex 1- or 2-syllabled Chirps

Often in series

SQUEEP or PITCHIT   *pp. 536, 586*

All year, in mild alarm. Variable and slightly plastic, but distinctive. Similar calls given by adults when feeding fledglings.

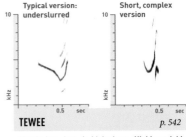

Typical version: underslurred

Short, complex version

TEWEE   *p. 542*

Mostly Apr.–Aug., in high alarm. Highly variable and rather plastic, but most versions high, clear, and 2-syllabled.

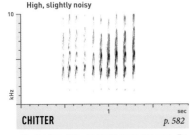

High, slightly noisy

CHITTER   *p. 582*

Mostly Apr.–Aug., in alarm near the nest. Extremely plastic; notes often longer than shown, polyphonic, grading into Jits.

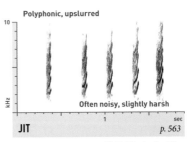

Polyphonic, upslurred

Often noisy, slightly harsh

JIT   *p. 563*

All year; most common call. Highly plastic. Often in series. Generally rougher than House Finch Jirp.

Northern Rough-winged Swallow

Stelgidopteryx serripennis

Forages over rivers and fields near water; nests singly in riverbank burrows or rock crevices. Named for tiny barbs on the leading edge of the wing.

Song (p. 599): faint, rarely heard clusters of hoarse whistled and gurgling phrases

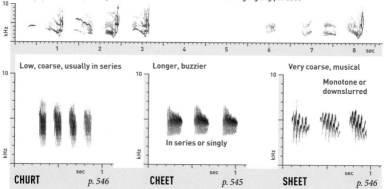

CHURT	CHEET	SHEET
Low, coarse, usually in series	Longer, buzzier / In series or singly	Very coarse, musical / Monotone or downslurred
p. 546	*p. 545*	*p. 546*

Churt and Cheet given all year. Cheet higher than Churt, more grating, and less variable; both are higher and more musical than Bank Swallow Churt. Sheet rarely heard; function unknown. Recalls Tree Swallow Sheet and certain House Sparrow Songs. Sometimes gives Snarls near nest site.

Bank Swallow

Riparia riparia

Our smallest swallow. Nests in colonies of 10 to 1,000 pairs, in holes in vertical riverbanks or in piles of sand at gravel quarries.

Song (p. 576): an accelerating chatter of Churtlike notes, ending in soft, complex, noisy gurgling

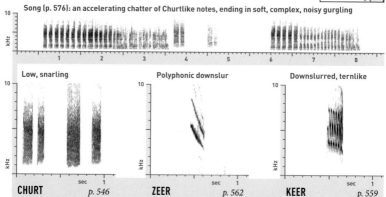

CHURT	ZEER	KEER
Low, snarling	Polyphonic downslur	Downslurred, ternlike
p. 546	*p. 562*	*p. 559*

Churt given all year; most common call. Lower, harsher than Rough-winged Churt. Zeer and Keer given all year, in alarm. Zeer is given in warning by first bird to spot danger; flock responds with Keers. Song given Apr.–July, in flight, at nest hole, and to drive away rivals.

TREE SWALLOW

Tachycineta bicolor

Breeds in tree cavities or nest boxes, usually in semi-open areas near water. Arrives earlier in spring migration than other swallows.

Dawn Song (p. 596): high, semimusical Chirps, sometimes in clusters of 2–3

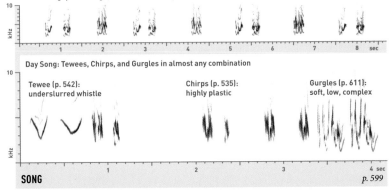

Day Song: Tewees, Chirps, and Gurgles in almost any combination

Tewee (p. 542): underslurred whistle

Chirps (p. 535): highly plastic

Gurgles (p. 611): soft, low, complex

SONG   p. 599

Mostly Apr.–July, by males. Females also reported to sing. Parts of Day Song given singly, sometimes with Sheets and other calls, or run together into long phrases. Dawn Song given prior to sunrise, often in flight; contains 1–7 (usually 2–3) different types of Chirps, often in repeating pattern.

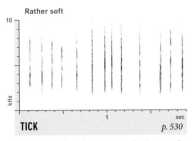

Rather soft

TICK   p. 530

Mostly Apr.–June, by male swooping low over female in close courtship; also sometimes given when diving at predators.

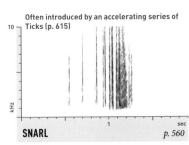

Often introduced by an accelerating series of Ticks (p. 615)

SNARL   p. 560

Mostly Apr.–June. Above version given by male swooping low over female in close courtship; single Snarls or series directed at predators.

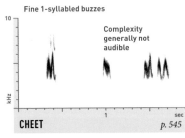

Fine 1-syllabled buzzes

Complexity generally not audible

CHEET   p. 545

All year, by both sexes, in response to predator or competitor for nest site. Plastic.

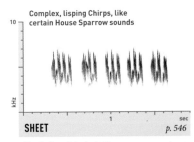

Complex, lisping Chirps, like certain House Sparrow sounds

SHEET   p. 546

Mostly Apr.–July, by both sexes, near nest; may function in defense of nest site against competitors.

VIOLET-GREEN SWALLOW

Tachycineta thalassina

A short-tailed, vividly colored swallow. Common nester in tree cavities and cliff crevices in areas of coniferous forest, sometimes under eaves of buildings.

Dawn Song (p. 596): long monotonous series of Cheetlike notes, often with repeating patterns of 4–5 notes

Day Song: hugely plastic, but often begins with Cheetlike notes and ends with musical rising phrases

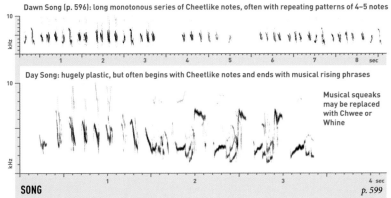

Musical squeaks may be replaced with Chwee or Whine

SONG   p. 599

Day Song poorly known, probably because it is so plastic that it is often difficult to distinguish from a sequence of calls. Dawn Song given May–Aug. long before first light, as early as 2:30 AM, by birds in high circling flight; becomes gradually more disjointed and less stereotyped, devolving into Cheet calls by dawn.

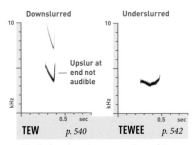

Downslurred   Upslur at end not audible

Underslurred

TEW   p. 540   **TEWEE**   p. 542

Mostly Apr.–Aug., in response to predators or disturbance near nest. Tew is the more common version, sometimes given in long plastic series.

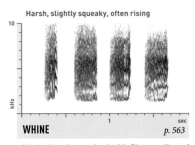

Harsh, slightly squeaky, often rising

WHINE   p. 563

Mostly Apr.–Aug., mixed with Chwee calls and grading smoothly into them. Also given when diving at predators, and sometimes in Day Song.

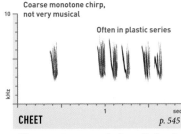

Coarse monotone chirp, not very musical

Often in plastic series

CHEET   p. 545

All year; most common call, given while foraging and in interactions. Like Tree Swallow Cheet but quicker, lower, coarser, often downslurred.

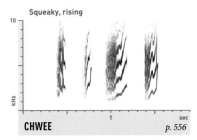

Squeaky, rising

CHWEE   p. 556

Mostly Apr.–Aug., apparently mostly in pair interactions, possibly also in aggression. Highly plastic; often in short series.

Purple Martin

Progne subis

Our largest swallow. In the East, nests almost exclusively in multi-compartment backyard birdhouses. Western populations still nest in tree cavities.

Dawn Song (p. 599): clusters of 1-syllabled burry, Veerlike notes; consecutive notes usually different

Complex musical, gurgling phrase

Usually contains Tews, and often Burrts

Most versions contain Ticks and ticking trills

DAY SONG

p. 602

Mostly Apr.–Aug. Both sexes give versions of Day Song in courtship. Female version simpler, with mostly Tews and Burrts, lacking ticking trills. First-year males, in femalelike plumage, sing like adult males. Dawn Song given by males only, prior to sunrise. Regional variation likely swamped by individual variation.

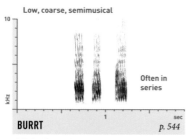

Low, coarse, semimusical

Often in series

BURRT

p. 544

Likely all year, in a variety of contexts, possibly indicating mild excitement. Often given in stereotyped songlike series of 2–6.

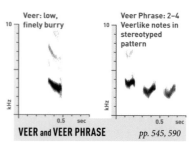

Veer: low, finely burry

Veer Phrase: 2–4 Veerlike notes in stereotyped pattern

VEER and VEER PHRASE

pp. 545, 590

Likely all year, in high alarm or excitement. Individuals may have multiple versions. More frequent than Burrts.

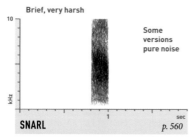

Brief, very harsh

Some versions pure noise

SNARL

p. 560

Mostly Apr.–July, in high alarm and aggression, at bottom of dive while swooping at potential predators or occasionally other bird species.

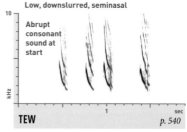

Low, downslurred, seminasal

Abrupt consonant sound at start

TEW

p. 540

Likely all year, in a variety of contexts. Most common call. Somewhat variable and plastic; some versions partly noisy.

CHICKADEES AND TITMICE (Family Paridae)

Birds in this family, known as "parids," are small, active, highly social species that include some of our most common and familiar visitors to backyard bird feeders. Vocal repertoires and social structures are highly complex. Most species gather in winter flocks that have well-defined dominance hierarchies; each member dominates all lower-ranked birds and submits to all higher-ranked birds. In spring, flocks split up, with the highest-ranked pairs claiming the best breeding territories.

Three different kinds of parid vocalization can show a level of complexity or a communicative function usually associated with "songs":

Whistled Songs are simple, musical, and learned. They are given by all species of titmice but only some chickadees. In our region, only Black-capped and Mountain Chickadees sing "classic" Whistled Songs; the distinction between Whistled Songs and Gurgle Songs in Mexican Chickadee is less clear.

Whistled Songs are strongly associated with breeding-season defense of territory by males, and may also help attract females. Male chickadees have repertoires of 2–3 (rarely 5) Whistled Songs, except in most Black-capped Chickadees, which create variety by shifting the pitch of a single songtype. Titmice have repertoires of 9–15 Whistled Songs.

Gurgle Songs are complex, musical, and learned, given by all chickadee species. They are strongly associated with aggression, although they can also be given in close-range courtship situations. Individual male Black-capped Chickadees have repertoires of 2–18 Gurgles (averaging 8), and other species are probably similar. The Tseet Songs of titmice may serve a similar function.

Chick-a-dee calls are the iconic sounds for which chickadees are named. Variable, versatile, and at least partly learned, they serve functions from maintaining flock contact to mobbing predators to giving the "all clear" after danger passes. In the species that have been well studied, Chick-a-dees consist of 3–6 different note types (usually four: A, B, C, D). All these may be repeated any number of times or left out altogether, but the order never varies: A always comes before B, B before C, etc. Only in Mexican Chickadee can B rarely precede A. In all chickadees, different numbers and combinations of notes apparently send different messages. Titmice make similar, but simpler, sounds, often with fewer note types.

BLACK-CRESTED TITMOUSE

Baeolophus atricristatus

In our area, restricted to West Texas, where it is common in mountain pine-oak and foothill pinyon-juniper forests. Formerly lumped with Tufted Titmouse.

Males have 9–11 songtypes, typically repeated many times before switching

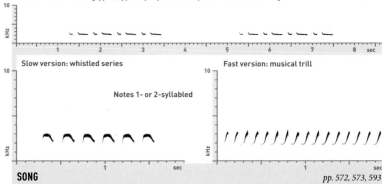

Slow version: whistled series

Notes 1- or 2-syllabled

Fast version: musical trill

SONG   *pp. 572, 573, 593*

Almost all year, but especially Jan.–June. Highly variable. Series average longer and faster than in Tufted Titmouse; trilled songtypes are fairly distinctive, but where ranges meet, Songs are rarely separable. Songs from the Big Bend region differ from those elsewhere, often including syncopated series (top).

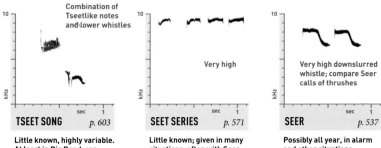

Combination of Tseetlike notes and lower whistles

Very high

Very high downslurred whistle; compare Seer calls of thrushes

TSEET SONG *p. 603*

SEET SERIES *p. 571*

SEER *p. 537*

Little known, highly variable. At least in Big Bend, can start with Chatters.

Little known; given in many situations, often with Seer and Tsip.

Possibly all year, in alarm and other situations. Sometimes in series.

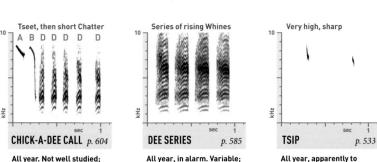
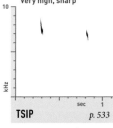

Tseet, then short Chatter
A B D D D D D

Series of rising Whines

Very high, sharp

CHICK-A-DEE CALL *p. 604*

DEE SERIES *p. 585*

TSIP *p. 533*

All year. Not well studied; C notes apparently rare or absent.

All year, in alarm. Variable; notes can be 2-syllabled.

All year, apparently to maintain contact with flockmates.

BRIDLED TITMOUSE

Baeolophus wollweberi

Fairly common in oak and pine-oak woods and some riparian areas. Often a key participant in mixed-species feeding flocks. Smaller than other titmice.

Males apparently average 3 songtypes; each repeated several times before switching

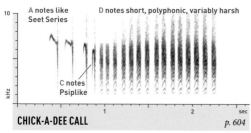

Some versions are triplet series (p. 594)

Typical version: simple series

Usually semimusical

Some versions are a low musical Trill

Common version: couplet series

SONG

pp. 572, 573, 593

Mostly Feb.–July, by males; female not known to sing. Song often difficult to distinguish from that of Juniper Titmouse; the 2 species overlap locally. In Bridled, couplets and triplets never include more than one note type. Bridled apparently lacks equivalent of Tseet Songs of other titmice.

A notes like Seet Series

D notes short, polyphonic, variably harsh

C notes Psiplike

CHICK-A-DEE CALL

p. 604

All year, in alarm, contact, and other contexts. Plastic; up to 3 note types (A, C, D). Ends in rapid Chatter like Juniper Titmouse, but opening notes usually longer, Seetlike, and slow enough to count.

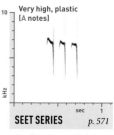

Very high, plastic (A notes)

SEET SERIES

p. 571

Possibly all year, likely in alarm, in courtship, and in other situations.

Unmusical, rattling version (C notes)

Polyphonic version (D notes)

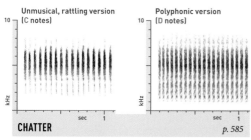

CHATTER

p. 585

All year, in alarm. Consists of C and/or D (Dee) notes from Chick-a-dee Call, and perhaps best considered a version of that call. Speed usually quite fast; notes always short, Psiplike or Chitlike.

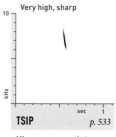

Very high, sharp

TSIP

p. 533

All year, apparently to maintain contact with flockmates.

OAK TITMOUSE

Baeolophus inornatus

Common in open woods dominated by oaks; rare in juniper and pine woods. Range barely overlaps with Juniper Titmouse in northeast California.

Males average 3–5 songtypes; each repeated several times before switching

Often just 2 renditions of a 3-syllable phrase

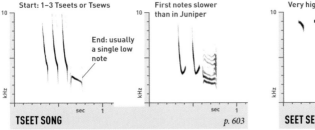

Infrequent version: simple series of musical whistles

Some versions are a low musical Trill

More common version: musical couplet or triplet series (p. 594)

SONG

pp. 572, 573, 593, 594

Mostly Feb.–June, by male; female not known to sing. Hugely variable but stereotyped. Averages slower and shorter than Juniper Titmouse Song (usually fewer than 8 syllables/second), often with more unique notes in each repeated phrase, but much overlap.

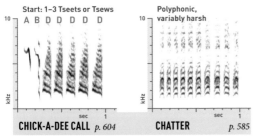

Start: 1–3 Tseets or Tsews

End: usually a single low note

First notes slower than in Juniper

TSEET SONG *p. 603*

Possibly all year. A confusing category of sounds, intermediate between whistled Song and Chick-a-dee Call, possibly grading into both. Function needs more study. Slower than in Juniper Titmouse.

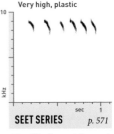

Very high, plastic

SEET SERIES *p. 571*

Possibly all year; infrequent. Given in courtship and likely in other situations.

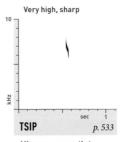

Start: 1–3 Tseets or Tsews

A B D D D D D

Polyphonic, variably harsh

Very high, sharp

CHICK-A-DEE CALL *p. 604*   **CHATTER** *p. 585*   **TSIP** *p. 533*

All year, in alarm, contact, and other contexts. Chick-a-dee Call not well studied in this species and Juniper Titmouse; additional note types may exist beyond A, B, and D.

All year, apparently to maintain contact with flockmates.

JUNIPER TITMOUSE

Baeolophus ridgwayi

Uncommon open woods dominated by pinyon and juniper. Almost identical to Oak Titmouse, and formerly considered the same species.

Males average 1–3 songtypes; each repeated several times before switching

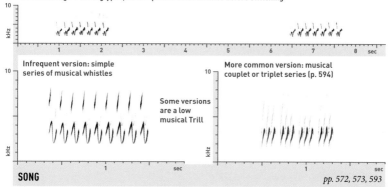

Infrequent version: simple series of musical whistles

Some versions are a low musical Trill

More common version: musical couplet or triplet series (p. 594)

SONG

pp. 572, 573, 593

Mostly Mar.–June, by male; one report of female song. Less variable than Oak Titmouse Song; averages longer and faster (often over 8 syllables/second). Most versions are musical couplet or triplet series; very rarely gives slow repeated triplets like Oak often does. Trilled songtypes much more common than in Oak.

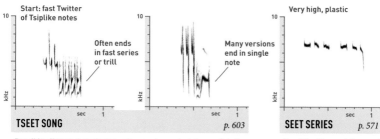

Start: fast Twitter of Tsiplike notes

Often ends in fast series or trill

Many versions end in single note

Very high, plastic

TSEET SONG
p. 603

SEET SERIES
p. 571

Possibly all year. A confusing category of sounds, intermediate between whistled Song and Chick-a-dee Call, possibly grading into both. Function needs more study. Faster than in Oak Titmouse.

Possibly all year; infrequent. Given in courtship and likely in other situations.

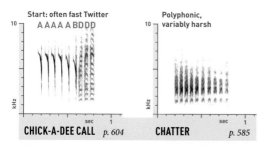

Start: often fast Twitter
A AAA A BDDD

Polyphonic, variably harsh

Very high, sharp

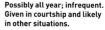

CHICK-A-DEE CALL *p. 604*

CHATTER *p. 585*

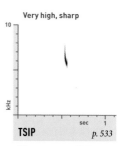

TSIP *p. 533*

All year, in alarm, contact, and other contexts. Chick-a-dee Call typically faster than in Oak Titmouse, with more and shorter D (Dee) notes and (especially) more introductory notes.

All year, apparently to maintain contact with flockmates.

BLACK-CAPPED CHICKADEE

Poecile atricapillus

Common and widespread in woodlands, parks, and towns; regular at bird feeders. Often flocks with other species, especially in winter.

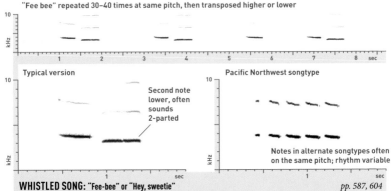

"Fee bee" repeated 30–40 times at same pitch, then transposed higher or lower

Typical version

Second note lower, often sounds 2-parted

Pacific Northwest songtype

Notes in alternate songtypes often on the same pitch; rhythm variable

WHISTLED SONG: "Fee-bee" or "Hey, sweetie"   pp. 587, 604

Mostly Feb.–June. In spring, often heard in "song duels" between rival males. In most of North America, the "Fee-bee" (left) is the only songtype, and dueling males vary its pitch rather than its pattern. In some areas (such as western Oregon and Washington, right) males instead deploy 2–3 similar songtypes.

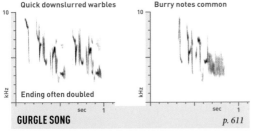

Quick downslurred warbles

Burry notes common

Very high, plastic (here, like A notes)

Ending often doubled

GURGLE SONG   p. 611

SEET SERIES   p. 571

All year, especially in winter flocks. Brief musical gurgles, starting high and ending low; in this species given mostly by males, who average 8 Gurgle types apiece.

All year. Highly plastic; given in response to danger and also in courtship.

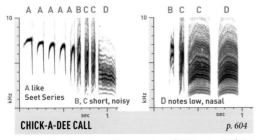

A A A A A B C C D

A like Seet Series

B, C short, noisy

B C   C   D

D notes low, nasal

Very high, quick

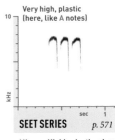
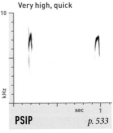

CHICK-A-DEE CALL   p. 604

PSIP   p. 533

All year. Plastic; can have all four note types (ABCD, above left), but more often only AD, BC, AC, or BCD. Chitters of C notes (p. 582) and series of D (Dee) notes common in agitation.

All year. The common contact call, given softly by calm birds.

Mountain Chickadee

Poecile gambeli

Common in coniferous forests; often visits bird feeders. Closely related to Black-capped Chickadee, and occasionally hybridizes with it.

Males have 4–7 songtypes, but use some more than others, and are rarely heard switching

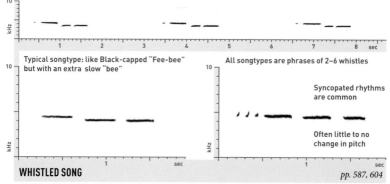

Typical songtype: like Black-capped "Fee-bee" but with an extra slow "bee"

All songtypes are phrases of 2–6 whistles

Syncopated rhythms are common

Often little to no change in pitch

WHISTLED SONG

pp. 587, 604

All year, but mostly Mar.–July. Males sing to defend territory and possibly to attract a mate; female Song evidently very rare. Much more variable than Black-capped Whistled Song; typically has more notes and less change in pitch. In any given area, the two species sing differently and ignore each other's songs.

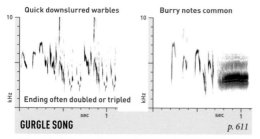

Quick downslurred warbles

Burry notes common

Ending often doubled or tripled

GURGLE SONG

p. 611

All year, by both sexes in aggression; also prior to copulation. Very similar to Black-capped Gurgle; may average longer. Repertoire size unknown, but probably like Black-capped.

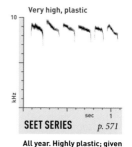

Very high, plastic

SEET SERIES

p. 571

All year. Highly plastic; given in response to danger and also in courtship.

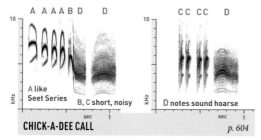

A A A A B D   D

A like Seet Series

B, C short, noisy

C C C C   D

D notes sound hoarse

CHICK-A-DEE CALL

p. 604

All year. Plastic; much like Black-capped Chick-a-dee, but Dee notes typically sound harsher, less musical. Chitters of C notes (p. 582) and series of D (Dee) notes common in agitation.

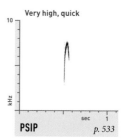

Very high, quick

PSIP

p. 533

All year. The common contact call, given softly by calm birds.

BOREAL CHICKADEE

Poecile hudsonicus

A tame, fluffy bird of northern forests, usually associated with spruces. Related to Chestnut-backed Chickadee, and like it, lacks a Whistled Song.

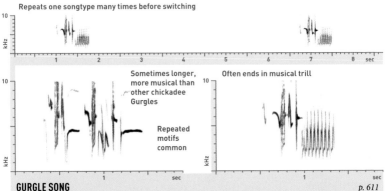

Repeats one songtype many times before switching

Sometimes longer, more musical than other chickadee Gurgles

Repeated motifs common

Often ends in musical trill

GURGLE SONG

p. 611

All year, especially during territory formation, Mar.–June. Apparently given only by males, usually in aggression, occasionally during copulation. Sometimes a complex series, with a short rapid phrase repeated 2–8 times. Repertoire size unknown, but probably each male has several Gurgles.

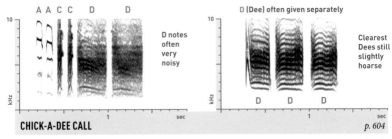

A A C C D D

D notes often very noisy

D (Dee) often given separately

Clearest Dees still slightly hoarse

D D D

CHICK-A-DEE CALL

p. 604

All year; plastic. Some noise and a hint of polyphony make the Dees sound hoarser, more strained than in Black-capped. Often abbreviated, with just 1–2 notes before a single Dee. C notes frequently given in separate Chitters or singly, as Chit. B notes rare or absent.

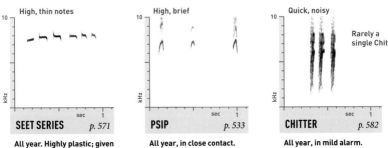

High, thin notes

SEET SERIES *p. 571*

All year. Highly plastic; given in response to danger and also in courtship.

High, brief

PSIP *p. 533*

All year, in close contact. Not very loud.

Quick, noisy

Rarely a single Chit

CHITTER *p. 582*

All year, in mild alarm. Same as C note of Chick-a-dee Call.

CHESTNUT-BACKED CHICKADEE

Poecile rufescens

Common in or near dense, humid coniferous forests. Also found in riparian willows and cottonwoods, especially in the southern part of its range.

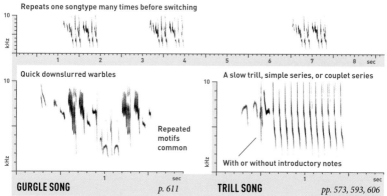

Repeats one songtype many times before switching

Quick downslurred warbles

Repeated motifs common

GURGLE SONG *p. 611*

A slow trill, simple series, or couplet series

With or without introductory notes

TRILL SONG *pp. 573, 593, 606*

The two types of song shown appear to have similar functions, and are probably both best considered versions of the Gurgle Song of other chickadees. Given in aggression, especially during territory formation, Mar.–June. Repertoire size unknown.

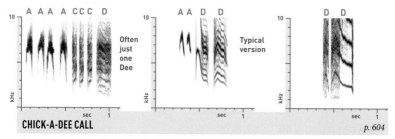

A A A A C C C D

Often just one Dee

A A D D

Typical version

D D

CHICK-A-DEE CALL *p. 604*

All year; plastic. D (Dee) notes are shorter, higher than in other chickadees, highly nasal, variably hoarse, and usually downslurred. Often just 1–2 notes given before a single Dee. Version at right (a short Dee followed by a long Dee) is used by courting females and by juveniles to beg for food. B notes rare or absent.

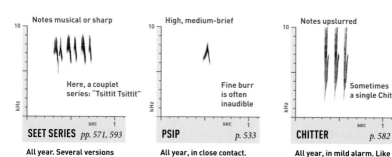

Notes musical or sharp

Here, a couplet series: "Tsittit Tsittit"

SEET SERIES *pp. 571, 593*

All year. Several versions have different functions; most common is shown.

High, medium-brief

Fine burr is often inaudible

PSIP *p. 533*

All year, in close contact. Not very loud.

Notes upslurred

Sometimes a single Chit

CHITTER *p. 582*

All year, in mild alarm. Like C note of Chick-a-dee Call.

Mexican Chickadee

Poecile sclateri

In its limited U.S. range, found in high-elevation pine or spruce forest. Note extensive black bib. Range does not overlap with other chickadees'.

Consecutive songs tend to be similar, with slight modifications

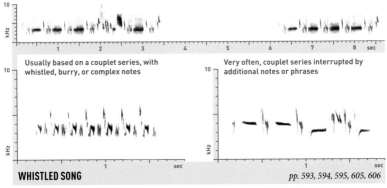

Usually based on a couplet series, with whistled, burry, or complex notes

Very often, couplet series interrupted by additional notes or phrases

WHISTLED SONG

pp. 593, 594, 595, 605, 606

All year. Songs given in spring average more complex; couplet series are more common in fall. Repertoire size unknown, in part because individuals often recombine parts of different songtypes. Simpler and more complex Songs may serve different functions; more study needed. Often hard to tell from Gurgle Song.

Quick warble, often buzzy

In altercations, repeats motifs

GURGLE SONG *p. 611*

All year, in many situations. Individuals have at least 3–10 versions.

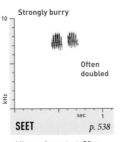

Strongly burry

Often doubled

SEET *p. 538*

All year, in contact. Often downslurred, like A note of Chick-a-dee call.

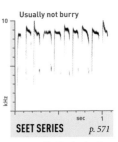

Usually not burry

SEET SERIES *p. 571*

All year. Highly plastic; given in courtship and by begging juveniles.

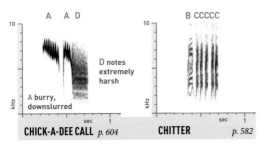

A   A D

D notes extremely harsh

A burry, downslurred

CHICK-A-DEE CALL *p. 604*

B CCCCC

CHITTER *p. 582*

All year. Plastic; four note types (ABCD), but mostly gives combinations AD, BC, or AC. Chitters of only C notes (p. 582) and series of D (Dee) notes common in agitation. B can rarely precede A.

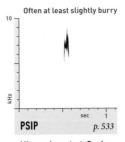

Often at least slightly burry

PSIP *p. 533*

All year, in contact. Grades into Seet. Also apparently a Tsip (p. 533) like titmice.

BUSHTIT (Family Aegithalidae)

The Bushtit is related to the long-tailed tits of Eurasia, which are only distantly related to chickadees and titmice. Vocalizations are simple and apparently innate.

VERDIN (Family Remizidae)

The Verdin is related to the penduline tits of Europe and Africa, in a sister group to chickadees and titmice. Song is likely learned.

NUTHATCHES (Family Sittidae)

Nuthatches cling to tree trunks like miniature woodpeckers, but are as likely to face downward as upward. Vocalizations are nasal or squeaky; songs may be learned.

CREEPERS (Family Certhiidae)

Creepers cling to tree trunks like nuthatches, but always facing skyward. Songs are complex, musical, and learned, with many local dialects.

BUSHTIT

Psaltriparus minimus

Forms large, active flocks, usually in dense bushes or low trees, sometimes in residential areas. Constant simple, quiet calls often announce its presence.

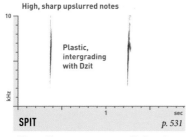

High, sharp upslurred notes

Plastic, intergrading with Dzit

SPIT *p. 531*

All year. Most common contact call, given almost constantly by flocks.

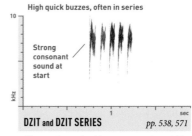

High quick buzzes, often in series

Strong consonant sound at start

DZIT and DZIT SERIES *pp. 538, 571*

All year, in territorial encounters, in mobbing predators, or in response to humans. Plastic.

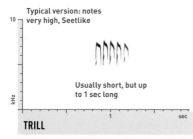

Typical version: notes very high, Seetlike

Usually short, but up to 1 sec long

TRILL

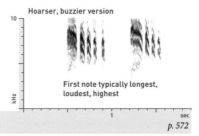

Hoarser, buzzier version

First note typically longest, loudest, highest

p. 572

Typical version of Trill given all year, by adult birds separated from the flock or warning of aerial predators. Some species of ground squirrel give very similar calls. Hoarse version given mostly May–Aug., by begging fledglings, but this or a similar call also given in various situations by highly agitated adults.

VERDIN

Auriparus flaviceps

Common and active, singly or in pairs, in deserts with thorny vegetation. Gives frequent simple, distinctive, rather loud vocalizations.

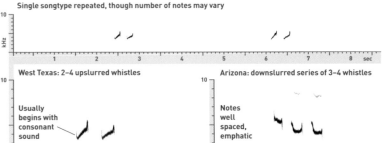

Single songtype repeated, though number of notes may vary

West Texas: 2–4 upslurred whistles

Usually begins with consonant sound

Arizona: downslurred series of 3–4 whistles

Notes well spaced, emphatic

SONG: "Twee twee twee" or "Teet toot toot"   *pp. 587, 588, 605*

Mostly Feb.–July. Shows regional dialects, indicating that Song is likely learned; each male may have only a single songtype, but consecutive renditions can vary slightly. Males apparently use Tew to attract females; Song begins only after pairs form, so may be territorial. More study needed.

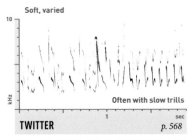

Soft, varied

Often with slow trills

TWITTER   *p. 568*

Feb.–July. Plastic; usually between mates, but sometimes by lone birds. May be learned; function unclear. Needs more study.

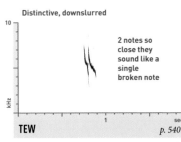

Distinctive, downslurred

2 notes so close they sound like a single broken note

TEW   *p. 540*

All year, but mostly Feb.–July, when males give it from high perches for long periods. Only slight geographic variation.

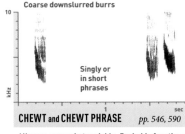

Coarse downslurred burrs

Singly or in short phrases

CHEWT and CHEWT PHRASE   *pp. 546, 590*

All year; somewhat variable. Probably functions as an alarm call, but more study needed.

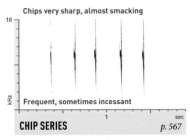

Chips very sharp, almost smacking

Frequent, sometimes incessant

CHIP SERIES   *p. 567*

All year. Most common call. Speed correlates with agitation; excited birds can accelerate into rapid chipping Twitters.

RED-BREASTED NUTHATCH

Sitta canadensis

Resident in northern coniferous forests; wanders irregularly south in winter. Often visits feeders. Usually excavates its own nesting cavities.

Slow Song continues at steady pace for long periods, occasionally a minute or more

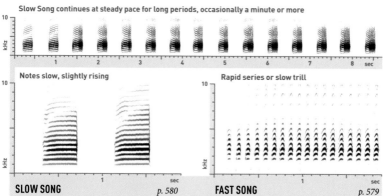

Notes slow, slightly rising

Rapid series or slow trill

SLOW SONG   *p. 580*   **FAST SONG**   *p. 579*

All year, by both sexes, especially Mar.–Aug. Uniform. Individuals apparently have 2 distinct song patterns. Slow Song given in courtship, mostly before pairing; Fast Song apparently territorial, mostly after pairing. Fast Song is 1–2 seconds long, repeated after a pause; always lower than in White-breasted Nuthatch.

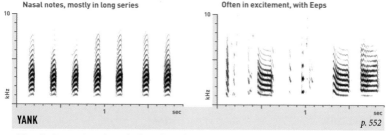

Nasal notes, mostly in long series

Often in excitement, with Eeps

YANK   *p. 552*

A broad category of nasal sounds given all year in many contexts; sometimes difficult to distinguish from Song. Short versions almost always given in long series. Longer versions, like individual notes of Slow Song, given mostly in high agitation, with Eeps.

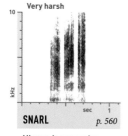

Very harsh

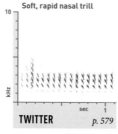

Soft, rapid nasal trill

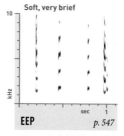

Soft, very brief

SNARL   *p. 560*   **TWITTER**   *p. 579*   **EEP**   *p. 547*

All year, in aggression. Plastic; often in series. Heard frequently at feeders.

All year, in excitement; made of Eeplike notes. Plastic.

All year. Soft, highly plastic contact calls, inaudible at a distance.

White-breasted Nuthatch (Pacific)

Sitta carolinensis (*aculeata* group)

Uncommon in deciduous and mixed woods, especially those dominated by oaks. One of three vocally distinctive populations (perhaps species).

Tewee Song and Fast Song both 1–2 seconds long, repeated after a pause

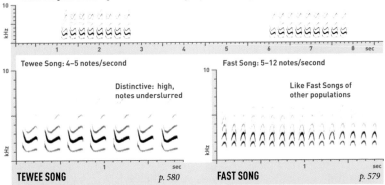

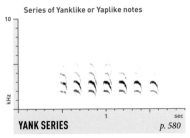

Tewee Song: 4–5 notes/second

Distinctive: high, notes underslurred

TEWEE SONG *p. 580*

Fast Song: 5–12 notes/second

Like Fast Songs of other populations

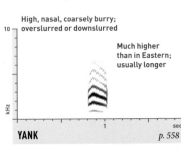

FAST SONG *p. 579*

All year, by males, but especially Jan.–May. Females not known to sing. Tewee Song and Fast Song seem quite distinct. Tewee Song is far more common and distinctive; no other population sounds similar. Fast Song averages higher than that of Montane; always higher than Fast Song of Eastern.

Series of Yanklike or Yaplike notes

YANK SERIES *p. 580*

Not well known; version shown may be considered a rapid series of Yanks or a variant of Fast Song. Also gives soft Chwit near nest (p. 531).

High, nasal, coarsely burry; overslurred or downslurred

Much higher than in Eastern; usually longer

YANK *p. 558*

All year; most common and distinctive call. No other population sounds similar. Almost never doubled or run into trill, even in high agitation.

Soft seminasal downslur

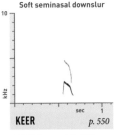

KEER *p. 550*

Soft, infrequent, by both sexes in courtship and high excitement.

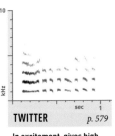

TWITTER *p. 579*

In excitement, gives high Eeplike notes, often in trills. Audible at close range.

Higher than in other populations

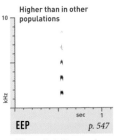

EEP *p. 547*

All year, by both sexes, in close contact. Plastic. Audible only at close range.

WHITE-BREASTED NUTHATCH (MONTANE)

Sitta carolinensis (*nelsoni* group)

Nearly identical in appearance to other forms of White-breasted Nuthatch, but vocally distinct. Found in deciduous, mixed, and coniferous forests.

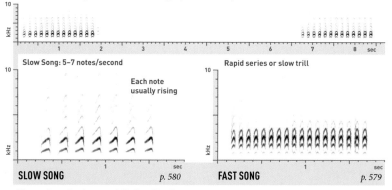

Fast and Slow Songs both 1–2 seconds long, repeated after a pause

Slow Song: 5–7 notes/second

Each note usually rising

Rapid series or slow trill

SLOW SONG   *p. 580*

FAST SONG   *p. 579*

All year, by males, but especially Jan.–May. Females not known to sing. Unlike in Red-breasted Nuthatch, Fast and Slow Songs are not well differentiated; given in similar contexts, perhaps ends of a continuum of variation. Pitch averages higher than Eastern, lower than Pacific, but overlaps both.

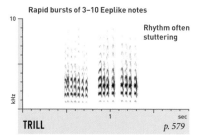

Rapid bursts of 3–10 Eeplike notes

Rhythm often stuttering

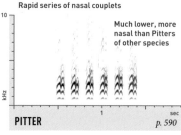

Rapid series of nasal couplets

Much lower, more nasal than Pitters of other species

TRILL   *p. 579*

PITTER   *p. 590*

All year; possibly an excited form of Pitter. Plastic; lower and louder than Twitter. Also gives soft low Chwut near nest (p. 555).

All year; most common and distinctive call. No other population sounds similar. Highly plastic, grading into Trill (p. 579).

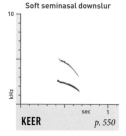

Soft seminasal downslur

KEER   *p. 550*

Soft, infrequent, by both sexes in courtship and high excitement.

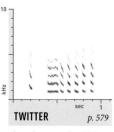

TWITTER   *p. 579*

In excitement, gives high Eeplike notes, often in trills. Audible at close range.

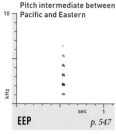

Pitch intermediate between Pacific and Eastern

EEP   *p. 547*

All year, by both sexes, in close contact. Plastic. Audible only at close range.

White-breasted Nuthatch (Eastern)

Sitta carolinensis (*carolinensis* group)

Most common in mature deciduous forests, including some residential areas. Vocally similar to Pacific population, but lower in pitch, without overlap.

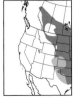

Fast and Slow Songs both 1–2 seconds long, repeated after a pause

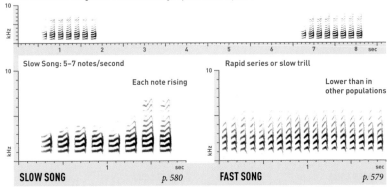

Slow Song: 5–7 notes/second

Each note rising

SLOW SONG *p. 580*

Rapid series or slow trill

Lower than in other populations

FAST SONG *p. 579*

All year, by males, but especially Jan.–May. Females not known to sing. Unlike in Red-breasted Nuthatch, Fast and Slow Songs are not well differentiated; given in similar contexts, apparently ends of a continuum of variation. Repertoire size unknown. Always higher than Red-breasted Nuthatch Song.

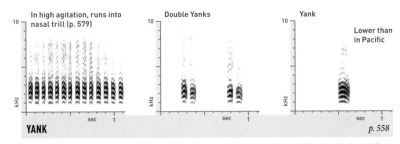

In high agitation, runs into nasal trill (p. 579)

Double Yanks

Yank

Lower than in Pacific

YANK *p. 558*

A broad category of nasal sounds given all year in many contexts; sometimes difficult to distinguish from Song. Trilled version given in high alarm; Double Yanks and Yanks in mild alarm and in interactions. Most versions slightly burry. Usually distinctive; other populations generally do not sound similar.

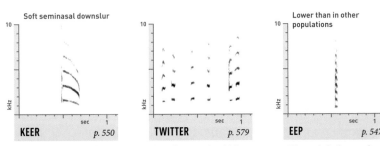

Soft seminasal downslur

Lower than in other populations

KEER *p. 550*

Soft, infrequent, by both sexes in courtship and high excitement.

TWITTER *p. 579*

In excitement, gives high Eeplike notes, but apparently not in trills.

EEP *p. 547*

All year, by both sexes, in close contact. Plastic. Audible only at close range.

PYGMY NUTHATCH

Sitta pygmaea

A tiny, active, social bird, associated primarily with ponderosa pine. Breeds cooperatively. Lowers body temperature to survive cold nights.

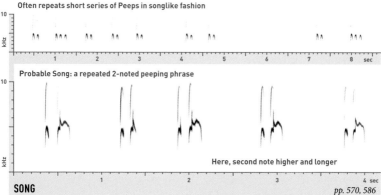

Often repeats short series of Peeps in songlike fashion

Probable Song: a repeated 2-noted peeping phrase

Here, second note higher and longer

SONG

pp. 570, 586

Apparently Feb.–June. Distinction between Song and calls in this species not entirely clear; almost all vocalizations are variations on a simple Peep. Peeping couplets or triplets, repeated at a steady pace for long periods from treetop, may function as Song. Single Peep repeated steadily may also be a type of Song.

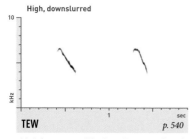

High, downslurred

TEW

p. 540

Little known; possibly given all year, perhaps in alarm or in excitement.

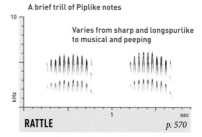

A brief trill of Piplike notes

Varies from sharp and longspurlike to musical and peeping

RATTLE

p. 570

All year. Not well known; likely given by adults feeding young, also possibly in aggression. Perhaps more common in coastal California.

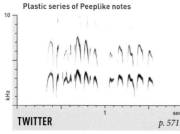

Plastic series of Peeplike notes

TWITTER

p. 571

High, clear, usually fairly musical

PEEP

p. 534

All year; most common calls. Feeding flocks are rarely silent. Single Peeps run often into Twitters. In Twitters, pitch and length of notes increase with excitement level; young birds beg with very high, long notes. Pitch and length of individual Peeps also plastic.

BROWN CREEPER

Certhia americana

Common in coniferous and mixed forests, but often inconspicuous. Slowly scales tree trunks like a woodpecker in upward spirals.

Dawn Song: stereotyped sequences of Tseews and Seets, often with complex Songs interspersed

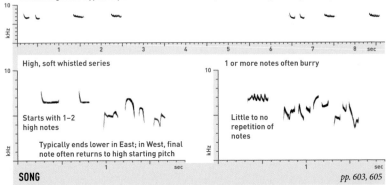

High, soft whistled series

Starts with 1–2 high notes

Typically ends lower in East; in West, final note often returns to high starting pitch

1 or more notes often burry

Little to no repetition of notes

SONG
pp. 603, 605

Apr.–July. Most males have only one songtype; some have two. Dawn Song begins with short repeated stereotyped phrases composed of Tseew- and Seetlike notes; as sun rises, these phrases become more variable and Song becomes more frequent. Dawn Song pattern may continue into late morning. This species shows subtle geographic variation in plumage, genetics, and vocalizations; more study needed.

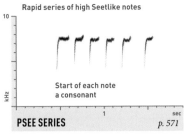

Rapid series of high Seetlike notes

Start of each note a consonant

PSEE SERIES
p. 571

Mostly Apr.–July. Given infrequently, by males during territorial encounters; also perhaps in some other situations.

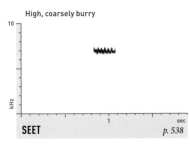

High, coarsely burry

SEET
p. 538

All year. Evidently given by both sexes in a variety of situations. One of the loudest calls.

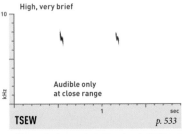

High, very brief

Audible only at close range

TSEW
p. 533

All year, in close contact. Also given in flight, but apparently not by night migrants.

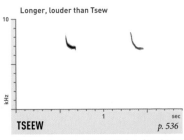

Longer, louder than Tsew

TSEEW
p. 536

All year, in alarm, during mobbing of predators, and during some territorial encounters.

WRENS (Family Troglodytidae)

The wrens are some of our most familiar songbirds, with one or more species common in almost every North American habitat. Most species are small and inconspicuously plumaged, but highly vocal, with loud, complex musical Songs. By contrast, the Cactus Wren of the desert Southwest is large and boldly patterned, with a monotonous, unmusical Song.

Subtle differences in vocalizations distinguish eastern and western populations of some species, including House Wren and Marsh Wren. Western populations of the Winter Wren were recently split into a separate species, Pacific Wren (*Troglodytes pacificus*), based in part on vocal differences. Field identification of these (sub-) species pairs can be difficult, sometimes requiring spectrographic analysis.

WINTER WREN

Troglodytes hiemalis

Tiny and shy, skulking in thick tangles and piles of downed wood. Breeds mostly in moist coniferous forests; winters in brushy woodlands.

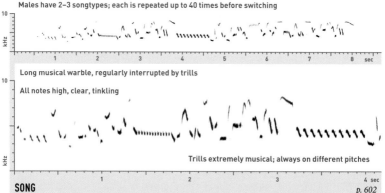

Males have 2–3 songtypes; each is repeated up to 40 times before switching

Long musical warble, regularly interrupted by trills

All notes high, clear, tinkling

Trills extremely musical; always on different pitches

SONG p. 602

Mostly Mar.–Aug. One of the most remarkable bird Songs in North America, sweetly musical and complex; rarely shorter than 4 seconds and sometimes longer than 30 seconds. Usually sings from a low exposed perch, but sometimes from high in trees; rarely in flight. Unknown whether female sings.

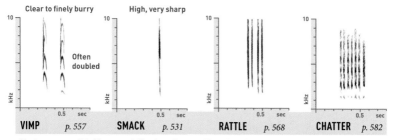

Clear to finely burry

Often doubled

High, very sharp

VIMP p. 557   **SMACK** p. 531   **RATTLE** p. 568   **CHATTER** p. 582

Most common call is Vimp, much like Song Sparrow Vimp, but often clearer. At times, possibly in alarm, Vimp grades into Smack, like call of Pacific Wren, which is sometimes run into rapid Rattle. In agitation, may also give a nasal Chatter with quality much like Vimp, as well as a soft Snarl (not shown).

Pacific Wren

Troglodytes pacificus

Common but skulking. Breeds in moist coniferous forests; winters in brushy woods. Recently split from Winter Wren; best identified by voice.

Males have 40–100 songtypes; consecutive songs often start the same but differ in ending

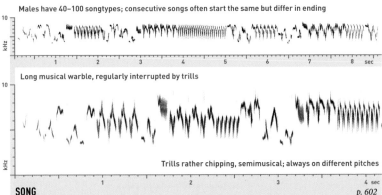

Long musical warble, regularly interrupted by trills

Trills rather chipping, semimusical; always on different pitches

SONG   p. 602

Mostly Mar.–Aug. Even more complex than Winter Wren Song; listen for the trills, which are less musical. Rarely shorter than 4 seconds and sometimes longer than 30 seconds. Usually sings from a low exposed perch, but sometimes from high in trees; rarely in flight. Unknown whether female sings.

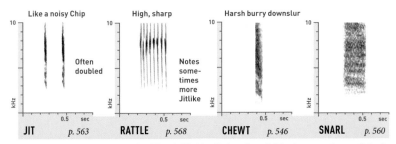

Like a noisy Chip | High, sharp | Harsh burry downslur |

Often doubled | Notes sometimes more Jitlike |

JIT p. 563 | **RATTLE** p. 568 | **CHEWT** p. 546 | **SNARL** p. 560

Most common call is Jit, much like Wilson's Warbler Jit, but often slightly sharper and more distinctly downslurred. In agitation, Jit runs into rapid Rattle, usually higher than Winter Wren's, but plastic. Chewt and Snarl infrequent in high alarm, grading smoothly into Jit and each other.

House Wren (Brown-throated) *Troglodytes aedon* (*brunneicollis* group)

The subspecies of House Wren in the mountains of Mexico is a warm brown below instead of grayish, with barred flanks and a faint buffy supercilium. Its song tends to be more complex than those of northern birds, often with fewer trills. Birds in the mountains of southeast Arizona are intermediate in appearance between Mexican and northern birds; their songs are variable but seem mostly indistinguishable from northern songs. Calls deserve study.

Southeast Arizona

HOUSE WREN

Troglodytes aedon

Common and widespread in brushy woodland edges, parks, and residential areas, where it nests in almost any type of small cavity.

Males estimated to sing 20–50 similar songtypes; consecutive songs usually same

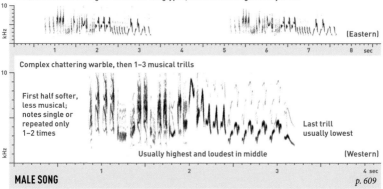

(Eastern)

Complex chattering warble, then 1–3 musical trills

First half softer, less musical; notes single or repeated only 1–2 times

Usually highest and loudest in middle

Last trill usually lowest

(Western)

MALE SONG   *p. 609*

Mostly Mar.–July; sings occasionally in winter. Variable; songs average shorter in West, with a Beertlike note near start. In close courtship or territorial disputes, given at rapid pace, one Song nearly running into the next. Subsong frequent fall through early spring; can suggest Winter Wren, but lower, more plastic.

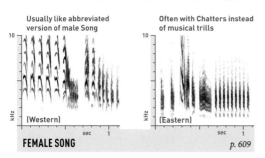
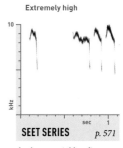

Usually like abbreviated version of male Song

(Western)

Often with Chatters instead of musical trills

(Eastern)

Extremely high

FEMALE SONG   *p. 609*

SEET SERIES   *p. 571*

Infrequent, by females who have lost track of their mates, and in female-female aggression. Hugely variable and plastic, ranging from 1-note Squeals to versions matching male Song.

In close courtship, often between Songs. Plastic; notes sometimes Chiplike.

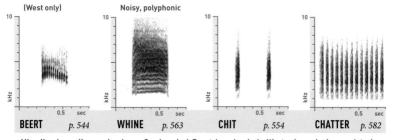

(West only)

Noisy, polyphonic

BEERT   *p. 544*   **WHINE**   *p. 563*   **CHIT**   *p. 554*   **CHATTER**   *p. 582*

All calls given all year, in alarm. Semimusical Beert heard only in West; given singly, run into long Sputters, or incorporated into Song. Plastic Whine recalls gnatcatchers or Gray Catbird. Chit varies from longer than shown to shorter, more like Dzik of Bewick's. Chatter noisy, variable in speed.

ROCK WREN

Salpinctes obsoletus

Found on open rocky slopes and eroded
bluffs. Nests in crevices; usually builds
a "pavement" of stones in front of the
nest, the purpose of which is unknown.

Individuals have 100 songtypes or more; consecutive songs well spaced and usually different

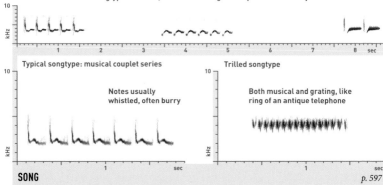

Typical songtype: musical couplet series

Notes usually
whistled, often burry

Trilled songtype

Both musical and grating, like
ring of an antique telephone

SONG p. 597

Mostly Mar.–Aug., apparently by males. Most songtypes are simple series or couplet series; some are single
trills; a very small number are burry or buzzy phrases. Birds choose from the same 3–6 songtypes repeatedly for a few minutes, then switch to a new set.

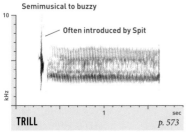

Semimusical to buzzy

Often introduced by Spit

TRILL p. 573

Mostly May–Aug., in defense of nest or fledged
young. Highly plastic, especially in pitch; tone
may recall chirping cricket.

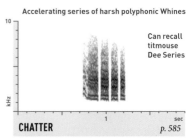

Accelerating series of harsh polyphonic Whines

Can recall
titmouse
Dee Series

CHATTER p. 585

Mar.–May, at least, in agitation and during pair
interactions. Variable and plastic. Juveniles beg
with similar notes.

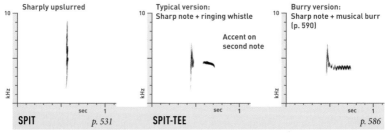

Sharply upslurred

SPIT p. 531

Typical version:
Sharp note + ringing whistle

Accent on
second note

SPIT-TEE

Burry version:
Sharp note + musical burr
(p. 590)

p. 586

All year, by both sexes. Spit given in high agitation, often with Chatter. Spit-tee given in a variety of circumstances, the bird vigorously bobbing its body with each call. Variable and plastic; different versions may
serve slightly different functions.

CANYON WREN

Catherpes mexicanus

Almost always found near cliffs or large open rock faces, usually in arid country. More often heard than seen, but can be approachable when found.

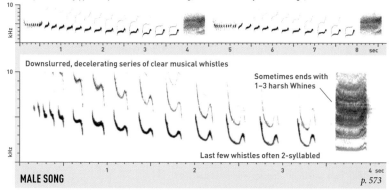

Repeats 1 songtype many times before switching; excited birds may double Songs, as here

Downslurred, decelerating series of clear musical whistles

Sometimes ends with 1–3 harsh Whines

Last few whistles often 2-syllabled

MALE SONG

p. 573

All year, but mostly Feb.–Aug. Variable, but distinctive for its musicality and decelerating, downslurred pattern. Individual notes may be upslurred, downslurred, or underslurred; speed at start varies greatly between songtypes. Repertoire size not fully known, but males appear to have at least 3 songtypes.

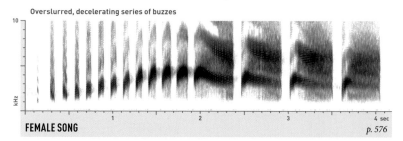

Overslurred, decelerating series of buzzes

FEMALE SONG

p. 576

All year. Distinctive but infrequent; first notes like Veet, last ones slightly harsher. Apparently given only by females, either in response to male Song or in territorial conflicts with other females. When duetting with mate, female often begins singing halfway through male's Song.

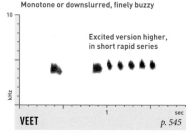

Monotone or downslurred, finely buzzy

Excited version higher, in short rapid series

VEET

p. 545

All year. Most common call, given in contact and mild alarm.

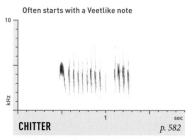

Often starts with a Veetlike note

CHITTER

p. 582

Mostly from agitated adults near the nest or fledged young. Highly plastic; sometimes shortened to a single Chit.

CAROLINA WREN

Thryothorus ludovicianus

Vocally conspicuous in many wooded or brushy habitats, including near human habitation. Range has expanded slowly northward in recent decades.

Males have 20–50 songtypes; consecutive songs are nearly always similar

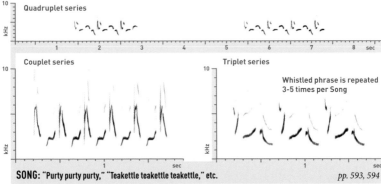

Quadruplet series

Couplet series

Triplet series

Whistled phrase is repeated 3-5 times per Song

SONG: "Purty purty purty," "Teakettle teakettle teakettle," etc.

pp. 593, 594

All year, by males; females not known to sing. Triplet series are most common, then couplet series, then quadruplet series; quintuplet series are rare. Males match songtypes during song duels; if one lacks a songtype, he may substitute Beert or equivalent. Songs in south Texas average lower, slower, simpler.

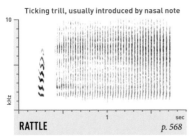

Ticking trill, usually introduced by nasal note

RATTLE *p. 568*

All year, reportedly only by females, often in response to and overlapping male Song. Also used in female-female aggression.

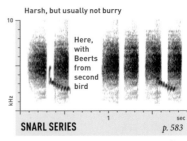

Harsh, but usually not burry

Here, with Beerts from second bird

SNARL SERIES *p. 583*

All year, in alarm; speed usually 2–3 notes/second. Agitated females may introduce series with 1–2 Pips; the result recalls chickadees.

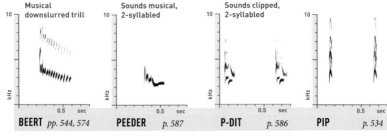

Musical downslurred trill

Sounds musical, 2-syllabled

Sounds clipped, 2-syllabled

BEERT *pp. 544, 574* **PEEDER** *p. 587* **P-DIT** *p. 586* **PIP** *p. 534*

All year, in mild alarm and contact. First 3 are male versions; Beert is most widespread and distinctive, but replaced in some areas by Peeder, P-dit, or other versions, especially in southern part of range. Pip is female version, occasionally run into short trill; compare Vimp of Winter Wren.

BEWICK'S WREN

Thryomanes bewickii

A prodigious singer found in brushy areas, open woods, and junipers. Formerly common in the East, but now rare east of the Mississippi.

Males have 10–20 songtypes; consecutive songs almost always similar

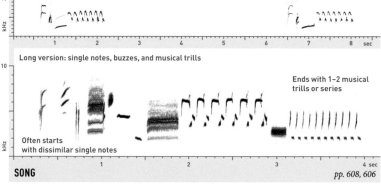

Long version: single notes, buzzes, and musical trills

Ends with 1–2 musical trills or series

Often starts with dissimilar single notes

SONG   *pp. 608, 606*

Nearly all year. Extremely varied, ranging from 1 to 4 seconds long, but always brightly musical. Long versions can be confused with Song Sparrow Song; short versions (top) have just 1–3 notes before a trill, recalling Black-throated Sparrow or Spotted Towhee. Female not known to sing.

High, polyphonic

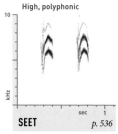

SEET   *p. 536*

Given continuously between Songs by some males; also singly in other situations.

Notes often Dziklike

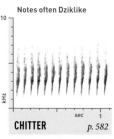

CHITTER   *p. 582*

All year, in agitation. Highly individual; some versions rather rattling.

Harsh, rasping

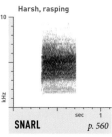

SNARL   *p. 560*

All year, in alarm. Sometimes given in series.

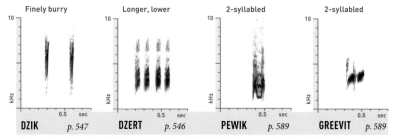

Finely burry

DZIK   *p. 547*

Longer, lower

DZERT   *p. 546*

2-syllabled

PEWIK   *p. 589*

2-syllabled

GREEVIT   *p. 589*

This species gives a bewildering array of 1- and 2-syllabled buzzy or noisy calls. Individuals usually give only one call type at a time, but one bird gave both Greevits and Dziks in mild agitation. Most common calls are 1-syllabled; some are intermediate between Dziks and Seets.

MARSH WREN (WESTERN)

Cistothorus palustris (paludicola group)

Locally common in freshwater cattail and bulrush marshes and coastal salt-marshes. Rather skulking, but vocally conspicuous, especially when breeding.

In excitement, some Songs are extended into complex warbles with many noisy notes (p. 602)

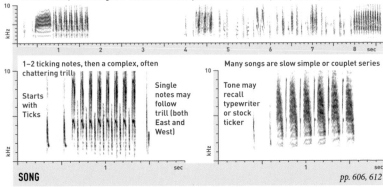

1–2 ticking notes, then a complex, often chattering trill

Starts with Ticks

Single notes may follow trill (both East and West)

Many songs are slow simple or couplet series

Tone may recall typewriter or stock ticker

SONG   *pp. 606, 612*

All year, especially Mar.–July, by males only, sometimes at night. Subsong common Aug.–Oct. and Jan.–Apr. Males have 100–220 similar songtypes; consecutive songs almost always differ. The Ticks can be followed by a buzz before the trill (top left), or by two consecutive trills. Some Songs resemble Sedge Wren's.

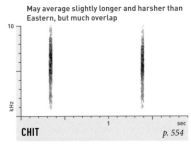

May average slightly longer and harsher than Eastern, but much overlap

CHIT   *p. 554*

All year, by both sexes, in contact and mild alarm; most common call. Fairly uniform. Juveniles give a sharp clear Chip (p. 531).

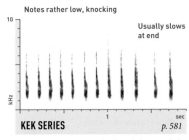

Notes rather low, knocking

Usually slows at end

KEK SERIES   *p. 581*

Little known; apparently given infrequently by territorial males. Western birds also give a Rasp like Eastern; no recordings available.

EASTERN AND WESTERN MARSH WRENS

Populations of Marsh Wren sort into two broad groups based on subtle differences in song forms and singing behavior. In addition to differences in introductory notes and the structure of terminal trills or series, Western males have larger repertoires and sing a greater variety of different-sounding songtypes. Call repertoires also apparently differ. In the zone of overlap between the two forms in the Great Plains, some interbreeding occurs, and some birds sing "mixed" songs with both Eastern and Western components. However, the majority of birds do not form mixed pairs. Outside the breeding season, subsinging birds may be identified with caution by general similarity of their songs to those of Eastern or Western breeding males.

MARSH WREN (EASTERN)

Cistothorus palustris (palustris group)

Nearly identical to the western group of subspecies, though some coastal populations are distinctively colored. Habits and habitat are similar.

Soft Twitters and musical warbles may precede or follow Songs, especially in coastal birds (p. 602)

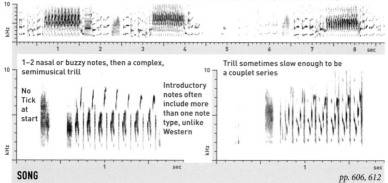

1–2 nasal or buzzy notes, then a complex, semimusical trill

No Tick at start

Introductory notes often include more than one note type, unlike Western

Trill sometimes slow enough to be a couplet series

SONG

pp. 606, 612

All year, especially Mar.–July, by males only, sometimes at night. Subsong common Aug.–Oct. and Jan.–Apr. Males have 30–70 similar songtypes; consecutive songs usually differ, but 2 songtypes may alternate, or 1 repeat a few times before switching. Usually lacks harsh Chatters; doubled trills rarer than in West.

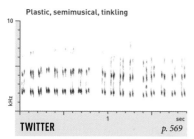

Plastic, semimusical, tinkling

TWITTER

p. 569

Little known; sometimes given in between songs, mostly by coastal birds. May recall distant Killdeer Trill.

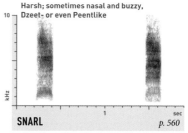

Harsh; sometimes nasal and buzzy, Dzeet- or even Peentlike

SNARL

p. 560

Similar to the notes that introduce Song; given separately by males during nest building and sometimes between Songs.

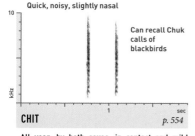

Quick, noisy, slightly nasal

Can recall Chuk calls of blackbirds

CHIT

p. 554

All year, by both sexes, in contact and mild alarm; most common call. Fairly uniform. Juveniles give a sharp clear Chip (p. 531).

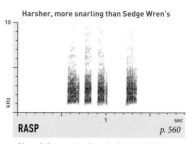

Harsher, more snarling than Sedge Wren's

RASP

p. 560

Given infrequently, in agitation or high alarm. Lower, harsher, and longer than Chit.

SEDGE WREN

Cistothorus platensis

A nomadic and opportunistic breeder in wet meadows of grass or sedges; may raise more than one brood per summer in widely separate locations.

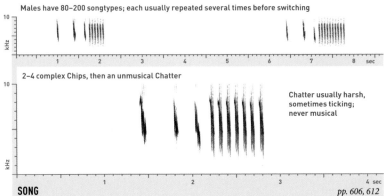

Males have 80–200 songtypes; each usually repeated several times before switching

2–4 complex Chips, then an unmusical Chatter

Chatter usually harsh, sometimes ticking; never musical

SONG *pp. 606, 612*

Mostly Feb.–July, by males. Females not known to sing. Unlike Marsh Wren, males usually repeat each songtype 10–20 times, but at dawn and in territorial conflicts, consecutive songs often differ. Many songtypes are very similar to the human ear.

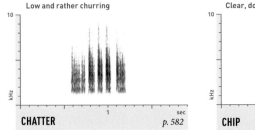

Low and rather churring

CHATTER *p. 582*

All year, by agitated birds, often with other calls. Highly plastic, but usually short (less than 2 seconds), unlike long Chatters of some other wrens.

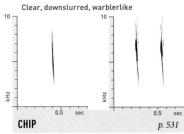

Clear, downslurred, warblerlike

CHIP *p. 531*

Apparently all year. Function and context little known. Generally higher than Chip of juvenile Marsh Wren.

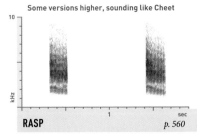

Some versions higher, sounding like Cheet

RASP *p. 560*

Often very short, sounding like Chup

CHURT *p. 546*

All year. Churt is the most common call; Rasp given in higher agitation. Both are plastic, intergrading with each other and variants. Compare to Chup and Churt of Common Yellowthroat (p. 546).

Cactus Wren

Campylorhynchus brunneicapillus

By far our largest wren, conspicuous both visually and vocally in southwestern deserts and arid shrublands, including near human habitation.

Males can have 14 or more similar songtypes; each usually repeated several times before switching

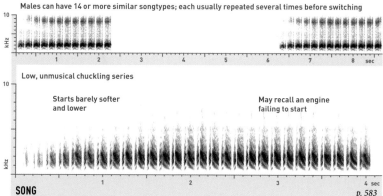

Low, unmusical chuckling series

Starts barely softer and lower

May recall an engine failing to start

SONG
p. 583

All year, mostly by males. Female Song reportedly infrequent, higher and softer than male's. Rather uniform rangewide; usually 3–4 seconds in length. Some versions fast enough to be called a slow trill. Some versions are couplet series.

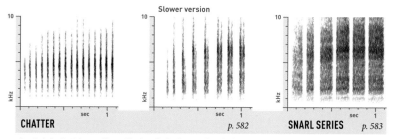

Slower version

CHATTER

p. 582

SNARL SERIES *p. 583*

A set of plastic, intergrading calls given by both sexes in agitation or alarm. Chatters infrequent; can recall oriole Chatters, but usually more rasping and plastic. Series of noisy Keklike notes given frequently in mild alarm, grading into longer Snarl Series when mobbing potential predators.

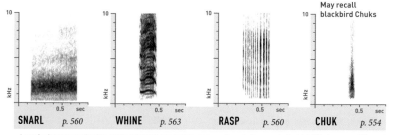

May recall blackbird Chuks

SNARL *p. 560*   **WHINE** *p. 563*   **RASP** *p. 560*   **CHUK** *p. 554*

A confusing array of calls, some of which grade into one another, but several of which likely serve different functions. Snarls and possibly Rasps given by mated pairs in greeting, Chuks in contact between adults and young. Whines may be given in high agitation.

GNATCATCHERS (Family Polioptilidae)

Some of our smallest and most active birds, gnatcatchers are slender, thin-billed, long-tailed, and subtly patterned. All species frequently flick their tails while gleaning insects from dense foliage.

Gnatcatcher vocal communication is highly complex and not well understood; this book only roughly outlines each species' repertoire. All vocalizations of a given species grade into one another, and nearly all occur in a variety of behavioral contexts. Many gnatcatchers, including Blue-gray and Black-capped but apparently not Black-tailed or California, include imitations of other bird species in their Complex Songs. All species snap their bills in alarm and aggression.

Black-tailed Gnatcatcher

Polioptila melanura

A bird of desert thorn scrub, especially mesquite and creosote thickets. The most vocally distinctive gnatcatcher in North America.

Complex Song (p. 597): harsh, snarling notes with orderly series of Chips and Tinks interspersed

Series of harsh, noisy notes

Length of notes and series variable, but tempo usually even

Faster series associated with higher levels of excitement

SIMPLE SONG   *p. 583*

Variable; Simple and Complex Song (p. 567) grade smoothly into one another, and into fast scolding Chatter (p. 582) that can be identical to House Wren's. Chip Series can strongly resemble Verdin's. More complex versions of Song exist with less repetition, more like Blue-gray Complex Song, but always dominated by harsh notes.

3–6 harsh downslurred Whines, each one lower and shorter

WHINE SERIES: Sheer-she-she   *pp. 563, 585*

Given in alarm, including in response to humans. Whinier, less noisy than other Black-tailed sounds; easy to confuse with Blue-gray.

Short burst of pure noise

Usually slightly downslurred; sometimes in series

SNARL   *p. 560*

All year, in contact; distinctively harsh, without any musicality. Like sound of letters "sh," but with abrupt start and end.

Blue-gray Gnatcatcher

Polioptila caerulea

Breeds in dry brush and in riparian, oak, and pinyon-juniper woods. Highly active, gleaning insects from leaves and occasionally catching them in midair.

Simple Song: highly variable, but short phrases sometimes repeat

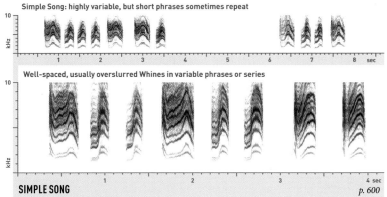

Well-spaced, usually overslurred Whines in variable phrases or series

SIMPLE SONG

p. 600

Given frequently, apparently serving functions including territorial advertisement. Often mistaken for a series of call notes; best distinguished from calls by repeating patterns, syncopated rhythms, and variety of note types, including overslurred notes. Grades smoothly into Complex Song; intermediates common.

Jumbled mix of varied high notes and short phrases

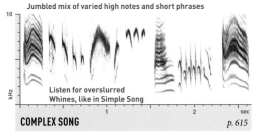

Listen for overslurred Whines, like in Simple Song

COMPLEX SONG

p. 615

Usually soft, not carrying far; by males in close courtship or in boundary disputes. May be continuous or given in short bursts; notes occasionally repeated. Often includes imitations.

High, downslurred Whine

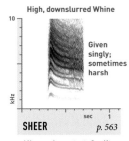

Given singly; sometimes harsh

SHEER

p. 563

All year, in contact. Quality similar to notes of Simple Song.

"EASTERN" BLUE-GRAY GNATCATCHER

Blue-gray Gnatcatchers breeding east of central Texas and western Oklahoma nest in moist deciduous woods and edges and differ slightly in vocalizations. Sheer calls average higher and less harsh, but may not be consistently identifiable. Simple Songs are higher and burrier, with most notes distinctly downslurred.

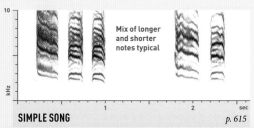

Mix of longer and shorter notes typical

SIMPLE SONG

p. 615

CALIFORNIA GNATCATCHER

Polioptila californica

Local and threatened in the U.S., where it breeds only in coastal sage scrub; in Baja California, also found in a range of other arid habitats.

Simple Song: harsh series and Mews often alternate

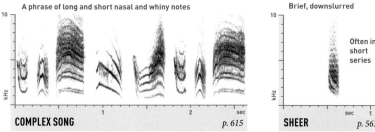

Series of 4–5 short harsh notes

Series of 2–3 Mews

SIMPLE SONG
pp. 583, 597

Mostly Jan.–June, reportedly by males only, in close pair interactions. Some versions consist of only series of short harsh notes, some contain only series of Mews, and some include both together, sometimes with other note types interspersed. Likely grades into Complex Song.

A phrase of long and short nasal and whiny notes

COMPLEX SONG
p. 615

Usually soft, not carrying far; by males in close courtship or in boundary disputes. Not known to include musical notes, series, or imitations. Stereotyped phrases sometimes repeat.

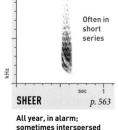

Brief, downslurred

Often in short series

SHEER
p. 563

All year, in alarm; sometimes interspersed with Mews.

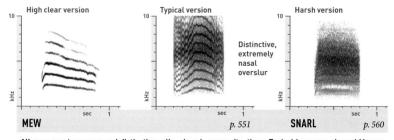

High clear version

Typical version

Harsh version

Distinctive, extremely nasal overslur

MEW
p. 551

SNARL
p. 560

All year; most common and distinctive calls, given in many situations. Typical long, overslurred Mew recalls Mew of Gray Catbird, but usually higher. Usually nasal, but some versions are polyphonic. Mews grade into harsher, more monotone Snarls, given in aggression and in high alarm.

BLACK-CAPPED GNATCATCHER

Polioptila nigriceps

Rare, local breeder in riparian thickets in deserts and arid canyons. Like Blue-gray, but breeding male has black cap; note long bill, graduated tail, voice.

Simple Song: Series of harsh notes, series of whiny notes, Chips, or a combination

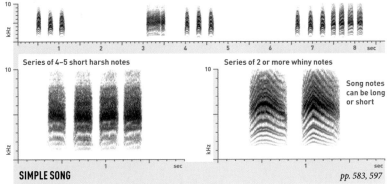

Series of 4–5 short harsh notes

Series of 2 or more whiny notes

Song notes can be long or short

SIMPLE SONG *pp. 583, 597*

Probably mostly Jan.–June, by males in close pair interactions; has also been recorded from solo males. Not well known, but usually consists of series, sometimes with Whine or Snarl calls interspersed, or variable Chips either singly or in series. Grades into Complex Song.

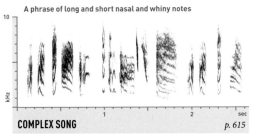

A phrase of long and short nasal and whiny notes

COMPLEX SONG *p. 615*

Usually soft; likely by males in close courtship or in boundary disputes. Some versions are more like Simple Song, but with greater variety, including musical notes, series, and imitations.

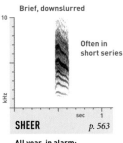

Brief, downslurred

Often in short series

SHEER *p. 563*

All year, in alarm; sometimes interspersed with Mews.

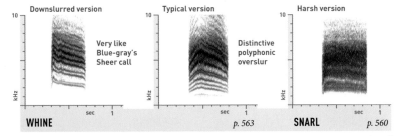

Downslurred version

Very like Blue-gray's Sheer call

WHINE

Typical version

Distinctive polyphonic overslur

p. 563

Harsh version

SNARL *p. 560*

All year. Whine is most common call, given in many situations. Most common versions shown; overslurred versions are distinctive, as are the rarer underslurred and upslurred versions. Whines grade into harsher, more monotone, churring Snarls, given in aggression and in high alarm.

DIPPERS (Family Cinclidae)

These remarkable, rotund birds can be found only along rushing streams and occasionally other types of shorelines. They fly only low over the water. When foraging they bob their entire bodies compulsively, hopping between exposed rocks and jumping into the current for aquatic insects, almost as though they recognize no boundary between water and air.

Even those who have known dippers for quite a while can be taken aback by the beauty and complexity of their songs, which are highly musical and likely learned. Calls are rather few and simple, and likely innate.

AMERICAN DIPPER

Cinclus mexicanus

Intimately tied to fast-moving streams with exposed rocks or logs, mostly in foothills and mountains. White eyelid is obvious when the bird blinks.

Consecutive series usually differ, often strung together without pause

Slow musical series mixed with single notes and trills

Not known to include imitations

Many notes burry, some whistled

SONG p. 601

All year, by both sexes, especially Mar.–July. Often quite loud, so as to be heard over river noise. Parents sing near nest after feeding chicks. Repertoire size unstudied; individuals seem to have at least 8 stereotyped phrases of 1–5 consecutive series or notes, which are then recombined into longer songs.

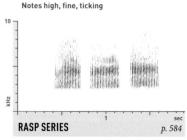

Notes high, fine, ticking

RASP SERIES p. 584

Little known. In sole available recording, call is given repeatedly during nest building, sometimes in response to mate's Song.

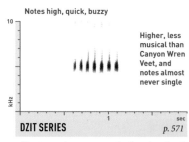

Notes high, quick, buzzy

Higher, less musical than Canyon Wren Veet, and notes almost never single

DZIT SERIES p. 571

All year; most common vocalization, given in agitation, in flight, and probably in contact.

KINGLETS (Family Regulidae)

Charismatic and active, these are the smallest birds in North America outside the hummingbird family. Our two species are not particuarly closely related. Both join mixed-species flocks in migration and winter. Songs appear to be learned.

WRENTIT (Family Sylviidae)

A unique bird, skulking but highly vocal; most closely related to the parrotbills and the *Sylvia* warblers of the Old World. It is not known whether sounds are learned.

GOLDEN-CROWNED KINGLET

Regulus satrapa

Found mostly in spruce trees year-round. Calls very like those of Brown Creeper, with which it often flocks outside the breeding season.

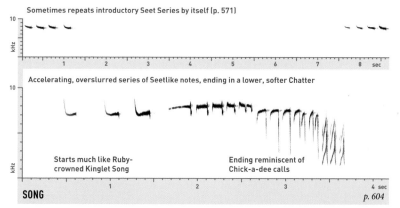

Sometimes repeats introductory Seet Series by itself (p. 571)

Accelerating, overslurred series of Seetlike notes, ending in a lower, softer Chatter

Starts much like Ruby-crowned Kinglet Song

Ending reminiscent of Chick-a-dee calls

SONG   *p. 604*

Mostly May–July. Slight geographic variation. Structure like the Chick-a-dee calls of parids (p. 604), with several different note types always in the same order, each repeated 0–6 times. Males appear to have only 1 songtype, but consecutive songs may vary considerably in number and type of notes.

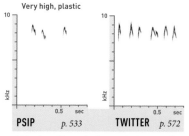

Very high, plastic

PSIP *p. 533*

Also singly

TWITTER *p. 572*

Psip given all year, in close contact; can run into plastic Twitter. 2-noted Psittip (not shown) given constantly by fledglings in late summer.

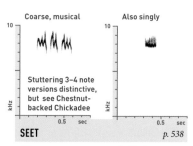

Coarse, musical

Stuttering 3–4 note versions distinctive, but see Chestnut-backed Chickadee

SEET *p. 538*

All year; most common call. Single-note versions, often given in alarm, average higher, less musical than Brown Creeper's.

Ruby-crowned Kinglet

Regulus calendula

Breeds in coniferous and mixed forests; found in various habitats in migration and winter. Male's red crown inconspicuous unless flared in agitation.

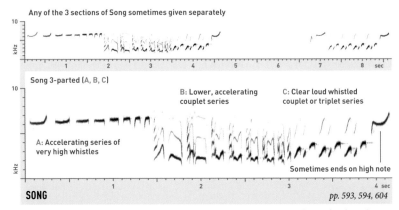

Any of the 3 sections of Song sometimes given separately

Song 3-parted (A, B, C)

B: Lower, accelerating couplet series

C: Clear loud whistled couplet or triplet series

A: Accelerating series of very high whistles

Sometimes ends on high note

SONG   *pp. 593, 594, 604*

Mostly Feb.–Aug.; versions on spring migration often plastic. Highly variable. Repertoire size unknown; individuals may have a single songtype. Female Song is shorter, apparently like first 2 parts of male Song; more study needed.

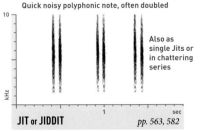

2-syllabled notes like in B section of Song

Singly or in rapid couplet series

In agitation, can speed up into complex trill

PITTER   *p. 590*

Mostly May–Aug. Given incessantly by alarmed or anxious birds. Highly variable; likely learned rather than innate. Second syllable may be higher or lower than first.

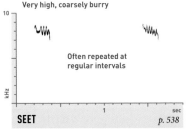

Quick noisy polyphonic note, often doubled

Also as single Jits or in chattering series

Very high, coarsely burry

Often repeated at regular intervals

JIT or JIDDIT   *pp. 563, 582*

SEET   *p. 538*

All year. Most common call, particularly in mixed-species foraging or mobbing flocks, but also from solo birds.

Mostly July–Aug., from begging juveniles; plastic. Very similar to burry Seets of Golden-crowned Kinglet and Brown Creeper.

WRENTIT

Chamaea fasciata

A small, skulking, sedentary resident of coastal scrub and chapparal, the sole North American representative of an Old World family.

Male Song sometimes extended with long series of Female Songlike notes, usually in duet with female

Here, female notes in red

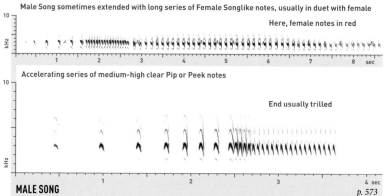

Accelerating series of medium-high clear Pip or Peek notes

End usually trilled

MALE SONG

p. 573

All year, but especially Apr. –Aug.; rather uniform rangewide. As far as is known, individual males have a single songtype, but number and speed of notes may vary with level of agitation; tone quality sometimes reported to vary as well. Purpose of extended version of Song unknown; more study needed.

Series of medium-high semimusical Chirps

Rhythm often slightly unsteady; speed somewhat variable

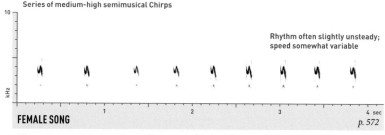

FEMALE SONG

p. 572

All year, but especially Apr. –Aug.; uniform rangewide. Less frequent than Male Song and almost always given at the same time, in unsynchronized duet, often when the pair is fairly far apart and from inside dense cover; easily overlooked.

Very soft quick nasal note

Does not carry far

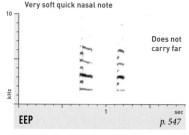

EEP

p. 547

Probably all year, in close contact; rarely heard owing to low volume. Highly plastic; some versions lower than shown, many much shorter.

Harsh but not loud; very fast, almost a Churr

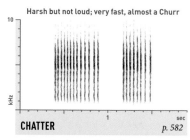

CHATTER

p. 582

All year, in alarm and agitation; most common call. Uniform, but highly plastic in length; some versions can continue for many seconds.

THRUSHES (Family Turdidae)

The thrushes are among our most celebrated singers. They are remarkable aesthetically, because unlike many other birds, their fluting, polyphonic voices often contain harmonies similar to those in traditional Western music, resulting in a beauty easily appreciated by humans. But thrushes are also remarkable behaviorally, displaying highly developed vocal communication systems with multiple Song categories and large call repertoires. They are favorites of night migration enthusiasts, thanks to their loud, frequent, and rather distinctive nocturnal flight calls.

Excited Songs of Thrushes

Most North American thrushes have at least two distinct modes of singing: a loud, musical song used for long-distance communication, and a softer, more complex, often faster song used in close-range courtship and/or aggressive encounters. In the genus *Catharus* (Veery, Hermit Thrush, and Swainson's Thrush), the Excited Song is much like the regular Song, but with additional components either in between the regular Song phrases, or appended to them. These additional components are often complex, polyphonic, and trilled.

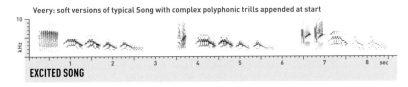

Veery: soft versions of typical Song with complex polyphonic trills appended at start

EXCITED SONG

Hermit Thrush: typical Song, with complex polyphonic phrases in between, or appended to end

EXCITED SONG

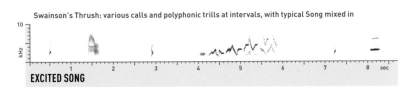

Swainson's Thrush: various calls and polyphonic trills at intervals, with typical Song mixed in

EXCITED SONG

Call Repertoires of Veery

The Veery is closely related to two species from eastern North America, the Bicknell's and Gray-cheeked Thrushes. All three species have similar complex and poorly understood repertoires of burry call notes. Each adult has at least 5–10 different versions of these calls, often giving long series of identical calls before switching to a different type. Mates and neighbors have a strong tendency to match call types. Sometimes consecutive calls may vary, or one type may slowly grade into another over the course of several renditions. Similar calls are given by all three species during nocturnal migration; they may be less variable than those given during the day, but the relationship between daytime and nighttime calls is poorly understood. More study needed.

MOUNTAIN BLUEBIRD

Sialia currucoides

A lovely bird of mountain meadows, sagebrush steppes and burns. Voice usually quite soft. Like other bluebirds, migrates early in spring and late in fall.

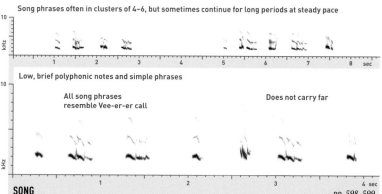

Song phrases often in clusters of 4–6, but sometimes continue for long periods at steady pace

Low, brief polyphonic notes and simple phrases

All song phrases resemble Vee-er-er call

Does not carry far

SONG

pp. 598, 599

Mostly Mar.–July, prior to sunrise; occasionally at dusk, rarely during the day. Not known whether female sings. Repertoire size unknown. Clustered and continuous versions of Song may serve different functions; more study needed. Various calls sometimes incorporated.

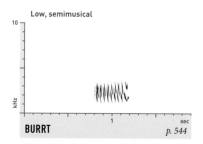

Low, semimusical

BURRT

p. 544

At least Apr.–June, apparently in excitement or during interactions. Associated with Songlike notes and with Chup.

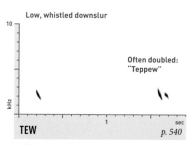

Low, whistled downslur

Often doubled: "Teppew"

TEW

p. 540

All year, by both sexes, in alarm, often with Chup and Chuppa-tew. Tew rarely run into series of more than 2 notes.

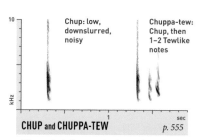

Chup: low, downslurred, noisy

Chuppa-tew: Chup, then 1–2 Tewlike notes

CHUP and CHUPPA-TEW

p. 555

All year, by both sexes, in alarm, often with Tew and Teppew. Chup very rarely run into Chatter as in Eastern and Western Bluebirds.

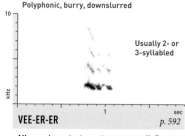

Polyphonic, burry, downslurred

Usually 2- or 3-syllabled

VEE-ER-ER

p. 592

All year, in contact; most common call. Does not carry far. Quite plastic; some versions are clear and/or single-syllabled. Distinctive.

WESTERN BLUEBIRD

Sialia mexicana

Frequents open pine forests, oak and pinyon-juniper woodlands, and burns. In winter, sometimes flocks with other bluebird species.

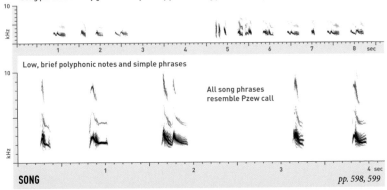

Song phrases usually given at fairly steady pace for long periods; rarely in clusters of 4–8

Low, brief polyphonic notes and simple phrases

All song phrases resemble Pzew call

SONG *pp. 598, 599*

Mostly Mar.–July, prior to sunrise; occasionally at dusk, rarely during the day. Not known whether female sings. Repertoire size unknown. Rare clustered version of Song can recall Eastern Bluebird, but pace of phrases considerably slower. Chups and short Chatters sometimes incorporated.

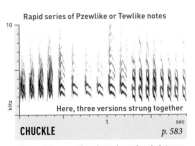

Rapid series of Pzewlike or Tewlike notes

Here, three versions strung together

CHUCKLE *p. 583*

Not well known; given in early spring, in interactions. Variable and plastic; individuals likely have multiple versions.

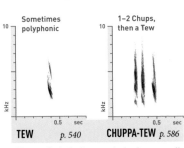

Sometimes polyphonic

1–2 Chups, then a Tew

TEW *p. 540*　　**CHUPPA-TEW** *p. 586*

These calls both given rarely, in alarm, usually with Chup or Chatter. Similar to corresponding calls of Mountain Bluebird.

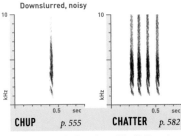

Downslurred, noisy

CHUP *p. 555*　　**CHATTER** *p. 582*

All year, in agitation, often while flicking wings. Quite plastic in rhythm. Chatters more common than single Chups.

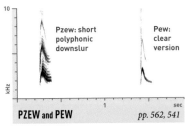

Pzew: short polyphonic downslur

Pew: clear version

PZEW and PEW *pp. 562, 541*

All year; most common call, in contact. Somewhat variable and plastic; most common variants shown.

EASTERN BLUEBIRD

Sialia sialis

Local in the West, frequenting meadows, orchards, and other open areas
with scattered trees. Readily nests in
boxes. An early spring migrant.

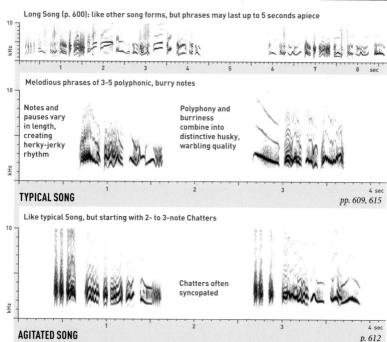

Long Song (p. 600): like other song forms, but phrases may last up to 5 seconds apiece

Melodious phrases of 3-5 polyphonic, burry notes

Notes and
pauses vary
in length,
creating
herky-jerky
rhythm

Polyphony and
burriness
combine into
distinctive husky,
warbling quality

TYPICAL SONG

pp. 609, 615

Like typical Song, but starting with 2- to 3-note Chatters

Chatters often
syncopated

AGITATED SONG

p. 612

Mostly Mar.–July, by both sexes. Typical Song given more often by males, loudly from prominent perch or
very softly, often in proximity to mate. Individual birds apparently sing at least 3–16 stereotyped phrase
types; consecutive songs almost always different. Grades into Agitated Song, given by both sexes in presence of potential predators, including humans, usually when mate is away from territory. Long Song little
known; may be given in excitement, possibly in courtship.

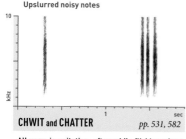

Upslurred noisy notes

CHWIT and CHATTER *pp. 531, 582*

All year, in agitation, often while flicking wings.
Quite plastic in rhythm. Chatters more common
than single Chwits.

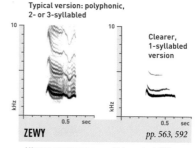

Typical version: polyphonic,
2- or 3-syllabled

Clearer,
1-syllabled
version

ZEWY *pp. 563, 592*

All year; most common call, in contact. Often underslurred. Somewhat variable and plastic;
male versions usually longer than female's.

Townsend's Solitaire

Myadestes townsendi

Slender and long-tailed; often perches
conspicuously on highest point in tree.
Breeds in western mountains; in winter, usually found near junipers.

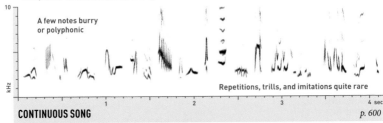

Complex musical warble, often lasting 30 seconds or more

A few notes burry
or polyphonic

Repetitions, trills, and imitations quite rare

CONTINUOUS SONG *p. 600*

All year, by breeding males in high, slow, circling flight, and by both sexes to establish summer and winter
territories. Extraordinarily long, complex, and beautiful. Pace rather slow; most notes are musical whistles, with a few chips, burrs, and creaks. Often soft, or fading slowly in or out.

In territorial disputes, may repeat the same phrase several times with only minor variation

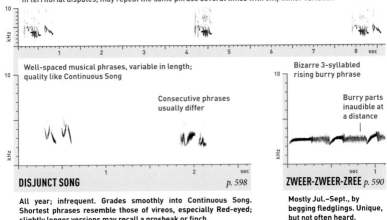

Well-spaced musical phrases, variable in length;
quality like Continuous Song

Consecutive phrases
usually differ

DISJUNCT SONG *p. 598*

All year; infrequent. Grades smoothly into Continuous Song.
Shortest phrases resemble those of vireos, especially Red-eyed;
slightly longer versions may recall a grosbeak or finch.

Bizarre 3-syllabled
rising burry phrase

Burry parts
inaudible at
a distance

ZWEER-ZWEER-ZREE *p. 590*

Mostly Jul.–Sept., by
begging fledglings. Unique,
but not often heard.

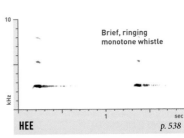

Lower, less
musical than
robin Cheep

Variably
harsh

Brief, ringing
monotone whistle

CHEEP *p. 535*

WHINE *p. 563*

HEE *p. 538*

These calls given in defense of winter territory
against waxwings and bluebirds (not robins or
other solitaires). Fairly plastic.

All year; most common call. Distinctive; uniform
and fairly stereotyped. Often repeated at 1- to
5-second intervals for long periods.

VARIED THRUSH

Ixoreus naevius

A retiring bird of dense, humid, mossy coniferous forest. Stunningly patterned, with a haunting voice. In winter, also found in parks and residential areas.

Consecutive songs given on different pitches, usually 5–10 seconds apart

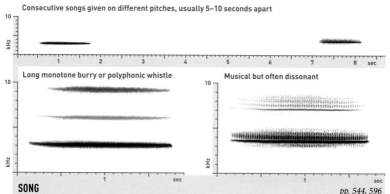

Long monotone burry or polyphonic whistle

Musical but often dissonant

SONG

pp. 544, 596

An ethereal sound, emblematic of the dark, dense, lush forests this species calls home. Mostly Mar.–July, by males; not known whether females sing. Individuals have 3–7 songtypes. Occasionally gives fast Song with only small pauses between notes, perhaps a type of Excited Song or juvenile practice.

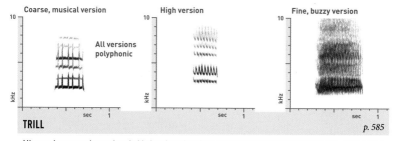

Coarse, musical version

All versions polyphonic

High version

Fine, buzzy version

TRILL

p. 585

All year, in aggression and probably in other situations. Individuals have multiple versions; winter flocks can chorus softly with a bewildering variety of similar calls. Repertoire size unknown. Usually given singly, but sometimes 2 or more different calls given in quick succession. A few versions are clear, not trilled.

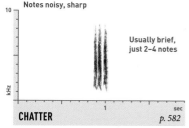

Notes noisy, sharp

Usually brief, just 2–4 notes

CHATTER

p. 582

All year, in moderate to high alarm, often with Chup. Harsh quality recalls Western Bluebird Chatter, but much faster.

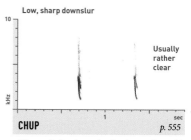

Low, sharp downslur

Usually rather clear

CHUP

p. 555

All year, in low to moderate alarm, often with Chatter. Usually clearer, more Tewlike than Chup of Hermit Thrush.

VEERY

Catharus fuscescens

Renowned for its ethereal, downward-spiraling song. In the East, breeds in dense deciduous forest understory; in the West, breeds in willow thickets.

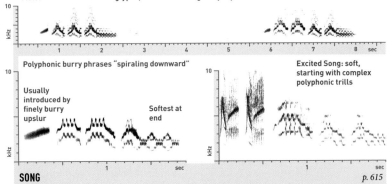

Males have 2–6 similar songtypes; consecutive songs may repeat or differ

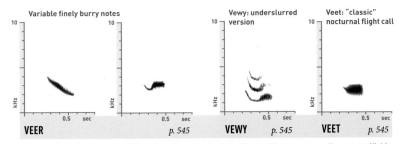

Polyphonic burry phrases "spiraling downward"

Usually introduced by finely burry upslur

Softest at end

Excited Song: soft, starting with complex polyphonic trills

SONG   *p. 615*

Mostly May–July. Distinctive and arrestingly beautiful. Overslurred phrases in downslurred series create "spiraling" sound; like Swainson's Thrush, but Song falls rather than rises in pitch. Excited Song (p. 366) often heard in altercations; much softer than typical Song, audible only at close range.

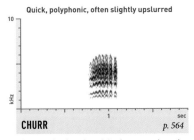

Variable finely burry notes

Vewy: underslurred version

Veet: "classic" nocturnal flight call

VEER   *p. 545*   **VEWY**   *p. 545*   **VEET**   *p. 545*

All year, by both sexes, especially on breeding grounds during the day. Most common call category. Highly variable; adults have at least 5–10 different versions, some of which may differ in function. Usually distinctive in the West, especially underslurred versions. Overslurred versions rare.

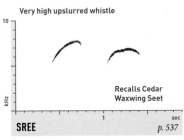

Quick, polyphonic, often slightly upslurred

Very high upslurred whistle

Recalls Cedar Waxwing Seet

CHURR   *p. 564*   **SREE**   *p. 537*

Apparently given by adults in aggression; also interspersed into bouts of Excited Song. Often introduced by soft, noisy Chuklike notes.

Usually upslurred, unlike similar calls of other thrushes, but sometimes nearly monotone. Given in altercations or in alarm.

HERMIT THRUSH

Catharus guttatus

Breeds in mixed and coniferous forests, where it gives extended Song performances at dawn and dusk. Like other thrushes, a vocal nocturnal migrant.

Males average 8–10 songtypes; consecutive songs always differ, often alternating high and low

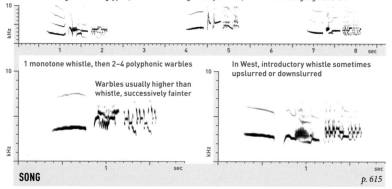

1 monotone whistle, then 2–4 polyphonic warbles

Warbles usually higher than whistle, successively fainter

In West, introductory whistle sometimes upslurred or downslurred

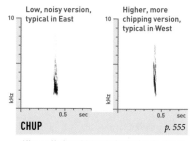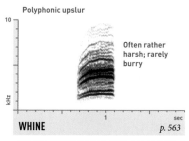

SONG   *p. 615*

Mostly May–Aug. A lovely, delicate song whose intricacy can be appreciated only at close range. Best distinguished from other thrushes by long monotone whistle at start of each song. Soft subsong and plastic song sometimes given in migration and winter. Excited Song (see p. 366) includes extra polyphonic warbles.

Low, noisy version, typical in East

Higher, more chipping version, typical in West

Polyphonic upslur

Often rather harsh; rarely burry

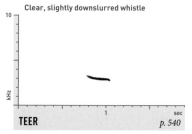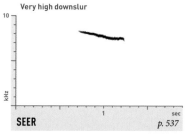

CHUP   *p. 555*

All year. Varies with level of agitation; in all populations, higher-pitched versions and longer low, burry versions associated with alarm.

WHINE   *p. 563*

All year. Distinctive; less harsh and burry than Spotted Towhee Wheeze. Often given with Chups; may indicate a lower level of alarm.

Clear, slightly downslurred whistle

Very high downslur

TEER   *p. 540*

All year. Given regularly by night migrants; also occasionally on breeding grounds, where finely burry versions have been recorded.

SEER   *p. 537*

All year, from motionless, often hidden birds; warns of aerial or other predators. Like other downslurred thrush Seers.

Swainson's Thrush (Russet-backed)

Catharus ustulatus (*ustulatus* group)

Breeds in mixed and coniferous forests and in deciduous riparian areas. Slightly redder above than Olive-backed form, like Veery, but note brownish flanks.

Males have 3–7 songtypes, differing only in details; consecutive songs differ

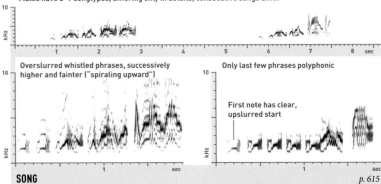

Overslurred whistled phrases, successively higher and fainter ("spiraling upward")

Only last few phrases polyphonic

First note has clear, upslurred start

SONG　　　　　　　　　　　　　　　*p. 615*

Mostly Mar.–July. Often sings in spring migration, but for a rather short period on the breeding grounds, mostly in June in many areas, and especially at dawn and dusk. First note consistently different from Olive-backed's on spectrogram, but this is hard to hear. Averages more repeated phrases than Olive-backed.

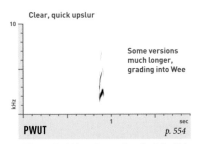

Clear, quick upslur

Some versions much longer, grading into Wee

PWUT　　　　　　　　　　　　*p. 554*

All year. Variable and somewhat plastic in pitch, length, and tone. May average slightly longer than Olive-backed's, but much overlap.

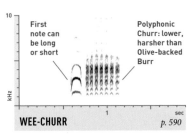

First note can be long or short

Polyphonic Churr: lower, harsher than Olive-backed Burr

WEE-CHURR　　　　　　　　　*p. 590*

Mostly May–Aug., with Pwuts in alarm near nest; also in territorial conflicts. Churr also given separately; sounds much like Veery's.

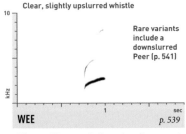

Clear, slightly upslurred whistle

Rare variants include a downslurred Peer (p. 541)

WEE　　　　　　　　　　　　*p. 539*

All year. Given regularly on breeding grounds and in nocturnal migration. Averages lower and longer than Olive-backed's.

Presumably like Olive-backed's; no recordings available

SEER　　　　　　　　　　　　*p. 537*

All year, from motionless, often hidden birds; warns of aerial or other predators. Like other downslurred thrush Seers.

SWAINSON'S THRUSH (OLIVE-BACKED)

Catharus ustulatus (*swainsoni* group)

Breeds in mixed and coniferous forests, and in deciduous riparian areas. Hybridizes with Russet-backed form where ranges meet.

Males have 3–7 songtypes, differing only in details; consecutive songs differ

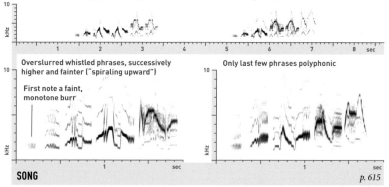

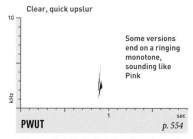

Overslurred whistled phrases, successively higher and fainter ("spiraling upward")

First note a faint, monotone burr

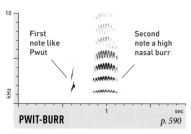

Only last few phrases polyphonic

SONG
p. 615

Mostly Mar.–July. Often sings in spring migration, but for a rather short period on the breeding grounds, mostly in June in many areas, and especially at dawn and dusk. No other thrush Song consistently rises in pitch. Excited Song adds trills at start of song phrases and/or call-like notes in between.

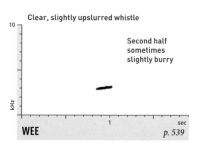

Clear, quick upslur

Some versions end on a ringing monotone, sounding like Pink

PWUT
p. 554

All year. Variable in pitch, length, and tone, but always at least vaguely upslurred. Most common call.

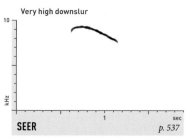

First note like Pwut

Second note a high nasal burr

PWIT-BURR
p. 590

Mostly May–Aug., with Pwuts in alarm and in territorial conflicts. Burr also given singly. Quite different from Russet-backed's Wee-Churr.

Clear, slightly upslurred whistle

Second half sometimes slightly burry

WEE
p. 539

All year. Given regularly on breeding grounds and in nocturnal migration. Much like the call of the Spring Peeper frog.

Very high downslur

SEER
p. 537

All year, from motionless, often hidden birds; warns of aerial or other predators. Like other downslurred thrush Seers.

AMERICAN ROBIN

Turdus migratorius

One of our most familiar birds, found in almost any habitat. Forms large flocks in migration and winter; famously pulls earthworms from lawns in summer.

After dawn, often lacks "hissely" phrases at end of each series

Short musical phrases, usually in series of 4–6

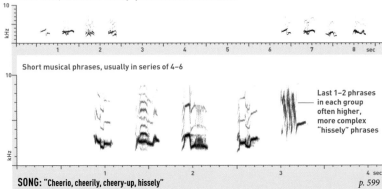

Last 1–2 phrases in each group often higher, more complex "hissely" phrases

SONG: "Cheerio, cheerily, cheery-up, hissely"

p. 599

Well known. All year, but especially Mar.–June. Both sexes reported to sing. Individuals have 6–20 "cheerio" phrases, which are medium-low, usually of 2–3 syllables, rarely burry or polyphonic. In a series, phrases usually switch constantly, but are sometimes repeated up to several times in a row.

A continuous string of "hissely" phrases, usually audible only at close range

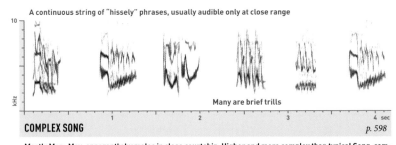

Many are brief trills

COMPLEX SONG

p. 598

Mostly Mar.–May, apparently by males in close courtship. Higher and more complex than typical Song, composed mostly of "hissely" phrases, of which an individual may have 50–100. Some louder Complex Songs mix "cheerio" phrases, "hissely" phrases, and multiple versions of Whinny; function unknown.

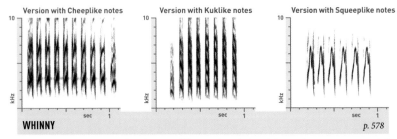

Version with Cheeplike notes

Version with Kuklike notes

Version with Squeeplike notes

WHINNY

p. 578

All year, likely by both sexes. Function not well known; given singly or in long series, or as a component of Complex Song. Individuals have several fairly stereotyped versions, each usually repeated before switching, but some birds switch frequently. First and last note often lower.

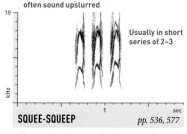

Notes higher, slightly longer than Cheep; often sound upslurred

Usually in short series of 2–3

SQUEE-SQUEEP *pp. 536, 577*

All year, by both sexes; function not well understood. May grade into both Whinny and Cheep. Often followed by a series of soft Kuks.

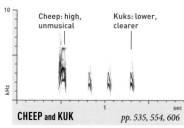

Cheep: high, unmusical

Kuks: lower, clearer

CHEEP and KUK *pp. 535, 554, 606*

All year, by both sexes, in alarm. Often in above pattern, with 1–2 Cheeps followed by 3–4 Kuks (p. 569), but both sounds also given separately.

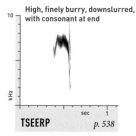

High, finely burry, downslurred, with consonant at end

TSEERP *p. 538*

All year. Sometimes in flight, including by night migrants. Can introduce Kuks.

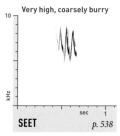

Very high, coarsely burry

SEET *p. 538*

All year. Often in flight, including by night migrants. Can introduce Kuks.

Very high, downslurred

SEER *p. 537*

From motionless, often hidden birds; warns of aerial or other predators.

REGIONAL VARIATION IN THRUSHES

In addition to the Russet-backed and Olive-backed forms of Swainson's Thrush, there are also distinctive regional forms of Hermit Thrush and Veery.

Hermit Thrush separates into three broad groups by plumage: a richly colored group in the East, a grayish group in the Interior West, and a group on the Pacific Coast like eastern birds, but with grayish flanks. Subtle differences in Song exist. In particular, eastern birds sing with a greater difference in pitch between their songtypes. Eastern birds almost always have at least one song in their repertoire with an introductory whistle below 2 kHz. These low songs may suggest Wood Thrush for those who are accustomed to Hermit Thrushes from the western groups, whose songtypes all cluster in a narrower range of high pitches.

The introductory whistles of eastern birds are almost always nearly monotone. These whistles tend to be slightly slurred in the Interior West, and can be strongly slurred and sometimes even slightly polyphonic or burry in birds along the Pacific Coast.

Chup calls also differ on average between eastern birds and other populations (see p. 373).

Veery also shows distinct regional variation in plumage. Eastern birds have rich brown upperparts, while Western birds tend to be significantly grayer. Differences in Song, if they exist, are apparently subtle.

CATBIRDS, THRASHERS, AND MOCKINGBIRDS
(Family Mimidae)

Members of this family, known as "mimids," are medium-sized, long-billed, long-tailed birds, generally plumaged in grays and browns, that tend to skulk in deep cover. Songs are learned, and famous for including vocal imitations of other bird species. All North American mimids incorporate at least some imitations into their Songs, and some, such as the Northern Mockingbird, even occasionally imitate non-bird sounds such as phone ringtones and car alarms. In addition to Songs, at least some call types may be learned.

Mimids tend to have huge Song repertoires. The Brown Thrasher is estimated to have the largest repertoire of any bird species yet studied, with individuals likely capable of singing 1,500–2,000 different phrase types, and possibly more. Northern Mockingbirds can learn new sounds throughout their lives, unlike most birds, and it has been suggested that some mimids may innovate new Song phrases continually. In these species, the patterns in which the sounds are arranged are more important for species identification than the types of sounds that are sung.

Regional Variation in Curve-billed Thrashers

In the U.S., eastern and western populations of Curve-billed Thrashers differ slightly in plumage, genetics, and some vocalizations. Eastern birds of the nominate subspecies group show paler breasts with more contrasting spots, more distinct wingbars, and larger pale spots on the tail corners. Western birds of the *palmeri* subspecies group (sometimes called "Palmer's Thrasher") have duskier breasts with less distinct spotting and wingbars, and smaller pale spots on the tail.

The typical whistled call notes of the western and eastern subspecies groups are distinct, but the situation is complicated by a third call type found around the Chiricahua Mountains in southeastern Arizona and southwest New Mexico. Birds giving this third call type appear to have plumage typical of the eastern subspecies group. However, the call type has occasionally been recorded well within the range of the *palmeri* group; the plumage of the birds giving it was not noted. In all groups, calls become longer and more plastic in agitation, complicating identification.

The whistled calls may be learned rather than innate. More study is needed.

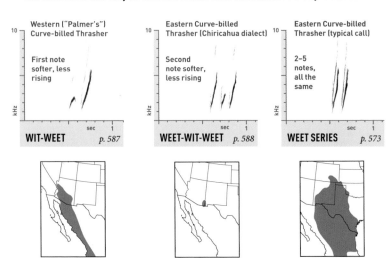

Western ("Palmer's") Curve-billed Thrasher
First note softer, less rising
WIT-WEET *p. 587*

Eastern Curve-billed Thrasher (Chiricahua dialect)
Second note softer, less rising
WEET-WIT-WEET *p. 588*

Eastern Curve-billed Thrasher (typical call)
2–5 notes, all the same
WEET SERIES *p. 573*

GRAY CATBIRD

Dumetella carolinensis

Common, but more often heard than seen. Skulks in dense deciduous undergrowth, especially along woodland edges. Named for its catlike Mew call.

Pauses between song phrases frequently quite long (p. 598)

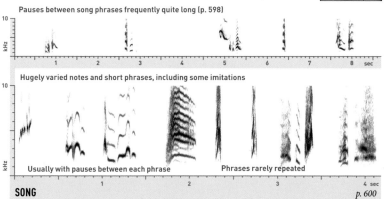

Hugely varied notes and short phrases, including some imitations

Usually with pauses between each phrase   Phrases rarely repeated

SONG *p. 600*

Mostly Apr.–Aug., by male; female sings rarely and softly. Males may have 100 or more song phrases. Soft subsong often heard in fall. Length of pauses between phrases varies greatly, but fairly constant within a singing bout. Occasionally repeats phrases several times, suggesting a mockingbird.

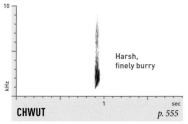

Harsh, finely burry

CHWUT *p. 555*

All year, in mild alarm. Like Hermit Thrush Chup, but rougher, slightly rising.

Very high, almost a Tsip

CHIP *p. 531*

High, brief, burry

SEET *p. 538*

Many types of high, brief calls given by begging juveniles, including a rising Chwit (p. 531). Adults sometimes give some of these calls.

Typical version: mostly clear

Harsh version: noisy, slightly grating

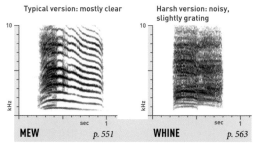

Loud, abrupt, harsh

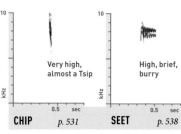

Usually sounds 2-noted, each "note" actually a rapid couplet

MEW *p. 551*   **WHINE** *p. 563*   **RATCHET** *p. 568*

All year; Mew is most common and distinctive call. Highly variable and plastic, becoming harsh Whine in excitement or alarm. Always at least partly polyphonic.

Given at dawn, in escape flight, or during chases. Plastic, but always short.

BROWN THRASHER

Toxostoma rufum

Found in dense deciduous brush and woodland edges. Named "thrasher" for its habit of foraging by tossing up leaf litter on the ground.

Pauses between song phrases quite plastic, but usually at least as long as the phrases themselves

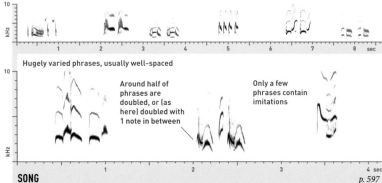

Hugely varied phrases, usually well-spaced

Around half of phrases are doubled, or (as here) doubled with 1 note in between

Only a few phrases contain imitations

SONG
p. 597

Mostly Mar.–July, by males, usually from conspicuous perch; females not known to sing. Individual males estimated to have more than 1,000 song phrases. Very soft versions of Song sometimes heard in close courtship; soft subsong also given in fall and winter.

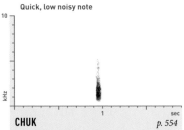

Quick, low noisy note

CHUK
p. 554

All year, likely in alarm. Variable, but all versions low, brief, and fairly soft.

Usually fairly brief

SNARL
p. 560

All year, in agitation or alarm.

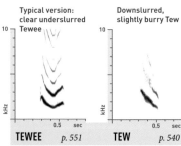

Typical version: clear underslurred Tewee

Downslurred, slightly burry Tew

TEWEE
p. 551

TEW
p. 540

All year, possibly in alarm or contact. Quite variable and plastic; can resemble certain calls of Veery, but usually clearer.

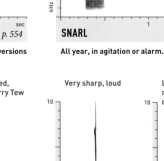

Very sharp, loud

Like Smack, but notes very quickly doubled

SMACK
p. 531

RATCHET
p. 568

All year. Smack is most common call, apparently given in alarm. Ratchet rare; notes may be more smacking than in Gray Catbird.

BENDIRE'S THRASHER

Toxostoma bendirei

Uncommon and local in desert scrub. Very like Curve-billed Thrasher, but bill shorter, eye slightly paler; spots below are more extensive and arrow-shaped.

Songs last 5–30 seconds, separated by long pauses

Varied short phrases, most of them repeated, without significant pauses

Low and musical like Sage Thrasher, but many notes slightly nasal and burry

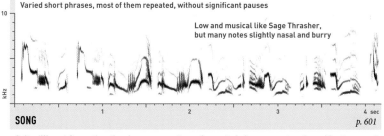

SONG *p. 601*

Quite different from other thrasher songs; pattern of repeated phrases run together without pauses is nearly diagnostic. Mostly Jan.–June. Not known whether female sings. Individual repertoire size unknown. Has not been reported to imitate other species.

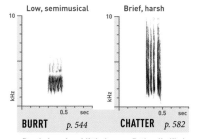

Low, semimusical | Brief, harsh

BURRT *p. 544*   **CHATTER** *p. 582*

Rarely heard and little known. Both calls likely given in agitation or in interactions. Chatter can resemble Ratchet calls of other thrashers.

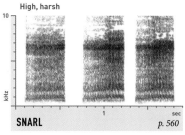

High, harsh

SNARL *p. 560*

Apparently given in high alarm; sole available example was recorded from adult defending nest.

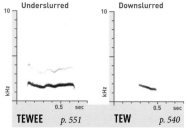

Underslurred | Downslurred

TEWEE *p. 551*   **TEW** *p. 540*

Rarely heard and little known. Plastic. In sole available recording, lower and more whistled (less nasal) than in other thrashers.

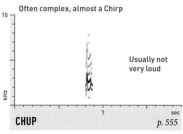

Often complex, almost a Chirp

Usually not very loud

CHUP *p. 555*

Likely all year; most common call, but still infrequent. Variable, plastic. Young Curve-billeds look like Bendire's and give a similar Chup call.

CALIFORNIA THRASHER

Toxostoma redivivum

Breeds in thick chapparal, where it is often difficult to see when not singing from an exposed perch. Note dark eye and buffy undertail coverts.

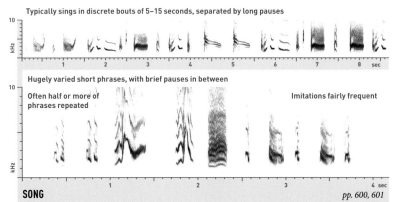

Typically sings in discrete bouts of 5–15 seconds, separated by long pauses

Hugely varied short phrases, with brief pauses in between

Often half or more of phrases repeated

Imitations fairly frequent

SONG　　　　　　　　　　　　　　　　　　　　　*pp. 600, 601*

All year, especially Nov.–June, by both sexes; mated pairs often countersing. Unlike other thrashers, but like Northern Mockingbird, sometimes repeats phrases 4 or more times in a row; sings with fewer pauses than mockingbirds. Phrases average noticeably lower and harsher than in other thrasher songs.

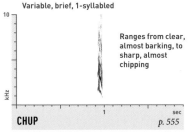

Variable, brief, 1-syllabled

Ranges from clear, almost barking, to sharp, almost chipping

CHUP　　　　　　　　*p. 555*

All year. Possibly belongs in the same category as Chweek and Weep-it; given in similar manner in similar situations.

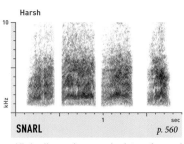

Harsh

SNARL　　　　　　　　*p. 560*

Likely all year, in aggressive interactions and possibly in alarm. Infrequent.

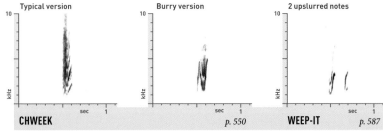

Typical version

Burry version

2 upslurred notes

CHWEEK　　　　　　　　*p. 550*　　　　　**WEEP-IT**　　*p. 587*

All year; most common calls. Highly variable; individuals may have more than one version, but the same version is usually repeated over and over at intervals, often from deep cover. Weep-it may recall Wit-weet of "Palmer's" Curve-billed Thrasher (see p. 378), but lower, softer, less musical.

LeConte's Thrasher

Toxostoma lecontei

Uncommon and local in sparsely vegetated desert scrub, where despite the
lack of cover it can be difficult to see.
Runs between bushes with tail cocked.

Songs last from 5 seconds to well over a minute

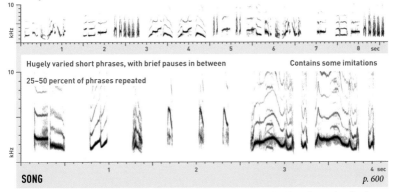

Hugely varied short phrases, with brief pauses in between

Contains some imitations

25–50 percent of phrases repeated

SONG   *p. 600*

Mostly Nov.–Mar., by males; females sing infrequently. Individual repertoire size unknown, but clearly very
large. Phrases average higher and clearer than in other thrasher songs, mostly without harsh notes; listen
for long clear slurred polyphonic phrases.

Quick series of 2–5 upslurred seminasal notes

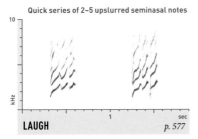

LAUGH   *p. 577*

Likely all year, in high alarm. Plastic. A rasping
call also reported to be given in high alarm,
probably similar to Snarl of other thrashers.

Very brief, ticking or smacking

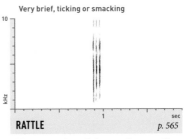

RATTLE   *p. 565*

Infrequent and poorly known; given sometimes
with Weep call, possibly in interactions. Recalls
Ratchet calls of other thrashers.

Vaguely 2-syllabled version   Simple version   Doubled version (Weep-weep)

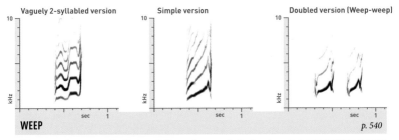

WEEP   *p. 540*

All year; most common calls. Single-syllable versions reportedly given in contact, doubled Weep-Weep in
mild alarm. Highly variable and plastic; the left and center spectrograms are from the same individual. Often given at regular intervals at dawn or dusk.

CRISSAL THRASHER

Toxostoma crissale

Usually skulks in dense desert scrub, especially in washes, in mesquite thickets, and near the interface with pinyon-juniper. Note pale eye, rufous vent.

Typically sings in loosely organized bouts of up to 60 seconds; occasionally phrases are up to 1 second apart

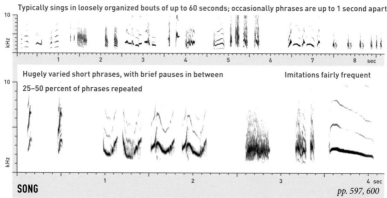

Hugely varied short phrases, with brief pauses in between

Imitations fairly frequent

25–50 percent of phrases repeated

SONG
pp. 597, 600

All year, but especially Jan.–Apr. Not known whether females sing. Generally intermediate in pitch and quality between California and LeConte's. Difficult to separate from Curve-billed Thrasher Song; averages clearer and more musical, with longer pauses between phrases and shorter pauses between bouts.

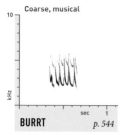

Coarse, musical

BURRT
p. 544

Little known; sole available example given with Churry-churry, likely in alarm.

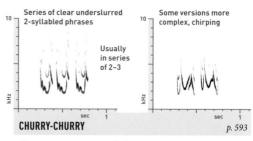

Series of clear underslurred 2-syllabled phrases

Usually in series of 2–3

Some versions more complex, chirping

CHURRY-CHURRY
p. 593

All year; most common call, often repeated for long periods. Distinctive, but highly variable and likely learned. "Churry" phrases rarely strung together into long series, likely in agitation.

SOUTHWESTERN THRASHER SONGS

The southwestern thrasher species are often more easily heard than seen, but can be difficult to identify by song. Two are highly distinctive and can always be identified by pattern:

- **Bendire's Thrasher** sings *repeated phrases without pauses*;
- **Sage Thrasher** sings *unique phrases without pauses*.

The other four (California, Crissal, LeConte's, and Curve-billed) tend to sing with broadly similar phrases and patterns. Tone quality is the most reliable indicator, but can be difficult to judge. Listen for species-specific call notes given during or between songs, but beware that each of these species may incorporate imitations of the others' calls into their own songs where ranges overlap.

Expert birders who have lived near Crissal and Curve-billed Thrashers for many years report being able to identify them by song with perhaps 80 percent confidence. In many cases unseen singers may best be left unidentified.

CURVE-BILLED THRASHER

Toxostoma curvirostre

Breeds in open arid habitats, including sparsely vegetated residential areas. Plumage and vocalizations regionally variable; see p. 378.

Typically sings in discrete bouts of 5–15 seconds, separated by long pauses

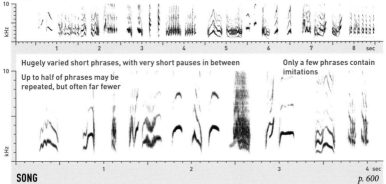

Hugely varied short phrases, with very short pauses in between

Up to half of phrases may be repeated, but often far fewer

Only a few phrases contain imitations

SONG *p. 600*

Nearly all year, but especially Mar.–July, by males, usually from conspicuous perch. Not known whether females sing. Individual repertoire size unknown, but clearly very large. Difficult to separate from Crissal Thrasher Song; averages faster and harsher, with phrases more clustered into discrete bouts.

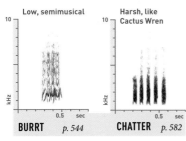

Low, semimusical

Harsh, like Cactus Wren

BURRT *p. 544*   **CHATTER** *p. 582*

Only recording of Burrt is from an agitated bird in early spring. Chatter given in alarm near nest and in altercations.

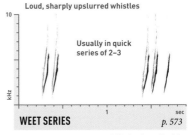

Loud, sharply upslurred whistles

Usually in quick series of 2–3

WEET SERIES *p. 573*

All year; most common and distinctive call of the eastern subspecies (see p. 378). Series becomes longer in agitation.

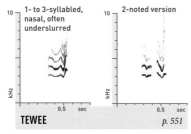

1- to 3-syllabled, nasal, often underslurred

2-noted version

TEWEE *p. 551*

Not well known; apparently given in courtship and possibly during or after territorial altercations. Highly variable and plastic.

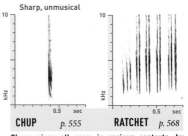

Sharp, unmusical

CHUP *p. 555*   **RATCHET** *p. 568*

Chups given all year, in various contexts, by adults and juveniles. Only recording of Ratchet is from two birds in a flight chase.

SAGE THRASHER

Oreoscoptes montanus

Breeds in sagebrush steppe. Small for a thrasher, with a short bill recalling a mockingbird's. Worn late-summer birds can be nearly unstreaked below.

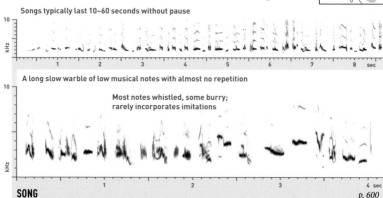

Songs typically last 10–60 seconds without pause

A long slow warble of low musical notes with almost no repetition

Most notes whistled, some burry; rarely incorporates imitations

SONG   *p. 600*

Mostly Mar.–June. Not known whether females sing. Individual repertoire size unknown, but clearly very large. Note types much less varied than in other thrasher songs. Sometimes sings at night. Single songs as long as 20 minutes have been recorded.

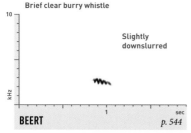

Brief clear burry whistle

Slightly downslurred

BEERT   *p. 544*

Likely all year, in alarm; infrequent. Also rarely gives other whistled calls, perhaps variants of Beert or notes from Song.

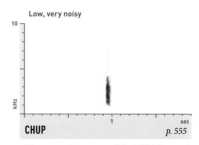

Low, very noisy

CHUP   *p. 555*

All year; most common call, but still infrequent. Lower, noisier than Brewer's Blackbird Chuk. Likely also gives a snarling call in altercations.

WHY DO MOCKINGBIRDS IMITATE OTHER BIRD SPECIES?

Northern Mockingbirds are not the only birds that imitate (see p. 30), but they are probably the most famous. The reasons for their prolific imitation are not entirely understood, but it is clear that their purpose is *not* to pass themselves off as members of the species they imitate. Being mistaken for another species would create two problems for a mockingbird: it might face a territorial challenge from the imitated species, and it might be ignored by other mockingbirds.

As far as is known, mockingbirds imitate other birds for more or less the same reason that human DJs use sample clips from other artists in making a remix. In essence, the mockingbirds are drawing from the sound-scape around them to show off their skill at assembling a completely original song.

Northern Mockingbirds that migrate may learn new sounds along the way and then bring those exotic voices back home with them, cataloguing their travels in their songs.

Northern Mockingbird

Mimus polyglottos

Visually and vocally conspicuous in open areas with scattered trees, including residential areas. In flight, shows white patches in the wings and tail.

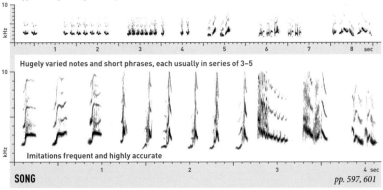

Typical Song has significant pauses between series

Hugely varied notes and short phrases, each usually in series of 3–5

Imitations frequent and highly accurate

SONG *pp. 597, 601*

Nearly all year. Both sexes sing, females much less often. Unmated males may sing all night long. Individual males have 50–200 song phrases, and can learn new phrases throughout their lives. Imitates many sounds, including nonbird sounds. Song best recognized by pattern of repetition.

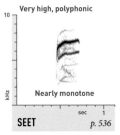

Very high, polyphonic

Nearly monotone

SEET *p. 536*

Given by begging juveniles; similar calls near nest by adults.

High, polyphonic

Burry version shown

SOFT CALLS *not indexed*

A varied set of highly plastic calls given by both sexes near the nest.

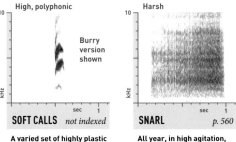

Harsh

SNARL *p. 560*

All year, in high agitation, especially in response to predators.

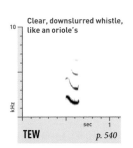

Clear, downslurred whistle, like an oriole's

TEW *p. 540*

Given infrequently, likely by both sexes, perhaps in alarm.

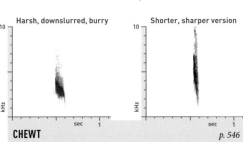

Harsh, downslurred, burry

CHEWT *p. 546*

Shorter, sharper version

Most common call, given by both sexes in mild alarm. Quite variable. Most versions quite abrupt and harsh. Often given in rapid series; rarely too fast to count.

STARLINGS AND MYNAS (Family Sturnidae)

This large and diverse family is native to the Old World; our species is introduced. Vocalizations are extremely complex and diverse. Songs are learned, often incorporating accurate imitations of other birds and environmental sounds.

EUROPEAN STARLING

Sturnus vulgaris

Released in New York City in 1890; now one of the most abundant birds in North America. A remarkable vocalist. Competes with native species for nest sites.

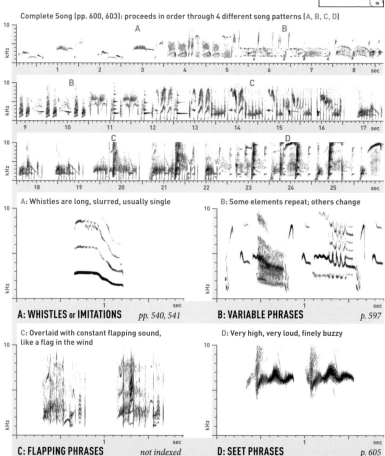

Complete Song (pp. 600, 603): proceeds in order through 4 different song patterns (A, B, C, D)

A: Whistles are long, slurred, usually single

A: WHISTLES or IMITATIONS *pp. 540, 541*

B: Some elements repeat; others change

B: VARIABLE PHRASES *p. 597*

C: Overlaid with constant flapping sound, like a flag in the wind

C: FLAPPING PHRASES *not indexed*

D: Very high, very loud, finely buzzy

D: SEET PHRASES *p. 605*

All year, but especially Apr.–June. Four distinct patterns make up the complete Song, up to 40 seconds total, but any pattern may be omitted or given by itself. Patterns A–C usually quite soft; Pattern D is far louder. Each male has 2–12 Whistles (A), 10–35 Variable Phrases (B), 2–14 Flapping Phrases (C), and up to 6 Seet Phrases (D), for a total of up to 70 phrase types per bird. On average, 18 of these phrases are imitations of other species, in patterns A or B (rarely C). Females also sing, but lack Seet Phrases (D).

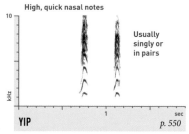

High, quick nasal notes

Usually singly or in pairs

YIP *p. 550*

All year, in response to hawks, in nest defense, and in chases of other starlings. Fairly uniform; slightly polyphonic or screechy.

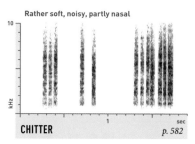

Rather soft, noisy, partly nasal

CHITTER *p. 582*

All year, when individuals join groups and in mild aggressive encounters. Fairly uniform. Sometimes shortened to a single Chit.

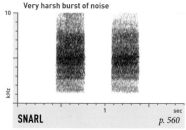

Very harsh burst of noise

SNARL *p. 560*

All year, but especially in alarm at nest, often with Yips. Varies in length; sometimes in series.

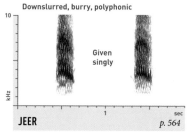

Downslurred, burry, polyphonic

Given singly

JEER *p. 564*

All year, by solo birds in flight and by small flocks. Juveniles beg with a similar call.

WAXWINGS (Family Bombycillidae) *next page*

The waxwings are handsome, medium-sized birds with crests and black masks, named for the red waxy tips on their secondary feathers. They eat mainly fruit, along with some insects, and are highly social and nomadic outside the breeding season, gathering in flocks that wander widely in search of fruiting trees. In flight silhouette they resemble starlings, with short tails and triangular wings. All their vocalizations are simple and apparently innate. Both species' vocal repertoires are traditionally considered to consist entirely of calls, lacking Songs altogether.

SILKY-FLYCATCHERS (Family Ptiliogonatidae) *p. 391*

Distant relatives of the waxwings, the silky-flycatchers also eat fruit as well as insects. They are named for their soft, silky plumage. Their vocal repertoires are complex, including learned songs and various calls.

OLIVE WARBLER (Family Peucedramidae) *p. 392*

Once considered a member of the warbler family (Parulidae), the Olive Warbler is now recognized as the sole surviving species of a very ancient lineage, perhaps most closely related to Old World families. Its songs are complex and learned; its call repertoire differs strikingly from those of parulid warblers.

Cedar Waxwing

Bombycilla cedrorum

A very social bird that spends much of the year in wandering flocks in search of fruiting trees. Vocalizations simple and very high-pitched.

Trills average shorter, higher, finer, and more monotone than Bohemian's

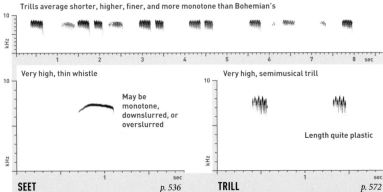

Very high, thin whistle

May be monotone, downslurred, or overslurred

SEET   *p. 536*

Very high, semimusical trill

Length quite plastic

TRILL   *p. 572*

Both calls given all year. Seet given frequently, especially during or prior to flight and in response to predators. Trill maintains flock contact, often in flight. Very short, soft versions given while feeding and in close courtship. Begging juveniles give longer, more Bohemian-like version.

Bohemian Waxwing

Bombycilla garrulus

An irregular winter visitor from the far north; invades towns and woodlands in flocks in search of fruiting trees, sometimes with Cedar Waxwings.

Trills lower than Cedar Waxwing's; longer, more downslurred on average

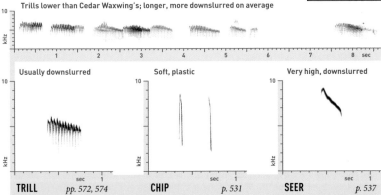

Usually downslurred

TRILL   *pp. 572, 574*

Soft, plastic

CHIP   *p. 531*

Very high, downslurred

SEER   *p. 537*

Trill is most common call all year; given frequently by winter flocks. Very short, soft versions given while feeding. Chips given all year, possibly in alarm, resembling single note of Trill, and grading into it. Seer given in response to predator near nest or young; given rarely, if ever, in winter.

Phainopepla

Phainopepla nitens

Locally common in desert washes, mesquite bosques, and sometimes riparian woodlands; closely associated with desert mistletoe, a parasitic plant.

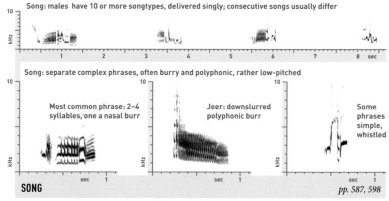

Song: males have 10 or more songtypes, delivered singly; consecutive songs usually differ

Song: separate complex phrases, often burry and polyphonic, rather low-pitched

Most common phrase: 2–4 syllables, one a nasal burr

Jeer: downslurred polyphonic burr

Some phrases simple, whistled

SONG

pp. 587, 598

Mostly Feb.–June, by unpaired males seeking mates. Usually loud and musical, but with a rather jangling, polyphonic tone. Phrases quite diverse, but typically 1–4 syllables; moderate geographic variation. Jeer phrases (p. 564) quite distinctive, but beware sounds of Common Grackle.

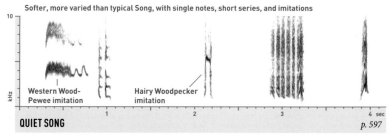

Softer, more varied than typical Song, with single notes, short series, and imitations

Western Wood-Pewee imitation

Hairy Woodpecker imitation

QUIET SONG

p. 597

Possibly all year; poorly understood. Imitations reported in response to predators, or while being banded; reasons for this behavior remain unclear. Soft, imitative song also apparently occurs in other contexts. Also gives quick bill Rattle in agitation (p. 566).

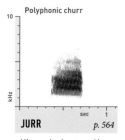

Polyphonic churr

JURR

p. 564

All year, in chases and in courtship. Harsh versions used to beg for food.

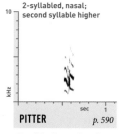

2-syllabled, nasal; second syllable higher

PITTER

p. 590

Possibly all year. Function unknown; may be a variant of Wert.

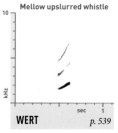

Mellow upslurred whistle

WERT

p. 539

All year; most common call, given in a variety of contexts. Some versions noisy.

OLIVE WARBLER

Peucedramus taeniatus

Breeds in mountain pine and pine-oak forests. Forages near tops of trees; fairly social. Black mask is muted and dusky in female and young birds.

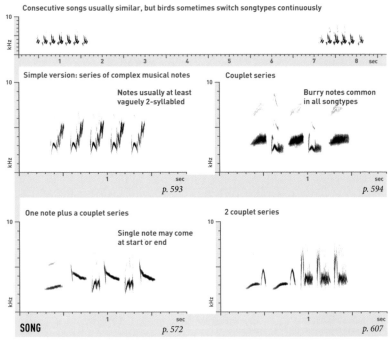

Consecutive songs usually similar, but birds sometimes switch songtypes continuously

Simple version: series of complex musical notes
Notes usually at least vaguely 2-syllabled
p. 593

Couplet series
Burry notes common in all songtypes
p. 594

One note plus a couplet series
Single note may come at start or end
p. 572

2 couplet series
p. 607

SONG

All year, by both sexes, especially Mar.–Aug. Repertoire size unknown, but at least some individuals have 5 or more songtypes. Some rapid simple versions recall a titmouse; many versions, especially complex ones, can be confused with Yellow-eyed Junco. Also has a very rare Complex Song (p. 606) consisting of many rapid consecutive musical series, recalling Continuous Songs of goldfinches. Function unknown; only available example given by a bird in the hand.

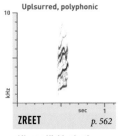

Uplsurred, polyphonic

ZREET *p. 562*

All year. Highly plastic. Function not well known. Young beg with similar calls.

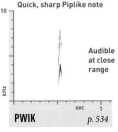

Quick, sharp Piplike note

Audible at close range

PWIK *p. 534*

All year. Highly plastic; some versions nearly a Whit. Function not well known.

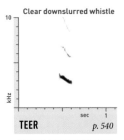

Clear downslurred whistle

TEER *p. 540*

All year, in many situations; most common call. Also a rare upslurred Wee (p. 539).

OLD WORLD SPARROWS (Family Passeridae)

The House Sparrow, native to Eurasia, is not closely related to New World sparrows.

WHYDAHS (Family Viduidae)

The whydahs are brood parasites, laying eggs in the nests of other species.

WAXBILLS, MANNIKINS, & MUNIAS (Family Estrildidae)

This diverse family of finchlike birds is closely related to the Old World Sparrows.

WEAVERS (Family Ploceidae)

Birds in this family, including the Northern Red Bishop, weave elaborate nests.

HOUSE SPARROW

Passer domesticus

Common near human habitation of all kinds. Has spread nearly everywhere on the continent since its introduction from Europe to New York in 1851.

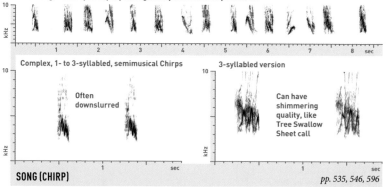

Fast Song: males cycle rapidly through many different Chirplike notes

Complex, 1- to 3-syllabled, semimusical Chirps

Often downslurred

3-syllabled version

Can have shimmering quality, like Tree Swallow Sheet call

SONG (CHIRP)

pp. 535, 546, 596

All year, by males in a variety of situations; rarely by females. Males reportedly have 4–12 different Chirp types; usually one is repeated for long periods, or two types are alternated, before switching to a new type. Fast Song may be associated with courtship, but also given by lone birds.

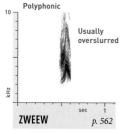

Polyphonic

Usually overslurred

ZWEEW *p. 562*

All year, by both sexes, in mild alarm. Somewhat plastic. Recalls House Finch.

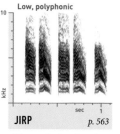

Low, polyphonic

JIRP *p. 563*

All year, singly or in series, in high alarm. Generally plastic.

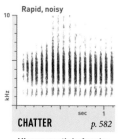

Rapid, noisy

CHATTER *p. 582*

All year, mostly by females, in interactions. Highly variable and plastic.

Pin-tailed Whydah

Vidua macroura

Native to Africa. Lays eggs in the nests of other species, like cowbirds; usual host in U.S. seems to be Scaly-breasted Munia, but more study needed.

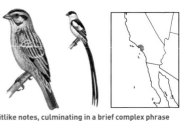

Displaying male gives long series of well-spaced Chwitlike notes, culminating in a brief complex phrase (p. 596)

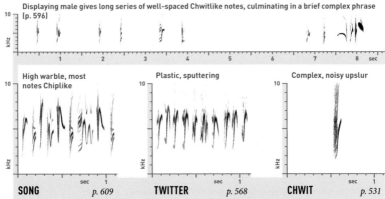

| High warble, most notes Chiplike | Plastic, sputtering | Complex, noisy upslur |

SONG *p. 609* **TWITTER** *p. 568* **CHWIT** *p. 531*

Vocal repertoire complex, but little studied. Display Song (top) may be different from the rapid complex warbles of 1–2 seconds (lower left) that may recall Barn Swallow Song. Twitter may also be a form of Song. Chwit is most common call; rather plastic, grading into notes of Display Song.

Orange-cheeked Waxbill *Estrilda melpoda*

Native to Africa; a few are feral in the Los Angeles area. Most sounds are like those shown; may also have a Song of similar notes.

High, variable

Plastic, often monotone

PSEEP *p. 534* **TWITTER** *p. 571*

Bronze Mannikin *Spermestes cucullata*

Native to Africa; a few breed in the Los Angeles area. Breet is most common call of adults; Chatter associated with begging of juveniles.

Recalls Evening Grosbeak

Much like House Sparrow's

BREET *p. 544* **CHATTER** *p. 582*

SCALY-BREASTED MUNIA

Lonchura punctulata

Also called Nutmeg Mannikin or Spice
Finch. Native to Asia; feral birds found
mostly in parks and residential areas.
Adult shown; juveniles are plain brown.

Occasionally gives 1-syllabled wheezy Cheeps (p. 535) along with Pi-peeps

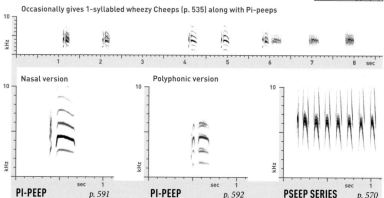

Nasal version	Polyphonic version	
PI-PEEP *p. 591*	**PI-PEEP** *p. 592*	**PSEEP SERIES** *p. 570*

Often rather quiet, but sometimes vocal in flocks. Most common call is a distinctive 2-noted Pi-peep, usually nasal or polyphonic. Clear whistled versions exist, and first note occasionally missing. Pseep Series
given loudly by begging juveniles.

NORTHERN RED BISHOP

Euplectes franciscanus

Also known as Orange Bishop. Native to
Africa. Feral in parts of Texas and California; has also been seen in Florida
and Arizona. Favors weedy floodplains.

Song (p. 593): a couplet series alternating Chwits and Tseets, often introduced by Tseets

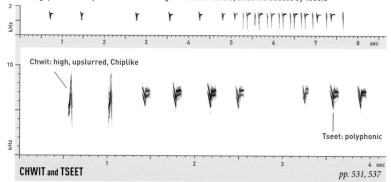

Chwit: high, upslurred, Chiplike

Tseet: polyphonic

CHWIT and TSEET *pp. 531, 537*

Most common calls appear to be Chwit and Tseet, given at least by males in many situations, often mixed
together as above. Chwit also given separately in mild alarm. Couplet series of Chwit- and Tseetlike notes
is apparent Song; also gives plastic Chirps and faint Whines, possibly a form of subsong.

PIPITS (Family Motacillidae)

Pipits are small birds of open country with generally cryptic, streaked plumage. They eat primarily insects and nest on the ground. They resemble larks and longspurs in their elongated hind claws and walking (instead of hopping) gait, and they sometimes flock with larks and longspurs in migration and winter, but the three families are not closely related.

Two species of pipit are regularly found in North America outside Alaska. Both perform long flight displays on the breeding grounds, and both show white outer tail feathers in flight. The bolder and more gregarious American Pipit wags its tail frequently, while the shyer and more solitary Sprague's Pipit does so rarely if at all. Songs in both species are apparently learned. Calls are probably innate, and call repertoires are quite simple.

AMERICAN PIPIT

Anthus rubescens

Breeds on arctic and alpine tundra. In migration and winter, found in flocks in fields and along shorelines. Bobs tail often. Note white outer tail feathers.

Series short to extremely long (30 seconds or more); consecutive series may repeat or differ

Simple or couplet series, often very long

Tone quality varies from unmusical and chirping to musical and whistled

SONG: "Chew chew chew chew" or "Teer teer teer teer" *pp. 572, 593*

Mostly May–Aug. Given in long, slow display flight or from ground. Repertoire size unknown; territorial birds may give many series consecutively without pauses. When excited, may switch series often, like American Goldfinch Song. Some series faster, like slow trills; some are brief, just a few repetitions.

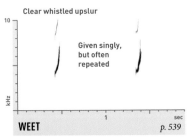

Clear whistled upslur

Given singly, but often repeated

WEET *p. 539*

May–Sept., in alarm near nest or fledglings; apparently rare away from breeding grounds. Compare Squeep of Sprague's Pipit.

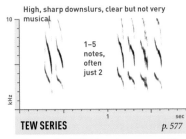

High, sharp downslurs, clear but not very musical

1–5 notes, often just 2

TEW SERIES *p. 577*

All year, often in flight; most common call. Plastic, but consecutive calls fairly consistent in pitch and pattern, unlike calls of Horned Lark.

Sprague's Pipit

Anthus spragueii

A local and declining bird of native prairies with a remarkable flight display. When flushed, rises steeply and flies far, then plummets back into the grass.

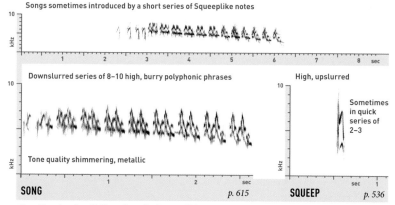

Songs sometimes introduced by a short series of Squeeplike notes

Downslurred series of 8–10 high, burry polyphonic phrases

High, upslurred

Sometimes in quick series of 2–3

Tone quality shimmering, metallic

SONG   *p. 615*

SQUEEP   *p. 536*

Song May–Aug. Recalls Veery Song, but longer, higher, and likely to be heard only from far overhead in native prairies. Display flights can last up to 3 hours. Repertoire size unknown, but consecutive songs similar. Squeep given all year, by flushed birds. Quite uniform; no other calls known from the species.

FINCHES (Family Fringillidae, Subfamily Carduelinae)   *next pages*

This group includes some of our most vocally and behaviorally complex birds. Songs and at least some calls are learned; songs tend to be long and complex, made up of large repertoires of notes. Many finches have more than one type of song. In a number of species, especially Cassin's Finch, Lesser Goldfinch, Lawrence's Goldfinch, Pine Siskin, and Pine Grosbeak, songs can incorporate excellent imitations of other bird species.

Most finches give several different types of calls, many of which are polyphonic, resulting in a characteristic "finchy" whining quality common to many species. The begging calls of many species are distinctively 2-noted, giving rise to transcriptions like "tea-cup," "chit-too," and "teer-weet."

Call Matching in Finches

Laboratory studies have shown that in many species of cardueline finches, mated birds change their calls to match their mate's. In most such cases the calls change only in minor details, but call matching can occur even among members of different finch species when they are kept in the same cage. Recordings suggest that in the redpolls, members of winter flocks may match calls with each other; more study is needed. It is possible that finches in the wild may in rare cases match the call notes of a different species.

GRAY-CROWNED ROSY-FINCH

Leucosticte tephrocotis

Breeds on arctic and alpine tundra, especially near rocky cliffs and permanent snowfields. Winters in flocks in mountains and in open country. Visits feeders.

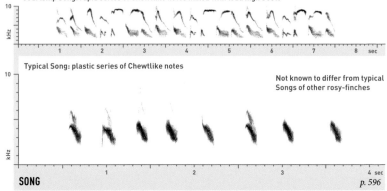

Courtship Song: rapid series of Chewtlike notes mixed with loud high Seets

Typical Song: plastic series of Chewtlike notes

Not known to differ from typical Songs of other rosy-finches

SONG *p. 596*

Distinction between Song and calls murky in all rosy-finches, but Chewtlike notes in long, loud series appear to function as typical Song. Courtship Song, given by potential mate at close range, at least sometimes involves simultaneous Chewts and Seets from the same individual.

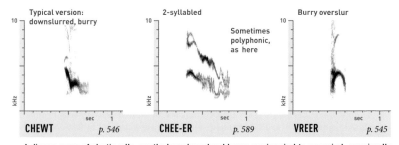

Typical version: downslurred, burry

2-syllabled

Sometimes polyphonic, as here

Burry overslur

CHEWT *p. 546* **CHEE-ER** *p. 589* **VREER** *p. 545*

A diverse group of plastic calls, mostly downslurred and burry, semimusical to unmusical, occasionally polyphonic, sometimes 2-syllabled. Given all year in pursuits, upon landing, and when mobbing predators. Different versions may serve different functions. Some versions perhaps best considered Song phrases.

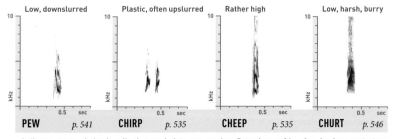

Low, downslurred

Plastic, often upslurred

Rather high

Low, harsh, burry

PEW *p. 541* **CHIRP** *p. 535* **CHEEP** *p. 535* **CHURT** *p. 546*

A diverse group of plastic calls that grade into one another. Pew given softly, often in close contact on ground. Chirp and Cheep often given in flight, with clearer versions in series upon takeoff and harsher versions in alarm. Churt given frequently, possibly in aggression.

BROWN-CAPPED ROSY-FINCH

Leucosticte australis

Breeds only on alpine tundra in Colorado and extreme southern Wyoming, near cliffs and permanent snowfields. Flocks with other rosy-finches in winter.

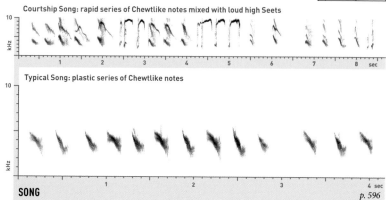

Courtship Song: rapid series of Chewtlike notes mixed with loud high Seets

Typical Song: plastic series of Chewtlike notes

SONG *p. 596*

Distinction between Song and calls murky in all rosy-finches, but Chewtlike notes in long, loud series appear to function as typical Song. No consistent differences from Gray-crowned Song have been reported, but more study needed.

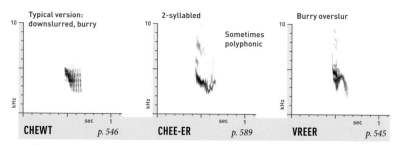

Typical version: downslurred, burry

2-syllabled

Sometimes polyphonic

Burry overslur

CHEWT *p. 546*

CHEE-ER *p. 589*

VREER *p. 545*

A diverse group of plastic calls, mostly downslurred and burry, semimusical to unmusical, occasionally polyphonic, sometimes 2-syllabled. Given all year in pursuits, upon landing, and when mobbing predators. Different versions may serve different functions. Some versions perhaps best considered Song phrases.

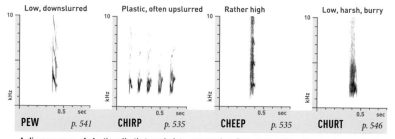

Low, downslurred

Plastic, often upslurred

Rather high

Low, harsh, burry

PEW *p. 541*

CHIRP *p. 535*

CHEEP *p. 535*

CHURT *p. 546*

A diverse group of plastic calls that grade into one another. Pew given softly, often in close contact on ground. Chirp and Cheep often given in flight, with clearer versions in series upon takeoff and harsher versions in alarm. Churt given frequently, possibly in aggression.

BLACK ROSY-FINCH

Leucosticte atrata

Breeds on alpine tundra near cliffs and permanent snowfields. Flocks with other rosy-finches in winter. Interbreeds with Gray-crowned where ranges meet.

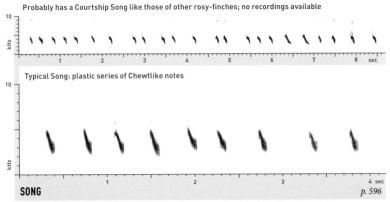

Probably has a Courtship Song like those of other rosy-finches; no recordings available

Typical Song: plastic series of Chewtlike notes

SONG *p. 596*

Distinction between Song and calls murky in all rosy-finches, but Chewtlike notes in long, loud series appear to function as typical Song. No consistent differences from Gray-crowned Song have been reported, but more study needed.

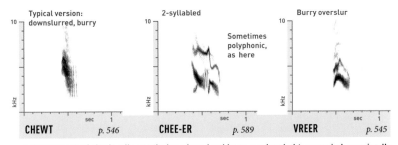

Typical version: downslurred, burry

2-syllabled

Sometimes polyphonic, as here

Burry overslur

CHEWT *p. 546* **CHEE-ER** *p. 589* **VREER** *p. 545*

A diverse group of plastic calls, mostly downslurred and burry, semimusical to unmusical, occasionally polyphonic, sometimes 2-syllabled. Given all year in pursuits, upon landing, and when mobbing predators. Different versions may serve different functions. Some versions perhaps best considered Song phrases.

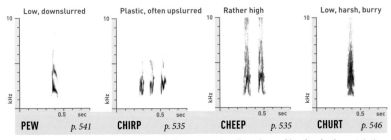

Low, downslurred

Plastic, often upslurred

Rather high

Low, harsh, burry

PEW *p. 541* **CHIRP** *p. 535* **CHEEP** *p. 535* **CHURT** *p. 546*

A diverse group of plastic calls that grade into one another. Pew given softly, often in close contact on ground. Chirp and Cheep often given in flight, with clearer versions in series upon takeoff and harsher versions in alarm. Churt given frequently, possibly in aggression.

PINE GROSBEAK

Pinicola enucleator

A large, tame, but often inconspicuous northern finch of open coniferous forests. In some winters, wanders well south of breeding range.

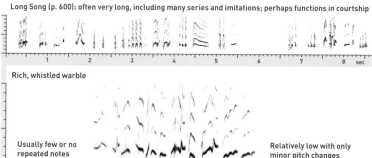

Long Song (p. 600): often very long, including many series and imitations; perhaps functions in courtship

Rich, whistled warble

Usually few or no repeated notes

Relatively low with only minor pitch changes

SONG
p. 609

Almost all year, mostly by males. Females apparently sing rarely; singers without red plumage may be first-spring males. Song grades into Long Song; some versions are short, but contain imitations. Repertoire size unknown; consecutive songs may start similarly, but then vary.

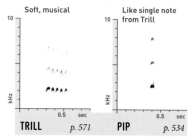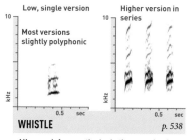

Soft, musical

Like single note from Trill

Low, single version

Most versions slightly polyphonic

Higher version in series

TRILL *p. 571* **PIP** *p. 534* **WHISTLE** *p. 538*

All year, by both sexes, in close contact. Generally audible only at close range, often mixed with other, louder calls. Plastic.

All year, infrequently, by both sexes, in alarm. Highly geographically variable and fairly plastic; most versions are nearly monotone.

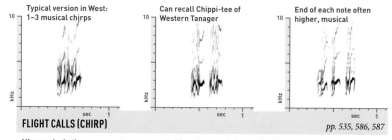

Typical version in West: 1–3 musical chirps

Can recall Chippi-tee of Western Tanager

End of each note often higher, musical

FLIGHT CALLS (CHIRP)
pp. 535, 586, 587

All year, by both sexes; most common call. Given frequently by perched birds. Extremely geographically variable and quite plastic; as in redpolls and Red Crossbills, flockmates appear to share a call type, and flocks with dissimilar call types reportedly do not mix. More study needed.

PURPLE FINCH

Haemorhous purpureus

Breeds in a variety of forest types and woodland edges. In winter, found in small flocks in many different habitats. Sometimes visits feeders.

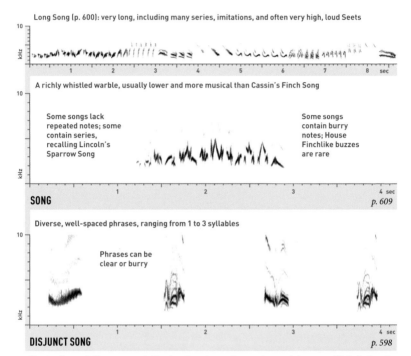

Long Song (p. 600): very long, including many series, imitations, and often very high, loud Seets

A richly whistled warble, usually lower and more musical than Cassin's Finch Song

Some songs lack repeated notes; some contain series, recalling Lincoln's Sparrow Song

Some songs contain burry notes; House Finchlike buzzes are rare

SONG *p. 609*

Diverse, well-spaced phrases, ranging from 1 to 3 syllables

Phrases can be clear or burry

DISJUNCT SONG *p. 598*

Nearly all year. Both sexes reported to sing, though many brown-plumaged singers are likely first-spring males. Typical Song is longer and more plastic in late winter, becoming shorter and more stereotyped by early summer; rarely but regularly includes imitations of other bird species. Individual repertoire size unknown; consecutive songs tend to start similarly but ends and middles often vary. Long Song directed at female, with crest raised and wings fluttering; often rather soft.

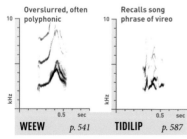

Overslurred, often polyphonic	Recalls song phrase of vireo
WEEW *p. 541*	**TIDILIP** *p. 587*

A confusing category of sounds, like phrases from the Disjunct Song, but given over and over, by both sexes and all ages. More study needed.

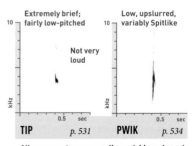

Extremely brief; fairly low-pitched	Low, upslurred, variably Spitlike
Not very loud	
TIP *p. 531*	**PWIK** *p. 534*

All year; most common calls, variable and possibly grading into one another. Pwik not known from populations east of the Canadian Rockies.

Cassin's Finch

Haemorhous cassinii

Breeds in open coniferous forests in mountains. Common in parts of range, uncommon in others. Sometimes visits feeders.

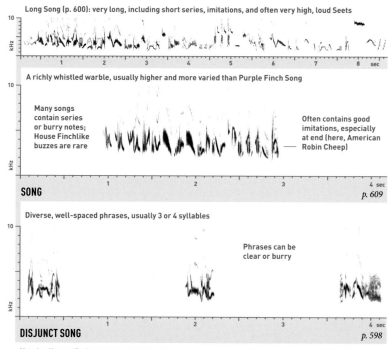

Long Song (p. 600): very long, including short series, imitations, and often very high, loud Seets

A richly whistled warble, usually higher and more varied than Purple Finch Song

Many songs contain series or burry notes; House Finchlike buzzes are rare

Often contains good imitations, especially at end (here, American Robin Cheep)

SONG
p. 609

Diverse, well-spaced phrases, usually 3 or 4 syllables

Phrases can be clear or burry

DISJUNCT SONG
p. 598

Nearly all year. Both sexes reported to sing, though many brown-plumaged singers are likely first-spring males. Typical Song is longer and more plastic in late winter, becoming shorter and more stereotyped by early summer; very frequently includes imitations of other bird species. Individual repertoire size unknown; consecutive songs tend to start similarly but ends and middles often vary. Long Song directed at female, with crest raised and wings fluttering; often rather soft.

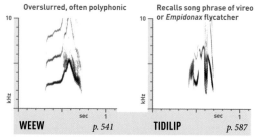

Overslurred, often polyphonic

Recalls song phrase of vireo or *Empidonax* flycatcher

Softer, more plastic than flight calls of goldfinches

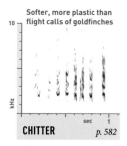

WEEW
p. 541

TIDILIP
p. 587

CHITTER
p. 582

A confusing category of sounds, like phrases from the Disjunct Song, but given over and over, by both sexes and all ages. More study needed. Weew like Purple Finch's; Tidilip often less musical.

Likely all year, in close contact and aggression, especially at feeders.

HOUSE FINCH

Haemorhous mexicanus

Common in semi-open habitats; often visits feeders. Originally a bird of the arid Southwest, introduced to the East in 1939; has expanded continent-wide.

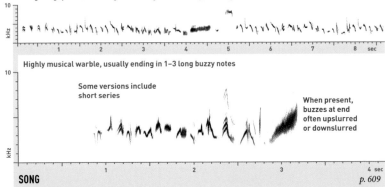

Long Song (p. 600): courting males sing continuously, with high loud Seets mixed in

Highly musical warble, usually ending in 1–3 long buzzy notes

Some versions include short series

When present, buzzes at end often upslurred or downslurred

SONG *p. 609*

All year, but mostly Mar.–Aug., by males. Females sing occasionally, mostly in spring. Individual males have 2–10 songtypes; consecutive songs may be same or different. Songs frequently truncated, lacking buzzy notes at end. Usually lacks repeated notes. Very rarely includes imitations of other species.

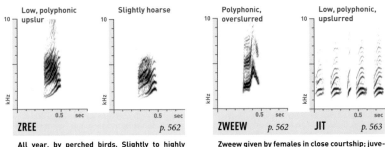

Low, polyphonic upslur

Slightly hoarse

Polyphonic, overslurred

Low, polyphonic, upslurred

ZREE *p. 562*

ZWEEW *p. 562*

JIT *p. 563*

All year, by perched birds. Slightly to highly plastic. One of the species' most common and characteristic calls.

Zweew given by females in close courtship; juveniles beg with similar notes. Jit poorly known.

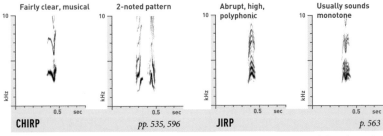

Fairly clear, musical

2-noted pattern

Abrupt, high, polyphonic

Usually sounds monotone

CHIRP *pp. 535, 596*

JIRP *p. 563*

All year, sometimes with other calls. Function poorly known. Individuals often alternate 2–3 versions; a few versions are 2-noted. Often highly plastic, but some versions fairly stereotyped. Several versions are easily confused with sounds of House Sparrow.

COMMON REDPOLL

Acanthis flammea

A finch of far-northern forest edges; winters in the northern part of our area, in some years irrupting farther south in numbers. Often visits feeders.

Disjunct Song (p. 597): various series and trills in any order, separated by pauses

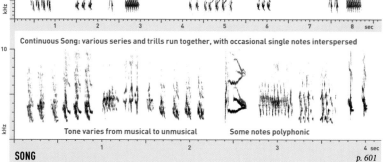
Continuous Song: various series and trills run together, with occasional single notes interspersed

Tone varies from musical to unmusical   Some notes polyphonic

SONG   *p. 601*

Mostly in breeding season. Continuous and Disjunct versions of Song intergrade, but Disjunct Song seems more common. Each bird uses several types each of series and trills; unclear how/whether these differ from Flight Calls and Buzzes.

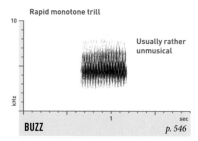
Rapid monotone trill

Usually rather unmusical

BUZZ   *p. 546*

All year. Variable; birds may have multiple versions. Often used in Song, but also given by winter flocks.

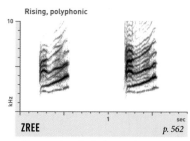
Rising, polyphonic

ZREE   *p. 562*

All year; rather plastic. Apparently serves to call groups together, and to signal excitement or alarm.

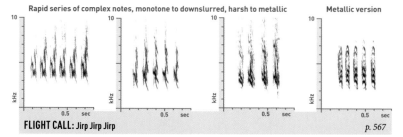
Rapid series of complex notes, monotone to downslurred, harsh to metallic   Metallic version

FLIGHT CALL: Jirp Jirp Jirp   *p. 567*

All year, in many contexts; most common call from winter flocks. Hugely variable; generally similar within flocks, but can differ greatly between them. Unknown whether individuals might have multiple call types or change call types over time. More study needed.

PINE SISKIN

Spinus pinus

Breeds in coniferous forests; wanders widely in winter, sometimes remaining to breed south of normal range. Forms vocal flocks; often visits feeders.

Disjunct Song (p. 598): various Zweewlike notes given singly, sometimes mixed with other calls

Continuous Song: long string of single notes and short series

Usually includes a few imitations of other species

Repeats most notes just once, or not at all

SONG *p. 600*

All year, especially Jan.–June, during breeding activity, reportedly only by males. Many notes resemble Zweew call; occasional diagnostic Zhree calls interspersed. Song often begins on high perch and continues during long circling flight; may conclude back on the perch. Continuous Song grades into Disjunct Song.

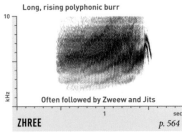

Long, rising polyphonic burr

Often followed by Zweew and Jits

ZHREE *p. 564*

All year; unmistakable. Given during Song or singly, usually while flashing yellow patches in wings and tail, perhaps in aggression.

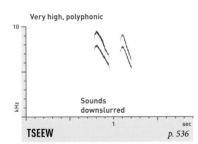

Very high, polyphonic

Sounds downslurred

TSEEW *p. 536*

Given near nest, and by individuals while foraging; function unclear. Fairly uniform; usually given in 1s and 2s.

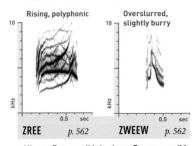

Rising, polyphonic

Overslurred, slightly burry

ZREE *p. 562*   **ZWEEW** *p. 562*

All year. Zree possibly in alarm; Zweew, possibly a variant of Zree, often with Flight Call. Phrases in Disjunct Song may resemble both calls.

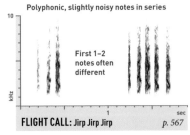

Polyphonic, slightly noisy notes in series

First 1–2 notes often different

FLIGHT CALL: Jirp Jirp Jirp *p. 567*

All year, in contact and flight. Only slightly variable. Lower, less musical than Lesser Goldfinch Flight Call.

AMERICAN GOLDFINCH

Spinus tristis

Common and widespread in weedy fields, second growth, and backyards; often visits feeders. Yellow plumage of male is present only in summer.

Disjunct Song (p. 598): 1- and 2-note Zreelike phrases mixed with call notes; often by countersinging males

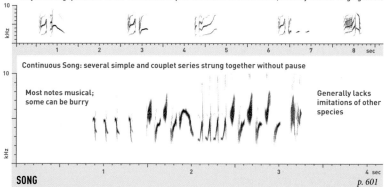

Continuous Song: several simple and couplet series strung together without pause

Most notes musical; some can be burry

Generally lacks imitations of other species

SONG *p. 601*

Mostly Apr.–Aug., from high perch or in circling flight. Short Songs of 2–4 series may be repeated at intervals, recalling Songs of Indigo Bunting or various warblers; at other times, consecutive songs may differ greatly in content and length. Most series contain 3–4 repetitions, but this varies greatly.

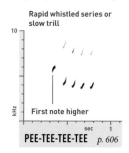

Rapid whistled series or slow trill

First note higher

PEE-TEE-TEE-TEE *p. 606*

Mostly June–July, by female in courtship. More musical than Flight Call.

2-note whistled phrase

Weet note sometimes repeated

TEW-WEET *p. 587*

Mostly July–Sept., by begging juveniles. Often polyphonic.

Short buzzes

DZIK SERIES *p. 576*

All year, in aggressive encounters. Highly plastic.

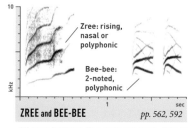

Zree: rising, nasal or polyphonic

Bee-bee: 2-noted, polyphonic

ZREE and BEE-BEE *pp. 562, 592*

All year, in alarm. Two main call types; Zree is more common, Bee-bee typically given in higher alarm, especially near nest.

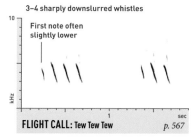

3–4 sharply downslurred whistles

First note often slightly lower

FLIGHT CALL: Tew Tew Tew *p. 567*

All year; most common call. Plastic in length, but otherwise uniform rangewide. Often given in flight. Sharp but fairly musical.

LAWRENCE'S GOLDFINCH

Spinus lawrencei

Breeds in dry oak woodlands and weedy or brushy areas near water; uncommon and local, often missing from areas where it was common the year before.

Disjunct Song (p. 598): various 1- and 2-note whistled and polyphonic phrases; less varied than in Lesser

Continuous Song: several series strung together without pause — Many notes are imitations

Often starts with Flight Call notes

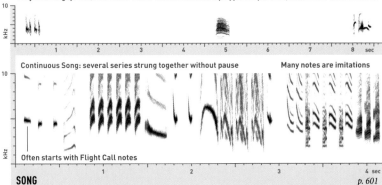

SONG *p. 601*

Mostly Feb.–July, by male; female sings occasionally. Continuous Song grades into Disjunct Song; very like Lesser's, but listen for Teet-tip calls, especially at start or between songs. Beware: Lesser and Lawrence's both sometimes use the other's Flight Calls in Continuous Song. Imitations of other species are excellent.

High, short, falling, finely buzzy

Polyphony can be hard to hear

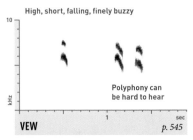

VEW *p. 545*

Not well known; sole available example given with Snarl, possibly by female in courtship. More study needed.

Harsh, buzzy, but high and soft

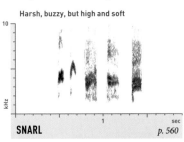

SNARL *p. 560*

All year, in aggressive encounters. Plastic. Juvenile reported to beg with 2-noted call like Lesser's Tea-cup; no recordings available.

1- to 2-syllabled broken whistle — Rising, polyphonic

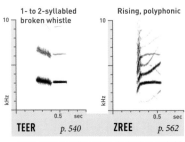

TEER *p. 540*   **ZREE** *p. 562*

All year. Teer easily confused with Lesser Goldfinch; clear, rising Zree is more common. Individuals may have multiple versions of both.

Soft, musical, very brief notes in 2- or 3-note patterns on slightly different pitches

Sometimes a single Tip (p. 531)

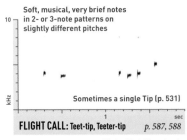

FLIGHT CALL: Teet-tip, Teeter-tip *p. 587, 588*

All year, in contact and flight. Variable, plastic. Distinctive, but soft and brief, easy to overlook. Young American Goldfinch can sound similar.

LESSER GOLDFINCH

Spinus psaltria

Common in a variety of habitats, from oak woods to pine forests to backyards; often visits feeders. Prefers more arid areas than American Goldfinch.

Disjunct Song (p. 598): various 1- and 2-note whistled and polyphonic phrases delivered at intervals

Continuous Song: several series strung together without pause

Many notes are imitations of other bird species

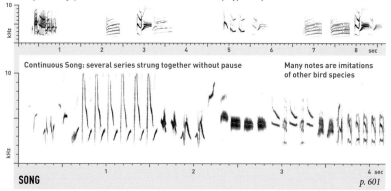

SONG *p. 601*

Mostly Feb.–July, from high perch or in circling flight. Continuous Song grades into Disjunct Song. Consecutive songs may differ greatly in content and length; most series contain 3–4 repetitions, but this varies. Imitations of other species are excellent, but can go unnoticed inside complex jumble of Song.

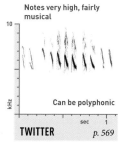

Notes very high, fairly musical

Can be polyphonic

TWITTER *p. 569*

Mostly June–July, by female in courtship. Highly plastic.

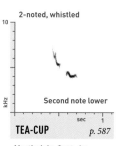

2-noted, whistled

Second note lower

TEA-CUP *p. 587*

Mostly July–Sept., by begging juveniles. Notes can be given separately.

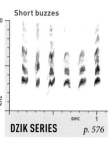

Short buzzes

DZIK SERIES *p. 576*

All year, in aggressive encounters. Highly plastic.

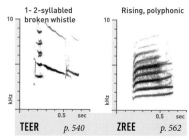

1- 2-syllabled broken whistle

Rising, polyphonic

TEER *p. 540*   **ZREE** *p. 562*

All year. Teer is most common and distinctive call. Zree lower, burrier than other finches'. Individuals likely have multiple versions of both.

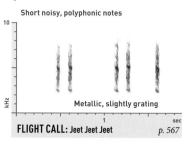

Short noisy, polyphonic notes

Metallic, slightly grating

FLIGHT CALL: Jeet Jeet Jeet *p. 567*

All year, in contact and flight. Somewhat variable, but distinctive. Higher, more metallic than Pine Siskin Flight Call.

EVENING GROSBEAK

Coccothraustes vespertinus

A social, somewhat nomadic finch of coniferous and mixed forests. In some winters, wanders far south of breeding range. Sometimes visits feeders.

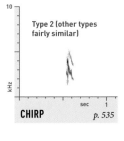

Evening Grosbeak Types

Five "types" have been identified among North American Evening Grosbeaks, corresponding to five different versions of the whistled Flight Calls. These types seem to represent regional variation, unlike Red Crossbill call types (pp. 412–417), which represent ecologically specialized populations that overlap in range.

Three Evening Grosbeak types (1, 2, and 4) are regular in the West. Type 3 breeds east of the Rockies in boreal Canada, and has not been recorded in the West; its Flight Call is a downslurred, polyphonic Pzeer, and it differs very slightly in structure and plumage. Type 5 is apparently found primarily in Mexico, and is rare north of the border.

Type 2 (other types fairly similar)

CHIRP   *p. 535*

Vocal Repertoire

In all Evening Grosbeaks, Song appears to consist of several different stereotyped Breetlike notes, sometimes mixed with Peerlike notes, given singly or in clusters for long periods, often from an exposed perch. As separate calls, the Breet and the Peer are loud and frequent. All types also have a soft Chirp given in close contact, often interspersed with or grading into a soft Breetlike note.

Type 1 Evening Grosbeak

The most widespread type in the West, and the only one recorded north of California and northwest of Wyoming. Occasionally irrupts in numbers southward and to low elevations.

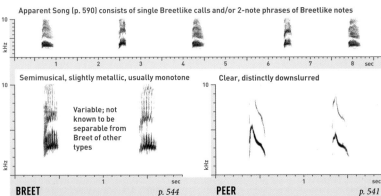

Apparent Song (p. 590) consists of single Breetlike calls and/or 2-note phrases of Breetlike notes

Semimusical, slightly metallic, usually monotone

Variable; not known to be separable from Breet of other types

BREET   *p. 544*

Clear, distinctly downslurred

PEER   *p. 541*

Apparent Song given mostly Apr.–June, by males; can include at least 3 different versions of Breet. Multi-note phrases apparently uncommon. Breet and Peer are common in many contexts. Although Peer begins at a higher frequency than Peer of other types, the shape of the call makes it sound lower.

Type 2 Evening Grosbeak

The common Evening Grosbeak type of the Sierra Nevada. Occasionally flocks with Type 1, which is an uncommon irruptive visitor to its range, where no other call types have been recorded.

Apparent Song (p. 590) consists mostly of 2- or 3-note phrases of Breetlike and sometimes Peerlike notes

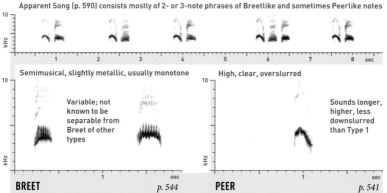

BREET — Semimusical, slightly metallic, usually monotone — Variable; not known to be separable from Breet of other types — *p. 544*

PEER — High, clear, overslurred — Sounds longer, higher, less downslurred than Type 1 — *p. 541*

Apparent Song given mostly Apr.–June, by males; can include at least 3 different versions of Breet and at least 2 versions of Peer. Multinote phrases may be the norm. Breet and Peer are most common calls, given in many contexts. Peer sounds longer, higher, less downslurred than Type 1.

Type 4 Evening Grosbeak

The common Evening Grosbeak type of the southern Rockies. Occasionally flocks with Type 1, which is an uncommon irruptive visitor to its range, where no other call types have been recorded.

Apparent Song (p. 544) consists mostly of single Breetlike and sometimes Peerlike calls

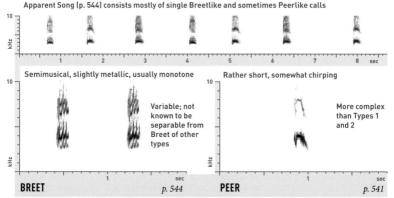

BREET — Semimusical, slightly metallic, usually monotone — Variable; not known to be separable from Breet of other types — *p. 544*

PEER — Rather short, somewhat chirping — More complex than Types 1 and 2 — *p. 541*

Apparent Song given mostly Apr.–June, by males; can include at least 3 different versions of Breet. Multinote phrases have not been documented. Breet and Peer are common in many contexts. Peer sounds slightly lower than Peer of Type 2, about the same pitch as Type 1, but with a distinctly different quality.

Type 5 Evening Grosbeak

Known only from a single recording. Believed to breed primarily in Mexico, occasionally wandering to the mountains of southeast Arizona.

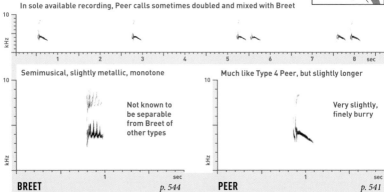

In sole available recording, Peer calls sometimes doubled and mixed with Breet

Semimusical, slightly metallic, monotone

Not known to be separable from Breet of other types

BREET *p. 544*

Much like Type 4 Peer, but slightly longer

Very slightly, finely burry

PEER *p. 541*

Very poorly known; apparently rather rare and local throughout its range. Sole available recording may represent an example of Song; more study needed.

RED CROSSBILL

Loxia curvirostra

A nomadic, social finch of coniferous forests, with a crossed bill specially adapted to extracting seeds from closed conifer cones. Exceedingly complex in biology and behavior.

Red Crossbill Types

Nine distinct groups of Red Crossbills, called "types," breed in North America. Eight are regular in the West. The types differ in vocalizations, bill structure, size, distribution, and ecology, but field identification requires experienced listening and often spectrographic analysis. The population formerly known as Type 9 was recently split into the Cassia Crossbill (*Loxia sinesciuris*), p. 417. Some have proposed that the other types are best recognized as separate species as well. All Red Crossbill types are nomadic; range maps on the following pages are courtesy of Matt Young.

Differences in size and bill structure apparently optimize different Red Crossbill types for feeding on different species of conifers, and different types usually do not interbreed where they overlap in range. However, field identification of the types depends on vocalizations that are learned, including the Flight Calls and the Excitement Calls. In rare cases, individuals have been shown to give calls of 2 different types. In one documented instance, a female that paired with a male of a different type changed her Flight Call type to match her mate's.

When identifying Red Crossbill types in the field, it is strongly recommended to make an audio recording, even if it is a poor one. Observers with recording equipment are needed to expand our knowledge of the status, distribution, behavior, and identification of these fascinating populations.

Vocal Repertoire of Red Crossbills

Flight Calls: These are the most common calls, variable but stereotyped, usually given in short series, from a perch or in flight. They are diagnostic for each type.

Excitement Calls: These are variable but stereotyped, given in agitation or alarm. Like the Flight Calls, they are usually given in short series, and are distinctive, sometimes diagnostic, for each type. When Red Crossbills in a flock give both Flight Calls and Excitement Calls, they are easy to mistake for two different Red Crossbill types.

Type 2 (other types similar)

CHIT-TOO *p. 592*

Songs: Red Crossbills may have two categories of Song. Short Songs usually consist of a couplet or triplet series, whistled or burry, with a few single notes at the start or end (pp. 596, 601). They grade into Long Songs, which are more plastic, often including strings of notes much like the Flight or Excitement Calls of the same or a different Red Crossbill type. To identify a Red Crossbill to type with confidence, even with a spectrogram, it is necessary to be sure that the bird is giving true Flight Calls or Excitement Calls, and not a variation of the Song.

Song repertoire size is unknown, but likely large. The spectrograms of Song shown in this book represent a fraction of the variation in each type. Song differences between types need more study.

Other calls: All crossbills give very soft, plastic calls when feeding. All juvenile crossbills beg with a distinctive Chit-too or Chit-too-too.

Type 1 Red Crossbill

One of the most genetically distinct crossbill types, found mostly in the Appalachians; apparently wanders sometimes to the Pacific Northwest. Medium bill and body size. Foraging habits in the West are poorly known.

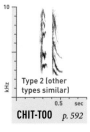

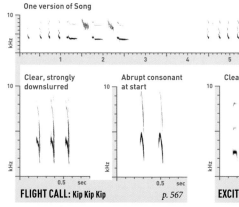

One version of Song

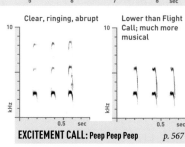

Clear, strongly downslurred

Abrupt consonant at start

FLIGHT CALL: Kip Kip Kip *p. 567*

Averages sharper than Type 2 Flight Call, but identification by ear is difficult; spectrogram shows downward hook at start (often subtle).

Clear, ringing, abrupt

Lower than Flight Call; much more musical

EXCITEMENT CALL: Peep Peep Peep *p. 567*

Averages slightly higher than Type 2 Excitement Call, but difficult to separate from it in the field. Notes on spectrogram show flatter shape.

Type 2 Red Crossbill

The most widespread crossbill type in North America, and the most common type in most parts of the West. One of our largest crossbills in body and bill size. Often feeds on large-seeded, hard-coned pine species such as ponderosa pine, but uses a wide variety of conifer species.

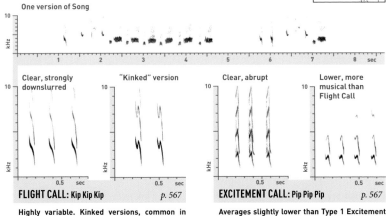

One version of Song

FLIGHT CALL: Kip Kip Kip *p. 567*

Clear, strongly downslurred

"Kinked" version

Highly variable. Kinked versions, common in West, sound only slightly different from nonkinked versions.

EXCITEMENT CALL: Pip Pip Pip *p. 567*

Clear, abrupt

Lower, more musical than Flight Call

Averages slightly lower than Type 1 Excitement Call; notes on spectrogram usually show more pronounced backward-N shape.

Type 3 Red Crossbill

Most common in the Pacific Northwest, but rarely wanders as far south as Arizona. Often found with Types 4 and 10, especially in winter. Small in bill and body size; often feeds on soft-coned conifers such as spruces and hemlocks.

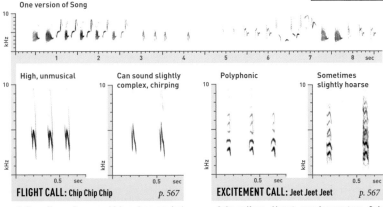

One version of Song

FLIGHT CALL: Chip Chip Chip *p. 567*

High, unmusical

Can sound slightly complex, chirping

Quite uniform. Averages higher, less musical than Types 1 and 2; usually sounds monotone. On spectrogram, note high peak, above 5 kHz.

EXCITEMENT CALL: Jeet Jeet Jeet *p. 567*

Polyphonic

Sometimes slightly hoarse

Quite uniform. Abrupt; sounds monotone. Only slightly musical. Averages higher than Excitement Calls of Types 5 and 6.

Type 4 Red Crossbill

Most common in the Pacific Northwest, but can be common in some years throughout the West, and wanders as far east as Maine. Medium bill and body size. May feed preferentially on Douglas-fir, but also forages on spruces and pines.

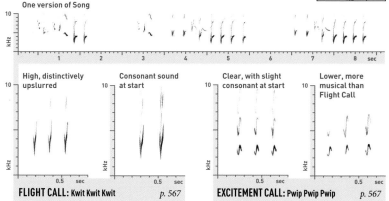

FLIGHT CALL: Kwit Kwit Kwit *p. 567*

High, distinctively upslurred

Consonant sound at start

Highly uniform. When heard well, can be confused only with Type 10; averages less musical, with initial consonant sound.

EXCITEMENT CALL: Pwip Pwip Pwip *p. 567*

Clear, with slight consonant at start

Lower, more musical than Flight Call

Fairly variable. Like Type 2 Excitement Call, but usually higher and slightly less musical. Often sounds slightly upslurred.

Type 5 Red Crossbill

Most common at high elevations in the southern Rockies. Occasionally wanders to low elevations, but rarely far from mountains. Medium-large in bill and body size; apparently specializes on cones of Lodgepole Pine.

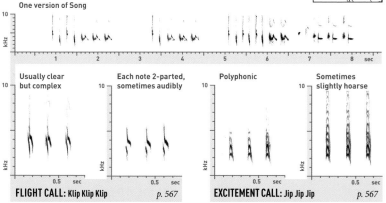

FLIGHT CALL: Klip Klip Klip *p. 567*

Usually clear but complex

Each note 2-parted, sometimes audibly

Fairly variable. Can sound 1-syllabled or vaguely 2-syllabled. Sounds most similar to Type 3, but usually more musical and complex.

EXCITEMENT CALL: Jip Jip Jip *p. 567*

Polyphonic

Sometimes slightly hoarse

Quite uniform. Averages slightly lower than Type 3 Excitement Call and more monotone than Type 6, but difficult to distinguish from both.

Type 6 Red Crossbill

The largest crossbill in bill and body size. Widespread in the mountains of Mexico; common in some years in southeast Arizona, apparently absent in others. Foraging habits need more study.

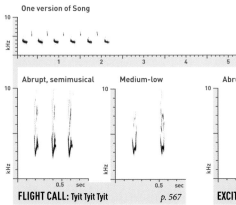

One version of Song

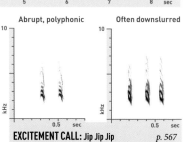

| Abrupt, semimusical | Medium-low | Abrupt, polyphonic | Often downslurred |

FLIGHT CALL: Tyit Tyit Tyit *p. 567*

Highly variable. Spectrogram can resemble Type 4, but sound is distinctly lower, more musical, usually rather monotone.

EXCITEMENT CALL: Jip Jip Jip *p. 567*

Fairly variable. Like Excitement Calls of Types 3 and 5, but averages more distinctly downslurred, lower, and less musical.

Type 7 Red Crossbill

Rarely recorded; poorly known. Medium bill and body size. Most records are from British Columbia, with a few from the interior Pacific Northwest. The main population of cross-bills on Kodiak Island, Alaska, may or may not be of this type; more study needed.

One version of Song

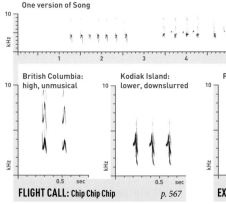

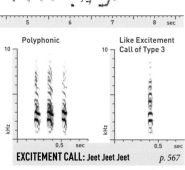

| British Columbia: high, unmusical | Kodiak Island: lower, downslurred | Polyphonic | Like Excitement Call of Type 3 |

FLIGHT CALL: Chip Chip Chip *p. 567*

Probably variable. "Classic" version, at left, is high, chirping, much like Type 3 Flight Call. Kodiak Island version sounds more like Tye 2.

EXCITEMENT CALL: Jeet Jeet Jeet *p. 567*

Not well known. Versions shown were given by captive birds in alarm, and thus may be hoarser than typical Excitement Calls in the wild.

Type 10 Red Crossbill

Locally common along the Pacific Coast. Small in bill and body size. Sometimes flocks with Types 3 and 4. Associated with Sitka spruce in the West; an Eastern population with similar Flight Calls may feed on other spruces and hemlock.

One version of Song

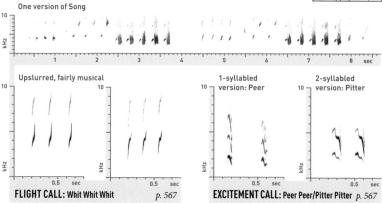

Upslurred, fairly musical

FLIGHT CALL: Whit Whit Whit p. 567

Variable, and more plastic than Flight Calls of other types. Recalls Whit of *Empidonax* flycatchers, but usually more musical.

1-syllabled version: Peer

2-syllabled version: Pitter

EXCITEMENT CALL: Peer Peer/Pitter Pitter p. 567

Distinctive, but some may match Excitement Calls of Types 4 and 2. Pitter may indicate high alarm; recalls Pitter of Ruby-crowned Kinglet.

CASSIA CROSSBILL

Loxia sinesciuris

Once known as "Type 9" Red Crossbill; now considered a full species. Sedentary, unlike Red Crossbill. Found only in the South Hills of Idaho.

Song (p. 576, 596, 601) highly variable, like Red Crossbill's, but most notes burry, series often longer

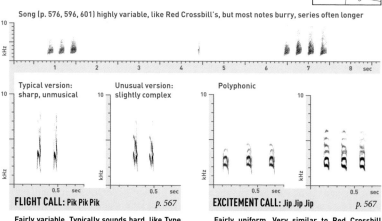

Typical version: sharp, unmusical

Unusual version: slightly complex

FLIGHT CALL: Pik Pik Pik p. 567

Fairly variable. Typically sounds hard, like Type 2 Red Crossbill, but less downslurred; vaguely recalls Pik of Downy Woodpecker, but lower.

Polyphonic

EXCITEMENT CALL: Jip Jip Jip p. 567

Fairly uniform. Very similar to Red Crossbill Type 3 and Type 5 Excitement Calls and probably not reliably distinguishable from them.

White-winged Crossbill

Loxia leucoptera

Breeds in northern coniferous forests; feeds primarily on cones of spruce and tamarack. In some winters, large numbers wander south of breeding range.

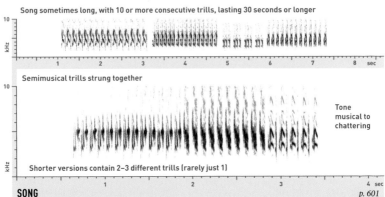

Song sometimes long, with 10 or more consecutive trills, lasting 30 seconds or longer

Semimusical trills strung together

Tone musical to chattering

Shorter versions contain 2–3 different trills (rarely just 1)

SONG *p. 601*

All year, by both sexes, but mostly Jan.–July, by males. Distinctive. Repertoire size unknown, but probably large; consecutive songs often differ in number and type of trills. Recalls Continuous Songs of redpolls, but trills longer, and single notes very rare.

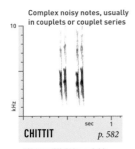

Complex noisy notes, usually in couplets or couplet series

CHITTIT *p. 582*

All year. Slightly variable; like redpoll Flight Calls, but averages harsher.

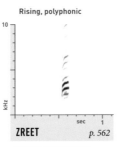

Rising, polyphonic

ZREET *p. 562*

All year; usually given singly. Variable, but shorter than goldfinch Zrees.

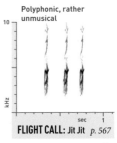

Polyphonic, rather unmusical

FLIGHT CALL: Jit Jit *p. 567*

All year; most common call. Variable; often upslurred.

LONGSPURS AND RELATIVES (Family Calcariidae) *next pages*

Formerly considered close relatives of the North American sparrows, longspurs and the Snow Bunting are now classified in their own family. Most species are found in open areas such as tundra, grasslands, agricultural fields, and beaches. Songs are complex and apparently learned; call repertoires are well developed.

In some species, individual birds have repertoires of several short whistled calls which appear to be learned. These calls show marked geographic variation. In the breeding season, in alarm near the nest or young, adults often cycle continuously through several or all of their whistled calls. Flocks also give them on the wintering grounds. The functions and patterns of variation of these calls need further study.

SNOW BUNTING

Plectrophenax nivalis

Breeds in rock cavities on Arctic tundra, often near human habitation; winters in open areas, often in flocks with Horned Larks and/or longspurs.

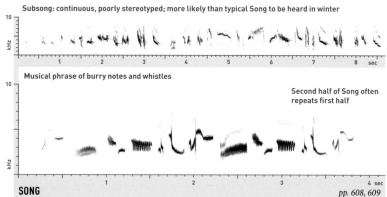

Subsong: continuous, poorly stereotyped; more likely than typical Song to be heard in winter

Musical phrase of burry notes and whistles

Second half of Song often repeats first half

SONG

pp. 608, 609

Mostly on breeding grounds, but subsong can be heard in late winter. Males have a single songtype, often truncated. Quality may recall Blue Grosbeak; generally lacks polyphony of Lapland Longspur song, and more likely to be constructed of repeated phrases, especially in flight display.

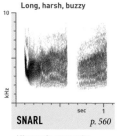

Long, harsh, buzzy

SNARL *p. 560*

All year, in aggressive interactions. Highly plastic.

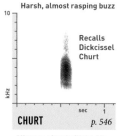

Harsh, almost rasping buzz

Recalls Dickcissel Churt

CHURT *p. 546*

All year, often in flight. No longspur has a similar call.

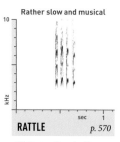

Rather slow and musical

RATTLE *p. 570*

All year; rather plastic. Averages slightly more musical than longspur Rattles.

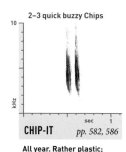

2–3 quick buzzy Chips

CHIP-IT *pp. 582, 586*

All year. Rather plastic; often replaces first 1–2 notes of Rattle.

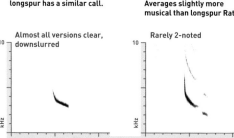

Almost all versions clear, downslurred

Rarely 2-noted

TEER

pp. 540, 587

All year; in alarm near nest, and when flocking in winter. At least some birds have repertoires of a few versions, but most versions quite similar.

LAPLAND LONGSPUR

Calcarius lapponicus

Breeds on Arctic tundra; winters in open areas, often in flocks with Horned Larks, Snow Buntings, and/or other longspur species.

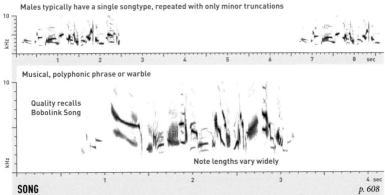

Males typically have a single songtype, repeated with only minor truncations

Musical, polyphonic phrase or warble

Quality recalls Bobolink Song

Note lengths vary widely

SONG *p. 608*

Almost exclusively on breeding grounds. Geographically variable, but males on adjacent territories tend to sing the same songtype. Usually about 2 seconds in length, but 2 or more songs sometimes strung together without pause, especially in courtship.

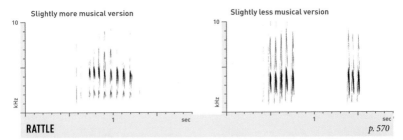

Slightly more musical version

Slightly less musical version

RATTLE *p. 570*

All year; frequently heard in winter. Somewhat plastic, especially in number of notes; most versions are 4–5 notes long, but shorter and longer versions, including single Piks, are not uncommon. Averages slightly longer and faster than McCown's, but overlaps with it.

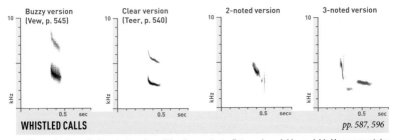

Buzzy version (Vew, p. 545)

Clear version (Teer, p. 540)

2-noted version

3-noted version

WHISTLED CALLS *pp. 587, 596*

All year, in alarm, flock contact, and possibly other contexts. Extremely variable; each bird has a repertoire of perhaps 6 versions. Tone may be clear, buzzy, or polyphonic; most versions 1-noted and downslurred, but some are monotone, and multinoted versions are fairly common.

Chestnut-collared Longspur

Calcarius ornatus

Breeds in native midgrass prairies, especially those that have been moderately mowed or grazed; winters in more arid grasslands, in flocks.

Males appear to have a single songtype, usually repeated at steady intervals

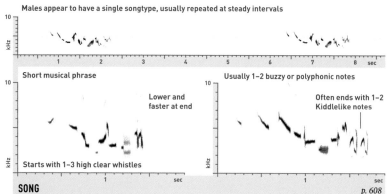

Short musical phrase

Lower and faster at end

Starts with 1–3 high clear whistles

Usually 1–2 buzzy or polyphonic notes

Often ends with 1–2 Kiddlelike notes

SONG   p. 608

Mostly Apr.–July. Given from ground or low perch, or in flight display as male glides to ground on spread wings and tail, sometimes with slow flaps. Pattern and quality can strongly resemble Western Meadowlark Song, but includes buzzy, polyphonic, and/or Kiddle-like notes.

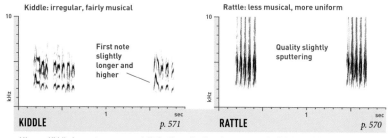

Kiddle: irregular, fairly musical

First note slightly longer and higher

Rattle: less musical, more uniform

Quality slightly sputtering

KIDDLE   p. 571   **RATTLE**   p. 570

All year. Kiddle is most common and distinctive call, given in contact and mild alarm. Highly plastic. Rattle is high, rapid, and rather unmusical; associated with male Song, responses to playback, and aggressive encounters. Both call types intergrade extensively.

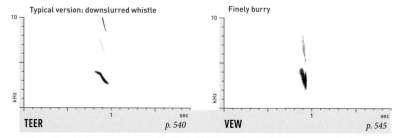

Typical version: downslurred whistle

Finely burry

TEER   p. 540   **VEW**   p. 545

All year. Given in alarm near nest, usually mixed with Kiddles or Rattles (p. 596); also given in winter flocks. Individuals may have multiple versions, but overall, less variable than Whistled Calls of Lapland and McCown's; the vast majority are clear Teerlike downslurs.

McCown's Longspur

Rhynchophanes mccownii

Breeds and winters in semiarid short-grass prairie. Generally prefers shorter grass than Chestnut-collared Longspur, but the two often found together.

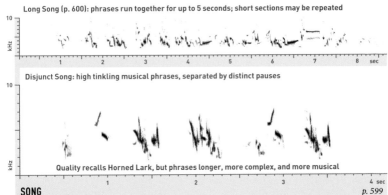

Long Song (p. 600): phrases run together for up to 5 seconds; short sections may be repeated

Disjunct Song: high tinkling musical phrases, separated by distinct pauses

Quality recalls Horned Lark, but phrases longer, more complex, and more musical

SONG p. 599

Mostly Apr.–June. Each male apparently has at least 10 phrase types. Disjunct Song often given from ground or low perch, Long Song in flight display as male glides to ground on spread wings and tail. The 2 song forms intergrade; pauses in Disjunct Song tend to become shorter as Song progresses.

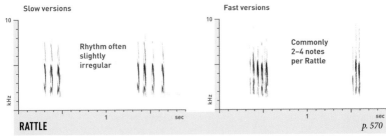

Slow versions

Rhythm often slightly irregular

Fast versions

Commonly 2–4 notes per Rattle

RATTLE p. 570

All year; most common call. Quite plastic in length, but averages shorter than other longspur Rattles, usually 4 or fewer notes. Slow, irregular rhythm fairly distinctive when present.

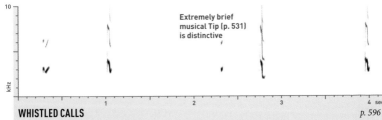

Various brief, musical whistles, upslurred, downslurred, or monotone

Extremely brief musical Tip (p. 531) is distinctive

WHISTLED CALLS p. 596

All year. Individuals have perhaps 4 versions. In alarm on breeding grounds, birds tend to cycle through several versions continuously, mixing them with short Rattles. Also heard from flocks in migration and winter. Almost always clear and 1-syllabled, but buzzy versions exist.

SPARROWS, JUNCOS, AND TOWHEES (Family Passerellidae)

This family includes mostly small to medium-sized birds with conical bills and brownish or grayish plumage. Representatives of the family can be found in almost any habitat, but most spend their time on or near the ground, often in dense cover. The towhees are larger than other sparrows, with long tails and distinctive calls.

Though they look and act quite different, the sparrows are closely related to the warblers (Parulidae), and this close affinity is evident in their vocalizations. Sparrow songs, which are apparently learned, can in many species be classified into typical Songs and Complex Songs, just as in some warbler species. In fact, the Complex Songs of warblers and sparrows often sound quite similar, and they occur in similar contexts, especially during infrequent flight displays, sometimes introduced by series of high-pitched call notes. Call repertoires are also similar in many warblers and sparrows, including a Seet or Dzeet call in contact or in flight (including during nocturnal migration), a high brief Tink in high alarm, and a Chip in mild alarm.

In addition, many sparrows also give various snarls, twitters, or whines in different kinds of interactions. The call repertoires of several species are poorly known, and it is likely that a number of sparrows give calls not described in this book. More study needed.

Pair Reunion Duets

In some sparrow species, mated pairs greet one another with complex, plastic "pair reunion duets." These duets, made up of call-like notes, tend to start with a series of high, Seetlike sounds and end with rapid, lower chips or twitters. Pair reunion duets are best known in the *Melozone* towhees, the Rufous-winged Sparrow, and the Rufous-crowned Sparrow. However, similar behaviors may occur at times in the *Pipilo* towhees and possibly a few other species.

Winter Singing

Many species of sparrow that winter in the U.S., including Song Sparrow, Fox Sparrow, Lark Sparrow, and the species in the genus *Zonotrichia*, can be heard singing nearly year-round. In fall and early winter, the majority of these are subsongs, presumably of first-year males. In late winter and early spring, the majority are plastic songs with a closer resemblance to the typical breeding songs of the species. Stereotyped breeding songs are rare before late spring.

BREWER'S SPARROW (TIMBERLINE)   *Spizella breweri taverneri*

The "Timberline" subspecies of Brewer's Sparrow is a highly local breeder in treeline scrub from Montana to Alaska. It is slightly larger-bodied and smaller-billed than the nominate subspecies, with marginally grayer and more contrasting plumage. Song averages more musical, often with many purely whistled series and fewer buzzy series, but overlap complicates identification. Calls include a downslurred Tsew (p. 533) like Field Sparrow's that is apparently rare or absent in the nominate race. Late-summer breeders at treeline in the southern Rockies are apparently nominate birds that have moved upslope after breeding at lower elevations.

LONG SONG   *not indexed*

EASTERN TOWHEE

Pipilo erythrophthalmus

Common in shrubby open woodlands in the East; rare at the western edge of its range, where it hybridizes with the closely related Spotted Towhee.

Complex Song (p. 596): various calls and call-like notes, singly or in series of 2–3

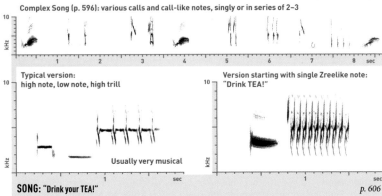

Typical version: high note, low note, high trill

Usually very musical

Version starting with single Zreelike note: "Drink TEA!"

SONG: "Drink your TEA!" *p. 606*

Mostly Apr.–July. Males have 2–5 songtypes (in Florida, up to 11), each usually repeated many times before switching. Females sing very rarely. Highly variable: 1–3 (rarely up to 5) intro notes are followed by 1 trill (rarely 2). Soft, disjointed Complex Song given at dawn, or after territorial encounters.

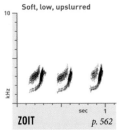

Soft, low, upslurred

ZOIT *p. 562*

One of several little-known calls, given with Zrees; may serve in courtship.

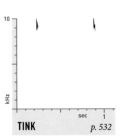

TINK *p. 532*

All year, by both sexes, in high alarm. Sometimes run into rapid Twitter.

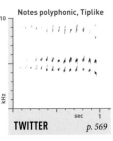

Notes polyphonic, Tiplike

TWITTER *p. 569*

Mostly by females in courtship, but also reported in male-male interactions.

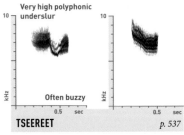

Very high polyphonic underslur

Often buzzy

TSEEREET *p. 537*

All year, by both sexes; second most common call. Plastic and variable, but usually long, underslurred, and polyphonic; often buzzy.

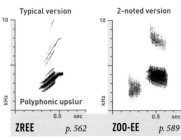

Typical version

Polyphonic upslur

ZREE *p. 562*

2-noted version

ZOO-EE *p. 589*

All year, by both sexes; most common call. Highly variable, but most versions are rising, metallic, 1-noted or only vaguely 2-noted.

SPOTTED TOWHEE

Pipilo maculatus

Formerly lumped with Eastern Towhee. Plumage and sounds vary geographically; birds at eastern end of range sound more similar to Eastern Towhee.

Complex Song (p. 596): various calls and call-like notes, singly or in series of 2–3

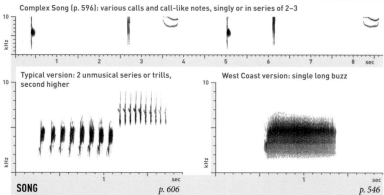

Typical version: 2 unmusical series or trills, second higher

West Coast version: single long buzz

SONG   *p. 606*   *p. 546*

Mostly Feb.–July. Males have 4–9 songtypes; 1 may be repeated many times before switching, or 2 may be alternated for long periods. Females sing very rarely. Highly variable. Hybrids with Eastern Towhee may sing like either parent. Soft, disjointed Complex Song given at dawn, or after territorial encounters.

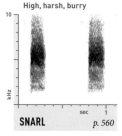
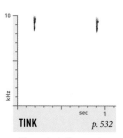
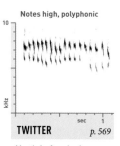

High, harsh, burry

Notes high, polyphonic

SNARL   *p. 560*

TINK   *p. 532*

TWITTER   *p. 569*

One of several little-known calls; this version given during midair tussle.

All year, by both sexes, in high alarm. Sometimes run into rapid Twitter.

Mostly by females in courtship; also in male-male interactions. Highly plastic.

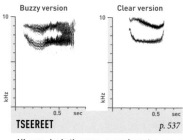
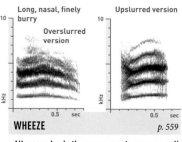

Buzzy version

Clear version

Long, nasal, finely burry

Overslurred version

Upslurred version

TSEEREET   *p. 537*

WHEEZE   *p. 559*

All year, by both sexes; second most common call. Plastic and variable, but usually long, underslurred, and polyphonic; often buzzy.

All year, by both sexes; most common call. Highly variable and plastic; some individuals give both over- and upslurred versions.

GREEN-TAILED TOWHEE

Pipilo chlorurus

Fairly common in semiarid habitats with diverse mature shrubs. Often skulks in deep cover, but can be quite conspicuous during breeding season.

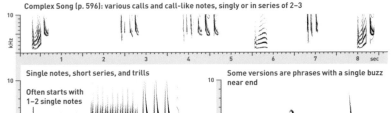

Complex Song (p. 596): various calls and call-like notes, singly or in series of 2–3

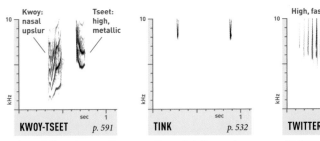

Single notes, short series, and trills

Often starts with 1–2 single notes

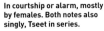

Some versions are phrases with a single buzz near end

SONG   p. 608

All year, especially Apr.–July, by males. Males have 5–12 songtypes; consecutive songs almost always differ. In California and Oregon, Song of "Thick-billed" Fox Sparrow (p. 456) averages fewer trills and more slurred whistles, but much overlap. Soft, disjointed Complex Song often given after territorial encounters.

Kwoy: nasal upslur   **Tseet:** high, metallic

KWOY-TSEET   p. 591

In courtship or alarm, mostly by females. Both notes also singly, Tseet in series.

TINK   p. 532

All year, by both sexes, in high alarm.

High, fast, umusical

TWITTER   p. 569

Mostly by courting females. More rattling, cowbirdlike than in other towhees.

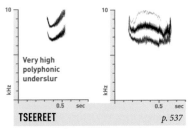

Very high polyphonic underslur

TSEEREET   p. 537

All year, by both sexes; second most common call. Plastic and variable, but usually long, underslurred, and polyphonic; often buzzy.

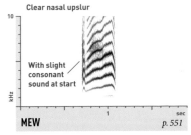

Clear nasal upslur

With slight consonant sound at start

MEW   p. 551

All year; most common call. Somewhat plastic, but rather uniform rangewide; rarely bury or noisy. A few versions are nearly monotone.

RUFOUS-CROWNED SPARROW

Aimophila ruficeps

Locally common on rocky hillsides with scattered dense shrubs. Usually nests on the ground. Not closely related to other North American sparrows.

Songs often introduced by 1–2 Keer notes

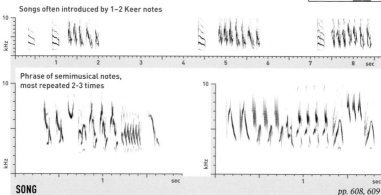

Phrase of semimusical notes, most repeated 2-3 times

SONG

pp. 608, 609

Mostly Mar.–Aug. May recall a fast Indigo Bunting Song, or the beginning of a House Wren Song. Males have repertoires of up to 14 songtypes; consecutive songtypes often similar, though number of repetitions in each series may vary. Typical Song is sometimes given during courtship flight display.

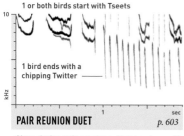

1 or both birds start with Tseets

1 bird ends with a chipping Twitter

PAIR REUNION DUET

p. 603

Given during pair reunions. May be given with and grade into Keer and Chatter.

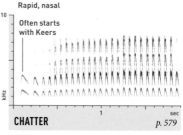

Rapid, nasal

Often starts with Keers

CHATTER

p. 579

All year, in alarm. Grades into Keers. Uniform but plastic; often unsteady in speed and pitch.

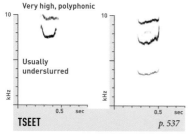

Very high, polyphonic

Usually underslurred

TSEET

p. 537

All year, likely in family contact; also perhaps in mild alarm. Quite plastic. Usually shorter, less burry than similar calls of towhees.

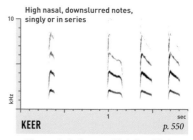

High nasal, downslurred notes, singly or in series

KEER

p. 550

All year, by both sexes, likely in mild alarm; most common call. Distinctive. Uniform but plastic; grades into Chatter.

CANYON TOWHEE

Melozone fusca

Common in rocky desert scrub, pinyon-juniper woodlands, and open riparian habitats, including residential areas. Note chestnut cap, spot on breast.

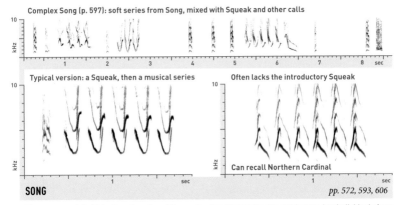

Complex Song (p. 597): soft series from Song, mixed with Squeak and other calls

Typical version: a Squeak, then a musical series

Often lacks the introductory Squeak

Can recall Northern Cardinal

SONG *pp. 572, 593, 606*

All year, especially Feb.–Aug. Female not known to sing. Repertoire size not known, but individuals have multiple songtypes, each usually repeated several times before switching, except in excitement, when songtype switches frequently. Some songtypes include couplet series, or rarely 2 consecutive series.

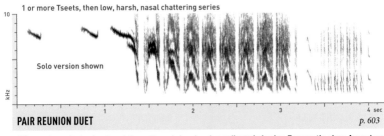

1 or more Tseets, then low, harsh, nasal chattering series

Solo version shown

PAIR REUNION DUET *p. 603*

All year, by mated pairs, to reinforce the pair bond and coordinate behavior. Frequently given from deep cover. Highly plastic; introductory Tseets are often absent, and chattering series may take many forms. Often given in display posture, bobbing head with scapulars raised.

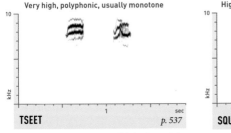

Very high, polyphonic, usually monotone

TSEET *p. 537*

All year, in contact and possibly in alarm. Quite plastic.

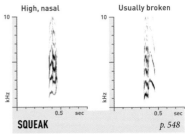

High, nasal

Usually broken

SQUEAK *p. 548*

All year; most common call. Variable but distinctive. Likely also gives a Tink in high alarm like other towhees; no recordings available.

CALIFORNIA TOWHEE

Melozone crissalis

Common in chaparral, riparian brush, and residential areas. Once lumped with Canyon Towhee, but more closely related to Abert's, as vocalizations show.

Individuals appear to have a single songtype, repeated with minor variations

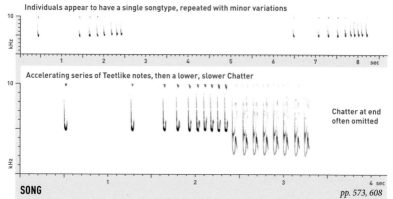

Accelerating series of Teetlike notes, then a lower, slower Chatter

Chatter at end often omitted

SONG

pp. 573, 608

Mostly Feb.–July, by unmated males. Female not known to sing. Some songtypes slightly more complex than shown, jumping several times from Teet series to Chatter and back. Complex Song not reported, but likely exists.

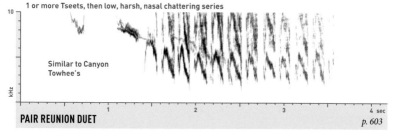

1 or more Tseets, then low, harsh, nasal chattering series

Similar to Canyon Towhee's

PAIR REUNION DUET

p. 603

All year, by mated pairs, to reinforce the pair bond and coordinate behavior. Frequently given from deep cover. Highly plastic; introductory Tseets are often absent, and chattering series may take many forms. Sometimes given in display posture with fluttering wings and open bills.

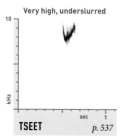

Very high, underslurred

TSEET *p. 537*

All year, in contact. Quite plastic. Underslur and polyphony can be hard to hear.

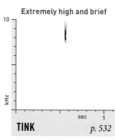

Extremely high and brief

TINK *p. 532*

Infrequent, in high alarm, usually with Teet, and perhaps grading into it.

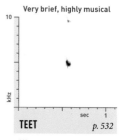

Very brief, highly musical

TEET *p. 532*

All year; most common and distinctive call. Audibly higher than Abert's Teet.

ABERT'S TOWHEE

Melozone aberti

Common in dense riparian brush and in some residential areas; not shy, but often stays within deep cover. Forages on the ground, like other towhees.

Individuals have multiple songtypes; consecutive songs may be same or different

Accelerating series of Teetlike notes, then a lower, slower Chatter

Sometimes ends with 2 consecutive series, or a partly repeating phrase

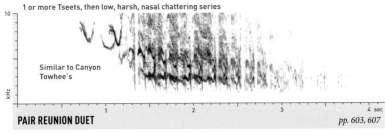

SONG
pp. 603, 608

Mostly Mar.–Aug. Repertoire size unknown; several songtypes from one individual may sound similar but show differences on the spectrogram. Some songtypes more complex than shown. Soft, infrequent Complex Song (p. 597) rather similar to that of Canyon Towhee.

1 or more Tseets, then low, harsh, nasal chattering series

Similar to Canyon Towhee's

PAIR REUNION DUET
pp. 603, 607

All year, by mated pairs, to reinforce the pair bond and coordinate behavior. Frequently given from deep cover. Highly plastic; introductory Tseets are often absent, and chattering series ranges from nasal to snarling.

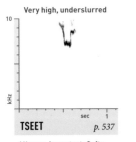 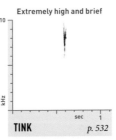 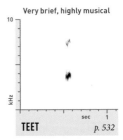

Very high, underslurred

Extremely high and brief

Very brief, highly musical

TSEET
p. 537

TINK
p. 532

TEET
p. 532

All year, in contact. Quite plastic. Underslur and polyphony can be hard to hear.

Infrequent, in high alarm, usually with Teet, and perhaps grading into it.

All year; most common and distinctive call. Rare variant is a clear Peep (p. 534).

Rufous-winged Sparrow

Peucaea carpalis

Uncommon, local bird of open, grassy desert scrub with hackberry, cholla, and often mesquite. Numbers north of Mexico have fluctuated.

Individuals have roughly 15–20 songtypes; consecutive songs usually similar

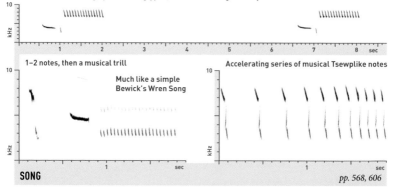

1–2 notes, then a musical trill

Much like a simple Bewick's Wren Song

Accelerating series of musical Tsewplike notes

SONG *pp. 568, 606*

All year, especially Feb.–Sept., by males; female not known to sing. Two basic song patterns, both shown above, are about equally common and seem to serve the same function. Neighboring males sometimes match songtypes.

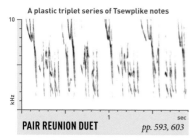

A plastic triplet series of Tsewplike notes

PAIR REUNION DUET *pp. 593, 603*

All year, by mated pairs or by male upon reuniting with female, likely to reinforce the pair bond. Not clear which notes are given by which sex.

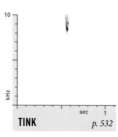

High, loud, monotone

More musical than Chip

SEET *p. 536*   **TSEWP** *p. 533*

Excited birds often give only the first note of Song for extended periods; most common forms are Seet and Tsewp, shown.

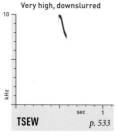

Very high, downslurred

TSEW *p. 533*

One of the more common calls; function little known. Grades into other calls.

High, rather musical

TINK *p. 532*

Likely in high alarm; grades into Tsew and Chip.

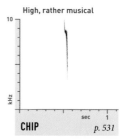

CHIP *p. 531*

Likely all year, in agitation; rather infrequent. Grades into other calls.

Cassin's Sparrow

Peucaea cassinii

Frequents arid grasslands dotted with mesquite, rabbitbrush, or other shrubs. Abundance can vary greatly from year to year at any one location.

1–3 whistles and a trill on nearly the same pitch, then 4 whistles in a high-low, high-low pattern

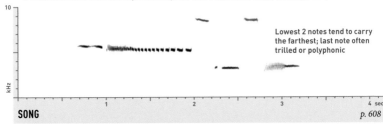

Lowest 2 notes tend to carry the farthest; last note often trilled or polyphonic

SONG *p. 608*

Highly distinctive and musical. Mostly Apr.–July in Texas, July–Sept. in Arizona, often in display flight as male glides downward, usually to a different perch. Individuals have 1–3 songtypes, all quite similar, each usually repeated 2–4 times before switching.

Complex Song often embedded in long twittering series of Psitlike notes

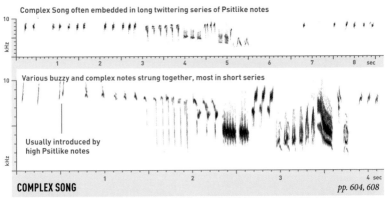

Various buzzy and complex notes strung together, most in short series

Usually introduced by high Psitlike notes

COMPLEX SONG *pp. 604, 608*

Mostly during breeding season by excited males, apparently in both close courtship and territorial encounters, sometimes in flight. Individuals may have a single Complex songtype, but repertoire size needs more study.

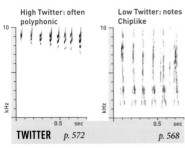

High Twitter: often polyphonic

Low Twitter: notes Chiplike

TWITTER *p. 572* *p. 568*

Mostly during interactions. Chipping version less frequent, usually introduced by and developing out of high version.

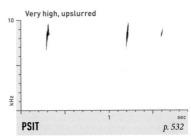

Very high, upslurred

PSIT *p. 532*

All year, in alarm and family contact; most common call. Highly plastic. May not be separable from Botteri's Psit.

Botteri's Sparrow

Peucaea botterii

Locally common in healthy semiarid grasslands, open oak woods, and open thorn scrub. Best distinguished from Cassin's Sparrow by voice.

A varied series of 1- or 2-syllabled notes, then a series of Chips accelerating into a trill

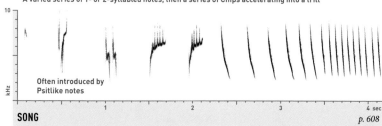

Often introduced by Psitlike notes

SONG

p. 608

Mostly prior to egg-laying, Mar.–June in Texas, May–July in Arizona, reportedly by males only. Sometimes given in low, level flight, especially during territorial disputes. Excited birds may double the accelerating trill at end, or omit it. Repertoire size unknown; consecutive songs tend to be similar.

Complex Song: highly varied, but certain short combinations of notes may be repeated

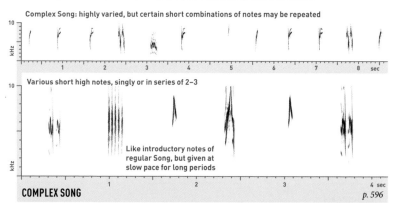

Various short high notes, singly or in series of 2–3

Like introductory notes of regular Song, but given at slow pace for long periods

COMPLEX SONG

p. 596

Mostly after egg-laying, May–July in Texas, July–Sept. in Arizona, reportedly by males only; if nest fails, male reverts to regular Song. Psits, or long Twitters of Psitlike notes, often interspersed. Number of phrase types per male unknown, but apparently large.

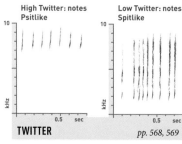

High Twitter: notes Psitlike

Low Twitter: notes Spitlike

TWITTER

pp. 568, 569

High Twitter given in alarm or with Complex Song. Low Twitter given in aggression, in high alarm, and during pair reunions.

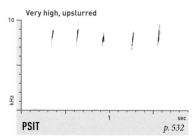

Very high, upslurred

PSIT

p. 532

All year, in alarm and family contact; most common call. Highly plastic. May not be separable from Cassin's Psit.

AMERICAN TREE SPARROW

Spizelloides arborea

A locally common winter visitor, found in flocks in weedy fields and open woods. Readily visits feeders. Note bi-colored bill and central breast spot.

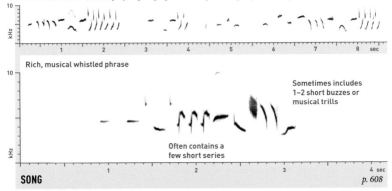

Most late-winter and early-spring singing is rambling subsong or plastic song

Rich, musical whistled phrase

Sometimes includes 1–2 short buzzes or musical trills

Often contains a few short series

SONG *p. 608*

Mostly on breeding grounds, but wintering birds often give subsong, plastic song, and occasionally typical Song before migrating north. Breeding males have a single songtype; easily confused with Song of Fox Sparrow, but higher and less likely to contain long, slurred whistles.

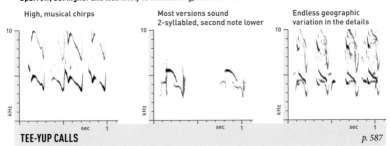

High, musical chirps

Most versions sound 2-syllabled, second note lower

Endless geographic variation in the details

TEE-YUP CALLS *p. 587*

Often given by flocks in migration or winter; apparently rather uniform within a given flock, but often different between flocks. Plastic; perhaps flocks or individuals change calls over time, or have multiple versions. More study needed.

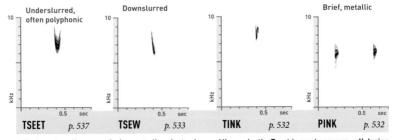

Underslurred, often polyphonic

Downslurred

Brief, metallic

TSEET *p. 537*   **TSEW** *p. 533*   **TINK** *p. 532*   **PINK** *p. 532*

Single-note calls are confusing; not all variants shown. All are plastic. Tseet is most common call during day and in night migration. Grades into Tsew, a common variant. Tink recorded from alarmed birds near nest in summer. Pink given infrequently all year, apparently in alarm. More study of calls needed.

BLACK-CHINNED SPARROW

Spizella atrogularis

Breeds in dry canyons, usually on rocky
hillsides with mixed shrubs. Winters at
slightly lower elevations. Female lacks
black face, recalling a junco.

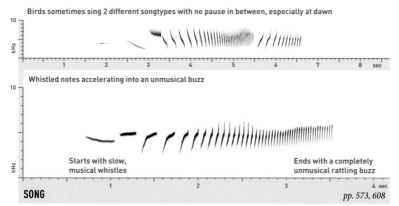

Birds sometimes sing 2 different songtypes with no pause in between, especially at dawn

Whistled notes accelerating into an unmusical buzz

Starts with slow,
musical whistles

Ends with a completely
unmusical rattling buzz

SONG   *pp. 573, 608*

Mostly Mar.–Aug., by males. Female not known to sing. Males have at least 2 songtypes, one with all the
notes upslurred, the other with all the notes downslurred; consecutive songs may be same or different,
but birds often alternate. Some birds add a few extra musical whistles at the start or end of the Song.

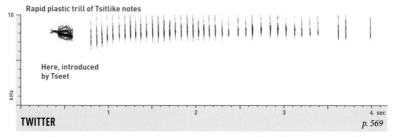

Rapid plastic trill of Tsitlike notes

Here, introduced
by Tseet

TWITTER   *p. 569*

Not well known. Version shown given by female in response to male Song. Likely gives other Twitter types
in other situations, like other members of the genus.

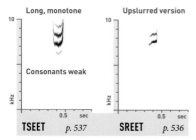

Long, monotone

Upslurred version

Consonants weak

TSEET *p. 537*   **SREET** *p. 536*

All year, in contact and in flight, including possibly by night migrants. Sometimes run into rapid
series. Quite plastic; rarely downslurred.

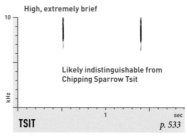

High, extremely brief

Likely indistinguishable from
Chipping Sparrow Tsit

TSIT   *p. 533*

All year, in alarm and possibly in other situations. Most common call; slightly plastic.

CHIPPING SPARROW

Spizella passerina

A common and familiar breeder in open woods, parks, and towns. Often nests in conifers. Outside the breeding season, often forms large flocks.

Dawn Song: songs delivered at very short intervals, often truncated

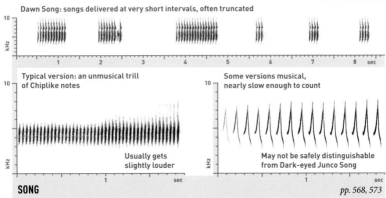

Typical version: an unmusical trill of Chiplike notes

Usually gets slightly louder

Some versions musical, nearly slow enough to count

May not be safely distinguishable from Dark-eyed Junco Song

SONG

pp. 568, 573

Mostly May–July. Highly variable. Females not known to sing. Most males have only 1 songtype; those with 2 never alternate. Rare variants include trills that change quality, pitch, or speed partway through. Dawn Song often given by multiple birds on ground at close range in complex interaction.

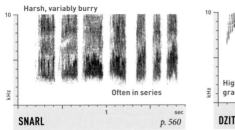

Harsh, variably burry

Often in series

SNARL

p. 560

All year, in close-range aggressive interactions. Highly variable and plastic. Other *Spizella* sparrows have similar calls.

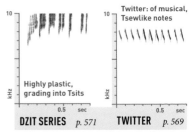

Twitter: of musical, Tsewlike notes

Highly plastic, grading into Tsits

DZIT SERIES *p. 571*   **TWITTER** *p. 569*

Mostly May–June. Dzit Series in courtship, in agitation, or by begging juvenile. Twitter before and during copulation, reportedly by female.

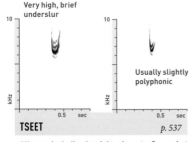

Very high, brief underslur

Usually slightly polyphonic

TSEET

p. 537

All year, including by night migrants. Somewhat variable and plastic, but fairly distinctive once learned.

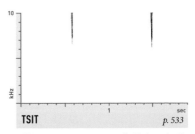

TSIT

p. 533

All year; most common call. Rather plastic, sometimes grading into Tseet. Often run into rapid Dzit Series by agitated birds.

CLAY-COLORED SPARROW

Spizella pallida

A common breeder in prairie shrubs, particularly snowberry, as well as second growth and thickets near water. Winters in dry, grassy or weedy areas.

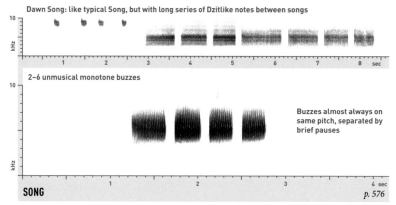

Dawn Song: like typical Song, but with long series of Dzitlike notes between songs

2–6 unmusical monotone buzzes

Buzzes almost always on same pitch, separated by brief pauses

SONG   *p. 576*

Mostly May–July, including on spring migration. Most males have a single songtype, but some males have up to 3; each songtype is repeated many times before switching. Most Songs contain only 1 type of buzz, but rarely the type of buzz will change partway through the Song.

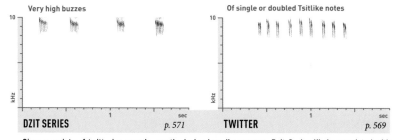

Very high buzzes

Of single or doubled Tsitlike notes

DZIT SERIES   *p. 571*

TWITTER   *p. 569*

Gives a variety of twittering sounds, mostly during breeding season. Dzit Series likely associated with alarm or agitation; similar sounds given in Dawn Song. Function of Twitter not clear. Likely also has musical version used in courtship as well as long snarling notes given in aggression.

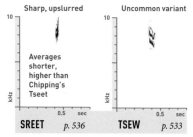

Sharp, upslurred

Averages shorter, higher than Chipping's Tseet

SREET   *p. 536*

Uncommon variant

TSEW   *p. 533*

All year, including by night migrants. Variable, plastic; most are Sreetlike.

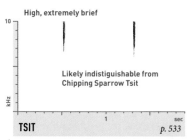

High, extremely brief

Likely indistiguishable from Chipping Sparrow Tsit

TSIT   *p. 533*

All year; most common call. Rather plastic. Very often given in rapid series; also run into rapid twitters by agitated birds.

BREWER'S SPARROW

Spizella breweri

Breeds in western sagebrush steppe. Forms flocks in migration, often with other *Spizella* species. See discussion of the "Timberline" subspecies on p. 423.

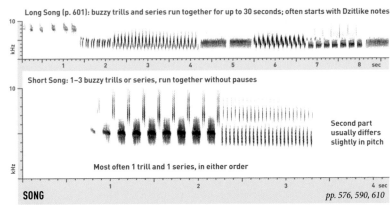

Long Song (p. 601): buzzy trills and series run together for up to 30 seconds; often starts with Dzitlike notes

Short Song: 1–3 buzzy trills or series, run together without pauses

Second part usually differs slightly in pitch

Most often 1 trill and 1 series, in either order

SONG

pp. 576, 590, 610

All year, especially May–June; subsong or plastic song frequent in winter. Most males have only 1 Short Song type, but some have 2, each repeated many times before switching. Short Song given at regular intervals during day by breeding males; Long Song at dawn, in courtship, and in territorial encounters.

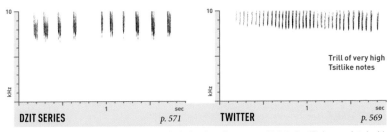

Trill of very high Tsitlike notes

DZIT SERIES *p. 571*

TWITTER *p. 569*

Gives a variety of twittering sounds, mostly during breeding season. Dzit Series likely associated with alarm or agitation; function of Twitter not clear. Likely also has musical version used in courtship as well as long snarling notes given in aggression, like Chipping Sparrow.

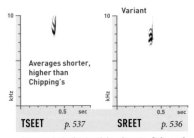

Variant

Averages shorter, higher than Chipping's

TSEET *p. 537*

SREET *p. 536*

All year, including by night migrants. Quite variable and somewhat plastic.

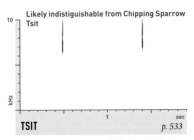

Likely indistiguishable from Chipping Sparrow
Tsit

TSIT *p. 533*

All year; most common call. Rather plastic. Very often given in rapid series; also run into rapid twitters by agitated birds.

FIELD SPARROW

Spizella pusilla

Fairly common in brushy fields, prairie edges, and early second growth. Nests at or near ground level. Note pinkish bill and legs.

Complex Song (p. 601): 1–2 series and 2–3 trills on nearly the same pitch; interspersed with call-like notes

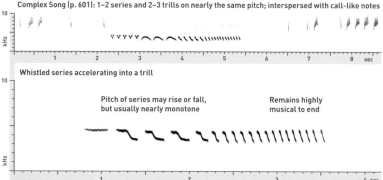

Whistled series accelerating into a trill

Pitch of series may rise or fall, but usually nearly monotone

Remains highly musical to end

SONG

p. 573

Mostly May–July. Simple, accelerating pattern distinctive in most of range. Males apparently have 1 regular and 1 Complex songtype, often shared by neighboring birds. Complex Song given mostly at dawn, especially later in season, and in territorial encounters. Females not known to sing.

Notes very high, downslurred, Tsewlike

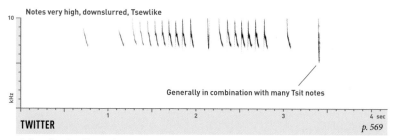

Generally in combination with many Tsit notes

TWITTER

p. 569

At least May–Aug. The version shown apparently used in male-male aggression; has been recorded in stereotyped form at regular intervals in late August, suggesting a type of "autumn song." Similar calls, likely representing several vocalizations, reported in courtship-related contexts.

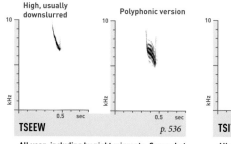

High, usually downslurred

Polyphonic version

TSEEW

p. 536

All year, including by night migrants. Somewhat variable and plastic, but fairly distinctive.

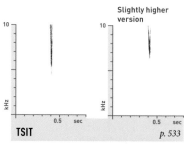

Slightly higher version

TSIT

p. 533

All year; most common call. Rather plastic.

VESPER SPARROW

Pooecetes gramineus

A bird of open grassy fields with a few low bushes; sometimes also cultivated fields. White outer tail feathers are an excellent field mark.

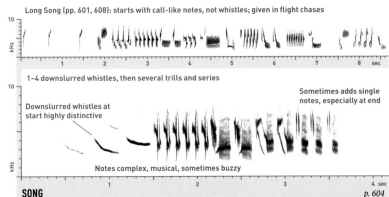

Long Song (pp. 601, 608): starts with call-like notes, not whistles; given in flight chases

1–4 downslurred whistles, then several trills and series

Downslurred whistles at start highly distinctive

Sometimes adds single notes, especially at end

Notes complex, musical, sometimes buzzy

SONG
p. 604

Mostly Apr.–Aug. Introductory whistles same in all Songs of 1 bird, and similar among neighbors, but regionally variable. Rest of Song highly variable; individuals combine 40 or more trills/series into 200 or more songtypes. Consecutive songtypes may have identical or different endings.

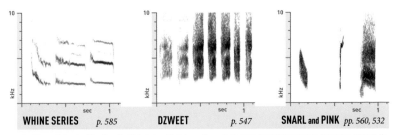

WHINE SERIES *p. 585*

DZWEET *p. 547*

SNARL and PINK *pp. 560, 532*

A wide variety of calls given in high-intensity interactions during the breeding season. Function of each little known; more study needed. Dzweets and Snarls likely intergrade; some very high-pitched snarllike calls have also been recorded.

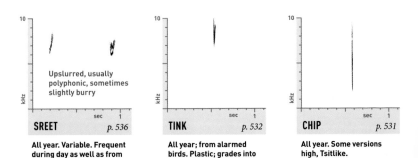

Upslurred, usually polyphonic, sometimes slightly burry

SREET *p. 536*

All year. Variable. Frequent during day as well as from night migrants.

TINK *p. 532*

All year; from alarmed birds. Plastic; grades into Chip.

CHIP *p. 531*

All year. Some versions high, Tsitlike.

LARK SPARROW

Chondestes grammacus

Found in open shrubby habitats, with or without scattered trees. Common in the West; quite local in the East. Note long, rounded tail with white corners.

Long Song (p. 601): like typical Song, but continues without pauses for long periods

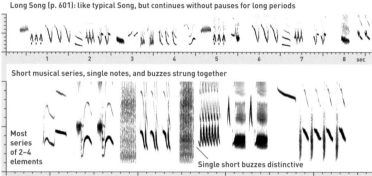

Short musical series, single notes, and buzzes strung together

Most series of 2–4 elements

Single short buzzes distinctive

SONG

pp. 601, 608

All year; especially Apr.–July, by males. Females not known to sing. Winter Songs often plastic. Repertoire size unknown; consecutive songs usually different, often with many of the same components rearranged. Courting males sing before females with wings drooped, tail raised and spread.

Twitter: extremely high, polyphonic

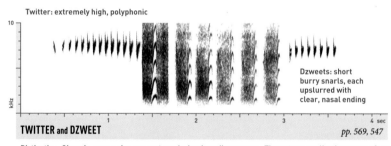

Dzweets: short burry snarls, each upslurred with clear, nasal ending

TWITTER and DZWEET

pp. 569, 547

Distinctive. Given in aggressive encounters during breeding season. These two vocalizations very often heard together, with Twitter immediately before or after Dzweets, but also given separately. Dzweets usually in short series, but sometimes given singly.

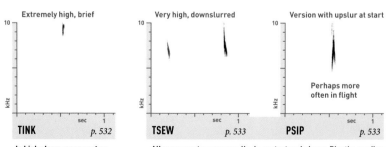

Extremely high, brief

TINK
p. 532

In high alarm, near nest or young. Grades into Tsip.

Very high, downslurred

TSEW
p. 533

Version with upslur at start

Perhaps more often in flight

PSIP
p. 533

All year; most common calls, in contact and alarm. Plastic, grading into Tink; 2 most common variants shown. No Seet call known; night migrants are not known to call.

Five-striped Sparrow

Amphispiza quinquestriata

Extremely local; only a few dozen pairs nest north of Mexico, on steep brushy hillsides in desert canyons. Secretive; difficult to find unless singing.

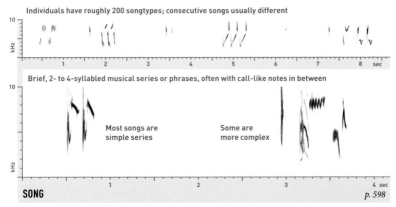

Individuals have roughly 200 songtypes; consecutive songs usually different

Brief, 2- to 4-syllabled musical series or phrases, often with call-like notes in between

Most songs are simple series

Some are more complex

SONG *p. 598*

Mostly Mar.–Aug, by males; female not known to sing. Distinctive; some songs recall Dickcissel. Pace of delivery often slower than shown, about 1 song every 4–5 seconds. Over short periods of time, males switch between only a few songtypes, often alternating just 2.

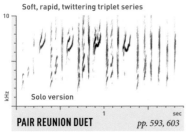

Soft, rapid, twittering triplet series

Solo version

PAIR REUNION DUET *pp. 593, 603*

All year, by mated pairs or by male upon reuniting with female, likely to reinforce the pair bond; also by males in territorial aggression.

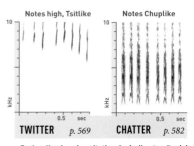

Notes high, Tsitlike

Notes Chuplike

TWITTER *p. 569*   **CHATTER** *p. 582*

Both calls given in agitation, including territorial aggression; Chatter also given by juveniles and in response to predators.

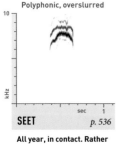

Polyphonic, overslurred

SEET *p. 536*

All year, in contact. Rather infrequent. Overslur distinctive when audible.

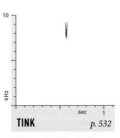

TINK *p. 532*

All year, in high agitation or alarm. Grades into Twitter.

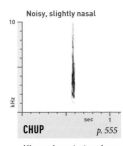

Noisy, slightly nasal

CHUP *p. 555*

All year, in contact or alarm. Plastic; some versions higher. Grades into Chatter.

BLACK-THROATED SPARROW

Amphispiza bilineata

Common in southwestern deserts. Breeds in a variety of arid shrubby habitats, particularly in creosote bush in the south and sagebrush in the north.

Long Song (p. 601): regular songs strung together without pause, songtypes often alternating

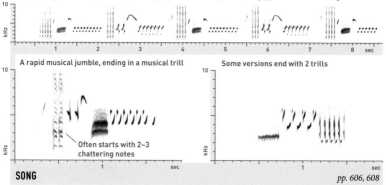

A rapid musical jumble, ending in a musical trill

Often starts with 2–3 chattering notes

Some versions end with 2 trills

SONG

pp. 606, 608

Mostly Mar.–Aug. High, tinkling, never very loud. Individuals have 5–9 songtypes; consecutive songtypes sometimes similar, but more often different; 2 songtypes often alternated for a while before switching to another 2. Truncated songs sometimes given for long periods.

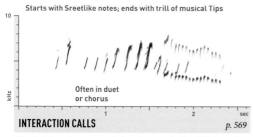

Starts with Sreetlike notes; ends with trill of musical Tips

Often in duet or chorus

INTERACTION CALLS
p. 569

Mostly in breeding season, during chases and other high-intensity aggressive interactions between two or more birds. Highly plastic.

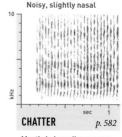

Noisy, slightly nasal

CHATTER
p. 582

Mostly in breeding season, during aggressive encounters.

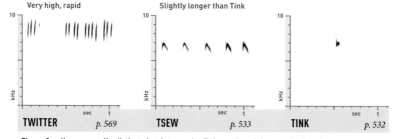

Very high, rapid

TWITTER
p. 569

Slightly longer than Tink

TSEW
p. 533

TINK
p. 532

These 3 calls are usually distinct, but intergrade. Twitter given infrequently; function unknown. Tsew usually given in short series, while Tink is usually given singly, but the two are often interspersed. Tsew and Tink are most common calls, given all year in a variety of situations.

SAGEBRUSH SPARROW

Artemisiospiza nevadensis

Breeds in sagebrush steppe; winters in fairly open arid scrub. Paler than nominate Bell's Sparrow, with streaked back and darker stripes bordering throat.

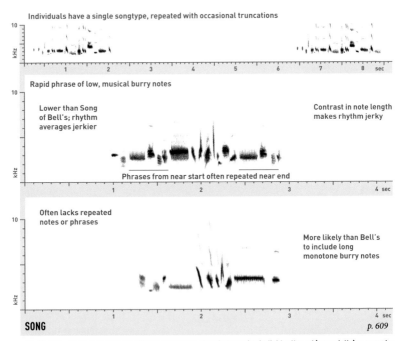

Individuals have a single songtype, repeated with occasional truncations

Rapid phrase of low, musical burry notes

Lower than Song of Bell's; rhythm averages jerkier

Contrast in note length makes rhythm jerky

Phrases from near start often repeated near end

Often lacks repeated notes or phrases

More likely than Bell's to include long monotone burry notes

SONG *p. 609*

Mostly Feb.–June, by males; females not known to sing. Song varies individually and (especially) geographically. On wintering grounds, subsong and plastic song are not uncommon. Typical versions are separable from those of Bell's Sparrow, but some versions approach Bell's in several characteristics. In most versions, relatively low pitch and highly musical burry notes contribute to distinctive "rich" tone quality.

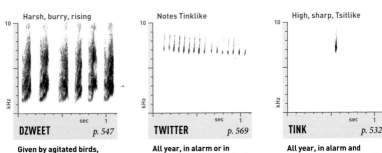

Harsh, burry, rising

DZWEET *p. 547*

Given by agitated birds, especially females, in territorial altercations.

Notes Tinklike

TWITTER *p. 569*

All year, in alarm or in agitation, grading into single Tinks.

High, sharp, Tsitlike

TINK *p. 532*

All year, in alarm and contact; most common call. Not very loud.

Bell's Sparrow (Nominate)

Artemisiospiza belli belli

Breeds in chaparral and sagebrush. Formerly lumped with Sagebrush Sparrow. Mojave Desert subspecies (next page) is intermediate in some ways.

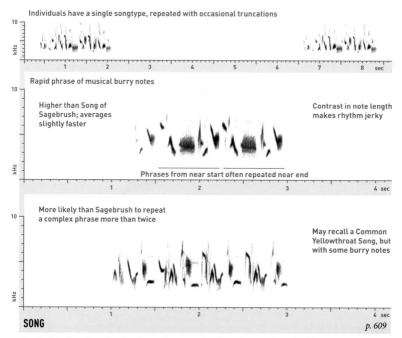

Individuals have a single songtype, repeated with occasional truncations

Rapid phrase of musical burry notes

Higher than Song of Sagebrush; averages slightly faster

Contrast in note length makes rhythm jerky

Phrases from near start often repeated near end

More likely than Sagebrush to repeat a complex phrase more than twice

May recall a Common Yellowthroat Song, but with some burry notes

SONG

p. 609

Mostly Feb.–June, by males; females not known to sing. Song shows moderate individual and geographical variation. On wintering grounds, subsong and plastic song are not uncommon. Songs of nominate *belli* are less variable than those of Sagebrush Sparrow, but see Mojave subspecies of Bell's (next page). In most versions by nominate subspecies, relatively high pitch and presence of numerous brief sharply whistled notes contribute to a "sweeter" tone quality than is typical of Sagebrush.

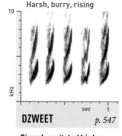

Harsh, burry, rising

DZWEET *p. 547*

Given by agitated birds, especially females, in territorial altercations.

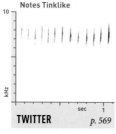

Notes Tinklike

TWITTER *p. 569*

All year, in alarm or in agitation, grading into single Tinks.

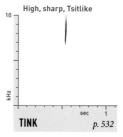

High, sharp, Tsitlike

TINK *p. 532*

All year, in alarm and contact; most common call. Not very loud.

BELL'S SPARROW (MOJAVE)

Artemisiospiza belli canescens

Appearance closer to Sagebrush Sparrow than to nominate Bell's Sparrow. Song like nominate Bell's, but more variable, often with more and longer burry notes, approaching Sagebrush Sparrow. Calls like Bell's.

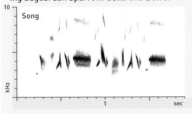

Song

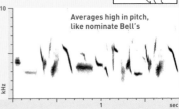

Averages high in pitch, like nominate Bell's

kHz — sec

SAVANNAH SPARROW (BELDING'S)

Passerculus sandwichensis beldingi

Year-round resident of coastal salt marshes in southern California. Dark and heavily streaked. Song variable, but not known to be consistently separable from those of other subspecies. Calls like other subspecies.

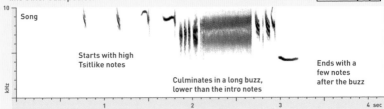

Song

Starts with high Tsitlike notes

Culminates in a long buzz, lower than the intro notes

Ends with a few notes after the buzz

kHz — 1 — 2 — 3 — 4 sec

SAVANNAH SPARROW (LARGE-BILLED)

Passerculus sandwichensis rostratus

Breeds in coastal salt marshes around the Gulf of California; a few winter in southern California. Pale and thick-billed. Song generally similar to that of other subspecies; the possibility of consistent differences needs more study. Calls apparently like other subspecies.

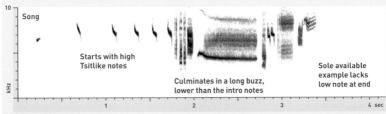

Song

Starts with high Tsitlike notes

Culminates in a long buzz, lower than the intro notes

Sole available example lacks low note at end

kHz — 1 — 2 — 3 — 4 sec

LARK BUNTING

Calamospiza melanocorys

A characteristic species of short-grass prairies. Nonbreeding males resemble females, but with more black in plumage. Forms flocks in winter.

Males have 1 to a few songtypes, each comprising several series and trills in a particular order

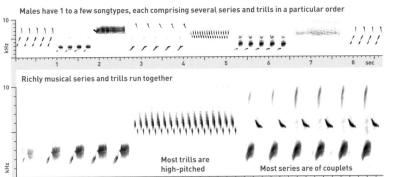

Richly musical series and trills run together

Most trills are high-pitched

Most series are of couplets

SONG *p. 601*

Mostly Apr.–June, by males in slow-flapping flight display. Consecutive songs usually similar, but many are truncated. Song often includes series of cardinal-like slurred whistles. In aggressive flight chases involving 2 or more males, Song reportedly contains fewer repeated elements.

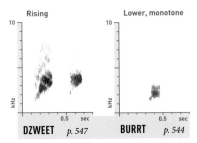

Rising

Lower, monotone

DZWEET *p. 547* **BURRT** *p. 544*

Possibly all year; function little known. Variable and/or plastic, apparently grading into Snarl.

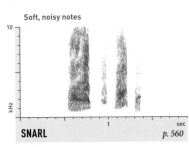

Soft, noisy notes

SNARL *p. 560*

Possibly all year; at least during nonbreeding season, by birds in flocks, perhaps in aggression. Highly plastic.

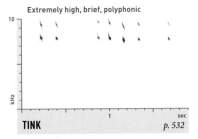

Extremely high, brief, polyphonic

TINK *p. 532*

Possibly all year; given in conjunction with Wert by birds in flocks, at least from spring to late summer. Slightly plastic.

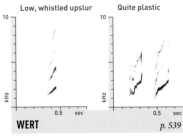

Low, whistled upslur

Quite plastic

WERT *p. 539*

All year; most common call, especially from flocks. Usually not very loud.

WHITE-THROATED SPARROW

Zonotrichia albicollis

Breeds in boreal forests; winters in brush. Head stripes may be white or tan in either sex; birds of one color morph prefer mates of the other morph.

Most males have 1 songtype; about 10 percent have 2, one of which is given more often than the other

Phrase of clear monotone whistles

First 1–2 notes longer than rest, usually on different pitch

Ends with syncopated, Morse-code whistles on a single pitch

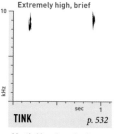

SONG: "Oh, Sweet Canada Canada Canada"
p. 604

A striking and easily recognized song, though quite variable. Mostly Mar.–July, by both sexes, though tan-striped females sing rarely. Female song tends to be shorter, with notes less steady in pitch. Subsong and plastic song given Aug.–Apr.

2–10 seminasal notes

Often introduced by Pink or Tseet

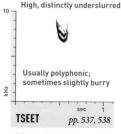

SQUEEP SERIES
p. 577

All year, in aggression, often with head extended forward and wings fluttering; also by mated pairs at the nest. Variable and plastic.

Rapid trill of Tsitlike notes

Often introduced by Tseets or Pinks

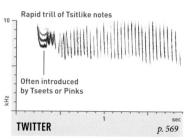

TWITTER
p. 569

Mostly Apr.–June, by females prior to copulation and by males in aggressive display, with head forward and wings fluttering. Quite plastic.

High, distinctly underslurred

Usually polyphonic; sometimes slightly burry

TSEET
pp. 537, 538

All year, in contact and in flight, including by night migrants. Plastic.

Extremely high, brief

TINK
p. 532

Mostly May–Aug., in alarm near nest, often with Pinks. Plastic.

Starts with consonant sound; some versions sound upslurred

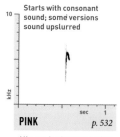

PINK
p. 532

All year, in alarm and other situations. Plastic.

Harris's Sparrow

Zonotrichia querula

Breeds at the interface between boreal forest and tundra; winters in brushy or weedy areas of the Great Plains, often flocking with White-crowned Sparrows.

Males have 1–3 songtypes; consecutive songtypes usually differ

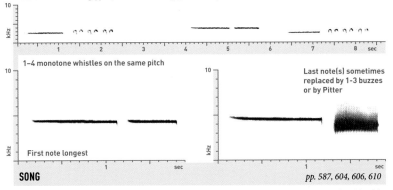

1–4 monotone whistles on the same pitch

First note longest

Last note(s) sometimes replaced by 1-3 buzzes or by Pitter

SONG

pp. 587, 604, 606, 610

Mostly Mar.–July, by males. Song by spring migrants is frequent and usually plastic. The relationship of the Pitter call to the Song needs more study; it apparently forms part of some songtypes, but can also be deployed by itself between Songs.

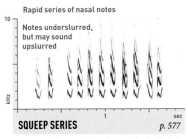

Rapid series of nasal notes

Notes underslurred, but may sound upslurred

SQUEEP SERIES

p. 577

Possibly all year; given frequently by birds in winter flocks. Quite plastic. Much like Squeep Series of other *Zonotrichia* species.

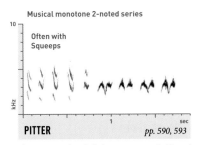

Musical monotone 2-noted series

Often with Squeeps

PITTER

pp. 590, 593

In summer, given in between or appended to end of Songs; in winter, given by flocks with Cheep Series, Pinks, and series of Pweews.

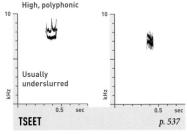

High, polyphonic

Usually underslurred

TSEET

p. 537

All year, including presumably by night migrants. Highly plastic. This species may also have a Tink, but no recordings are available.

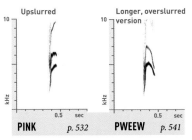

Upslurred

Longer, overslurred version

PINK

p. 532

PWEEW

p. 541

All year. Variable, but apparently rather stereotyped within individuals. Longer, more polyphonic than other *Zonotrichia* Pinks.

WHITE-CROWNED SPARROW

Zonotrichia leucophrys

Breeds in shrubby subalpine meadows and stunted conifers near treeline. In migration and winter, found in brush and weedy fields, often in flocks.

Most males have 1 songtype, shared with males breeding nearby, but geographic variation is great

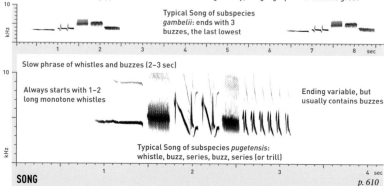

Typical Song of subspecies *gambelii*: ends with 3 buzzes, the last lowest

Slow phrase of whistles and buzzes (2–3 sec)

Always starts with 1–2 long monotone whistles

Ending variable, but usually contains buzzes

Typical Song of subspecies *pugetensis*: whistle, buzz, series, buzz, series (or trill)

SONG *p. 610*

All year, especially Mar.–July, by males; females occasionally sing softly and variably. Subsong and plastic song frequent Aug.–Apr. Males breeding at dialect boundaries may sing both dialects. Song of subspecies *gambelii* of Alaska and Canada varies relatively little. Songs of Pacific Coast subspecies *pugetensis* and *nuttalli* more variable, and those of Interior West subspecies *oriantha* highly variable.

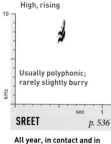

2–10 seminasal or slightly noisy notes

Often introduced by Pink or Seet

SQUEEP SERIES *p. 577*

All year, by both sexes, in mild agitation, often from flocks under feeding stations. Variable and highly plastic.

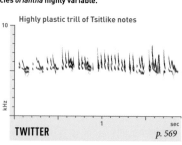

Highly plastic trill of Tsitlike notes

TWITTER *p. 569*

Mostly May–July, by both sexes, often with fluttering wings. By females in courtship, males in territorial defense.

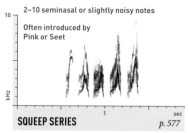

High, rising

Usually polyphonic; rarely slightly burry

SREET *p. 536*

All year, in contact and in flight, including by night migrants. Plastic.

Extremely high, brief

TINK *p. 532*

Mostly May–Aug., in alarm near nest, often with Pinks. Plastic.

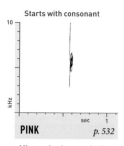

Starts with consonant

PINK *p. 532*

All year, in alarm and other situations; plastic. Some versions sound upslurred.

Golden-crowned Sparrow

Zonotrichia atricapilla

Similar to White-crowned Sparrow in breeding habitat, and flocks with it in migration and winter, but tends to prefer chaparral and dense riparian brush.

Males have 1 songtype, shared with males breeding nearby; several major geographic dialects

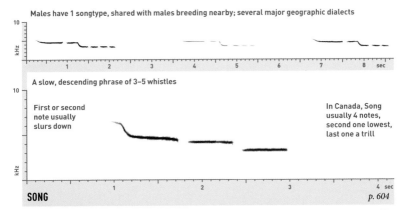

A slow, descending phrase of 3–5 whistles

First or second note usually slurs down

In Canada, Song usually 4 notes, second one lowest, last one a trill

SONG   *p. 604*

Mostly Mar.–July, by males; female not known to sing. Subsong and plastic song given Aug.–Apr. The 3-note songtype immediately above is typical of coastal Alaskan breeders; the songtype shown at top is typical of interior Alaskan breeders.

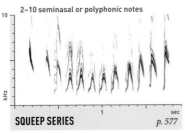

2–10 seminasal or polyphonic notes

SQUEEP SERIES   *p. 577*

All year, by both sexes, in mild agitation, often from flocks under feeding stations. Variable and highly plastic. Often introduced by Tew.

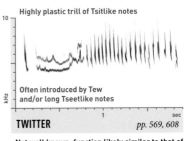

Highly plastic trill of Tsitlike notes

Often introduced by Tew and/or long Tseetlike notes

TWITTER   *pp. 569, 608*

Not well known; function likely similar to that of Twitter of other *Zonotrichia* species.

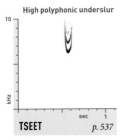

High polyphonic underslur

TSEET   *p. 537*

All year, in contact and in flight, including by night migrants. Plastic.

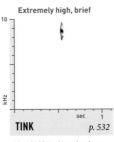

Extremely high, brief

TINK   *p. 532*

Mostly May–Aug., in alarm near nest, often with Pinks. Plastic.

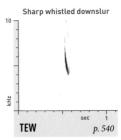

Sharp whistled downslur

TEW   *p. 540*

All year, in alarm and other situations; plastic. Most common call. Distinctive.

LINCOLN'S SPARROW

Melospiza lincolnii

Breeds in swampy willow thickets and boggy areas. Rather secretive in migration and winter, skulking in a variety of brushy habitats, usually near water.

1–6 songtypes, repeated many times before switching, though endings frequently vary

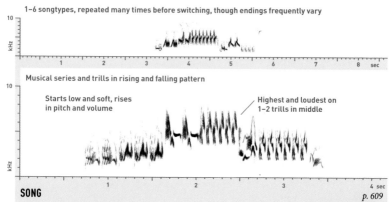

Musical series and trills in rising and falling pattern

Starts low and soft, rises in pitch and volume

Highest and loudest on 1–2 trills in middle

SONG
p. 609

Mostly May–July. Most likely to be mistaken for House Wren Song, but richer and more rigidly structured, especially at start. Sometimes given in flight display after introductory series of Dzeets, but no separate Complex Song as in Swamp Sparrow.

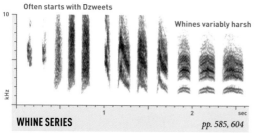

Often starts with Dzweets

Whines variably harsh

WHINE SERIES
pp. 585, 604

Mostly May–July; highly plastic. Similar to Swamp Sparrow Whine Series and likely used in similar circumstances. Sometimes starts with slow twitter of Tsitlike notes, much like Song Sparrow.

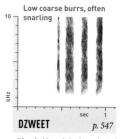

Low coarse burrs, often snarling

DZWEET
p. 547

Mostly May–July, by agitated birds on territory. Plastic.

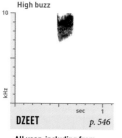

High buzz

DZEET
p. 546

All year, including from night migrants. Rather variable and plastic.

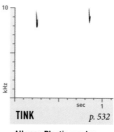

TINK
p. 532

All year. Plastic; grades smoothly into Chip.

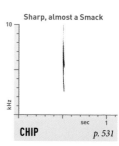

Sharp, almost a Smack

CHIP
p. 531

All year; most common call. Highly plastic in pitch and quality, but distinctive.

SWAMP SPARROW

Melospiza georgiana

Breeds in cattail marshes as well as salt marshes and cedar swamps. Can be rather shy in migration and winter, frequenting brushy habitats near water.

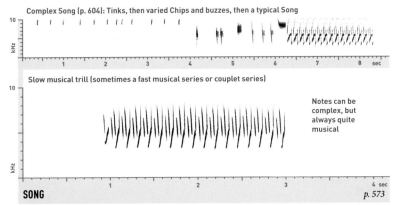

Complex Song (p. 604): Tinks, then varied Chips and buzzes, then a typical Song

Slow musical trill (sometimes a fast musical series or couplet series)

Notes can be complex, but always quite musical

SONG *p. 573*

Mostly Apr.–Aug. Males sing 1–4 songtypes, each repeated several times before switching. More musical and slower than most other single-trill songs. Complex Song rarely heard, mostly in early spring flight displays when males are establishing territories.

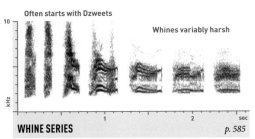

Often starts with Dzweets

Whines variably harsh

WHINE SERIES *p. 585*

Mostly Apr.–July, by males and possibly females, in territorial disputes. No equivalent of Song Sparrow Twitter known in Swamp, though females leaving the nest give quick series of Chips or Tinks.

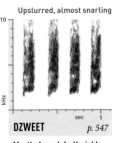

Upslurred, almost snarling

DZWEET *p. 547*

Mostly Apr.–July. Variable; like Lincoln's Sparrow Dzweets.

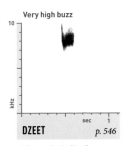

Very high buzz

DZEET *p. 546*

All year, including from night migrants. Monotone or barely underslurred.

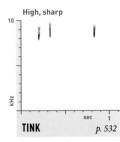

High, sharp

TINK *p. 532*

All year. Plastic; grades smoothly into Chip.

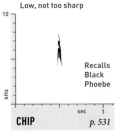

Low, not too sharp

Recalls Black Phoebe

CHIP *p. 531*

All year; most common call. Variable and plastic but fairly distinctive.

Song Sparrow

Melospiza melodia

Widespread and familiar, frequenting a wide variety of brushy habitats and wetlands. Plumage and size vary geographically.

Males have 6–12 songtypes, each repeated 5–15 times before switching

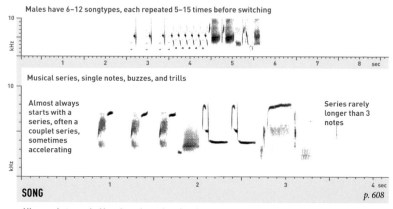

Musical series, single notes, buzzes, and trills

Almost always starts with a series, often a couplet series, sometimes accelerating

Series rarely longer than 3 notes

SONG *p. 608*

All year, but mostly Mar.–Aug., by males; females rarely sing. Extremely variable, but recognizable rangewide once learned. Alternating high and low notes are typical, as are alternating buzzy and musical tones; occasionally some parts are polyphonic. Song length 2–4 seconds.

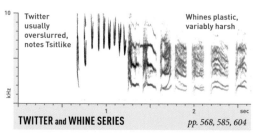

Twitter usually overslurred, notes Tsitlike

Whines plastic, variably harsh

TWITTER and WHINE SERIES *pp. 568, 585, 604*

All year, in high agitation. Twitters and Whine Series often given separately, mostly by females during breeding season, by males on winter territories. Single Tsitlike or Tsewlike notes may be given.

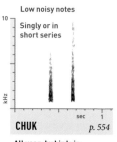

Low noisy notes

Singly or in short series

CHUK *p. 554*

All year; by birds in aggressive interactions. Also a harsh Dzweet (p. 547).

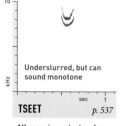

Underslurred, but can sound monotone

TSEET *p. 537*

All year; in contact and on nocturnal migration. Very like Fox Sparrow Tseet.

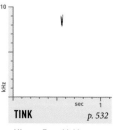

TINK *p. 532*

All year. From highly alarmed birds, often with Twitters and Whine Series.

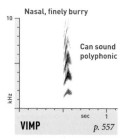

Nasal, finely burry

Can sound polyphonic

VIMP *p. 557*

All year. Most common and distinctive call; variable but easily recognizable.

Fox Sparrow (Sooty)

Passerella iliaca (*unalaschcensis* group)

Breeds from sea level to treeline in dense deciduous thickets, especially near water. Populations that breed the farthest north winter the farthest south.

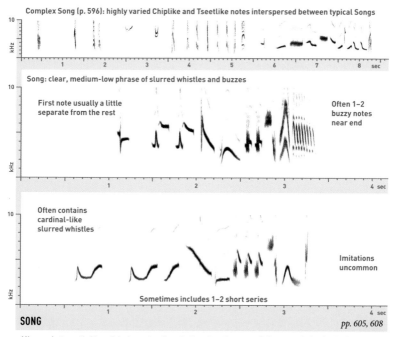

Complex Song (p. 596): highly varied Chiplike and Tseetlike notes interspersed between typical Songs

Song: clear, medium-low phrase of slurred whistles and buzzes

First note usually a little separate from the rest

Often 1–2 buzzy notes near end

Often contains cardinal-like slurred whistles

Imitations uncommon

Sometimes includes 1–2 short series

SONG
pp. 605, 608

All year, but mostly Mar.–July, by males; female Song less frequent. Subsong and plastic singing common in early spring. Individual males have 1–3 songtypes; in birds with multiple songtypes, consecutive songs usually different. Also gives a soft, rather nasal Chitter all year during altercations (p. 582). Generally quite similar to Song of Red form of Fox Sparrow, which breeds from Alaska to Newfoundland and is rare in winter in the West.

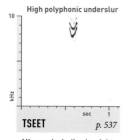

High polyphonic underslur

TSEET
p. 537

All year, including by night migrants. Very like Song Sparrow Tseet.

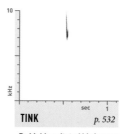

TINK
p. 532

By highly agitated birds near nest or in response to playback, often with Smack.

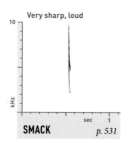

Very sharp, loud

SMACK
p. 531

All year, in alarm; most common call. Like Smack of Red Fox Sparrow of the East.

Fox Sparrow (Thick-billed)

Passerella iliaca (*megarhyncha* group)

Nests in montane shrublands and open coniferous forests with dense brush at mid- to high elevation. All Fox Sparrow forms are highly variable.

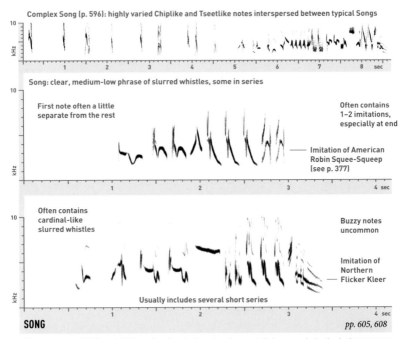

Complex Song (p. 596): highly varied Chiplike and Tseetlike notes interspersed between typical Songs

Song: clear, medium-low phrase of slurred whistles, some in series

First note often a little separate from the rest

Often contains 1–2 imitations, especially at end

Imitation of American Robin Squee-Squeep (see p. 377)

Often contains cardinal-like slurred whistles

Buzzy notes uncommon

Imitation of Northern Flicker Kleer

Usually includes several short series

SONG *pp. 605, 608*

All year, but mostly Mar.–July, by males; female Song less frequent. Subsong and plastic singing common in early spring. Individual males have 3–7 songtypes; consecutive songs usually different. Also gives a soft, rather nasal Chitter all year during altercations (p. 582). The most distinctive Song among Fox Sparrow forms, incorporating the most imitations. Easily confused with Song of Green-tailed Towhee (p. 426), which often breeds in the same locations.

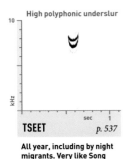

High polyphonic underslur

TSEET *p. 537*

All year, including by night migrants. Very like Song Sparrow Tseet.

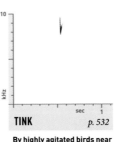

TINK *p. 532*

By highly agitated birds near nest or in response to playback, often with Teenk.

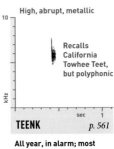

High, abrupt, metallic

Recalls California Towhee Teet, but polyphonic

TEENK *p. 561*

All year, in alarm; most common call. Distinctive among Fox Sparrow forms.

Fox Sparrow (Slate-colored)

Passerella iliaca (*schistacea* group)

Nests in riparian willow thickets, usually at high elevation, especially in southern part of range. Resembles Thick-billed, but note smaller bill, call.

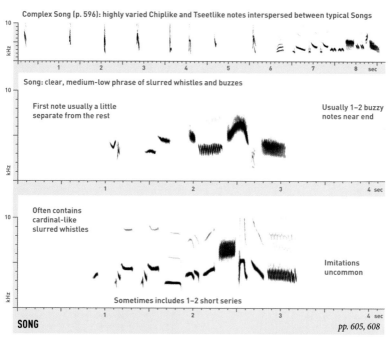

Complex Song (p. 596): highly varied Chiplike and Tseetlike notes interspersed between typical Songs

Song: clear, medium-low phrase of slurred whistles and buzzes

First note usually a little separate from the rest

Usually 1–2 buzzy notes near end

Often contains cardinal-like slurred whistles

Imitations uncommon

Sometimes includes 1–2 short series

SONG

pp. 605, 608

All year, but mostly Mar.–July, by males; female Song less frequent. Subsong and plastic singing common in early spring. Individual males have 2–4 songtypes; consecutive songs usually different. In agitation, runs Chips into a Twitter (p. 568); may also give an aggressive Chitter (p. 582) like other forms. Quite similar to Song of Sooty and Red forms of Fox Sparrow, but includes more buzzes on average.

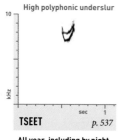

High polyphonic underslur

TSEET   *p. 537*

All year, including by night migrants. Very like Song Sparrow Tseet.

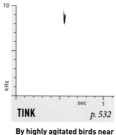

TINK   *p. 532*

By highly agitated birds near nest or in response to playback, often with Tew.

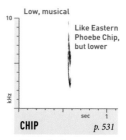

Low, musical

Like Eastern Phoebe Chip, but lower

CHIP   *p. 531*

All year, in alarm; most common call. Subtly distinct from Smack of Sooty form.

DARK-EYED JUNCO (SLATE-COLORED)

Junco hyemalis (*hyemalis* group)

A familiar "snowbird" that readily visits feeders in winter. Breeds in various wooded and brushy habitats. Note white outer tail feathers.

Complex Song (p. 602): typical Song trills mixed with variety of calls and call-like notes

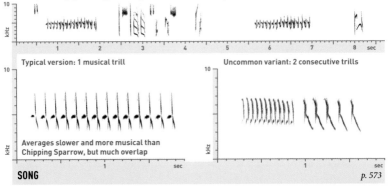

Typical version: 1 musical trill

Averages slower and more musical than Chipping Sparrow, but much overlap

Uncommon variant: 2 consecutive trills

SONG *p. 573*

Mostly Feb.–July, by males. Stereotyped but highly variable. Individuals have 1–7 songtypes, repeated many times before switching. Complex Song softer, by both sexes in close courtship during early breeding season. Subsong frequent Jan.–Mar.; like Complex Song but less stereotyped.

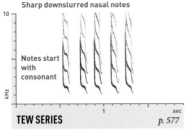

Sharp downslurred nasal notes

Notes start with consonant

TEW SERIES *p. 577*

All year, in aggressive encounters. Often given with other calls, especially Twitter.

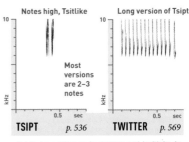

Notes high, Tsitlike

Most versions are 2–3 notes

Long version of Tsipt

TSIPT *p. 536*   **TWITTER** *p. 569*

Tsipt given all year, in contact and in flight, including by night migrants. Twitter given in aggression, and by female prior to copulation.

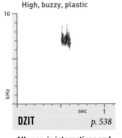

High, buzzy, plastic

DZIT *p. 538*

All year, in interactions and short flights; not known from night migrants.

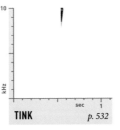

TINK *p. 532*

Not well known; given in response to playback and possibly in high alarm.

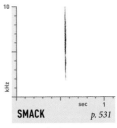

SMACK *p. 531*

All year, in alarm. Distinctively sharp, almost ticking.

DISTINCTIVE SUBSPECIES OF DARK-EYED JUNCO

These four groups are highly distinctive in plumage, but have songs and calls identical to those of the Slate-colored group, as far as is known. "Oregon" and "Slate-colored" Juncos intergrade where ranges meet in the Canadian Rockies; intergrades between other forms are uncommon.

Dark-eyed Junco (Oregon)
Junco hyemalis (*oreganus* group)

Dark-eyed Junco (White-winged)
Junco hyemalis aikeni

Dark-eyed Junco (Pink-sided)
Junco hyemalis mearnsi

Dark-eyed Junco (Gray-headed)
Junco hyemalis caniceps

DARK-EYED JUNCO (RED-BACKED)

Junco hyemalis dorsalis

Like "Gray-headed" subspecies, but bill bicolored and song slower and often more complex, thus approaching Yellow-eyed in both appearance and voice.

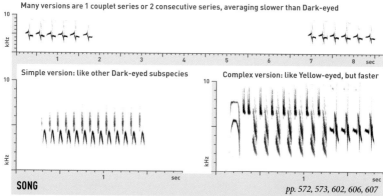

Many versions are 1 couplet series or 2 consecutive series, averaging slower than Dark-eyed

Simple version: like other Dark-eyed subspecies

Complex version: like Yellow-eyed, but faster

SONG

pp. 572, 573, 602, 606, 607

Song intermediate in many respects between those of Dark-eyed and Yellow-eyed, with variation that bridges the gap between them. Never matches the most complex songs of Yellow-eyed. Calls need more study; they include Tink and Smack like Dark-eyed, Tseet (p. 537) and Tsip (p. 533) like Yellow-eyed.

Yellow-eyed Junco

Junco phaeonotus

Common resident of mountain pine-oak forests, moving to lower elevations in winter, when it sometimes flocks with Dark-eyed Juncos.

Repertoire size unknown; individuals can have at least 6 songtypes, each usually repeated before switching

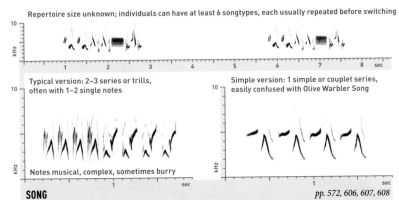

Typical version: 2–3 series or trills, often with 1–2 single notes

Notes musical, complex, sometimes burry

Simple version: 1 simple or couplet series, easily confused with Olive Warbler Song

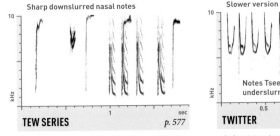

SONG *pp. 572, 606, 607, 608*

Mostly Mar.–Aug., by males; females not known to sing. Very unlike Dark-eyed's Song; may recall Bewick's Wren, Spotted Towhee, or Lazuli Bunting. Version shown at top is unusually complex. Song elements often truncated or transposed from one Song to the next. Soft Complex Song (p. 606) much like Dark-eyed's.

Sharp downslurred nasal notes

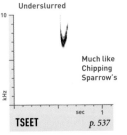

TEW SERIES *p. 577*

All year, in interactions. Often given with other calls, especially Twitter. Here, with Pinklike and Tseetlike calls of unknown function.

Slower version   Like Dark-eyed's

Notes Tseetlike, underslurred

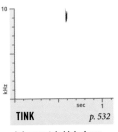

TWITTER *p. 569*

At least two types of Twitter given in apparent aggression; the version shown at left appears to be more common.

Underslurred

Much like Chipping Sparrow's

TSEET *p. 537*

Apparently rare; associated with Tew Series on breeding grounds. Plastic.

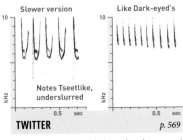

TINK *p. 532*

Infrequent, in high alarm. Plastic; grades into Tsip.

High, sharp downslur

Sometimes as sharp as Dark-eyed's Smack; never as low

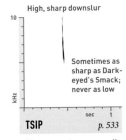

TSIP *p. 533*

All year; most common call. Highly variable and plastic; grades into Tink and Tseet.

SAVANNAH SPARROW

Passerculus sandwichensis

Breeds in grassy fields, wet meadows, salt marshes, and other well-watered open habitats. Some subspecies are distinctive; see p. 446.

Males have 1 songtype (very rarely 2), repeated with little variation

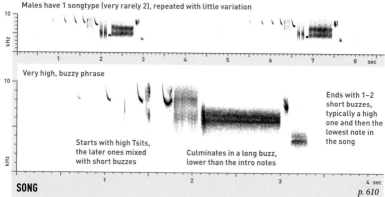

Very high, buzzy phrase

Ends with 1–2 short buzzes, typically a high one and then the lowest note in the song

Starts with high Tsits, the later ones mixed with short buzzes

Culminates in a long buzz, lower than the intro notes

SONG *p. 610*

Mostly Apr.–July. Fairly variable rangewide, but usually adhering closely to the pattern shown. Long buzz near end is the loudest part of Song and carries farthest; at a distance, beware confusion with LeConte's Sparrow Song.

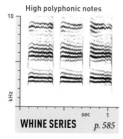

High polyphonic notes

WHINE SERIES *p. 585*

By both sexes in same-sex aggression, or in response to predators.

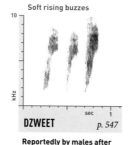

Soft rising buzzes

DZWEET *p. 547*

Reportedly by males after losing a fight, or by females rejecting a male.

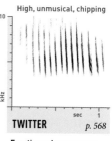

High, unmusical, chipping

TWITTER *p. 568*

Function unknown; more study needed.

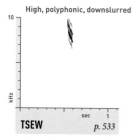

High, polyphonic, downslurred

TSEW *p. 533*

All year, including by night migrants. Most versions clear, but some are buzzy.

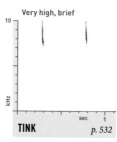

Very high, brief

TINK *p. 532*

All year, in high alarm, reportedly more often by male. Grades into Chip.

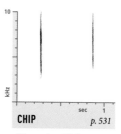

CHIP *p. 531*

All year, in mild alarm, reportedly more often by female. Grades into Tink.

GRASSHOPPER SPARROW

Ammodramus savannarum

Locally common but inconspicuous in
tall grass with patches of bare ground.
All vocalizations can be inaudible to
those with high-pitched hearing loss.

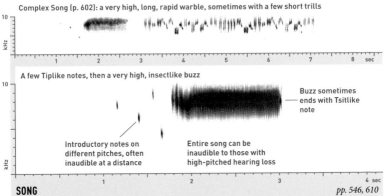

Complex Song (p. 602): a very high, long, rapid warble, sometimes with a few short trills

A few Tiplike notes, then a very high, insectlike buzz

Buzz sometimes
ends with Tsitlike
note

Introductory notes on
different pitches, often
inaudible at a distance

Entire song can be
inaudible to those with
high-pitched hearing loss

SONG   *pp. 546, 610*

Mostly Apr.–July, by males. Quite uniform. Each male has 2 songtypes, 1 simple and 1 Complex; simple
song mostly territorial. Complex Song may strengthen the pair bond; given regularly, especially late in
season, often right after typical Song or alternating with it, but rare from unpaired males.

Trill of high Tsitlike notes

Often starts with a lower note

TWITTER   *p. 569*

Mostly May–July, by both sexes, to maintain pair bond and possibly to signal approach to nest. Sometimes
considered a type of Song. Twitter of female often associated with Complex Song of male. Male and female
versions may differ slightly; more study needed.

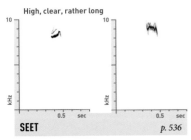

High, clear, rather long

SEET   *p. 536*

All year, including by night migrants. Plastic.
Often sounds slighty upslurred, sometimes
monotone.

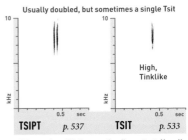

Usually doubled, but sometimes a single Tsit

High,
Tinklike

TSIPT   *p. 537*   **TSIT**   *p. 533*

All year, in mild alarm or family contact. Usually
delivered in fairly distinctive pattern of rapid
doublets or triplets.

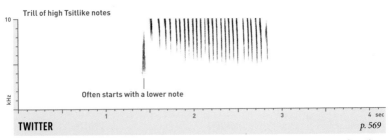

BAIRD'S SPARROW

Centronyx bairdii

Uncommon and local in native prairies with dense grass and few or no shrubs. Secretive and difficult to detect when not singing.

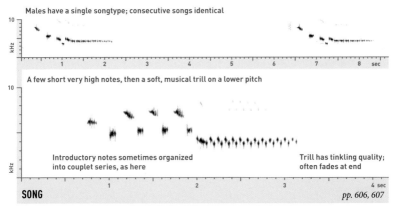

Males have a single songtype; consecutive songs identical

A few short very high notes, then a soft, musical trill on a lower pitch

Introductory notes sometimes organized into couplet series, as here

Trill has tinkling quality; often fades at end

SONG *pp. 606, 607*

Mostly May–Aug., by males. Females not known to sing. Slightly variable but stereotyped; much more musical than any other sparrow Song in its range. Some versions end in single lower note after trill. Rare flight song reported: a few Tsiplike notes as bird ascends, then a typical Song as it descends.

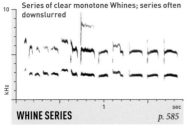

Series of clear monotone Whines; series often downslurred

WHINE SERIES *p. 585*

May–Aug., likely by both sexes, in intense interactions and territorial disputes. Rather plastic; grades into Tsweet Series.

Rapid series of upslurred notes

Notes plastic, sometimes partly buzzy

TSWEET SERIES *p. 568*

May–Aug., possibly by both sexes, in intense interactions and territorial disputes. Agitated birds occasionally give single Tsweets.

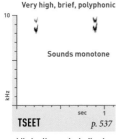

Very high, brief, polyphonic

Sounds monotone

TSEET *p. 537*

Likely all year, including by night migrants. Plastic; rarely longer than shown.

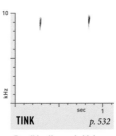

TINK *p. 532*

Possibly all year, in high alarm. Grades into Chip.

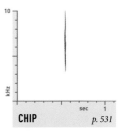

CHIP *p. 531*

Possibly all year, in mild alarm. Grades into Tink.

LeConte's Sparrow

Ammospiza leconteii

Rather secretive and subtly plumaged, but strikingly beautiful when seen up close. Frequents wet meadows, sedge fields, and marsh edges.

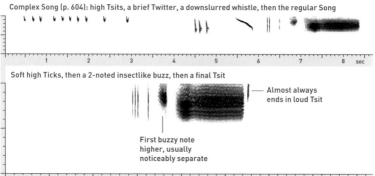

Complex Song (p. 604): high Tsits, a brief Twitter, a downslurred whistle, then the regular Song

Soft high Ticks, then a 2-noted insectlike buzz, then a final Tsit

Almost always ends in loud Tsit

First buzzy note higher, usually noticeably separate

SONG *pp. 546, 610*

Mostly May–July, reportedly only by males, sometimes at night. Fairly soft and easily overlooked, but highly uniform. Each male apparently has a single simple Song and a single Complex Song, given infrequently by excited birds, often during flight display.

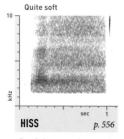

Quite soft

HISS *p. 556*

Poorly known; sole example is from agitated male responding to playback.

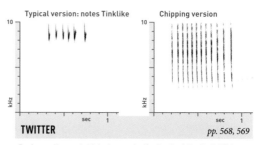

Typical version: notes Tinklike

Chipping version

TWITTER *pp. 568, 569*

Perhaps all year, in high alarm; plastic. Grades into single Tinks. Sole example of chipping version (right) is from agitated male responding to playback. Also gives a Whine Series (p. 585).

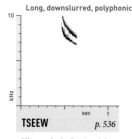

Long, downslurred, polyphonic

TSEEW *p. 536*

All year, including by night migrants. Distinctive.

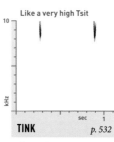

Like a very high Tsit

TINK *p. 532*

All year, in alarm; most common call, at least on breeding grounds.

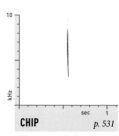

CHIP *p. 531*

All year; may indicate greater alarm than Tink.

NELSON'S SPARROW

Ammospiza nelsoni

Rather shy and local. Breeds in grassy freshwater marshes of the northern Great Plains, in sedges ringing Hudson Bay, and in North Atlantic salt marshes.

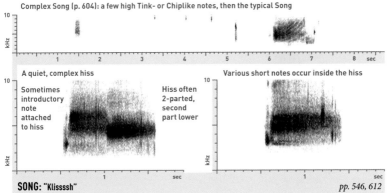

Complex Song (p. 604): a few high Tink- or Chiplike notes, then the typical Song

A quiet, complex hiss

Sometimes introductory note attached to hiss

Hiss often 2-parted, second part lower

Various short notes occur inside the hiss

SONG: "Klissssh"   *pp. 546, 612*

Mostly May–July, by males, including at night. Females rarely sing. Distinctive, but does not carry far; only slightly variable between populations. Males likely sing a single regular and a single Complex Song. Complex Song usually given in display flight.

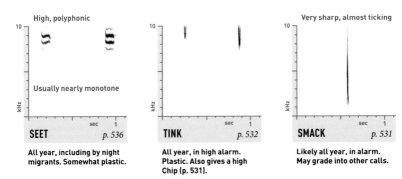

High, polyphonic

Usually nearly monotone

SEET   *p. 536*

All year, including by night migrants. Somewhat plastic.

TINK   *p. 532*

All year, in high alarm. Plastic. Also gives a high Chip (p. 531).

Very sharp, almost ticking

SMACK   *p. 531*

Likely all year, in alarm. May grade into other calls.

YELLOW-BREASTED CHAT (Family Icteriidae)   *next page*

For many years, the Yellow-breasted Chat was classified as a particularly odd member of the warbler family, but recent DNA studies have clarified that it is actually a unique bird, the sole surviving member of an ancient lineage. Its new family name, Icteriidae, is exceedingly similar to the name of the oriole and meadowlark family, Icteridae, but the two should not be confused. It is much larger than a warbler, with a longer tail, a distinctive mimid-like voice, and a slow-flapping flight display.

YELLOW-BREASTED CHAT

Icteria virens

Loud but skulking, more often heard than seen. Nests in second growth and clear-cuts in the East and riparian thickets in the West.

Song: widely spaced single notes, series, and trills, each with a very different tone quality

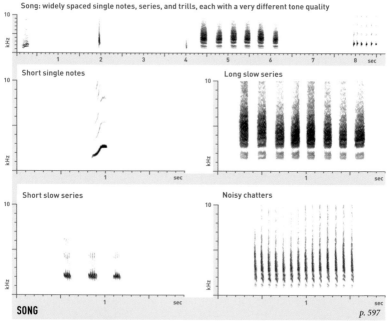

Short single notes

Long slow series

Short slow series

Noisy chatters

SONG

p. 597

Mostly Apr.–July, reportedly only by males. Each male has about 60 songtypes on average, but only 4–6 are typically used in any given five minutes of singing. Each songtype comprises either 1 note or a series of similar notes. Consecutive songs differ so greatly, and are separated by such long pauses, that they are easily mistaken for the sounds of different species; some are in fact imitations of other species. Tone quality may be whistled, nasal, noisy, or polyphonic. Song in slow-flapping display flight like regular Song, but with slightly shorter intervals filled with the audible beating of wings.

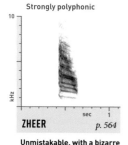

Strongly polyphonic

ZHEER *p. 564*

Unmistakable, with a bizarre polyphonic dissonance. All year, in alarm.

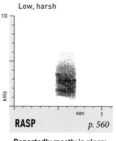

Low, harsh

RASP *p. 560*

Reportedly mostly in alarm near nest.

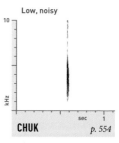

Low, noisy

CHUK *p. 554*

All year; most common call, in alarm. Sometimes doubled.

BLACKBIRDS, MEADOWLARKS, GRACKLES, COWBIRDS, AND ORIOLES (Family Icteridae)

These birds, often called the "icterids," comprise a highly diverse New World family of several distinct lineages. The blackbirds and grackles are dark-plumaged, highly social birds of wetlands and woodland edges. The orioles are colorful arboreal birds that hang their intricately woven nests from tree branches. The meadowlarks and the Bobolink are ground-nesting birds of fields and prairies. And the cowbirds are brood parasites, laying their eggs in the nests of other species.

The icterids have astonishingly varied vocal repertoires. Songs are learned and highly complex, usually whistled or polyphonic. A few species of oriole incorporate imitations of other birds into their songs. In some icterids, different categories of song serve different functions: for example, the Complex Songs of meadowlarks and the Whistled Songs of cowbirds are given mostly in excitement or in flight. In a few icterids, including Rusty and Yellow-headed Blackbirds, certain categories of song can form part of the alarm response to aerial predators.

Songlike Calls

In this family, the boundary between songs and calls is particularly blurry. Many calls are learned, and individuals often have repertoires of multiple call types. For example, the orioles have repertoires of short whistled or burry calls that they give for long periods, often switching call types, much as singing birds might switch songtypes. The function of these calls is not entirely clear.

Other icterid vocalizations that share characteristics of songs and calls include the whistled and burry calls of Bobolink, the Alert Calls of Red-winged Blackbird, and the series calls of Great-tailed Grackle.

Female Songs

In most oriole species, both sexes sing. In nonmigratory southern species, the sexes tend to look and sound fairly similar, and females sing frequently, in some cases more frequently than males. In migratory northern species, the sexes tend to look different, and female songs tend to be slightly less complex and less frequent.

In the meadowlarks, the cowbirds, and Red-winged Blackbird, male-female pairs often perform synchronized duets, the female's Rattle starting halfway through the male's Song. These duets can be a good way to identify mated pairs. By some criteria, the Rattles that females use in these situations can be considered a type of song.

RED-WINGED BLACKBIRD (BICOLORED)

Red-winged Blackbirds of the Central Valley of California and the central California coast lack the yellow stripe adjoining the red wing patch. Their songs are variable; some versions resemble those of typical populations, while some versions are more complex and nasal, strongly recalling Tricolored Blackbird. "Bicolored's" songs are always higher-pitched and never match the strangled, frog-like quality of Tricolored.

Interbreeding between "Bicolored" and typical forms is apparently extensive. Note that typical Red-wings winter in the range of "Bicolored" in large numbers.

Agelaius phoeniceus (californicus group)

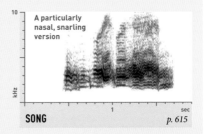

A particularly nasal, snarling version

SONG   p. 615

BOBOLINK

Dolichonyx oryzivorus

Nests in mid- to tall-grass prairie and hayfields. Leaves as early as July for marshlands where birds molt before continuing south to South America.

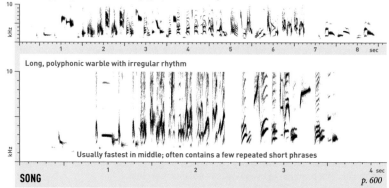

Males typically have 2 similar songtypes, delivered in any order

Long, polyphonic warble with irregular rhythm

Usually fastest in middle; often contains a few repeated short phrases

SONG *p. 600*

Mostly Apr.–June, often from highest perch in territory, or during song flights of up to 1 minute in length. Full Songs are generally 4–6 seconds long, but birds very frequently repeat just the first few notes, sometimes for long periods. Neighboring birds tend to share songtypes.

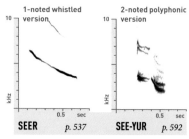

1-noted whistled version

2-noted polyphonic version

SEER *p. 537*   **SEE-YUR** *p. 592*

High, soft, snarling

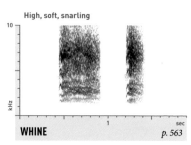

WHINE *p. 563*

Mostly Apr.–July, by singing males, singly or appended to start of Song. Highly variable; may simply represent extremely abbreviated Song.

Reportedly given by males, directed at females, usually after a song flight. Plastic; usually not very loud.

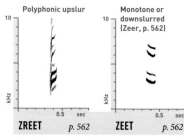

Polyphonic upslur

Monotone or downslurred (Zeer, p. 562)

ZREET *p. 562*   **ZEET** *p. 562*

Variable, noisy

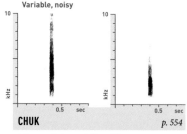

CHUK *p. 554*

Zreet all year, often in flight; most common call during migration, including at night. Zeet reportedly by breeding female.

All year. Quite variable; different versions reported in different contexts, but much variation is likely individual or geographic.

TRICOLORED BLACKBIRD

Agelaius tricolor

Biology and voice very different from Red-winged Blackbird. Nests in very dense itinerant colonies in marshes. Population has declined steeply.

Male Flight Song (p. 601): like Red-winged's, but composed of lower squeaky or nasal notes

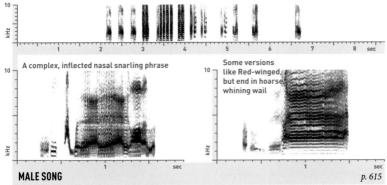

A complex, inflected nasal snarling phrase

Some versions like Red-winged, but end in hoarse whining wail

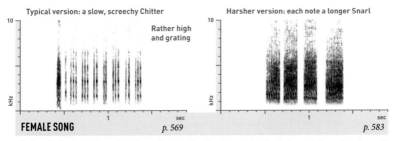

MALE SONG   *p. 615*

All year, especially Jan.–Apr., by males. May sing from exposed perch, but display postures used mostly deep inside cattails. Song usually recalls Yellow-headed Blackbird more than Red-winged, but beware Song of "Bicolored" Red-wingeds in California (see p. 467). Flight Song infrequent, by males leaving nest.

Typical version: a slow, screechy Chitter

Rather high and grating

Harsher version: each note a longer Snarl

FEMALE SONG   *p. 569*   *p. 583*

By females in the nesting colony, generally only when leaving or returning to the nest. Sometimes immediately follows Male Song, in apparent duet like that of Red-winged Blackbird. Highly plastic; the two forms shown grade into one another. Males also give Snarl Series like that shown at right, in aggression.

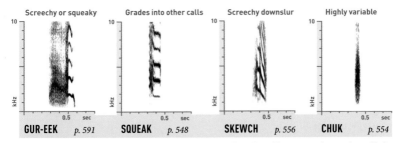

Screechy or squeaky | Grades into other calls | Screechy downslur | Highly variable

GUR-EEK *p. 591*   **SQUEAK** *p. 548*   **SKEWCH** *p. 556*   **CHUK** *p. 554*

Gives a wide variety of calls; a few of the most common are shown here. Most are nasal, squeaky, or Chuk-like; apparently lacks Seer-type, Pink-type, and Jenk-type calls of Red-winged (see p. 471). Tricolored colonies often fall suddenly silent upon sighting a hawk, unlike Red-winged colonies.

RED-WINGED BLACKBIRD

Agelaius phoeniceus

One of our most abundant native birds, especially in wetland habitats, where it forms dense breeding colonies. Vocal communication unique and complex.

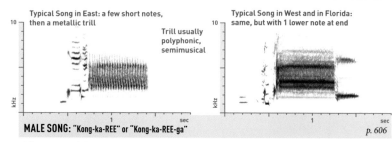

Typical Song in East: a few short notes, then a metallic trill

Trill usually polyphonic, semimusical

Typical Song in West and in Florida: same, but with 1 lower note at end

MALE SONG: "Kong-ka-REE" or "Kong-ka-REE-ga" *p. 606*

All year, especially Apr.–July. Territorial males spread wings slightly during each Song to display red shoulders. Males have 3–7 songtypes, each usually repeated many times before switching, but males in close courtship may cycle through their Song repertoire, interspersing Twitter calls.

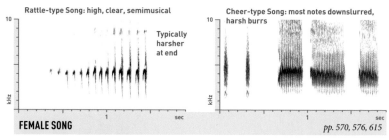

Rattle-type Song: high, clear, semimusical

Typically harsher at end

Cheer-type Song: most notes downslurred, harsh burrs

FEMALE SONG *pp. 570, 576, 615*

Mostly Apr.–July. Rattle-type Songs, often started halfway through mate's Song, tend to reinforce pair bond. Cheer-type Songs more aggressive, territorial. Many intermediate Songs begin with Rattles and end with Cheers. Chuklike notes often incorporated.

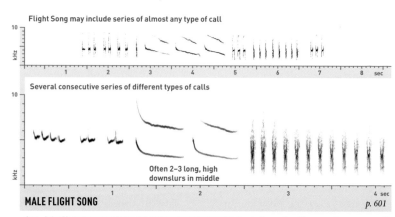

Flight Song may include series of almost any type of call

Several consecutive series of different types of calls

Often 2–3 long, high downslurs in middle

MALE FLIGHT SONG *p. 601*

Apr.–July. Given by males flying out of their breeding territories, and occasionally upon return, apparently to notify mates of departure and arrival. Extremely geographically variable; level of variation within individual males unknown.

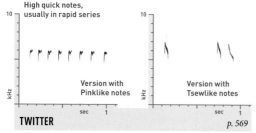

High quick notes, usually in rapid series

Version with Pinklike notes

Version with Tsewlike notes

TWITTER
p. 569

Variable; by both sexes before copulation, by males between Songs at dawn and when other blackbirds fly over. Female version usually lower and clearer.

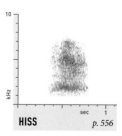

HISS
p. 556

By males confronting intruders or showing potential nest sites to females.

ALERT CALLS OF RED-WINGED BLACKBIRDS

Individual male Red-winged Blackbirds have repertoires of 12–20 different Alert Calls (p. 538), including some from each category below. Birds in a breeding colony tend to share call types, but regional variation is tremendous. Neighboring males call almost continuously, all giving the same call type. An individual switches call types upon noticing potential danger, and the others change calls to match. Thus, it is the change in call rather than the type of call that propagates the danger signal. Highly agitated males may switch frequently between 2 or 3 different call types. Females give only 2–3 such calls.

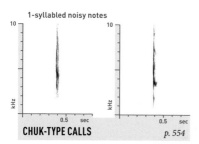

1-syllabled noisy notes

CHUK-TYPE CALLS
p. 554

Brief and usually low, though pitch may vary. Different versions serve as pair contact calls, in mobbing predators, etc.

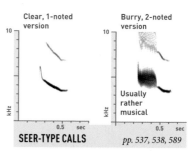

Clear, 1-noted version

Burry, 2-noted version

Usually rather musical

SEER-TYPE CALLS
pp. 537, 538, 589

Males have 3-7 of these calls, usually given in high alarm. May be 1- or 2-syllabled, clear or burry. Usually downslurred.

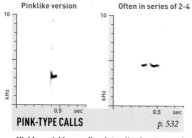

Pinklike version

Often in series of 2-4

PINK-TYPE CALLS
p. 532

Highly variable, grading into all other types of Alert Call. Some versions burry.

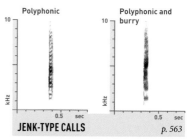

Polyphonic

Polyphonic and burry

JENK-TYPE CALLS
p. 563

Calls in this category are abrupt and polyphonic. May be clear or burry; grade into Chuck-type calls.

Eastern Meadowlark (Lilian's)

Sturnella magna (*lilianae* group)

Breeds in arid grasslands. Very like Western, with more white in tail; paler than Eastern Meadowlarks in the East, with a lower-pitched song.

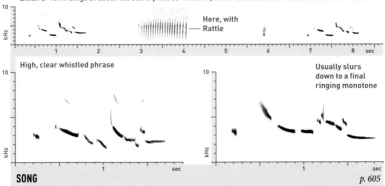

"Lilian's" form sings at about the same pitch as Western, lower than Eastern Meadowlarks in the East

Here, with Rattle

High, clear whistled phrase

Usually slurs down to a final ringing monotone

SONG *p. 605*

Mostly Mar.–Aug., by males; female not known to sing. Individuals have 50–100 songtypes, each usually repeated before switching. Simpler than Western Meadowlark Song, lacking warble at end. Complex Song (p. 600) given infrequently by excited males, usually in flight.

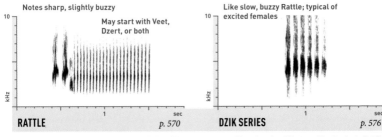

Notes sharp, slightly buzzy

May start with Veet, Dzert, or both

Like slow, buzzy Rattle; typical of excited females

RATTLE *p. 570*

DZIK SERIES *p. 576*

All year. Variable, innate. Generally slightly higher, slower, and buzzier than Western Meadowlark Rattle. Often given by female halfway through her mate's Song, but also from both sexes in response to threats. Sometimes shortened to just Veet or Dzert plus 2–3 rattling notes.

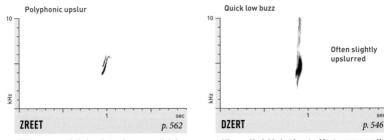

Polyphonic upslur

Quick low buzz

Often slightly upslurred

ZREET *p. 562*

DZERT *p. 546*

All year, especially in winter. Averages slightly higher than Western's. Also likely gives a learned Veet (p. 545, 589), like Easterns in the East.

All year. Variable but innate. Most common call, heard in many contexts. Higher, buzzier than Western's Chup.

WESTERN MEADOWLARK

Sturnella neglecta

The voice of the western prairie. Looks nearly identical to Eastern, but prefers shorter grass and drier habitats, and vocalizations differ markedly.

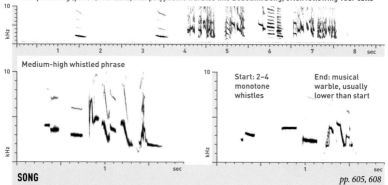

Complex Song (p. 600): variable, rich polyphonic warbles like Bobolink Song, often following Teer calls

Medium-high whistled phrase

Start: 2–4 monotone whistles

End: musical warble, usually lower than start

SONG

pp. 605, 608

Mostly Mar.–Aug., by males; females not known to sing. Very loud; can carry for over a mile. Males average 7 songtypes, each usually repeated. Rarely a Western may learn some Eastern songtypes. Complex Song (top) given infrequently by excited males, usually in flight; faster, lower than Eastern's.

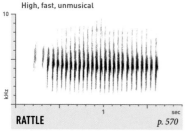

High, fast, unmusical

RATTLE

p. 570

All year. Variable; innate. Often given by female halfway through her mate's Song. Rare from adult males. Often introduced by Chup.

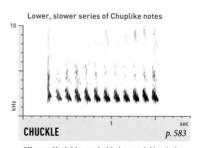

Lower, slower series of Chuplike notes

CHUCKLE

p. 583

All year. Variable; probably learned. Mostly from adult males. May sometimes intergrade with Rattle. Also a little-known harsh Churr (p. 558).

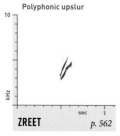

Polyphonic upslur

ZREET

p. 562

All year. Variable; juveniles in late summer give a high short monotone version.

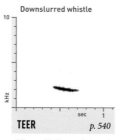

Downslurred whistle

TEER

p. 540

All year. Mellow and clear; in response to predators or prior to Complex Song.

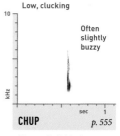

Low, clucking

Often slightly buzzy

CHUP

p. 555

All year. Variable; innate. Longest, buzziest versions can approach Eastern Dzert.

GREAT-TAILED GRACKLE

Quiscalus mexicanus

A Mexican border bird as late as 1960, but now a widespread fixture of suburbs, parks, and wetlands. Gathers in deafening flocks, often in city parks.

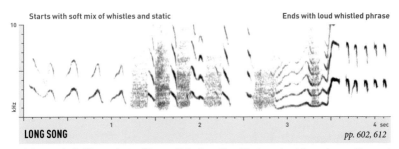

Starts with soft mix of whistles and static | Ends with loud whistled phrase

LONG SONG

pp. 602, 612

Infrequent in the West, and not well known. Males in northeast Mexico and east Texas share a widespread Long Song dialect, but in other populations it is not clear to what extent Long Song is distinct from Short Song. An example from Arizona is shown.

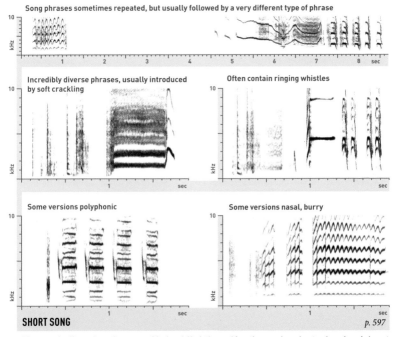

Song phrases sometimes repeated, but usually followed by a very different type of phrase

Incredibly diverse phrases, usually introduced by soft crackling

Often contain ringing whistles

Some versions polyphonic

Some versions nasal, burry

SHORT SONG

p. 597

Given nearly all year by males rangewide. Loud. Variation and function poorly understood; each male has at least 5 Short Songs that can be alternated or combined with each other or with Series. Most Short Songs begin with whispered "stick-breaking" sounds or whistling audible only at close range.

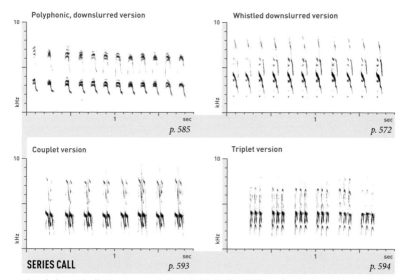

Polyphonic, downslurred version

p. 585

Whistled downslurred version

p. 572

Couplet version

SERIES CALL   *p. 593*

Triplet version

p. 594

A confusing catch-all category; more study needed. By feather-fluffing males during song bouts or in close courtship of females; also upon leaving a song perch and in other situations. Individuals have many series types, which vary geographically. One series type may switch to another without pause.

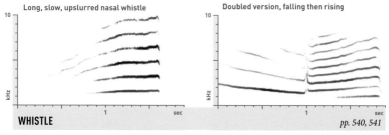

Long, slow, upslurred nasal whistle

Doubled version, falling then rising

WHISTLE   *pp. 540, 541*

Distinctive. Given rangewide by males during song bouts, often accompanied by feather fluffing; sometimes appended to the start of a Short Song. Most versions are upslurred, but some are downslurred or combinations.

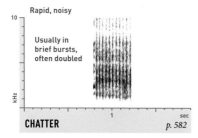

Rapid, noisy

Usually in brief bursts, often doubled

CHATTER   *p. 582*

Versions given by both sexes, but more often by females, at the approach of a courting male and in disputes with other females.

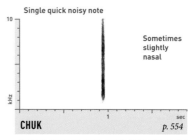

Single quick noisy note

Sometimes slightly nasal

CHUK   *p. 554*

By both sexes, in alarm and in flight. Somewhat variable; versions of males and females reportedly differ.

RUSTY BLACKBIRD

Euphagus carolinus

Breeds in bogs; winters in swamps, often in large flocks. In migration, found in wetlands and riparian woods. Uncommon and apparently declining.

Gurgle-creaks and Gurgles sometimes alternated, but either may be repeated many times

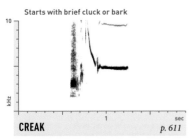

Starts with short hissing warble

Ends with high monotone whistle

GURGLE-CREAK SONG *p. 611*

Hissing warble

No whistle at end

GURGLE SONG *pp. 611, 612*

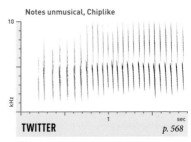

Individuals of both sexes apparently have one Gurgle-creak and one Gurgle. Both types of Song given all winter, plastic in fall, becoming stereotyped by May. Gurgle apparently not given after first 2 weeks on breeding grounds; may serve courtship function. Gurgle-creak continues throughout summer.

Starts with brief cluck or bark

CREAK *p. 611*

Apparently only on breeding grounds, in territorial disputes, in alarm, and when mobbing predators, often with Gurgle-Creaks.

Notes unmusical, Chiplike

TWITTER *p. 568*

Function unknown; given on breeding grounds in response to playback of Song, and sometimes in spring migration. More study needed.

Harsh, snarling

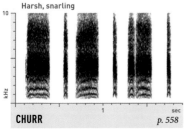

CHURR *p. 558*

Given by both sexes in high alarm near nest. Variable, especially in length. Usually interspersed with Chuks.

Noisy

CHUK *p. 554*

Lower, softer

CHUP *p. 555*

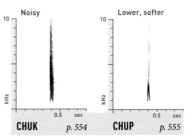

Chuk given all year, by both sexes, in mild alarm or contact. Some versions clearer, more nasal than shown. Chup often given prior to flight.

BREWER'S BLACKBIRD

Euphagus cyanocephalus

Closely related to Rusty Blackbird, but favors more open country at all seasons. Originally a western species; has greatly expanded range eastward.

Single songtype usually repeated, often with Chuks interspersed, but Creak and Gurgle may alternate

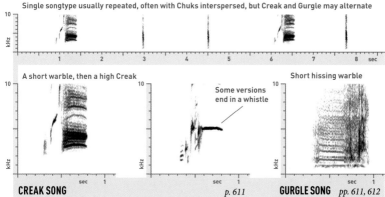

A short warble, then a high Creak

Some versions end in a whistle

Short hissing warble

CREAK SONG *p. 611* **GURGLE SONG** *pp. 611, 612*

All year, by both sexes. Males reported to have one Creak and one Gurgle each; both given with wing-spreading display in summer, mostly without display in winter. Songs have both territorial and courtship functions; differences between Creak and Gurgle need more study.

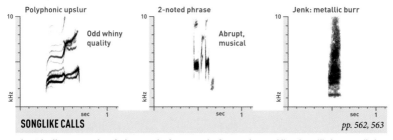

Polyphonic upslur

Odd whiny quality

2-noted phrase

Abrupt, musical

Jenk: metallic burr

SONGLIKE CALLS *pp. 562, 563*

A catch-all category of confusing sounds. Some may be Song variants, while others likely serve distinct functions; more study needed. All appear stereotyped within individuals, but variable between birds. Not illustrated: a downslurred 2-syllabled whistle reportedly given in response to predators.

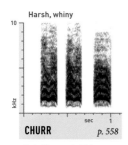 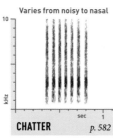 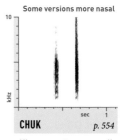

Harsh, whiny

Varies from noisy to nasal

Some versions more nasal

CHURR *p. 558* **CHATTER** *p. 582* **CHUK** *p. 554*

By both sexes in high agitation; often in series. Plastic.

By courting males or fighting females; speed and quality vary situationally.

All year; most common call. Variable, but rather like other blackbird Chuks.

COMMON GRACKLE

Quiscalus quiscula

Abundant in open woods, fields, towns, parks, and lawns; has expanded range westward in recent years. Forms large, loud flocks, often with other blackbirds.

In close courtship, male bows head, circles female, and rapidly alternates Whistles and Songs

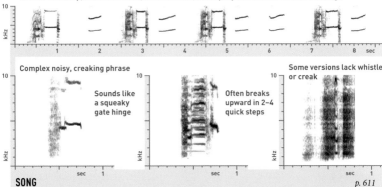

Complex noisy, creaking phrase

Sounds like a squeaky gate hinge

Often breaks upward in 2–4 quick steps

Some versions lack whistle or creak

SONG   *p. 611*

All year, especially Mar.–June, by both sexes, but primarily males. Female sings mostly in response to mate, both birds fluffing feathers and spreading wings. Each individual reportedly has 1 unique songtype, given repeatedly or mixed with Jeers and other calls.

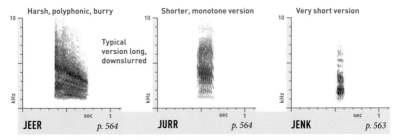

Harsh, polyphonic, burry

Typical version long, downslurred

Shorter, monotone version

Very short version

JEER *p. 564*   **JURR** *p. 564*   **JENK** *p. 563*

Given frequently by females, rarely by males; highly plastic. Versions shown are common, but the full range of intermediates may be given by any bird. Given in a variety of situations, including upon arrival and departure from a colony and during disputes.

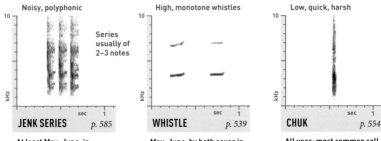

Noisy, polyphonic

Series usually of 2–3 notes

High, monotone whistles

Low, quick, harsh

JENK SERIES *p. 585*   **WHISTLE** *p. 539*   **CHUK** *p. 554*

At least May–June, in courtship. Function little known.

May–June, by both sexes in courtship, often in series.

All year; most common call. Uniform. Other grackle Chuks can be similar.

Yellow-headed Blackbird

Xanthocephalus xanthocephalus

Visually and vocally distinctive. Breeds in tall marshes, nesting over deeper water than Red-winged Blackbird. Often forages in agricultural areas.

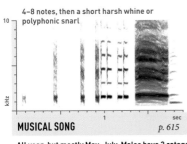

4–8 notes, then a short harsh whine or polyphonic snarl

MUSICAL SONG   *p. 615*

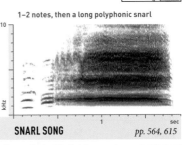

1–2 notes, then a long polyphonic snarl

SNARL SONG   *pp. 564, 615*

All year, but mostly May–July. Males have 2 categories of song. During Musical Song, male usually lifts head and both wings. During Snarl Song, used especially during territorial disputes, lifts wings slightly and turns head to left. Snarl sometimes omitted from both types of song, or given by itself, sounding like Jurr (p. 564).

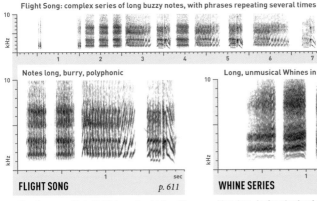

Flight Song: complex series of long buzzy notes, with phrases repeating several times

Notes long, burry, polyphonic

FLIGHT SONG   *p. 611*

May–Aug., usually in flight, by males driving off other males, or by groups of males mobbing hawks or herons. Also during copulation.

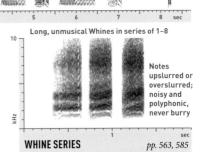

Long, unmusical Whines in series of 1–8

Notes upslurred or overslurred; noisy and polyphonic, never burry

WHINE SERIES   *pp. 563, 585*

May–Aug., by females leaving nest or confronting other females. Head and wings sometimes lifted. Rarely begun during mate's Song.

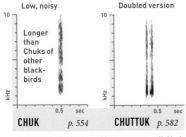

Low, noisy

Longer than Chuks of other blackbirds

CHUK   *p. 554*

Doubled version

CHUTTUK   *p. 582*

All year, by both sexes. Most common call. Variable; breeding males often give doubled version Chuttuk (right).

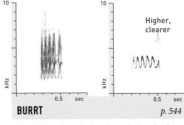

Brief, rather noisy

Juvenile version

Higher, clearer

BURRT   *p. 544*

Mostly May–Aug., by females or by begging juveniles. Function little known; more study needed.

BROWN-HEADED COWBIRD

Molothrus ater

Once limited to following bison herds on the high plains; now common in many habitats. A brood parasite, laying its eggs in other species' nests.

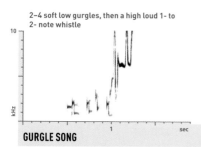

2–4 soft low gurgles, then a high loud 1- to 2- note whistle

GURGLE SONG
p. 611

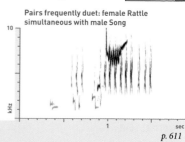

Pairs frequently duet: female Rattle simultaneous with male Song

Nearly all year; gives plastic songs in winter, becoming stereotyped by spring; sings little in fall. Individual males average 4 Gurgle songtypes in the East, more in the West; consecutive songs may be same or different. When directed at other cowbirds, given in display with bowed head, arched wings.

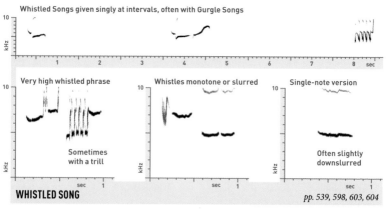

Whistled Songs given singly at intervals, often with Gurgle Songs

Very high whistled phrase

Sometimes with a trill

Whistles monotone or slurred

Single-note version

Often slightly downslurred

WHISTLED SONG
pp. 539, 598, 603, 604

Nearly all year, mostly or exclusively by males; given prior to and during flight, in songlike contexts, and in alarm. A learned sound, like Gurgle Song, with local dialects. Most males have 2 versions, 1 multinoted and 1 single-noted; the multinoted version is often truncated.

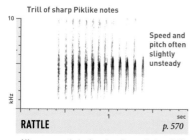

Trill of sharp Piklike notes

Speed and pitch often slightly unsteady

RATTLE
p. 570

All year, mostly by females. Somewhat variable and quite plastic, especially in female; male's version is quite monotone.

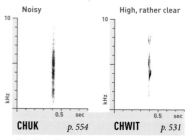

Noisy

High, rather clear

CHUK
p. 554

CHWIT
p. 531

Possibly all year; rarely heard. Chuk by both sexes; Chwit little known. Juveniles beg with high, coarse Cheet (p. 545).

BRONZED COWBIRD

Molothrus aeneus

Found in semi-open country. Note red eye and thick ruff of feathers on back of male's neck. A brood parasite, laying its eggs in the nests of other bird species.

Soft, slow series of rising musical whines, each ending in a gurgle

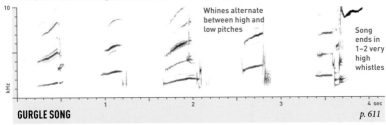

Whines alternate between high and low pitches

Song ends in 1–2 very high whistles

GURGLE SONG — *p. 611*

All year, especially May–July, apparently only by males. Given from perch, often in display with bowed head, arched wings, and fluffed neck feathers; sometimes preceded by hovering display and followed by Whistled Song in circling flight. Repertoire size unknown.

Gurgle Song sometimes culminates in much louder Whistled Song

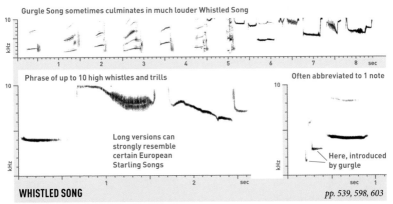

Phrase of up to 10 high whistles and trills

Long versions can strongly resemble certain European Starling Songs

Often abbreviated to 1 note

Here, introduced by gurgle

WHISTLED SONG — *pp. 539, 598, 603*

All year, mostly or exclusively by males; given in courtship during both hover display and circling flight display, as well as from perch. Likely also given in other contexts. A learned sound, with regional dialects; individuals have a single version, frequently truncated.

Trill of sharp Piklike notes

Quick, rather noisy

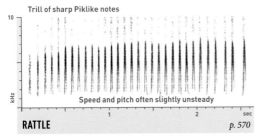

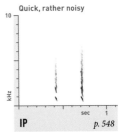

Speed and pitch often slightly unsteady

RATTLE — *p. 570*

IP — *p. 548*

All year, mostly by females. Somewhat variable and quite plastic. Very similar to Brown-headed Cowbird Rattle.

Possibly all year; rare. Juveniles beg with high polyphonic Zeer (p. 562).

ORCHARD ORIOLE

Icterus spurius

Our smallest oriole. Nests in shelter-belts and woodland edges, especially near water. Young males resemble females, but with black bib.

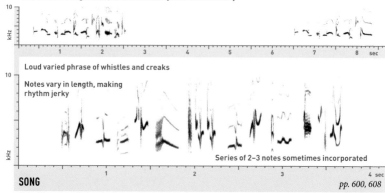

Consecutive songs differ, or start similarly but end differently

Loud varied phrase of whistles and creaks

Notes vary in length, making rhythm jerky

Series of 2–3 notes sometimes incorporated

SONG   *pp. 600, 608*

Mostly Mar.–June. Highly variable; consecutive songs can vary tremendously. Chuklike noisy notes occasionally included. Apparently incorporates imitations only rarely, and possibly only in quiet subsong or whisper song. First-year males sing often.

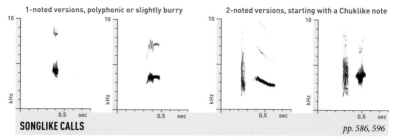

1-noted versions, polyphonic or slightly burry

2-noted versions, starting with a Chuklike note

SONGLIKE CALLS   *pp. 586, 596*

All year. Variable but stereotyped; individuals appear to have 3–4 versions, usually given singly, in random order, often interspersed with Chuks. Components of 2-noted versions may also be given separately, but quite distinctive when heard together.

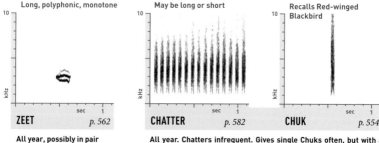

Long, polyphonic, monotone

May be long or short

Recalls Red-winged Blackbird

ZEET *p. 562*

CHATTER *p. 582*

CHUK *p. 554*

All year, possibly in pair contact. Somewhat variable and plastic.

All year. Chatters infrequent. Gives single Chuks often, but with a complete range of intergrades in between.

HOODED ORIOLE

Icterus cucullatus

Breeds in arid regions with scattered trees, including suburbs; particularly fond of ornamental palms. Young males resemble females, but with black bib.

Consecutive songs tend to differ greatly, though first 1–2 notes are often the same

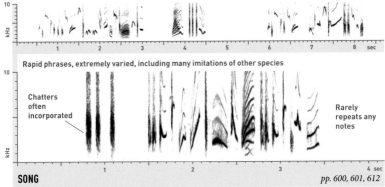

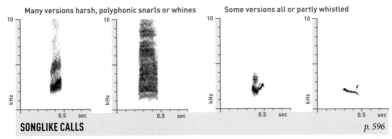

Rapid phrases, extremely varied, including many imitations of other species

Chatters often incorporated

Rarely repeats any notes

SONG   *pp. 600, 601, 612*

Apparently mostly Mar.–May. Hooded Orioles have a reputation for singing less than other orioles, but song is perhaps overlooked because it is rather soft and jumbled. Imitations of other bird species often make up the majority of song. Short Chatters often given between or within songs.

Many versions harsh, polyphonic snarls or whines   Some versions all or partly whistled

SONGLIKE CALLS   *p. 596*

All year. Variable but stereotyped; individuals appear to have 2–4 versions, usually given singly, in random order, interspersed with many Chatters and sometimes Zreets. Burry or noisy polyphonic versions typical and fairly distinctive; clear whistled versions generally less common.

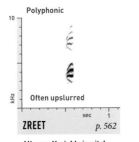

Polyphonic

Often upslurred

ZREET   *p. 562*

All year. Variable in pitch, and rather plastic. Some versions monotone (p. 562).

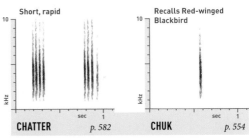

Short, rapid

Recalls Red-winged Blackbird

CHATTER   *p. 582*   **CHUK**   *p. 554*

All year. Slightly variable; plastic, especially in length. Chatter averages shorter than in other orioles; given frequently. Single Chuks infrequent, usually mixed with Chatters.

BALTIMORE ORIOLE

Icterus galbula

Familiar and highly popular for its bright plumage and cheerful song. Nests in woodland edges and mature shade trees in urban parks.

Long Song (p. 601): soft, continuous, and highly varied, with occasional Chatters; likely in close courtship

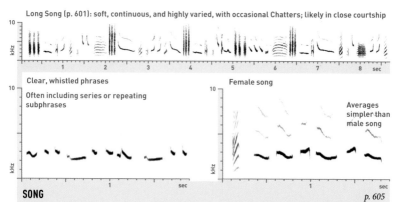

Clear, whistled phrases

Often including series or repeating subphrases

Female song

Averages simpler than male song

SONG
p. 605

Mostly Apr.–Aug. Males have 2–12 similar songtypes, rarely repeated; some males sing mostly short versions of 4–6 notes, others longer versions of 12–14 notes. Female repertoire size unknown. Females sing mostly near mate; most Songs are short series of downslurs, but some are as complex as males'.

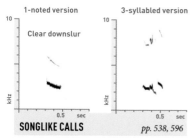

1-noted version

Clear downslur

3-syllabled version

Burry upslurs

Clear downslurs

SONGLIKE CALLS
pp. 538, 596

KREE SERIES p. 576   **TWITTER** p. 571

Possibly all year. Individuals have perhaps 3 versions, given singly in random order; most are clear and whistled, 1- to 3-syllabled.

Softly, by breeding birds; Kree Series in chases and pair interactions, Twitter possibly in pair contact.

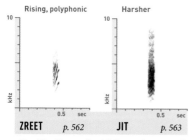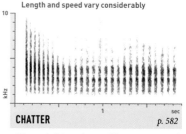

Rising, polyphonic

Harsher

Length and speed vary considerably

ZREET p. 562   **JIT** p. 563

CHATTER p. 582

Possibly all year. Zreet infrequent; apparently used rarely by night migrants. Jit mixed with Songlike Calls or Chatters.

All year, in interactions and in response to various types of threat. Plastic; perhaps rarely shortened to a single Chuk.

BULLOCK'S ORIOLE

Icterus bullockii

Breeds in riparian woods. Interbreeds with Baltimore where ranges overlap; hybrids may have any combination of parental plumage and vocalizations.

Long Song (p. 601): soft, continuous, and highly varied, with occasional Chatters; likely in close courtship

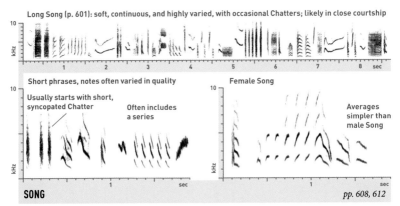

Short phrases, notes often varied in quality

Usually starts with short, syncopated Chatter

Often includes a series

Female Song

Averages simpler than male Song

SONG   *pp. 608, 612*

Mostly Apr.–July. Notes may have almost any tone quality, but whistles or slightly nasal notes usually predominate. Females sing regularly, mostly near mate; repertoire size unknown in both sexes, but at least some males have multiple songs, usually repeated many times before switching.

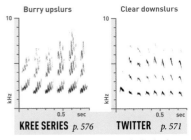

Brief whistled version

More nasal version

SONGLIKE CALLS   *pp. 538, 596*

Possibly all year. Individuals have perhaps 3 versions, given singly in random order; some are nasal or noisy, 1- or 2-noted.

Burry upslurs

Clear downslurs

KREE SERIES *p. 576*   **TWITTER** *p. 571*

Softly, by breeding birds; Kree Series in chases and pair interactions, Twitter possibly in pair contact.

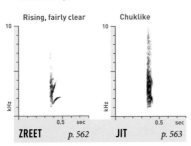

Rising, fairly clear

Chuklike

ZREET *p. 562*   **JIT** *p. 563*

Not well known. Zreet and Jit both given mixed with Songlike Calls, and possibly best classified in that category.

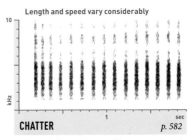

Length and speed vary considerably

CHATTER *p. 582*

All year, in interactions and in response to various types of threat. Plastic; rarely shortened to a single Chuk.

SCOTT'S ORIOLE

Icterus parisorum

Found in arid habitats, especially where desert transitions into pinyon-juniper or oak woodlands. Often nests in palms, tall yuccas, or junipers.

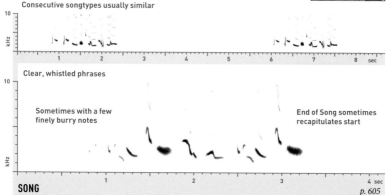

Consecutive songtypes usually similar

Clear, whistled phrases

Sometimes with a few finely burry notes

End of Song sometimes recapitulates start

SONG   *p. 605*

All year. Males likely have multiple songtypes, but repertoire size unknown. Females also sing, sometimes from the nest; not known whether female Song differs consistently from male Song. Sometimes contains 1–2 repeated notes.

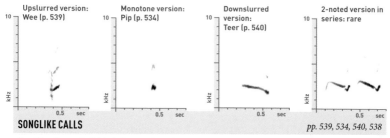

Upslurred version: Wee (p. 539)

Monotone version: Pip (p. 534)

Downslurred version: Teer (p. 540)

2-noted version in series: rare

SONGLIKE CALLS   *pp. 539, 534, 540, 538*

Possibly all year. Individuals have perhaps 1–2 versions; consecutive versions tend to be the same. Very short, 1-noted, upslurred or monotone versions are typical. The 2-noted version in series is unusual; may be part of an aberrant songtype, or an undescribed vocalization.

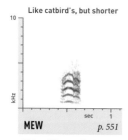

Like catbird's, but shorter

MEW   *p. 551*

Not well known. Variable and plastic; some versions fairly harsh.

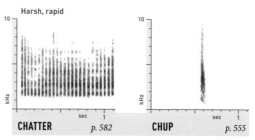

Harsh, rapid

CHATTER   *p. 582*   **CHUP**   *p. 555*

All year. Chatter apparently rare; sole available example shown. Chup much more common; quite variable, ranging from quite harsh to a polyphonic Jit (p. 563). May grade into Mew.

WARBLERS (Family Parulidae)

Warblers are small, highly active, and often brightly colored. Their beauty and diversity make them very popular with birders. Most are highly migratory.

Warblers learn their songs. They sort into 3 broad groups by repertoire type:

Warblers with Primary and Complex Songs
(*Ovenbird, Northern Waterthrush, MacGillivray's Warbler, Common Yellowthroat*)

In this group, vocal repertoires are very similar to those of many sparrows. Males tend to have a single short **primary song** as well as a single longer **complex song**, which is often given during a flight display and introduced by a series of call notes. Complex songs are given infrequently.

Warblers with "First-Category" and "Second-Category" Songs
(*Black-and-white Warbler, most* Setophaga)

Many species in this group also have two different types of singing behaviors. **First-category singing** is done mostly during the day by unpaired males, often directed at females. **Second-category singing** is done mostly at dawn by unpaired males or during the day by paired males, often directed at other males. Second-category singing tends to be characterized by long series of call-like notes in between songs, by shorter pauses between songs, and/or by a greater tendency for consecutive song-types to differ.

In species such as the Black-throated Green Warbler of the East, First- and Second-Category Songs are easy to distinguish in the field, because they differ consistently in form. In other species, songs in the two categories may be difficult to distinguish. In fact, in species such as the Yellow Warbler, the very same songtype may be used in first-category singing by one individual and in second-category singing by another individual (or even the same individual). In these cases it is not the form of the songs that determines their category, but the pattern in which they are sung.

Warblers with No Clear Distinction Between Song Categories
(*Wilson's Warbler, Red-faced Warbler, Painted Redstart, all* Oreothlypis)

These warblers have not been documented to sing two different categories of songs, but more study may reveal that the categories do in fact exist.

Warbler Calls

The call repertoires of most warblers include a Seetlike or Dzitlike call given in contact and in flight (including by night migrants); a high brief Tink given in high alarm; and a Chiplike note given in mild alarm. In addition, many species give Twitters or Rattles in alarm or aggression (see p. 570).

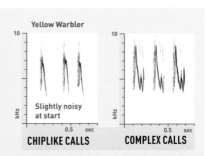

DAWN SONG CALLS OF WARBLERS

At dawn, especially in the second half of the breeding season, several warbler species give Second-Category Songs with strings of call-like notes in between. The calls between the songs are variable and slightly plastic; they may resemble Chips, Tinks, Seets, or the begging calls of juveniles, but often they are subtly distinct from any of these. Two different versions from Yellow Warbler Dawn Song are shown. Many species of sparrows also give call-like notes between songs at dawn.

Yellow Warbler

Slightly noisy at start

CHIPLIKE CALLS

COMPLEX CALLS

OVENBIRD

Seiurus aurocapilla

A very distinctive, thrushlike warbler. Breeds in mature deciduous or mixed woods, foraging mostly in leaf litter; builds dome-shaped nest on ground.

Loud, unmusical couplet series, increasing in volume throughout

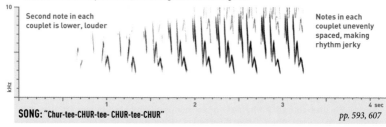

Second note in each couplet is lower, louder

Notes in each couplet unevenly spaced, making rhythm jerky

SONG: "Chur-tee-CHUR-tee- CHUR-tee-CHUR" *pp. 593, 607*

Mostly Apr.–Aug., by males, sometimes from high perches; females not known to sing. Somewhat variable; individuals have a single version. Rare versions are triplet series, or 2 consecutive couplet series. Often transliterated as "Teacher Teacher," but the accent is always on the lower "CHUR" notes.

Often, 2–3 phrases from typical Song included near start or end of complex phrase

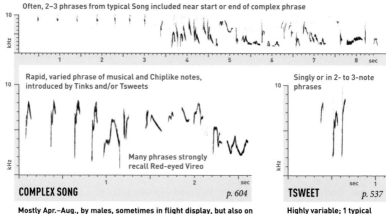

Rapid, varied phrase of musical and Chiplike notes, introduced by Tinks and/or Tsweets

Many phrases strongly recall Red-eyed Vireo

COMPLEX SONG *p. 604*

Mostly Apr.–Aug., by males, sometimes in flight display, but also on ground; often at dusk. Infrequent, but not rare; partial flight songs common in courtship and aggressive interactions.

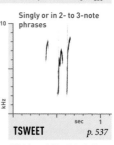

Singly or in 2- to 3-note phrases

TSWEET *p. 537*

Highly variable; 1 typical version shown. By males, in interactions.

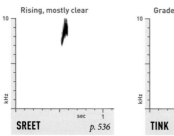

Rising, mostly clear

SREET *p. 536*

All year, often by night migrants. Some versions grade into Tsweet.

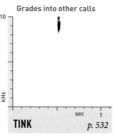

Grades into other calls

TINK *p. 532*

Apr.–Aug., by both sexes in interactions. Sometimes run into rapid Twitter.

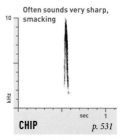

Often sounds very sharp, smacking

CHIP *p. 531*

All year, in alarm. Highly plastic, especially in pitch; becomes lower in alarm.

Northern Waterthrush

Parkesia noveboracensis

Large, stocky, and short-tailed. Breeds in boggy thickets and along wooded streams. Walks along the water's edge, frequently bobbing tail up and down.

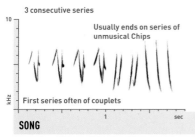

3 consecutive series

Usually ends on series of unmusical Chips

First series often of couplets

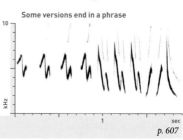

Some versions end in a phrase

SONG
p. 607

Mostly Apr.–July, reportedly only by males. Individuals apparently have a single songtype. Pitch distinctively low, tone emphatic. A few Songs have 4 sections, or only 2. Consecutive series tend to increase slightly in speed and descend slightly in pitch.

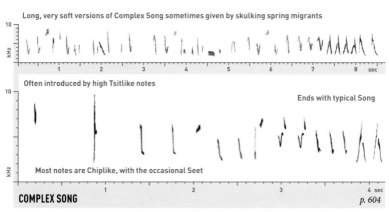

Long, very soft versions of Complex Song sometimes given by skulking spring migrants

Often introduced by high Tsitlike notes

Ends with typical Song

Most notes are Chiplike, with the occasional Seet

COMPLEX SONG
p. 604

Given infrequently, mostly on breeding grounds, often in display flight. Repertoire size unknown. Becomes louder and more recognizable toward end, like Complex Song of Ovenbird.

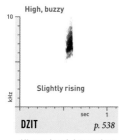

High, buzzy

Slightly rising

DZIT
p. 538

All year, in pair interactions and in flight, including by night migrants.

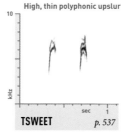

High, thin polyphonic upslur

TSWEET
p. 537

Little known; given in agitation or alarm, sometimes with Complex Song.

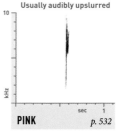

Usually audibly upslurred

PINK
p. 532

All year; most common call. Fairly distinctive; sometimes run into a Sputter.

Black-and-white Warbler

Mniotilta varia

Found in mature forests and woodland edges. Unlike other warblers,
creeps along tree trunks and large
branches, much like a nuthatch.

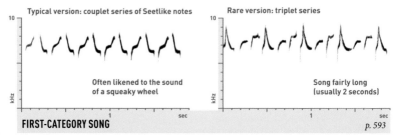

Typical version: couplet series of Seetlike notes

Often likened to the sound
of a squeaky wheel

Rare version: triplet series

Song fairly long
(usually 2 seconds)

FIRST-CATEGORY SONG p. 593

Mostly Apr.–Aug. Little studied, but likely functions like First-Category Songs of other warblers. Repertoire size unknown, but males may have a single version. Consecutive songs usually similar, but males some-times switch frequently between a First-Category and a Second-Category songtype.

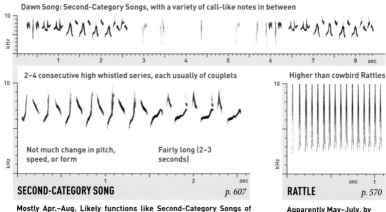

Dawn Song: Second-Category Songs, with a variety of call-like notes in between

2–4 consecutive high whistled series, each usually of couplets

Not much change in pitch,
speed, or form

Fairly long (2–3
seconds)

SECOND-CATEGORY SONG p. 607

Mostly Apr.–Aug. Likely functions like Second-Category Songs of other warblers, but more study needed. Individuals may have a sin-gle version; consecutive songs usually similar.

Higher than cowbird Rattles

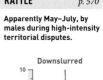

RATTLE p. 570

Apparently May–July, by males during high-intensity territorial disputes.

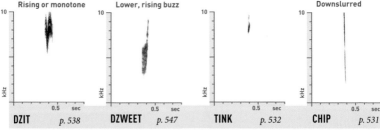

Rising or monotone

DZIT p. 538

Lower, rising buzz

DZWEET p. 547

TINK p. 532

Downslurred

CHIP p. 531

Gives a variety of plastic calls; most common versions shown. Dzit given all year, often in flight, including by night migrants. Dzweet apparently given in aggressive encounters. Tink given in high alarm. Chip given all year in mild alarm; sometimes has initial upslur like single note from Rattle.

TENNESSEE WARBLER

Oreothlypis peregrina

An abundant breeder in boreal forests, especially during outbreaks of spruce budworm. Like Orange-crowned, but undertail always whiter than breast.

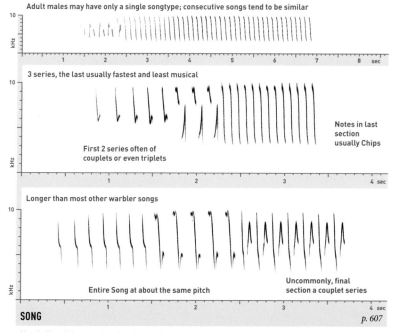

Adult males may have only a single songtype; consecutive songs tend to be similar

3 series, the last usually fastest and least musical

First 2 series often of couplets or even triplets

Notes in last section usually Chips

Longer than most other warbler songs

Entire Song at about the same pitch

Uncommonly, final section a couplet series

SONG

p. 607

Mostly May–July, by males; females not known to sing. Variable, but usually fairly distinctive. Can be confused with Northern Waterthrush Song; note higher pitch and less musical quality. Some males sing 2-parted Song that can be confused with Nashville Warbler. Multiple songtypes and Complex Songs have not been documented.

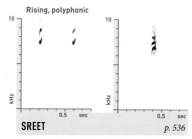

Rising, polyphonic

SREET

p. 536

Given frequently all year, including by both night and daytime migrants. Somewhat plastic; resembles other warbler Sreets.

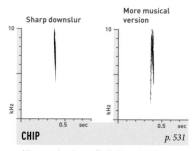

More musical version

Sharp downslur

CHIP

p. 531

All year, in alarm. Typical version is sharp, downslurred; more musical overslurred version may serve a different function.

ORANGE-CROWNED WARBLER

Oreothlypis celata

Common, especially in the West, nesting in open shrubby woods. Western populations are greener, eastern populations grayer; undertail coverts always yellowish.

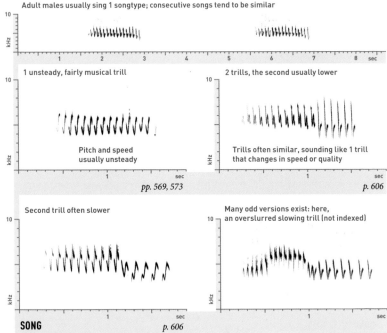

Adult males usually sing 1 songtype; consecutive songs tend to be similar

1 unsteady, fairly musical trill

Pitch and speed usually unsteady

pp. 569, 573

2 trills, the second usually lower

Trills often similar, sounding like 1 trill that changes in speed or quality

p. 606

Second trill often slower

Many odd versions exist: here, an overslurred slowing trill (not indexed)

SONG   *p. 606*

Mostly Mar.–July, by males; females not known to sing. Highly variable. Song of spring migrants is frequently plastic; becomes stereotyped shortly after arrival on breeding grounds. Most males apparently have a single songtype; 1 male documented singing 2 songtypes. Complex Songs have not been documented. Typically more musical and faster than Wilson's, but some songs are very similar. Trills reportedly average slightly faster in West (10–25 notes/second) than in East (7–18 notes/second).

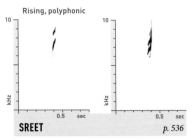

Rising, polyphonic

SREET   *p. 536*

Given frequently all year, including by both night and daytime migrants. Somewhat plastic; resembles other warbler Sreets.

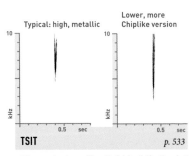

Typical: high, metallic   Lower, more Chiplike version

TSIT   *p. 533*

All year, in alarm. Usually fairly distinctive, but Nashville can give similar calls.

COLIMA WARBLER

Oreothlypis crissalis

Extremely local in the U.S., breeding in scrubby high-elevation oak woodlands. Much like Virginia's Warbler, but larger and browner.

Adult males apparently sing 1 songtype; consecutive songs tend to be similar, sometimes with truncations

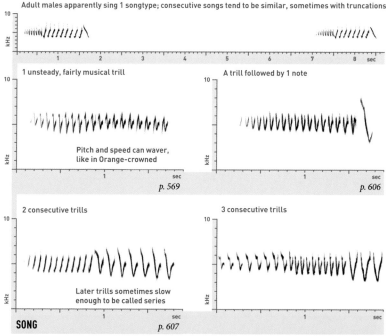

1 unsteady, fairly musical trill

Pitch and speed can waver, like in Orange-crowned

p. 569

A trill followed by 1 note

p. 606

2 consecutive trills

Later trills sometimes slow enough to be called series

SONG   *p. 607*

3 consecutive trills

Mostly Mar.–July, by males; females not known to sing. Variable; many songs are quite similar to those of Orange-crowned; others resemble Virginia's Warbler in pattern, but are always faster. Complex Songs have not been documented. Most typical song patterns shown here; rare variants have a single different note at the start of the song or in between trills or series.

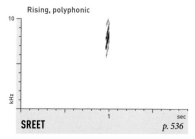

Rising, polyphonic

SREET   *p. 536*

All year, including possibly by night migrants. Somewhat plastic; some versions are intergrades with Pink.

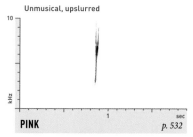

Unmusical, upslurred

PINK   *p. 532*

All year, in alarm. Generally similar to Pink of other *Oreothlypis* warblers.

Lucy's Warbler

Oreothlypis luciae

Breeds in mesquite bosques and riparian cottonwoods. One of our few cavity-nesting warblers. Very small and pale; red rump usually hidden by wings.

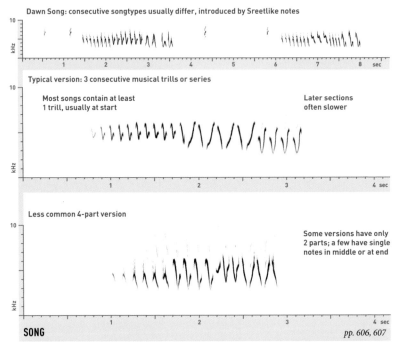

Dawn Song: consecutive songtypes usually differ, introduced by Sreetlike notes

Typical version: 3 consecutive musical trills or series

Most songs contain at least 1 trill, usually at start

Later sections often slower

Less common 4-part version

Some versions have only 2 parts; a few have single notes in middle or at end

SONG *pp. 606, 607*

Mostly Mar.–Aug., by males; females not known to sing. Unlike other members of the genus *Oreothlypis*, has a distinct Dawn Song pattern, and most males apparently have at least two songtypes, which are often alternated. The songtypes of an individual often have similar beginnings. Generally quite similar to Virginia's Warbler Song, but almost always faster and often longer. Rarely comprises mostly Chiplike notes, recalling Grace's Warbler. Complex Song has not been documented.

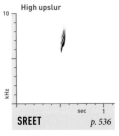

High upslur

SREET *p. 536*

All year, including possibly by night migrants. Like other warbler Sreets.

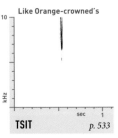

Like Orange-crowned's

TSIT *p. 533*

Possibly all year, likely in high alarm.

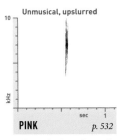

Unmusical, upslurred

PINK *p. 532*

All year, in alarm. Generally similar to Pink of other *Oreothlypis* warblers.

NASHVILLE WARBLER

Oreothlypis ruficapilla

Breeds in mixed forests and second growth. Eastern population sometimes pumps tail; western population, called "Calaveras Warbler," pumps tail often.

Adult males may have only a single songtype; consecutive songs tend to be similar

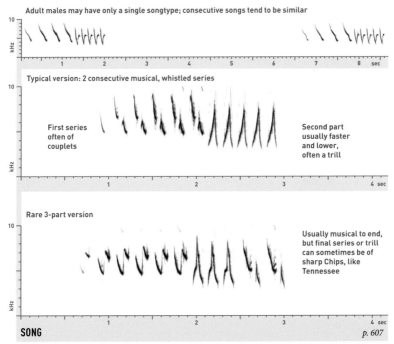

Typical version: 2 consecutive musical, whistled series

First series often of couplets

Second part usually faster and lower, often a trill

Rare 3-part version

Usually musical to end, but final series or trill can sometimes be of sharp Chips, like Tennessee

SONG   *p. 607*

Mostly May–July, by males; females not known to sing. Variable; birds occasionally sing only first series, sometimes for up to a few minutes at a time. Some versions difficult to separate from certain songs of Wilson's and Tennessee Warblers. Multiple songtypes and Complex Songs have not been documented. Songs and calls of western subspecies, shown, apparently do not differ significantly from those of eastern birds.

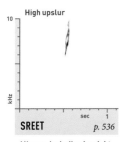

High upslur

SREET   *p. 536*

All year, including by night migrants. Like other warbler Sreets.

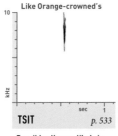

Like Orange-crowned's

TSIT   *p. 533*

Possibly all year, likely in high alarm. Not shown is a sharp Chip (p. 531).

Unmusical, upslurred

PINK   *p. 532*

All year, in alarm. Generally similar to Pink of other *Oreothlypis* warblers.

VIRGINIA'S WARBLER

Oreothlypis virginiae

Breeds in dry mixed forests, especially where oaks mix with pinyon and juniper or with pines. Closely related to Nashville Warbler, with similar vocalizations.

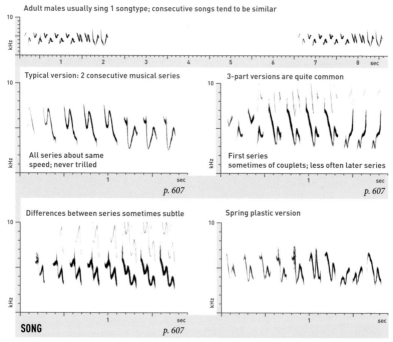

Adult males usually sing 1 songtype; consecutive songs tend to be similar

Typical version: 2 consecutive musical series

All series about same speed; never trilled

p. 607

3-part versions are quite common

First series sometimes of couplets; less often later series

p. 607

Differences between series sometimes subtle

Spring plastic version

SONG   *p. 607*

Mostly May–July, by males; females not known to sing. Variable. Complex Songs have not been documented. Similar to Nashville Warbler Song, but typically a little lower and shorter; 3-part versions are much more common, and a few are 4-parted. Final series frequently shortened to just one note. All parts of Song usually highly musical and clear, lacking burry richness of MacGillivray's. Some versions easy to mistake for certain Yellow Warbler songtypes.

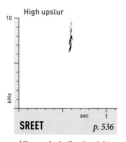

High upslur

SREET   *p. 536*

All year, including by night migrants. Like other warbler Sreets.

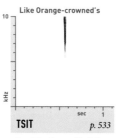

Like Orange-crowned's

TSIT   *p. 533*

Possibly all year, likely in high alarm.

Unmusical, upslurred

PINK   *p. 532*

All year, in alarm. Generally similar to Pink of other *Oreothlypis* warblers.

MacGillivray's Warbler

Geothlypis tolmiei

Breeds in riparian willow thickets, dense second growth, and regenerating clear-cuts. Skulking. Closely related to Mourning Warbler of the East.

Males have a single songtype, repeated at intervals

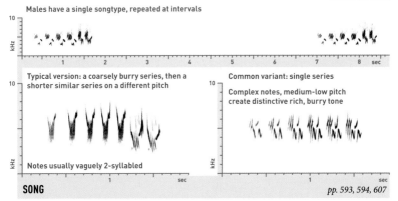

Typical version: a coarsely burry series, then a shorter similar series on a different pitch

Notes usually vaguely 2-syllabled

Common variant: single series

Complex notes, medium-low pitch create distinctive rich, burry tone

SONG   *pp. 593, 594, 607*

Mostly May–July, by males; females not known to sing. Variable but stereotyped. Uncommon versions have 3 or 4 series. Tone quality generally distinctive, but parts of some songs are simpler and more musical, creating potential for confusion with Virginia's Warbler.

Complex series, warbles, and Chwitlike notes

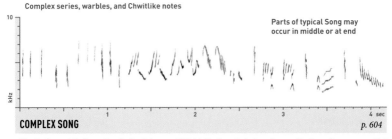

Parts of typical Song may occur in middle or at end

COMPLEX SONG   *p. 604*

Mostly June–July, by males during infrequent flight displays, but also given softly from dense cover. Not well known; only available recording is shown. Begins with series of Chwitlike notes. Not known whether individuals may have multiple versions.

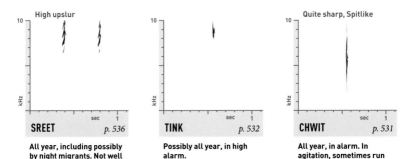

High upslur

SREET   *p. 536*

All year, including possibly by night migrants. Not well known.

TINK   *p. 532*

Possibly all year, in high alarm.

Quite sharp, Spitlike

CHWIT   *p. 531*

All year, in alarm. In agitation, sometimes run into rapid Chatter.

COMMON YELLOWTHROAT

Geothlypis trichas

Common and widespread in wetlands and dense undergrowth near water. Plumage varies geographically, especially in amount of yellow on belly.

Males have 1 typical Song, given repeatedly, and 1 Complex Song, given rarely

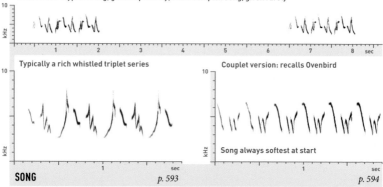

Typically a rich whistled triplet series

Couplet version: recalls Ovenbird

Song always softest at start

SONG   *p. 593*

SONG   *p. 594*

Mostly Apr.–July. Distinctive, but highly variable; western birds more likely to sing quadruplet or quintuplet series. About 3 percent of males sing a 2-part Song of consecutive complex series. Details of each male's Song are unique, but broadly similar songtypes may occur across entire regions.

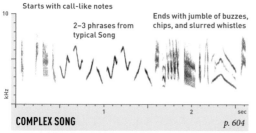

Starts with call-like notes

2–3 phrases from typical Song

Ends with jumble of buzzes, chips, and slurred whistles

Notes smacking

COMPLEX SONG   *p. 604*

RATTLE   *p. 568*

Given infrequently by males on breeding grounds, usually in flight display. May serve a territorial function, but possibly also given in response to potential predators.

Frequent from aggressive breeding males. Highly plastic in length.

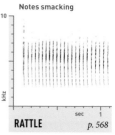

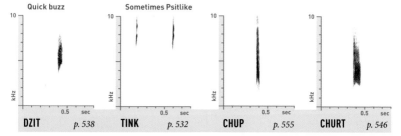

Quick buzz

Sometimes Psitlike

DZIT   *p. 538*

TINK   *p. 532*

CHUP   *p. 555*

CHURT   *p. 546*

Gives a variety of calls. Dzit given all year, including by night migrants; somewhat plastic, but fairly distinctive. Most common call is Chup, given all year in alarm. Tink given in high alarm; Churt is longer, burrier version of Chup. Both Tink and Churt highly plastic, grading into Chup.

RUFOUS-CAPPED WARBLER

Basileuterus rufifrons

A primarily tropical species; recently established as a rare, local breeder in southeast Arizona in brushy oak woods in open canyons, usually near water.

Sometimes switches songtypes constantly, with call-like notes in between

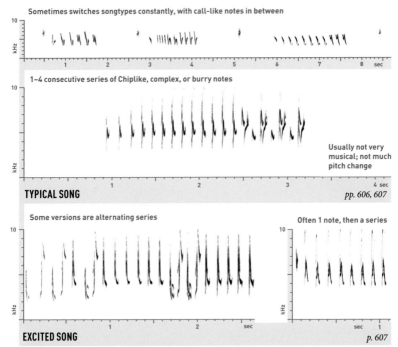

1–4 consecutive series of Chiplike, complex, or burry notes

Usually not very musical; not much pitch change

TYPICAL SONG   *pp. 606, 607*

Some versions are alternating series

Often 1 note, then a series

EXCITED SONG   *p. 607*

Typical Song given all year, especially Mar.–Aug. Not known whether both sexes sing. Repertoire size unknown, but birds apparently have multiple songtypes; consecutive songs may be same or different. Pattern shown at top resembles Second-Category Song of other warblers; more study needed. Songtypes extremely variable. Most songs in the U.S. seem to be 1- and 2-parted versions. Burry notes and couplet series, if present, usually at the end. Excited Song given in interactions; poorly known. Can be difficult to distinguish from Twitter or rapid series of Pink calls.

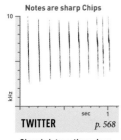

Notes are sharp Chips

TWITTER   *p. 568*

Given in interactions along with Excited Song. Highly plastic.

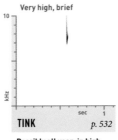

Very high, brief

TINK   *p. 532*

Possibly all year, in high alarm.

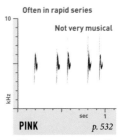

Often in rapid series

Not very musical

PINK   *p. 532*

All year; most common call. Variable; some versions more complex than shown.

AMERICAN REDSTART

Setophaga ruticilla

Common in deciduous woods. Highly active from ground level to treetops, often flycatching. Fans tail frequently, possibly to startle insect prey.

Males have 1 First-Category (repeated) songtype; 2–8 Second-Category (alternated) songtypes

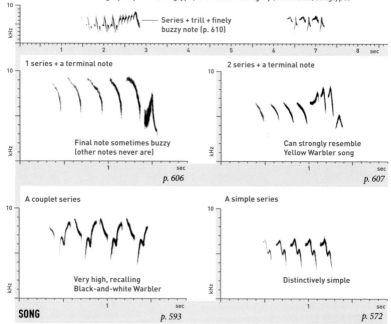

Series + trill + finely buzzy note (p. 610)

1 series + a terminal note

Final note sometimes buzzy (other notes never are)

p. 606

2 series + a terminal note

Can strongly resemble Yellow Warbler song

p. 607

A couplet series

Very high, recalling Black-and-white Warbler

p. 593

A simple series

Distinctively simple

p. 572

SONG

Perhaps the most variable warbler song. Mostly Apr.–Aug.; stereotyped in breeding males, often plastic in migrants. Femalelike birds that sing are likely young males. Two song categories: First-Category Songs are always repeated; Second-Category Songs are continually switched. First-Category Songs have accented endings in about 80 percent of males; these recall Yellow Warbler, but tend to be simpler and often higher and slower. Second-Category Songs are diverse and easily confused with many other warbler species, but the constant switching of songtypes is a good identification clue.

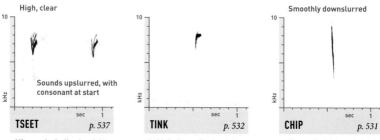

High, clear

Sounds upslurred, with consonant at start

TSEET *p. 537*

All year, including by night migrants. Variable and plastic, but fairly distinctive.

TINK *p. 532*

In high alarm. Grades into Chip.

Smoothly downslurred

CHIP *p. 531*

All year. Variable and plastic, but generally high and clear.

YELLOW WARBLER

Setophaga petechia

Our most common and widespread warbler, found in a variety of habitats from woodland edges, parks, and second growth to suburban backyards.

Dawn Song: consecutive songtypes differ, introduced by strings of call-like notes

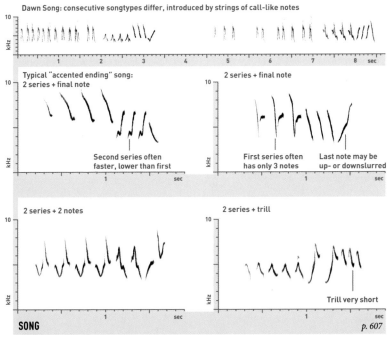

Typical "accented ending" song:
2 series + final note

Second series often
faster, lower than first

2 series + final note

First series often
has only 3 notes

Last note may be
up- or downslurred

2 series + 2 notes

2 series + trill

Trill very short

SONG   *p. 607*

Mostly Apr.–Aug.; stereotyped in breeding males, often plastic in migrants. Two song categories: First-Category Songs are always repeated; usually, Second-Category Songs are continually switched. Males have 1, rarely 2 First-Category Songs, 5–17 different Second-Category Songs. First-Category Songs usually have 2 series plus 1 terminal note, but identical songtypes sometimes used by different males in different categories.

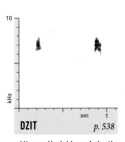

DZIT   *p. 538*

All year. Variable and plastic, but tends to be loud, fairly musical.

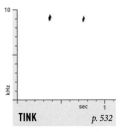

TINK   *p. 532*

In high alarm. May grade into Chip.

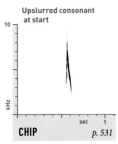

Upslurred consonant
at start

CHIP   *p. 531*

All year; most common call. Plastic; some versions higher, without initial upslur.

YELLOW-RUMPED WARBLER (MYRTLE)

Setophaga coronata (coronata group)

Breeds in northern coniferous forests. Uncommon to fairly common in migration and winter in the West, usually outnumbered by Audubon's.

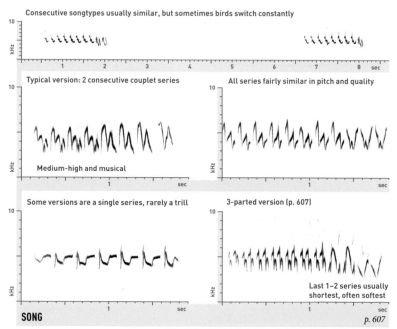

Consecutive songtypes usually similar, but sometimes birds switch constantly

Typical version: 2 consecutive couplet series

Medium-high and musical

All series fairly similar in pitch and quality

Some versions are a single series, rarely a trill

3-parted version (p. 607)

Last 1–2 series usually shortest, often softest

SONG   *p. 607*

Mostly Apr.–Aug., by males. Females not known to sing. Not known whether males have 2 song categories. Repertoire size unknown, but males apparently have multiple songtypes. Much overlap with Audubon's song, but usually faster, sometimes trilled at end; series more often simple.

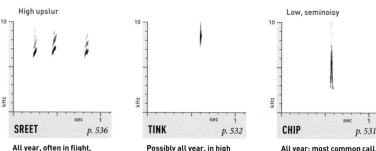

High upslur

SREET   *p. 536*

All year, often in flight, including by night migrants. Like Audubon's Sreet.

TINK   *p. 532*

Possibly all year, in high alarm. Plastic; grades into Sreet.

Low, seminoisy

CHIP   *p. 531*

All year; most common call. Fairly distinctive.

Yellow-rumped Warbler (Audubon's)

Setophaga coronata (*auduboni* group)

Breeds in mountain coniferous forests. Formerly considered a separate species; hybridizes with Myrtle in a narrow contact zone in Canada.

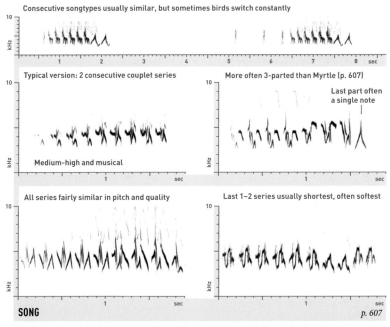

Consecutive songtypes usually similar, but sometimes birds switch constantly

Typical version: 2 consecutive couplet series

Medium-high and musical

More often 3-parted than Myrtle (p. 607)

Last part often a single note

All series fairly similar in pitch and quality

Last 1–2 series usually shortest, often softest

SONG   *p. 607*

Mostly Apr.–Aug., by males. Females not known to sing. Not known whether males have 2 song categories. Repertoire size unknown, but males apparently have multiple songtypes. Averages slower than Myrtle's Song, rarely including trills, most series usually of couplets; almost never a single series.

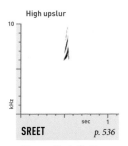

High upslur

SREET   *p. 536*

All year, often in flight, including by night migrants. Like Myrtle Sreet.

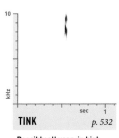

TINK   *p. 532*

Possibly all year, in high alarm. Plastic; grades into Sreet.

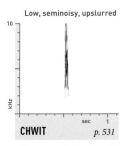

Low, seminoisy, upslurred

CHWIT   *p. 531*

All year; most common call. Subtly but distinctively different from Myrtle's Chip.

GRACE'S WARBLER

Setophaga graciae

A pine forest specialist, common in open stands of ponderosa pine and similar species. Forages near the tops of tall trees, where it can be hard to see.

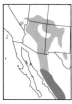

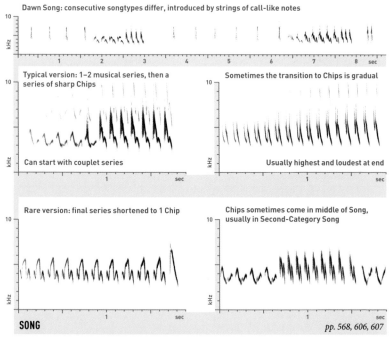

Dawn Song: consecutive songtypes differ, introduced by strings of call-like notes

Typical version: 1–2 musical series, then a series of sharp Chips

Can start with couplet series

Sometimes the transition to Chips is gradual

Usually highest and loudest at end

Rare version: final series shortened to 1 Chip

Chips sometimes come in middle of Song, usually in Second-Category Song

SONG

pp. 568, 606, 607

Mostly Apr.–Aug., by males. Females not known to sing. Two song categories: First-Category Songs are always repeated; usually, Second-Category Songs are continually switched. Males have 1 or more First-Category Songs, 2 or more different Second-Category Songs, for a typical total of 6 or more Songs per male. First-Category Songs are generally more stereotyped than Second-Category Songs, but identical songtypes are sometimes used by different males in different categories. In high agitation, gives plastic Song with Rattles of Chiplike notes. Unmusical, chipping series at end of Song is usually distinctive in range.

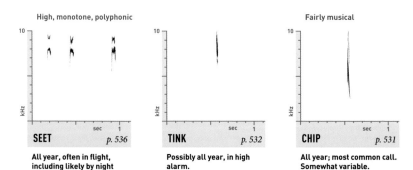

High, monotone, polyphonic

Fairly musical

SEET *p. 536*

All year, often in flight, including likely by night migrants.

TINK *p. 532*

Possibly all year, in high alarm.

CHIP *p. 531*

All year; most common call. Somewhat variable.

BLACK-THROATED GRAY WARBLER

Setophaga nigrescens

Breeds in pinyon-juniper and arid oak scrub, sometimes with an open coniferous overstory. Related to Townsend's and Hermit Warblers.

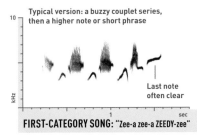
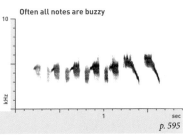

Typical version: a buzzy couplet series, then a higher note or short phrase

Last note often clear

Often all notes are buzzy

FIRST-CATEGORY SONG: "Zee-a zee-a ZEEDY-zee"   *p. 595*

Highly variable. Mostly Mar.–May; stereotyped in breeding males, often plastic in migrants. Repertoire size unknown. Given mostly prior to pairing; reportedly also in territorial conflicts, unlike First-Category Songs of other warblers, but more study needed. Categorization of songs on this page is tentative.

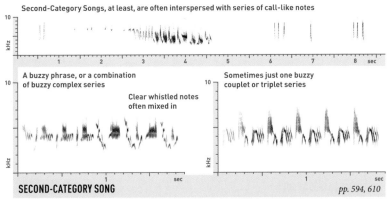

Second-Category Songs, at least, are often interspersed with series of call-like notes

A buzzy phrase, or a combination of buzzy complex series

Clear whistled notes often mixed in

Sometimes just one buzzy couplet or triplet series

SECOND-CATEGORY SONG   *pp. 594, 610*

Highly variable and poorly understood. Apparently given mostly at dawn and dusk, and after males are paired. Repertoire size unknown; at least some males likely have 2 or more Second-Category Songs.

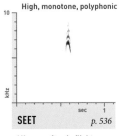

High, monotone, polyphonic

SEET   *p. 536*

All year, often in flight, including likely by night migrants.

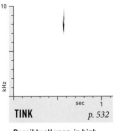

TINK   *p. 532*

Possibly all year, in high alarm.

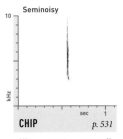

Seminoisy

CHIP   *p. 531*

All year; most common call. Rather low and noisy, recalling Myrtle Warbler.

TOWNSEND'S WARBLER

Setophaga townsendi

Breeds in tall, dense coniferous forests, primarily of fir and spruce. Closely related to Hermit Warbler, and hybridizes with it in southern Washington.

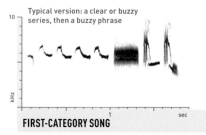

Typical version: a clear or buzzy series, then a buzzy phrase

Last note often highest

FIRST-CATEGORY SONG

p. 610

Mostly Mar.–July, by males; females not known to sing. Repertoire size unknown. Extremely variable; few consistent differences from Hermit Warbler. Spring songs often highly plastic. First-Category Song probably given mostly prior to pairing, but more study needed. Categorization of songs on this page is tentative.

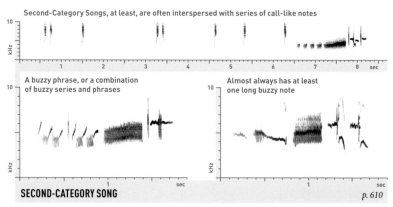

Second-Category Songs, at least, are often interspersed with series of call-like notes

A buzzy phrase, or a combination of buzzy series and phrases

Almost always has at least one long buzzy note

SECOND-CATEGORY SONG

p. 610

Highly variable and poorly understood. Apparently given mostly at dawn and dusk, and after males are paired. Repertoire size unknown; at least some males likely have 2 or more Second-Category Songs.

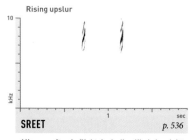

Rising upslur

SREET

p. 536

All year, often in flight, including likely by night migrants. Likely also gives a Tink (p. 532) in high alarm; no recordings available.

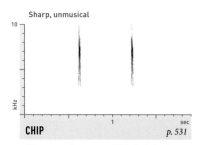

Sharp, unmusical

CHIP

p. 531

All year; most common call. Typically begins with very quick upslur, though this is difficult to hear. Recalls Myrtle Warbler, but higher.

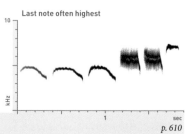

Hermit Warbler

Setophaga occidentalis

Breeds in tall, dense coniferous forests. Forages near the tops of tall trees. Range apparently contracting as that of Townsend's Warbler expands.

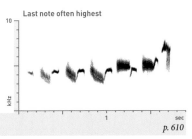

Typical version: a clear or buzzy series, then a buzzy phrase

Last note often highest

FIRST-CATEGORY SONG

p. 610

Mostly Mar.–July, by males; females not known to sing. Reportedly 1 First-Category Song per male, likely given mostly prior to pairing; more study needed. Extremely variable; few consistent differences from Townsend's Warbler. Spring songs often highly plastic. Categorization of songs on this page is tentative.

Second-Category Songs, at least, are often interspersed with series of call-like notes

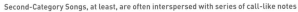

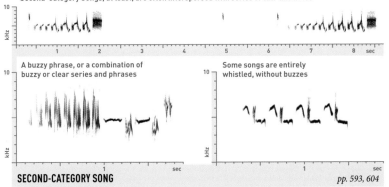

A buzzy phrase, or a combination of buzzy or clear series and phrases

Some songs are entirely whistled, without buzzes

SECOND-CATEGORY SONG

pp. 593, 604

Highly variable and poorly understood. Apparently given mostly at dawn and dusk, and after males are paired. Many males reportedly have 2 or more Second-Category Songs.

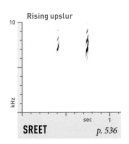

Rising upslur

SREET
p. 536

All year, often in flight, including likely by night migrants.

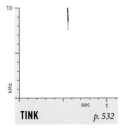

TINK
p. 532

Possibly all year, in high alarm. May grade into other calls.

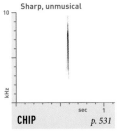

Sharp, unmusical

CHIP
p. 531

All year; most common call. Much like Townsend's Warbler Chip.

WILSON'S WARBLER

Cardellina pusilla

Breeds in swampy thickets in bogs and wet meadows. Female like female Yellow Warbler, but longer-tailed, with more contrast between crown and face.

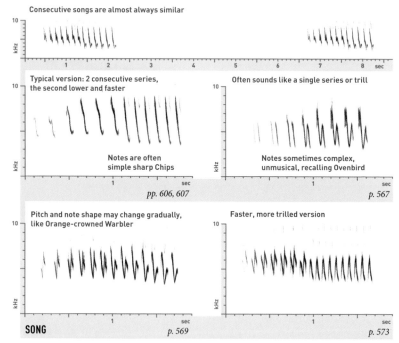

Consecutive songs are almost always similar

Typical version: 2 consecutive series, the second lower and faster

Notes are often simple sharp Chips

pp. 606, 607

Often sounds like a single series or trill

Notes sometimes complex, unmusical, recalling Ovenbird

p. 567

Pitch and note shape may change gradually, like Orange-crowned Warbler

Faster, more trilled version

SONG

p. 569

p. 573

Mostly Apr.–July, by males. At least 1 female has been recorded singing a slower, rising series of Seetlike notes. Quite variable. Males reported to have a single songtype, but 1 male has been recorded alternating 2 songtypes with call-like notes in between. Rarely adds a third series or single note at start or end. Usually sounds less musical than Nashville Warbler; never includes couplet series. Orange-crowned Warbler is almost always faster (trilled from the start).

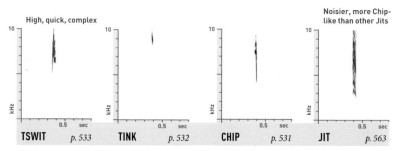

High, quick, complex

Noisier, more Chip-like than other Jits

TSWIT *p. 533*

TINK *p. 532*

CHIP *p. 531*

JIT *p. 563*

Tswit given all year, often in flight, including by night migrants. Tink apparently rare, in high alarm. High Chip given at least in late summer by juveniles, likely also sometimes in fall. Jit is most common and distinctive call; recalls Ruby-crowned Kinglet Jit but higher, briefer, sharper.

RED-FACED WARBLER

Cardellina rubrifrons

A spectacular bird, locally common in coniferous and mixed forests, or riparian areas within them. Nests on the ground, in a hole or natural shelter.

Consecutive songs may be similar or different

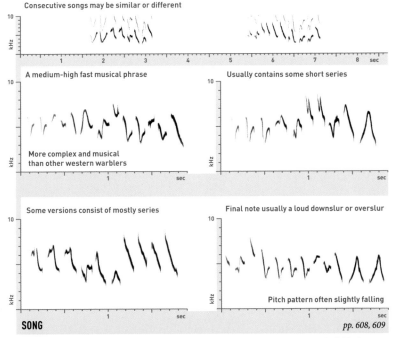

A medium-high fast musical phrase

More complex and musical than other western warblers

Usually contains some short series

Some versions consist of mostly series

Final note usually a loud downslur or overslur

Pitch pattern often slightly falling

SONG

pp. 608, 609

Mostly Apr.–July. Not known whether females sing. Repertoire size unknown, but most individuals apparently have multiple similar songtypes. Rapid switching of songtypes with relatively short pauses between songs may constitute Second-Category Song; more study needed. Distinctive, but beware of plastic songs of other warbler species in spring migration. Painted Redstart Song is usually lower and less varied, with few single notes.

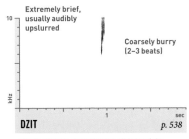

Extremely brief, usually audibly upslurred

Coarsely burry (2–3 beats)

DZIT

p. 538

Likely all year, including possibly by night migrants. Fairly distinctive but apparently infrequent.

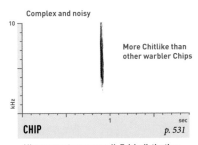

Complex and noisy

More Chitlike than other warbler Chips

CHIP

p. 531

All year; most common call. Fairly distinctive; sounds longer than most Chips, sometimes giving the impression of a second "beat."

PAINTED REDSTART

Myioborus pictus

Breeds in oak and pine-oak woodlands. Highly active, often flicking wings and fanning tail while foraging to startle prey with white plumage patches.

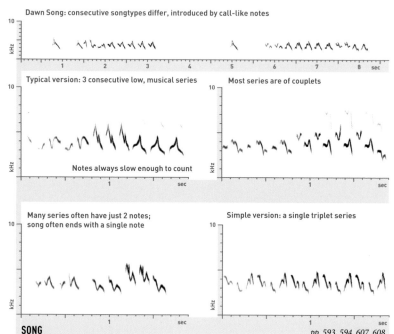

Dawn Song: consecutive songtypes differ, introduced by call-like notes

Typical version: 3 consecutive low, musical series

Notes always slow enough to count

Most series are of couplets

Many series often have just 2 notes; song often ends with a single note

Simple version: a single triplet series

SONG

pp. 593, 594, 607, 608

All year, especially Mar.–Sept., by both sexes. Mated pairs sometimes duet. Individuals have 10 or more songtypes. When countersinging or defending territory against other individuals, consecutive songs are more likely to be different. Pairing status may also affect song switching, suggesting two song categories, as in other warblers, but more study is needed. Notes always clear and highly musical; usually not much change in pitch of Song. A rare, soft jumbled song (not indexed) may represent Complex Song or subsong.

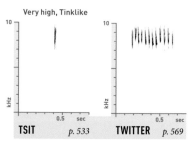

Very high, Tinklike

TSIT *p. 533*

TWITTER *p. 569*

Possibly all year, in high alarm. As agitation increases, Tsit becomes doubled Tsipt (p. 537) and then brief rapid bursts of Twitter.

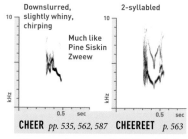

Downslurred, slightly whiny, chirping

Much like Pine Siskin Zweew

2-syllabled

CHEER *pp. 535, 562, 587*

CHEEREET *p. 563*

All year; most common call. Highly variable but stereotyped; likely learned. Cheer and Cheereet about equally common.

PLASTIC SONGS OF SPRING MIGRANT WARBLERS

In spring migration, warblers less than a year old (and perhaps some older birds) often sing highly plastic songs. This is typical of many passerines, but it can be particularly confusing in spring warblers. Young birds' attempts at trills may sound more like warbles, and their attempts at series may sound more like phrases. Hints of the tone quality and pattern of adult birds are usually present, but some songs at an early stage of development may be unidentifiable.

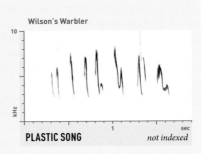

Wilson's Warbler

PLASTIC SONG *not indexed*

AGGRESSIVE TWITTERS AND RATTLES OF WARBLERS

Several species of warblers give plastic Twitters or cowbirdlike Rattles during aggressive interactions or in extreme agitation. Such sounds are heard particularly often from Black-and-white Warbler (and in the East, from Black-throated Blue Warbler, Magnolia Warbler, and Pine Warbler), but they have been recorded or reported in a number of other species, and it is likely that most or all warblers occasionally give similar sounds.

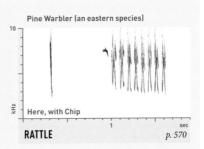

Pine Warbler (an eastern species)

Here, with Chip

RATTLE *p. 570*

CARDINALS, GROSBEAKS, BUNTINGS, AND THEIR ALLIES (Family Cardinalidae) *next pages*

This diverse family is closely related to the sparrows (family Emberizidae) and the finches (family Fringillidae), and taxonomists have shuffled some species among these families more than once. Recently, the North American tanagers were added to the Cardinalidae. The Dickcissel, long a taxonomic enigma, is now known to belong here as well.

Like sparrows and finches, most species in this family have thick, often conical bills, used in many species for cracking hard seeds. Sexes usually differ in plumage.

Most species in the family share the traits of bright coloration, at least in breeding males, and complex, musical Songs, which are apparently learned. The *Piranga* tanagers and *Pheucticus* grosbeaks give a variety of soft, plastic calls, while the *Passerina* buntings, Blue Grosbeak, and Dickcissel have call repertoires that resemble those of many warblers and sparrows. Female Song is common in some species.

HEPATIC TANAGER

Piranga flava

Breeds in dry, open pine forests, sometimes mixed with pinyon-juniper or with
oak. Note smaller, darker bill than
Summer Tanager, indistinct face mask.

At dawn, phrases given at steady pace (about 2 phrases/second) for long periods (p. 598)

Several short musical phrases in clusters

Most phrases brief, clear; clusters
often long, with 6–10 phrases

Usually slower
than Black-headed
Grosbeak Song

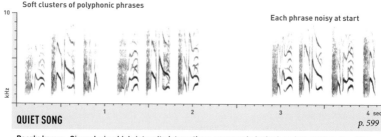

SONG

p. 599

Mostly Apr.–July. Not known whether both sexes sing. Repertoire size unknown. Burry phrases rare. Long
clusters of phrases can strongly recall Black-headed Grosbeak Song, but pace usually steadier and notes
simpler, averaging more musical and less varied.

Soft clusters of polyphonic phrases

Each phrase noisy at start

QUIET SONG

p. 599

Poorly known. Given during high-intensity interactions, apparently by both males and females. Phrases
sometimes fairly stereotyped, as shown above; sometimes plastic, grading into Zreetlike calls. Other tanager species may give similar sounds; more study needed.

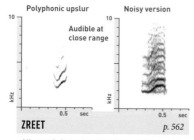

Polyphonic upslur

Noisy version

Audible at
close range

ZREET

p. 562

All year. Soft versions like those at left given frequently in close contact. Louder versions given
by females and likely by begging juveniles.

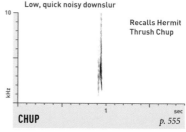

Low, quick noisy downslur

Recalls Hermit
Thrush Chup

CHUP

p. 555

All year; most common call, given frequently in
alarm. Fairly uniform. Given singly, unlike calls
of other tanagers.

SUMMER TANAGER

Piranga rubra

In West, breeds in riparian woods, mesquite, and tamarisk. Adult male is red all year. First-year males and some females are mottled red and green.

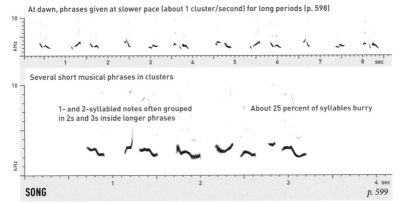

At dawn, phrases given at slower pace (about 1 cluster/second) for long periods (p. 598)

Several short musical phrases in clusters

1- and 2-syllabled notes often grouped in 2s and 3s inside longer phrases

About 25 percent of syllables burry

SONG *p. 599*

Mostly May–July, by males. Female sings infrequently; Song reportedly shorter and softer than male's, with less pause between notes. Easy to confuse with American Robin Song, but averages slightly burrier, with more herky-jerky rhythm created by groupings of notes inside phrases.

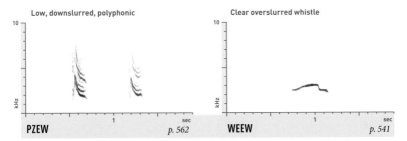

Low, downslurred, polyphonic

Clear overslurred whistle

PZEW *p. 562*

WEEW *p. 541*

A little-known set of calls. In West, Pzew appears to be common in close contact. In nominate subspecies of East, underslurred Pewy (p. 542) and Vewy (pp. 545, 513) seem more common. Weew recorded from a begging juvenile in West, Vewy from juveniles in East. Courting female may give Whine Series.

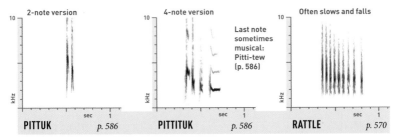

2-note version

4-note version

Last note sometimes musical: Pitti-tew (p. 586)

Often slows and falls

PITTUK *p. 586*

PITTITUK *p. 586*

RATTLE *p. 570*

All year; most common call. Usually 3–4 notes in descending series, but highly plastic; number of notes tends to increase with agitation level, grading into Rattle. First note usually highest, but tendency decreases in high alarm near nest or young. Single-note Pik calls very rare.

WESTERN TANAGER

Piranga ludoviciana

A stunning bird of open coniferous and mixed woodlands. Common and widespread, but not always conspicuous. Amount of red on face variable in males.

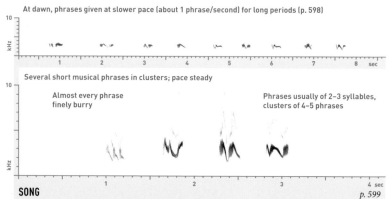

At dawn, phrases given at slower pace (about 1 phrase/second) for long periods (p. 598)

Several short musical phrases in clusters; pace steady

Almost every phrase finely burry

Phrases usually of 2–3 syllables, clusters of 4–5 phrases

SONG   *p. 599*

Mostly Apr.–July, by male. Female sings infrequently. Males have 7 individual phrases on average, which can appear in any order within a given cluster. Very like American Robin Song, but note burry quality. Chippi-tee and Chippituk sometimes incorporated into Song.

Polyphonic downslur

Polyphonic upslur

Polyphonic and/or buzzy underslur

ZEER   *p. 562*

ZREET   *p. 562*

ZEWY   *p. 563*

A confusing set of plastic, intergrading calls. Soft versions given in close contact and possibly by night migrants; louder versions, especially of Zewy or similar calls, sometimes repeated by females, especially in courtship, or by begging juveniles. Function of various versions not well known.

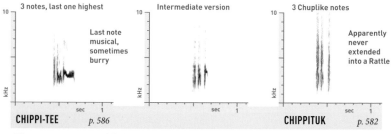

3 notes, last one highest

Last note musical, sometimes burry

Intermediate version

3 Chuplike notes

Apparently never extended into a Rattle

CHIPPI-TEE   *p. 586*

CHIPPITUK   *p. 582*

All year; most common and distinctive call. Individual birds give both Chippi-tee and Chippituk, frequently switching between them, often via intermediate versions. Plastic, but almost always 3 notes, sometimes 2. Chippituk may indicate higher alarm; more study needed.

DICKCISSEL

Spiza americana

A seminomadic bunting of weedy prairies and agricultural areas. In any area, may be common one year and absent the next. Forms large flocks in winter.

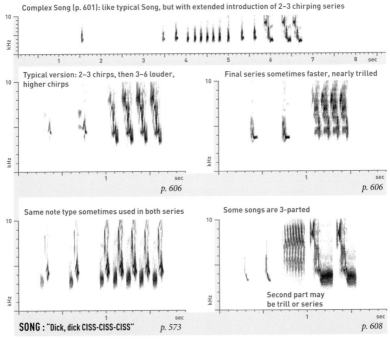

Complex Song (p. 601): like typical Song, but with extended introduction of 2–3 chirping series

Typical version: 2–3 chirps, then 3–6 louder, higher chirps

p. 606

Final series sometimes faster, nearly trilled

p. 606

Same note type sometimes used in both series

Some songs are 3-parted

Second part may be trill or series

SONG : "Dick, dick CISS-CISS-CISS" p. 573

p. 608

Mostly Apr.–Aug., by males. Females not known to sing. Repertoire size unknown; most males repeat a single songtype for long periods, but at least some males can alternate 2 songtypes. Number of notes, and number of types of notes, may increase with excitement level, grading smoothly into Complex Song, sometimes given in flight. Not very musical.

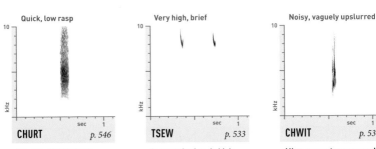

Quick, low rasp

CHURT p. 546

All year, often in flight, including by night migrants. Distinctive.

Very high, brief

TSEW p. 533

Apparently given in high alarm, possibly near nest.

Noisy, vaguely upslurred

CHWIT p. 531

All year; most common call. May sound vaguely 2-syllabled.

Northern Cardinal

Cardinalis cardinalis

A familiar, beloved resident of shrubby woods, deserts, and backyards. Often visits feeders. Eastern range has expanded over the past century.

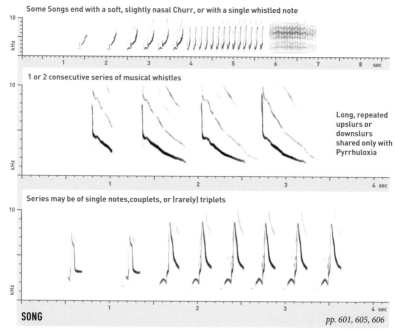

Some Songs end with a soft, slightly nasal Churr, or with a single whistled note

1 or 2 consecutive series of musical whistles

Long, repeated upslurs or downslurs shared only with Pyrrhuloxia

Series may be of single notes, couplets, or (rarely) triplets

SONG

pp. 601, 605, 606

Mostly Feb.–Aug., by both sexes. Males have 8–10 songtypes; female repertoire apparently similar. Females sometimes sing from the nest. Each Song usually repeated several times before switching, but Songs are often truncated. Highly variable; some Songs have 3 or more consecutive series, especially when birds are excited. Each series usually faster than the last, with later series often trills. First series sometimes accelerating.

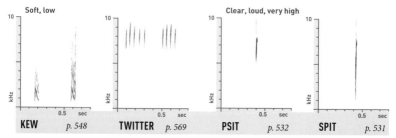

Soft, low

Clear, loud, very high

KEW *p. 548*   **TWITTER** *p. 569*   **PSIT** *p. 532*   **SPIT** *p. 531*

Kew given softly in family contact, and between mates during encounters with rival males. Twitter given in response to hawks or other potential predators; also sometimes in response to playback. Psit is most common call; grades into lower Spit, which is rarer, and possibly more aggressive.

PYRRHULOXIA

Cardinalis sinuatus

Found in scrubby southwestern deserts and arid backyards, often side by side with the related Northern Cardinal; the two species rarely interact.

Some Songs end with a soft, slightly nasal Churr, or with a single whistled note

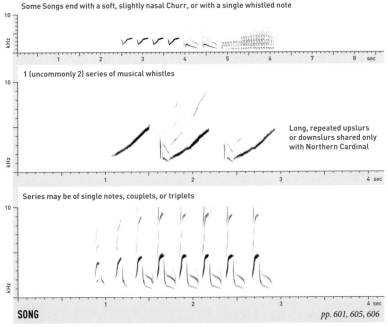

1 (uncommonly 2) series of musical whistles

Long, repeated upslurs or downslurs shared only with Northern Cardinal

Series may be of single notes, couplets, or triplets

SONG   *pp. 601, 605, 606*

Mostly Feb.–Aug., by males; females sing much less often than female Northern Cardinals. Males have 10–14 songtypes. Each Song usually repeated several times before switching, but Songs are often truncated. Songs generally simpler and slower than Northern Cardinal: usually one series, rarely faster than 5 notes/second, often just 2–3 notes total; but Northern Cardinal often difficult or impossible to rule out by Song alone.

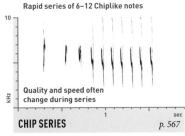

Rapid series of 6–12 Chiplike notes

Quality and speed often change during series

CHIP SERIES   *p. 567*

All year, by both sexes, in contact, alarm, and aggression. Quite plastic, but distinctive; Northern Cardinal rarely sounds similar.

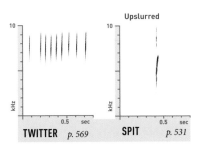

Upslurred

TWITTER *p. 569*   **SPIT** *p. 531*

Twitter like Northern Cardinal's. Spit plastic but generally lower, less sharp, more obviously whistled than cardinal's Psit.

ROSE-BREASTED GROSBEAK

Pheucticus ludovicianus

Breeds in deciduous and mixed forests, woodland edges, and residential areas with mature trees. Regularly visits feeders. Female has pale pinkish bill.

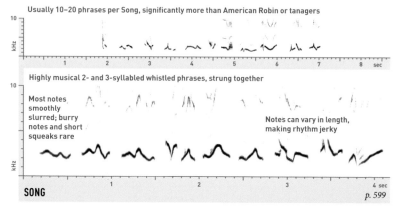

Usually 10–20 phrases per Song, significantly more than American Robin or tanagers

Highly musical 2- and 3-syllabled whistled phrases, strung together

Most notes smoothly slurred; burry notes and short squeaks rare

Notes can vary in length, making rhythm jerky

SONG　　*p. 599*

Mostly Apr.–July, by both sexes, sometimes from the nest. Individual birds use about 15–25 different phrases in their Songs. In excitement, Song becomes longer, sometimes including softer, varied portions with cardinal-like whistles. Reported to give Complex Song in flight display; no recordings available.

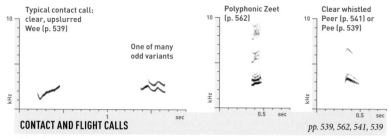

Typical contact call: clear, upslurred Wee (p. 539)

One of many odd variants

Polyphonic Zeet (p. 562)

Clear whistled Peer (p. 541) or Pee (p. 539)

CONTACT AND FLIGHT CALLS　　*pp. 539, 562, 541, 539*

A hugely variable and plastic set of calls given softly during the day, apparently by both sexes, in close pair contact. One common night flight call is a monotone whistled Pee like the version shown at right, like Hermit Thrush Teer but shorter; some nocturnal versions apparently a burry Veet.

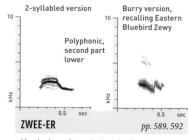

2-syllabled version

Burry version, recalling Eastern Bluebird Zewy

Polyphonic, second part lower

ZWEE-ER　　*pp. 589, 592*

Mostly June–Sept., by begging juveniles. Likely also sometimes by adult females in courtship. Quite variable.

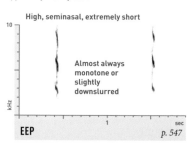

High, seminasal, extremely short

Almost always monotone or slightly downslurred

EEP　　*p. 547*

All year; most common call. Fairly uniform but slightly plastic.

Black-headed Grosbeak

Pheucticus melanocephalus

Breeds in riparian deciduous woods or drier coniferous forests. Sometimes hybridizes with Rose-breasted Grosbeak. Female has grayish, bicolored bill.

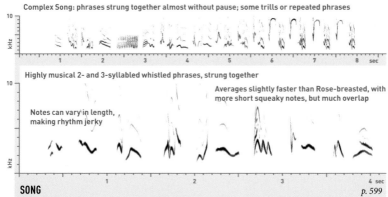

Complex Song: phrases strung together almost without pause; some trills or repeated phrases

Highly musical 2- and 3-syllabled whistled phrases, strung together

Averages slightly faster than Rose-breasted, with more short squeaky notes, but much overlap

Notes can vary in length, making rhythm jerky

SONG *p. 599*

Mostly Apr.–July, by both sexes, sometimes from the nest. Males have about 25 different phrases, females about 13. In excitement, Song becomes longer, sometimes including softer, varied portions with cardinal-like whistles. Complex Song infrequent, in flight display with exaggerated wingbeats.

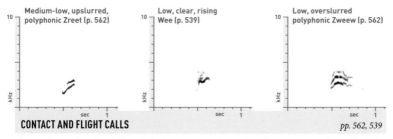

Medium-low, upslurred, polyphonic Zreet (p. 562)

Low, clear, rising Wee (p. 539)

Low, overslurred polyphonic Zweew (p. 562)

CONTACT AND FLIGHT CALLS *pp. 562, 539*

A poorly known, hugely variable and plastic category of calls. Versions like those shown given softly during the day, apparently by both sexes in close pair contact. Nocturnal flight calls little known, but likely similar to Rose-breasted Grosbeak's; perhaps more reliably upslurred.

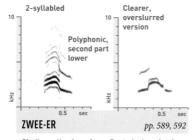

2-syllabled

Clearer, overslurred version

Polyphonic, second part lower

ZWEE-ER *pp. 589, 592*

Similar calls given June–Sept., by begging juveniles and sometimes by adult females, likely in courtship. Quite variable.

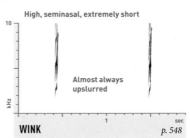

High, seminasal, extremely short

Almost always upslurred

WINK *p. 548*

All year; most common call. Fairly uniform, slightly plastic. Difference from Rose-breasted Eep is subtle but consistent.

INDIGO BUNTING

Passerina cyanea

Common in overgrown fields and wood-land edges. Nonbreeding males become mostly brown. Female like female Lazuli, but slightly streaked below.

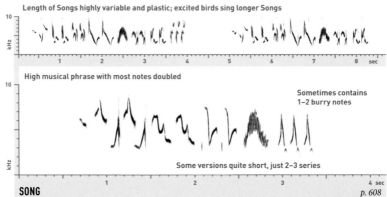

Length of Songs highly variable and plastic; excited birds sing longer Songs

High musical phrase with most notes doubled

Sometimes contains 1–2 burry notes

Some versions quite short, just 2–3 series

SONG *p. 608*

Apr.–Aug., by males only. Most adult males have a single songtype, which they repeat with minor variations, usually truncations; a few have 2 songtypes. Doubling of most phrases is fairly distinctive. Quality more musical than most warbler Songs; like American Goldfinch, but slower with shorter series.

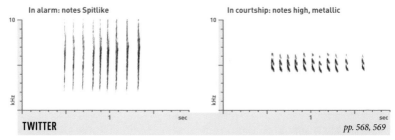

In alarm: notes Spitlike

In courtship: notes high, metallic

TWITTER *pp. 568, 569*

Twitter of Spitlike notes given mostly June–Aug., in high alarm near nest or young, usually interspersed among many single Spit calls. Twitter of polyphonic Tsewlike notes given Apr.–Aug., by females in close courtship, and by highly agitated males in territorial encounters, usually with fluttering wings.

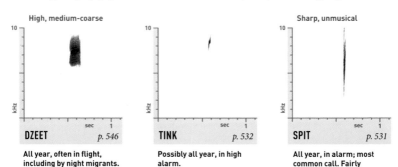

High, medium-coarse

Sharp, unmusical

DZEET *p. 546*

TINK *p. 532*

SPIT *p. 531*

All year, often in flight, including by night migrants. Rarely a Dzweet (p. 547).

Possibly all year, in high alarm.

All year, in alarm; most common call. Fairly uniform.

LAZULI BUNTING

Passerina amoena

Breeds in shrubby areas, riparian woods, and chapparal. Nonbreeding males much like breeding males, with some patchy brown above.

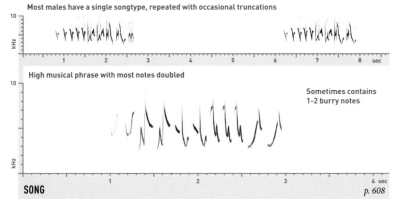

Most males have a single songtype, repeated with occasional truncations

High musical phrase with most notes doubled

Sometimes contains 1-2 burry notes

SONG   *p. 608*

Apr.–Aug., by males only. Very similar to Indigo Bunting's song, but averages higher and faster, and thus more likely to be mistaken for a warbler; somewhat more likely than Indigo Bunting to include longer series of 3–5 notes. Some versions quite short, just 2–3 series.

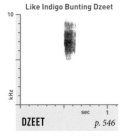

Like Indigo Bunting Dzeet

DZEET   *p. 546*

All year, often in flight. May average barely higher than Indigo's.

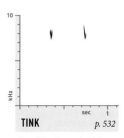

TINK   *p. 532*

Possibly all year, in high alarm.

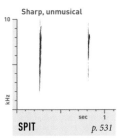

Sharp, unmusical

SPIT   *p. 531*

All year, in alarm; most common call. Much like Indigo Bunting's Spit.

SONGS OF INDIGO BUNTING, LAZULI BUNTING, AND HYBRIDS

The Songs of Indigo and Lazuli Bunting are very similar, but in areas where only one species occurs, the Songs contain only species-specific note types, and each species usually ignores playback of the other's Song. However, in areas where both species occur, they respond aggressively to each other's Songs, and some individuals sing with note types from both species. Hybrids, which occur frequently in the zone of overlap, may also sing with both species' note types; more study needed. For these reasons, it is frequently difficult or impossible to identify Indigo and Lazuli Buntings by sound alone, at least in areas where both occur.

VARIED BUNTING

Passerina versicolor

Breeds in dense arid scrub, in washes or adjacent to riparian areas. Females resemble other female buntings, but lack wingbars; bill slightly decurved.

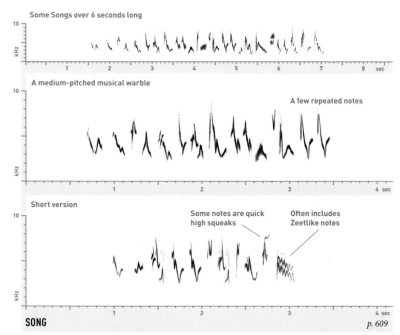

Some Songs over 6 seconds long

A medium-pitched musical warble

A few repeated notes

Short version

Some notes are quick high squeaks

Often includes Zeetlike notes

SONG   *p. 609*

Mostly Apr.–Aug., by males only. Males appear to have a single basic songtype, which they deliver with minor variations, usually truncations, at any time of day. Much like Blue Grosbeak Song, but slightly higher-pitched; combination of high squeaks and lower burry notes may recall western Warbling Vireo Song. Long, soft versions of Song reportedly given during courtship.

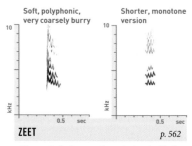

Soft, polyphonic, very coarsely burry

Shorter, monotone version

ZEET   *p. 562*

All year, by both sexes, in contact and while foraging, often mixed with Pinks or Song. Highly plastic, and not very loud.

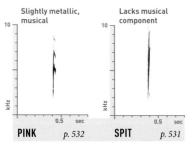

Slightly metallic, musical

Lacks musical component

PINK   *p. 532*   **SPIT**   *p. 531*

All year, by both sexes, in contact and alarm. Varies from Pink to Spit; slightly plastic. Higher Tinklike note reported in high alarm.

PAINTED BUNTING

Passerina ciris

Unmistakable. Found in semi-open brushy fields, coastal scrub, and sometimes residential areas. First-year males resemble females.

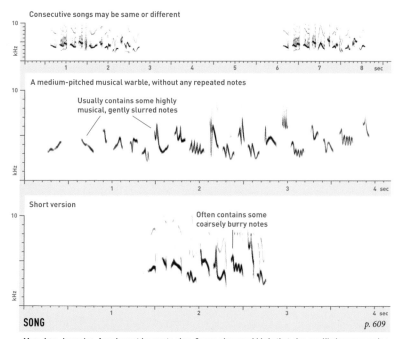

Consecutive songs may be same or different

A medium-pitched musical warble, without any repeated notes

Usually contains some highly musical, gently slurred notes

Short version

Often contains some coarsely burry notes

SONG *p. 609*

Mar.–Aug., by males; females not known to sing. Green-plumaged birds that sing are likely young males. Individuals estimated to have at least 3–5 songtypes, each of which may be delivered with minor variations, especially in the middle of the song. Length of Song quite variable; young birds may sing shorter songs on average. Similar to Varied Bunting Song, but most notes more musical, more obviously whistled; some versions recall House Finch Song.

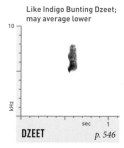

Like Indigo Bunting Dzeet; may average lower

DZEET *p. 546*

All year, likely including by night migrants, but apparently infrequent.

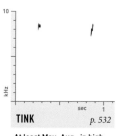

TINK *p. 532*

At least May–Aug., in high agitation. Plastic.

Pink: slightly metallic

PINK and SPIT *pp. 532, 531*

All year, in contact and alarm. Varies from Pink to Spit, like in Varied Bunting.

BLUE GROSBEAK

Passerina caerulea

Found in overgrown fields and scrub with scattered trees. Larger than Indigo Bunting, with chestnut wingbars and a much thicker, bicolored bill.

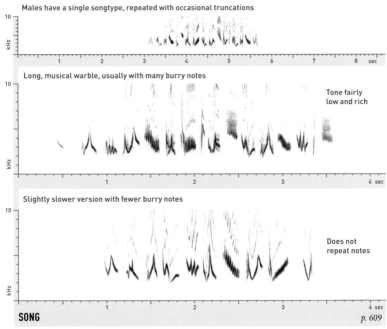

Males have a single songtype, repeated with occasional truncations

Long, musical warble, usually with many burry notes

Tone fairly low and rich

Slightly slower version with fewer burry notes

Does not repeat notes

SONG *p. 609*

Mostly Apr.–Sept., by males. Females not known to sing; brown-plumaged singers may be first-spring males. Older males average longer songs than younger males. Much like Songs of Varied Bunting and western Warbling Vireo, but lower-pitched; lacks squeaky notes. Lower, slower, and burrier than Song of Purple Finch; notes often slow enough to count.

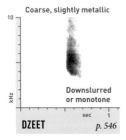

Coarse, slightly metallic

Downslurred or monotone

DZEET *p. 546*

All year, including by night migrants. Distinctly lower than Indigo Bunting Dzeet.

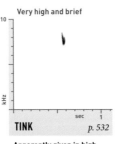

Very high and brief

TINK *p. 532*

Apparently given in high alarm and in response to playback.

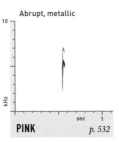

Abrupt, metallic

PINK *p. 532*

All year, by both sexes, in alarm; most common call.

INDEX

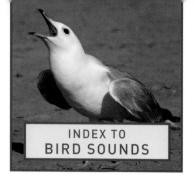

INDEX TO
BIRD SOUNDS

HOW TO USE THE VISUAL INDEX

This visual index lists every sound in the species accounts in a group with similar sounds. Each group is indicated by a name, a description, and a symbol that indicates roughly how the sounds in the group appear on the spectrogram. Page numbers in the index refer to the species accounts; page numbers in the species accounts refer to the index.

How to look up a sound

- **By characteristics of the sound:** Use the Quick Index on the inside back cover of the book. Find the description that best matches the sound, and then turn to the corresponding pages in the visual index.
- **By similarity with a known sound of a familiar species.** Turn to the species account of the familiar bird. Find the similar sound in the species account, and then use the page number listed with it to turn to the corresponding pages in the visual index.

Organization of the index

The index is split into seven major parts by sound pattern:
1. Single-note sounds (p. 530)
2. Sounds made of repeated similar notes (p. 565)
3. 2- and 3-syllabled phrases (p. 586)
4. Complex series (p. 593)
5. Songs of different separate notes, series, or phrases (p. 596)
6. Very long complex songs (p. 600)
7. Medium to long phrases and combinations (p. 603)

Within these categories, sounds are generally listed according to the following principles:
1. Short to long (or simple to complex)
2. High to low
3. Slow to fast

The basic tone qualities (p. 14) generally appear in the following order:

1. Ticking or mechanical
2. Whistled
3. Hooting, cooing, or growling
4. Burry or buzzy
5. Nasal
6. Harsh
7. Combinations of burry/buzzy, nasal, and harsh
8. Polyphonic

The basic pitch patterns (p. 7) generally appear in the following order:

1. Monotone
2. Upslurred
3. Downslurred
4. Overslurred
5. Underslurred

Bird sounds are complex and variable; many exceptions to these guidelines occur.

INDEX PART I: A SINGLE-NOTE SOUND

A single note given by itself, or repeated after a pause

A CLICK, TAP, OR SNAP
A nearly instantaneous, wholly unmusical note that shows on the spectrogram as a completely vertical line.

TICK
Extremely brief, noisy clicklike note without any change in pitch.

Soft
Mountain Plover, p. 133
Buff-breasted Sandpiper,
   p. 144

Louder, often in irregular series
Forster's Tern, p. 182
Violet-crowned Humming-
   bird, p. 116
Tree Swallow, p. 326

TOCK
Like Tick, but more resonant and sometimes more musical. Often quite loud.

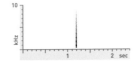

American Coot, p. 125
Common Gallinule, p. 124

Sora, p. 123

A CHIPLIKE NOTE
An extremely brief, extremely sharp note, slightly more musical than a Click.

WHIT
Extremely brief, sharply upslurred single note. Pitch medium to high.

Solitary Sandpiper, p. 151
Least Flycatcher, p. 271
Willow Flycatcher, p. 267
Dusky Flycatcher, p. 269
Gray Flycatcher, p. 270

Buff-breasted Flycatcher,
   p. 274

Lower
Wild Turkey (juvenile), p. 65

SPIT

Like Whit, but sharper, nearly a snapping sound.

Costa's Hummingbird, p. 110
Bushtit, p. 339
Rock Wren, p. 350
Northern Cardinal, p. 516

Pyrrhuloxia, p. 517
Indigo Bunting, p. 520
Lazuli Bunting, p. 521
Varied Bunting, p. 522

Painted Bunting, p. 523

CHWIT

Like Whit, but slightly noisier and longer.

White-breasted Nuthatch
(Pacific), p. 342
Gray Catbird (juvenile), p. 379
Pin-tailed Whydah, p. 394
Yellow-rumped Warbler
(Audubon's), p. 503
Dickcissel, p. 515

Sharper, more Spitlike
MacGillivray's Warbler,
p. 497

Averages squeakier
Northern Red Bishop, p. 395

Lower, mostly noisy
Eastern Bluebird, p. 369

Slightly more ringing
Brown-headed Cowbird,
p. 480

CHIP

An extremely brief, sharply downslurred note. Pitch medium to high.

Fairly sharp, rather high
Sedge Wren, p. 356
Ovenbird, p. 488
Tennessee Warbler, p. 491
Nashville Warbler, p. 495
Black-and-white Warbler,
p. 490
Townsend's Warbler, p. 506
Hermit Warbler, p. 507
American Redstart, p. 500
Yellow Warbler, p. 501
Grace's Warbler, p. 504

High, slightly more musical
Blue-throated Hummingbird,
p. 105
Gray Catbird (juvenile), p. 379
Rufous-winged Sparrow,
p. 431
Wilson's Warbler (juvenile),
p. 508

High, ringing/metallic
(*see also* Tsit, p. 533)

Lower, slightly musical
Rivoli's Hummingbird, p. 106
Lucifer Hummingbird, p. 107
Calliope Hummingbird, p. 111
Rufous Hummingbird, p. 112
Allen's Hummingbird, p. 113
Broad-tailed Hummingbird,
p. 114
Sharp-shinned Hawk, p. 201
Eastern Phoebe, p. 277
Swamp Sparrow, p. 453
Fox Sparrow (Slate-colored),
p. 457

Lower, fairly sharp
Marsh Wren (juvenile),
pp. 354–355

Fairly low, slightly complex
Scaled Quail, p. 67
Yellow-rumped Warbler
(Myrtle), p. 502
Black-throated Gray Warbler,
p. 505

Low, complex, noisy, Chitlike
Red-faced Warbler, p. 509

Very sharp, almost smacking
Anna's Hummingbird, p. 109
Bohemian Waxwing, p. 390
Vesper Sparrow, p. 440
Lincoln's Sparrow, p. 452
Savannah Sparrow, p. 461
Baird's Sparrow, p. 463
LeConte's Sparrow, p. 464
Nelson's Sparrow, p. 465

SMACK

Like Chip, but sharper, almost clicking, like the smacking of lips.

Rather high, thin
Winter Wren, p. 347
Ovenbird, p. 488
Fox Sparrow (Sooty), p. 455
Dark-eyed Junco, p. 458

Lower, slightly harsher, Chuklike
Brown Thrasher, p. 380

Slightly lower
Nelson's Sparrow, p. 465

TIP

Extremely brief whistled note with no change in pitch. Pitch medium.

McCown's Longspur, p. 422
Purple Finch, p. 402

Lawrence's Goldfinch, p. 408

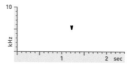

TEET

Brief whistled note with no change in pitch. Pitch medium. Slightly longer and more musical than Tip; more abrupt than Hee, with distinct consonant sounds.

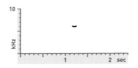

California Towhee, p. 429
Abert's Towhee, p. 430

(California Ground Squirrel call)

TINK

Like Tip (p. 531), but extremely high-pitched. More musical than Chip (p. 531) or Tsit. Most versions are quite plastic and variable.

Very high
Ovenbird, p. 488
Common Yellowthroat, p. 498
MacGillivray's Warbler, p. 497
Rufous-capped Warbler, p. 499
Black-and-white Warbler, p. 490
Townsend's Warbler (possibly), p. 506
Hermit Warbler, p. 507
American Redstart, p. 500
Yellow Warbler, p. 501
Yellow-rumped Warbler (Myrtle), p. 502
Yellow-rumped Warbler (Audubon's), p. 503
Grace's Warbler, p. 504
Black-throated Gray Warbler, p. 505
Wilson's Warbler, p. 508
Eastern Towhee, p. 424

Spotted Towhee, p. 425
Green-tailed Towhee, p. 426
California Towhee, p. 429
Abert's Towhee, p. 430
Rufous-winged Sparrow, p. 431
American Tree Sparrow, p. 434
Vesper Sparrow, p. 440
Lark Sparrow, p. 441
Five-striped Sparrow, p. 442
Black-throated Sparrow, p. 443
Sagebrush Sparrow, p. 444
Bell's Sparrow, p. 445
Lark Bunting, p. 447
White-throated Sparrow, p. 448
White-crowned Sparrow, p. 450
Golden-crowned Sparrow, p. 451

Song Sparrow, p. 454
Lincoln's Sparrow, p. 452
Swamp Sparrow, p. 453
Fox Sparrow (all forms), pp. 455–457
Dark-eyed Junco, p. 458
Yellow-eyed Junco, p. 460
Savannah Sparrow, p. 461
Baird's Sparrow, p. 463
LeConte's Sparrow, p. 464
Nelson's Sparrow, p. 465
Indigo Bunting, p. 520
Lazuli Bunting, p. 521
Painted Bunting, p. 523
Blue Grosbeak, p. 524

Slightly lower
Pacific-slope Flycatcher, p. 272
Cordilleran Flycatcher, p. 273

PSIT

Very high, very sharp upslurred whistle. Like Spit (p. 531), but higher.

Cassin's Sparrow, p. 432
Botteri's Sparrow, p. 433

Even sharper
Sharp-shinned Hawk, p. 201
Northern Cardinal, p. 516

PINK

Extremely brief single note, sharply upslurred at start, but main part of note monotone, slightly ringing, often metallic. Pitch medium to high. See also Teenk (p. 561).

Solitary Sandpiper, p. 151
Vesper Sparrow, p. 440
White-throated Sparrow, p. 448
White-crowned Sparrow, p. 450
Varied Bunting, p. 522
Painted Bunting, p. 523
Blue Grosbeak, p. 524

Averages longer
Harris's Sparrow, p. 449

Often higher, sharper, Spitlike
Northern Waterthrush, p. 489
Colima Warbler, p. 493
Lucy's Warbler, p. 494
Nashville Warbler, p. 495
Virginia's Warbler, p. 496
Rufous-capped Warbler, p. 499

High, metallic, Tinklike
American Tree Sparrow, p. 434

Lower, more musical
Red-winged Blackbird, pp. 470–471

PSIP

Like Psit, but more musical. Shorter than Pseep (p. 534).

Singly or in series
Boreal Chickadee, p. 336
Black-capped Chickadee,
  p. 334
Mountain Chickadee, p. 335

Chestnut-backed Chickadee,
  p. 337
Mexican Chickadee, p. 338

Slightly sharper
Lark Sparrow, p. 441

Very high
Golden-crowned Kinglet,
  p. 363

TSIT

Higher than Chip (p. 531), sharp, like an extremely high Smack (p. 531). Can be confused with Tink or with Psit.

Orange-crowned Warbler,
  p. 492
Lucy's Warbler, p. 494
Nashville Warbler, p. 495
Virginia's Warbler, p. 496
Black-chinned Sparrow,
  p. 435

Chipping Sparrow, p. 436
Clay-colored Sparrow, p. 437
Brewer's Sparrow, p. 438
Field Sparrow, p. 439
Grasshopper Sparrow, p. 462
Painted Redstart, p. 510

Slightly lower
Orange-crowned Warbler,
  p. 492,

TSWIT

Like Sreet (p. 536), but with stronger consonant sound at start; generally lower, sharper, and more Chiplike. See Tseet (p. 537), which is slightly longer.

Wilson's Warbler, p. 508

TSIP

Like Chip (p. 531), but higher. Sharper than Tsew. Some versions of Tink are similar.

Black-crested Titmouse,
  p. 330
Bridled Titmouse, p. 331
Oak Titmouse, p. 332

Juniper Titmouse, p. 333
Mexican Chickadee, p. 338
Dark-eyed Junco (Red-
  backed), p. 459

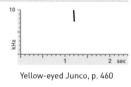

Yellow-eyed Junco, p. 460

TSEW

Like Tsip, but less sharp, more musical. Shorter than Tseew (p. 536).

Rufous-winged Sparrow,
  p. 431
American Tree Sparrow,
  p. 434
Lark Sparrow, p. 441
Dickcissel, p. 515

Slightly longer, more Seetlike
Brown Creeper, p. 346

Slightly longer, polyphonic (though this is difficult to hear)
Savannah Sparrow, p. 461
Clay-colored Sparrow, p. 437

Often in series
Black-throated Sparrow,
  p. 443

TSEWP

Like Tsew, but sharper, with lower and more abrupt ending. Less sharp and more musical than Chip (p. 531); sharper and less musical than High Tew (p. 540).

Rufous-winged Sparrow, p. 431

A PEEP- OR CHIRPLIKE NOTE
Very brief notes that start and end abruptly lower, creating the impression that they begin and/or end with the consonants "p" or "k."

PIP (WHISTLED)
Very brief, sharply overslurred whistle or seminasal note.

High, sharp
Black Swift, p. 117

Less sharp at start, almost a Wip
Willow Flycatcher, p. 267

Medium pitch
Black-bellied Plover, p. 135
Ash-throated Flycatcher,
p. 280
Scott's Oriole, p. 486

Pine Grosbeak, p. 401

Low, rarely single
Eared Grebe, p. 87
Red Knot, p. 141
Red-necked Phalarope,
p. 156

Medium pitch, seminasal
Scissor-tailed Flycatcher,
p. 290

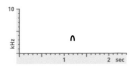

Western Kingbird, p. 288
Couch's Kingbird, p. 285

Briefer, slightly more nasal; compare Vimp, p. 557
Carolina Wren, p. 352

WIP (WHISTLED)
Like Pip, but with a weaker initial consonant sound. See Nasal Wip, p. 550.

Medium-high
Dusky-capped Flycatcher,
p. 279

Medium pitch
Great Crested Flycatcher,
p. 281
Brown-crested Flycatcher,
p. 282

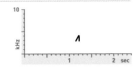

PEEP
Higher than Pip, and often longer. Clearer than Cheep.

Very high, short
Red Phalarope, p. 156

High, short
Buff-breasted Sandpiper,
p. 144
Long-billed Dowitcher,
p. 148
Alder Flycatcher, p. 266

Hammond's Flycatcher,
p. 268
Gray Flycatcher, p. 270
Pygmy Nuthatch, p. 345
Abert's Towhee, p. 430

Medium-high to high, slightly longer
Black-legged Kittiwake,
p. 162

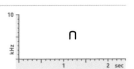

Black Oystercatcher, p 128
Semipalmated Plover, p. 130
Piping Plover, p. 131
Rock Sandpiper, p. 143
Wild Turkey (juvenile), p. 65

PWIK
Like Pip, but much sharper and less musical.

Olive Warbler, p. 392

Purple Finch, p. 402

PSEEP
Like Peep, but higher, less sharp, with weaker consonant sounds. Longer than Psip (p. 533).

Blue-winged Teal (male),
p. 46
Ring-necked Duck (male),
p. 53
Pigeon Guillemot, p. 157

Lewis's Woodpecker, p. 234
Vermilion Flycatcher, p. 278
Orange-cheeked Waxbill,
p. 394

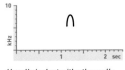

Usually in duet with other calls
Couch's Kingbird, p. 285

PIK

Like Pip, but sharper, with stronger consonant sounds at start and end. Lower than Peek.

Downy Woodpecker, p. 242
Ladder-backed Woodpecker,
    p. 240

Sharper, briefer
    Nuttall's Woodpecker, p. 241

Lower
    American Three-toed
        Woodpecker, p. 246

PEEK

Like Pik, but higher and/or longer.

Hairy Woodpecker, p. 243
Arizona Woodpecker, p. 244

KIP

Like Pik, but sharper and much briefer. Like Eep (p. 547), but with a much stronger consonant sound at start.

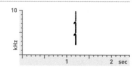

High
    Common Tern, p. 180
    Merlin, p. 255

Lower
    Black-backed Woodpecker,
        p. 247
    Boreal Owl, p. 224

CHIRP (TYPICAL)

A brief semimusical note, like Peep or Cheep but more complex. Pitch medium to high. Usually downslurred. Lacks polyphonic quality of Jirp (p. 563).

High, long, sometimes multisyllabled
    Ancient Murrelet, p. 158

Medium pitch, stereotyped
    Hutton's Vireo, p. 297
    Painted Redstart Cheer,
        p. 510

Medium pitch, stereotyped or plastic
    Buff-breasted Flycatcher,
        p. 274
    House Finch, p. 404
    House Sparrow, p. 393
    Gray-crowned Rosy-Finch,
        p. 398
    Brown-capped Rosy-Finch,
        p. 399

Black Rosy-Finch, p. 400
Pine Grosbeak, p. 401

Medium pitch, stereotyped or plastic, sometimes finely buzzy
    Tree Swallow, p. 326

Medium-low, finely buzzy
    Evening Grosbeak,
        pp. 410–412

CHEEP

High, plastic, often slightly screechy complex notes. Higher than Chirp; usually harsher and more complex than Peep. Higher than screechy Bark (p. 555).

High, plastic
    White-rumped Sandpiper,
        p. 145
    Horned Lark, p. 320

Medium-high, plastic
    Snowy Plover, p. 129
    Surfbird, p. 141
    Semipalmated Sandpiper,
        p. 147
    Western Sandpiper, p. 147
    Red Phalarope, p. 156
    Scaly-breasted Munia, p. 395

Medium-high, abrupt, often stereotyped
    Townsend's Solitaire, p. 370
    American Robin, pp. 376–377

Medium pitch, plastic
    Ruddy Turnstone, p. 140
    Gray-crowned Rosy-Finch,
        p. 398
    Brown-capped Rosy-Finch,
        p. 399
    Black Rosy-Finch, p. 400

Medium to low pitch, almost a Squeak or screechy Bark
    White-winged Scoter, p. 56
    Stilt Sandpiper, p. 142
    Rock Sandpiper, p. 143
    Buff-breasted Sandpiper,
        p. 144
    Red-necked Phalarope,
        p. 156

A PEEP- OR CHIRPLIKE NOTE, CONTINUED

SQUEEP
A high Cheeplike note, but clearer and more stereotyped. Often sounds upslurred.

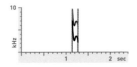

Dusky Flycatcher, p. 269
Sprague's Pipit, p. 397
American Robin, pp. 376–377

Some versions vaguely 2-syllabled
Barn Swallow, p. 324

A HIGH SEETLIKE NOTE
Very high musical whistled notes. Sometimes polyphonic or buzzy, though this can be difficult to hear.

SREET
Extremely high, short, slightly upslurred whistled or polyphonic note.

Tennessee Warbler, p. 491
Orange-crowned Warbler, p. 492
Colima Warbler, p. 493
Lucy's Warbler, p. 494
Nashville Warbler, p. 495
Virginia's Warbler, p. 496
MacGillivray's Warbler, p. 497
Townsend's Warbler, p. 506
Hermit Warbler, p. 507
Clay-colored Sparrow, p. 437
Brewer's Sparrow, p. 438

Slightly longer
Black-chinned Sparrow, p. 435
Vesper Sparrow, p. 440
White-crowned Sparrow, p. 450

Sometimes slightly buzzy (though this can be difficult to hear); see Dzit, p. 538
Ovenbird, p. 488

Slightly lower
Yellow-rumped Warbler (Myrtle), p. 502

Yellow-rumped Warbler (Audubon's), p. 503

Even lower, longer
Pacific-slope Flycatcher, p. 272
Cordilleran Flycatcher, p. 273

SEET (NOT BURRY)
Extremely high, nearly monotone whistled or polyphonic note.

Short, polyphonic (though this can be difficult to hear)
Grace's Warbler, p. 504
Black-throated Gray Warbler, p. 505

Short, not polyphonic
Blue-throated Hummingbird, p. 105

Fairly long
Grasshopper Sparrow, p. 462

Like Grasshopper Sparrow, but often polyphonic (though this can be difficult to hear)
Nelson's Sparrow, p. 465

Similar but even longer, lower, polyphony easier to hear
Northern Mockingbird (juvenile), p. 387
Bewick's Wren, p. 353

Long, overslurred
Five-striped Sparrow, p. 442

Very long, very high, not polyphonic
Cedar Waxwing, p. 390

Highly variable
Rufous-winged Sparrow, p. 431

TSEEW
Extremely high, short, slightly downslurred whistled or polyphonic note. Like Tsew (p. 533), but slightly longer.

Broad-billed Hummingbird, p. 115
Brown Creeper, p. 346
Field Sparrow, p. 439

More downslurred, often doubled
Pine Siskin, p. 406

Longer, more monotone, usually polyphonic (though this can be difficult to hear)
LeConte's Sparrow, p. 464

TSEET
Extremely high, short, slightly underslurred whistled or polyphonic note. Like Seet, but with stronger consonant sound at start. Shorter than Tseereet.

Shortest
Brewer's Sparrow, p. 438
Baird's Sparrow, p. 463

Slightly longer, with particularly strong consonant sounds
American Tree Sparrow, p. 434
Chipping Sparrow, p. 436
Dark-eyed Junco (Red-backed), p. 459
Yellow-eyed Junco, p. 460

Slightly longer still, with weaker consonant sounds
American Redstart, p. 500
Northern Red Bishop, p. 395

Even longer, more monotone, with weaker consonant sounds
Black-chinned Sparrow, p. 435
White-throated Sparrow, p. 448
Golden-crowned Sparrow, p. 451

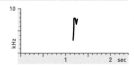

Harris's Sparrow, p. 449
Song Sparrow, p. 454
Fox Sparrow (all forms), pp. 455–457

Very long, clear
Rufous-crowned Sparrow, p. 427
Canyon Towhee, p. 428
California Towhee, p. 429
Abert's Towhee, p. 430

TSWEET
High, rising complex note; starts with consonant, sounds squeaky.

Hutton's Vireo, p. 297
Ovenbird, p. 488

Northern Waterthrush, p. 489

SREE
Very high upslurred whistle.

Long, very high
Veery, p. 372

Short, very high
Pigeon Guillemot, p. 157

Short, medium high
Cordilleran Flycatcher, p. 273

SEER
Extremely high, long, downslurred whistle.

Swainson's Thrush (Russet-backed), p. 374
Swainson's Thrush (Olive-backed), p. 375
Hermit Thrush, p. 373
American Robin, pp. 376–377
Bohemian Waxwing, p. 390

Sometimes in series
Black-crested Titmouse, p. 330

Lower, often longer
Bobolink, p. 468
Red-winged Blackbird, pp. 470–471

Finely burry
Pigeon Guillemot, p. 157

VERY HIGH OVERSLURRED WHISTLE
Like Seet, but longer, and distinctly overslurred.

Costa's Hummingbird Dive Display, p. 110

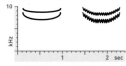

TSEEREET
Like Tseet, but longer, and often more noticeably underslurred.

Clear to finely buzzy
Eastern Towhee, p. 424

Spotted Towhee, p. 425
Green-tailed Towhee, p. 426

TSIPT
Very brief coarse note made of 2–3 high sharp Tsit- or Tsiplike notes. Shorter, coarser, and less musical than Dzit.

Dark-eyed Junco), pp. 458–459
Grasshopper Sparrow, p. 462

Painted Redstart, p. 510

A HIGH SEETLIKE NOTE, CONTINUED

DZIT

Very short, very high buzzy note. Shorter and usually finer than Burry Seet.

Yellow Warbler, p. 501
Dark-eyed Junco,
pp. 458–459

Black-and-white Warbler,
p. 490
Red-faced Warbler, p. 509

Often slightly rising; see Sreet,
p. 536
Northern Waterthrush, p. 489

Slightly lower
Bushtit (usually in rapid
series), p. 339

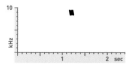

Lower still
Common Yellowthroat, p. 498

TSEERP

Like Tseet (p. 537), but longer, burry, and downslurred.

American Robin, pp. 376–377

SEET (BURRY)

Very coarse, very high, musical burry whistles. Longer than Dzit.
Compare Very High Trills (p. 572).

Mexican Chickadee, p. 338
Ruby-crowned Kinglet
(juvenile), p. 364
Brown Creeper, p. 346
American Robin, pp. 376–377

Gray Catbird (juvenile),
p. 379
White-throated Sparrow
Tseet, p. 448

Rhythm sometimes stuttering
Golden-crowned Kinglet,
p. 363

A WHISTLE

Medium- to high-pitched, rather musical notes.

VARIABLE SINGLE WHISTLES

These species give a huge variety of single whistled notes, usually 1-syllabled, which vary individually and geographically. In this group, consecutive whistles are usually the same. See also the group "Mostly Single 1-Syllabled Notes," p. 596, in which consecutive sounds are usually different.

Medium-low, brief, usually
monotone or downslurred
Scott's Oriole Songlike Calls,
p. 486
Baltimore Oriole Songlike
Calls, p. 484

Bullock's Oriole Songlike
Calls, p. 485

Medium-low, usually monotone
or upslurred, slightly polyphonic
Pine Grosbeak, p. 401

Medium to high, usually
monotone or downslurred
Red-winged Blackbird Alert
Calls, pp. 470–471

HEE

A brief clear, ringing monotone whistle, without any strong consonant sounds. Less abrupt than Teet (p. 532).

High, breathy, often preceded or
followed by rustling or quacking
Gadwall, p. 48
Mallard, p. 52
Mexican Duck (likely), p. 53
Northern Pintail, p. 50
Ruddy Duck (female), p. 62

Breathy, slightly downslurred
Anna's Hummingbird Dive
Display, p. 109

High, extremely plastic
Sabine's Gull, p. 163

High, clear, ringing
Townsend's Solitaire, p. 370

High, slightly burry
Green-winged Teal, p. 51

Lower, slightly burry
Northern Pintail, p. 50

Even lower, fairly mellow
Lesser Prairie-Chicken,
p. 77
Pied-billed Grebe, p. 90

PEE

Like Hee, but with stronger consonant sound at start. See also *Pip or Peep Series (p. 570).*

Short-billed Dowitcher,
  p. 148

Rose-breasted Grosbeak,
  p. 518

A LONGER MONOTONE WHISTLE

Long clear monotone whistles, like Hee but longer.

High, thin
  Common Grackle, p. 478
  Brown-headed Cowbird,
    p. 480
  Bronzed Cowbird, p. 481

Medium-high
  Western Wood-Pewee, p. 263
  Say's Phoebe, p. 275
    (occasionally truncated songs
    from other species)

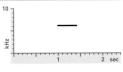

Low, mellow, highly plastic
  Western Screech-Owl, p. 226
  Whiskered Screech-Owl,
    p. 228

WEET

Like Whit (p. 530), but higher and more musical. Higher and much sharper than Wee.

American Pipit, p. 396

PSEWEET

Like Weet, but slightly longer; vaguely 2-syllabled and/or slightly underslurred.

Pacific-slope Flycatcher, p. 272

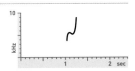

WEE

A short, slightly upslurred whistle, sometimes nearly monotone. Medium pitch.

Medium-high
  Swainson's Thrush
    (Russet-backed), p. 374
  Swainson's Thrush
    (Olive-backed), p. 375

Olive Warbler, p. 392
Rose-breasted Grosbeak,
  p. 518
Black-headed Grosbeak,
  p. 519

Medium-low
  Scott's Oriole, p. 486

WERT

A short upslurred whistle, lower and sharper than Wee. Longer and more musical than Pwut (p. 554).

Mountain Plover, p. 133
Snowy Plover, p. 129

Phainopepla, p. 391
Lark Bunting, p. 447

PWEE OR DEET

Like Wee, but with stronger consonant sound at start; often slightly underslurred.

Short
  Alder Flycatcher, p. 266

Longer, higher, shriller
  Killdeer Deet, p. 132

Low, variable
  American Golden-Plover,
    p. 134
  Pacific Golden-Plover, p. 134

A WHISTLE, CONTINUED

WEEP
Like Wee (p. 539), but with stronger consonant sound at end. Longer and higher than Wip (p. 534).

*Medium-low, often in series
(Weep-weep)*
  LeConte's Thrasher, p. 383

Medium-high
  Dusky-capped Flycatcher,
    p. 279

Great Crested Flycatcher,
  p. 281
Olive-sided Flycatcher,
  p. 265

*High, often in series
(Weep-weep)*
  Hudsonian Godwit, p. 139

Baird's Sandpiper, p. 145
Stilt Sandpiper, p. 142

Very high, shrill
  Sharp-shinned Hawk, p. 201

A LONGER UPSLURRED WHISTLE
A rising whistle, longer than Weet or Weep.

Piping Plover Weet, p. 131
Whimbrel Curree, p. 137

More nasal
  European Starling,
    pp. 388–389

Great-tailed Grackle,
  pp. 474–475

*Sometimes broken. See also
Rising Broken Nasal Note
(p. 553).*
  Snowy Plover Ter-weet, p. 129

TEW (HIGH)
A high, sharply downslurred whistle; less sharp and more musical than Chip (p. 531) or Tsewp (p. 533).

Very high, sharp
  Northern Goshawk, p. 203

*Longer, with subtle break in
middle*
  Verdin, p. 340

High, not very sharp
  Black Phoebe, p. 276

Violet-green Swallow, p. 327
Pygmy Nuthatch, p. 345
Golden-crowned Sparrow,
  p. 451

*Lower, varying in pitch and
sharpness*
  White-tailed Kite, p. 200

Sharp-shinned Hawk, p. 201
Osprey, p. 211
Golden Eagle, p. 210

TEW (LOW)
Like sounds in the preceding group, but lower and more musical.

Purple Martin, p. 328

Sharper
  Northern Bobwhite, p. 70
  Mountain Bluebird, p. 367
  Western Bluebird, p. 368

Less sharp
  Northern Mockingbird,
    p. 387
  Brown Thrasher, p. 380
  Bendire's Thrasher, p. 381

TEER
Musical downslurred whistle, longer and less sharp than Tew.

*Quite high, long, often vaguely
multisyllabic*
  Lawrence's Goldfinch, p. 408
  Lesser Goldfinch, p. 409

High, thin, mostly rather sharp
  Black Phoebe, p. 276
  Horned Lark, p. 320
  Olive Warbler, p. 392

Chestnut-collared Longspur,
  p. 421
Lapland Longspur, p. 420
Snow Bunting, p. 419

Medium pitch, nearly monotone
  Hermit Thrush, p. 373

Low, loud, often nearly monotone
  Western Meadowlark, p. 473
  Scott's Oriole, p. 486
  (Songlike Calls of some
    orioles)

PEW

Like Tew, but with different consonant sound at start.

Black-bellied Whistling-Duck, p. 37
Lesser Yellowlegs, p. 153
Stilt Sandpiper, p. 142
Western Bluebird, p. 368

Gray-crowned Rosy-Finch, p. 398
Brown-capped Rosy-Finch, p. 399
Black Rosy-Finch, p. 400

PEER

Lower and longer than Tew, much less sharp, with different consonant sound at start.

Gray Flycatcher, p. 270
Buff-breasted Flycatcher, p. 274
Evening Grosbeak (Types 1, 2, 4, 5), pp. 410–412
Rose-breasted Grosbeak, p. 518

Much longer, slightly higher
Black-bellied Plover, p. 135
Say's Phoebe, p. 275

Long, quite high
Northern Beardless-Tyrannulet, p. 262

Quite low
Swainson's Thrush (Russet-backed), p. 374

A LONGER DOWNSLURRED WHISTLE

A long downslurred whistle, longer than Peer or Teer, usually without strong consonant sounds.

High, clear, variable
European Starling, pp. 388–389

Great-tailed Grackle, pp. 474–475

Low, mellow, not very loud
Montezuma Quail, p. 71

PWEEW

Longer, distinctly overslurred version of Pip (p. 534); lower than Pseep (p. 534).

Alder Flycatcher, p. 266
Willow Flycatcher, p. 267

Yellow-billed Magpie, p. 315
Harris's Sparrow, p. 449

PWEER

Like Pweew, but more drawn out at end.

Alder Flycatcher, p. 266

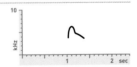

WEEW (WHISTLED)

Medium-high overslurred whistle, more drawn out than Pweew. Clearer than Wheew (p. 542). Has less polyphonic quality than Zweew (p. 562). Sometimes vaguely multisyllabic.

Purple Finch, p. 402
Cassin's Finch, p. 403

Summer Tanager, p. 513

A LONGER OVERSLURRED WHISTLE

A long overslurred whistle, longer than Weew, usually without strong consonant sounds. See also Mellow Wail Series (p. 575).

Medium-high, clear
Dusky-capped Flycatcher Weer, p. 279

High, often broken
Black-bellied Whistling-Duck Sweeoo, p. 37

Variable, often mellow
Gray Jay, p. 304

PEWY

Brief underslurred whistle, usually with consonant at start. Medium pitch.

American Golden-Plover,
  p. 134

Alder Flycatcher, p. 266
Summer Tanager, p. 513

PYOOWEE OR TEWEE (WHISTLED)

Like Pewy, but longer. See also Seminasal Tewee, p. 551.

Medium-low, long (Pyoowee)
  Black-bellied Plover, p. 135
  Long-billed Curlew, p. 136

High, thin (Tewee)
  Barn Swallow, p. 324
  Tree Swallow, p. 326
  Violet-green Swallow, p. 327

VARIABLE SHORT BREATHY WHISTLES

A catch-all category of variable Wheewlike notes, including short low Whews and Whips, upslurred Wheeps, downslurred Wheers, and burry Whirrs. Each species listed here gives several different sounds in this category.

Fulvous Whistling-Duck,
  p. 36

Black-bellied Whistling-
  Duck, p. 37
Wood Duck, p. 45

Greater Scaup, p. 54
Lesser Scaup, p. 54
Ring-necked Duck, p. 53

WHEEP

High, breathy, nasal upslur. Longer and more nasal than Weep (p. 540).

Short
  Mandarin Duck, p. 44

Long
  Wood Duck, p. 45

WHEEW

High, breathy, nasal or seminasal overslur; like Wow, but much higher. Shriller or more nasal than seminasal Weew (p. 551).

High, variable, clear to shrill
  Northern Harrier, p. 199

High, sometimes burry at start or end
  Wood Duck, p. 45

Eurasian Wigeon, p. 49

Medium pitch
  Northern Pintail, p. 50

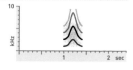

Quite variable
  Black-bellied Whistling-
    Duck, p. 37

A LOW-PITCHED NOTE

A low whistle, such as a Whoop, Hoot, or Coo, or a low burry or harsh note, such as a Growl or Groan.

OOIT

A low, mellow, upslurred whistle, longer than Pwut (p. 554), very slightly nasal.

Sora, p. 123

Softer, shorter, slightly lower
  Boreal Owl, p. 224

Even shorter
  Long-eared Owl, p. 222
  Greater Prairie-Chicken,
    p. 76

WHOOP

A low, mellow whistle or slightly nasal cry, medium-long.

Usually loud
Tundra Swan, p. 43

Loud, vaguely 2-syllabled
Mountain Quail Song, p. 66

Rather soft
Common Loon, p. 184
Yellow-billed Loon, p. 185
Pacific Loon, p. 186

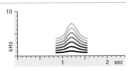

WOW

Longer than Wheew; lower, more nasal, and often longer than Whoop.

Redhead, p. 55

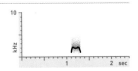

WHUP

Low, short whistled note, often slightly complex, like a low-pitched Chirp (p. 535). Usually plastic.

Slightly screechy
Canvasback, p. 55

HOOT/COO

Low, clear, medium to long notes.

Very deep, monotone
Dusky Grouse Hoomch, p. 82
Sooty Grouse Hoomch, p. 83

Higher, nearly monotone
Sharp-tailed Grouse, pp. 74–75
Long-eared Owl, p. 222
Flammulated Owl, p. 229

Upslurred
Common Ground-Dove, p. 95
Great Gray Owl, p. 219

Overslurred
Snowy Owl, p. 214

Overslurred, broken
Mourning Dove Growl Song, p. 96

Downslurred
Barred Owl Hoo-wah, p. 220

GRIP

Like Pwut (p. 554), but burry. Mostly audible at close range.

Mexican Whip-poor-will, p. 103

Common Poorwill, p. 104

GROWL

A very low, nearly monotone burry note. Usually audible only at close range.

Greater Roadrunner, p. 100

Mexican Whip-poor-will, p. 103

GROAN

A very low, nearly monotone, partly noisy note. Usually audible only at close range. See also Croaking Groan (p. 561).

Short, very low
Montezuma Quail, p. 71
Common Nighthawk, p. 101

Long, higher, more nasal
Emperor Goose, p. 38

Domestic Chicken, p. 63
Greater Sage-Grouse, p. 78
Gunnison Sage-Grouse, p. 79
American White Pelican, p. 188

Brown Pelican, p. 188
Little Blue Heron, p. 194

A BURRY OR BUZZY NOTE

A note that rises and falls very rapidly in pitch, creating a trilled effect. More musical versions are called burry; less musical versions are called buzzy. Both burry and buzzy notes can be either coarse (meaning the up-and-down changes in pitch are slower) or fine (meaning the changes are faster).

TREMOLO

A musical, mellow whistled note that repeatedly rises and falls in pitch, slowly enough that each change in pitch is clearly audible.

Medium-high, seminasal
Long-billed Curlew, p. 136
Least Tern, p. 176

Low, monotone or upslurred
Common Loon, p. 184
Yellow-billed Loon, p. 185

Even lower, long, downslurred
Eastern Screech-Owl, p. 227

BREET

A medium-high, coarsely burry, nearly monotone musical note. Can also be considered a short trill.

Often slightly rising
Pectoral Sandpiper, p. 144
Baird's Sandpiper, p. 145
Least Sandpiper, p. 146

Higher, usually downslurred
Western Sandpiper, p. 147

High, coarse, musical
Sandhill Crane (juvenile), p. 126

Slightly lower, less musical
Great Crested Flycatcher,
p. 281
Evening Grosbeak (all types),
pp. 410–412
Bronze Mannikin, p. 394

Slightly lower, often slightly downslurred
Ash-throated Flycatcher,
p. 280

BURRT

Like Breet, but lower.

Montezuma Quail, p. 71
Hammond's Flycatcher,
p. 268
Dusky Flycatcher, p. 269
Least Flycatcher, p. 271
Gray Flycatcher, p. 270
Pacific-slope Flycatcher,
p. 272
Cordilleran Flycatcher,
p. 273
Buff-breasted Flycatcher,
p. 274

Great Crested Flycatcher,
p. 281
Brown-crested Flycatcher,
p. 282
Loggerhead Shrike, p. 292
Northern Shrike, p. 293
Mountain Bluebird, p. 367

Lower, less musical, Churtlike
Snowy Plover, p. 129
Pectoral Sandpiper, p. 144
Eurasian Skylark, p. 321
Bendire's Thrasher, p. 381

Curve-billed Thrasher, p. 385
Yellow-headed Blackbird,
p. 479

Even lower, quite musical
Greater Pewee, p. 264
Purple Martin, p. 328
Crissal Thrasher, p. 384
Lark Bunting, p. 447

A LONG BURRY NOTE

A long, burry, monotone musical note. Can also be considered a trill.

Varied Thrush Song, p. 371

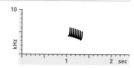

BEERT

Like Burrt, but downslurred. See also Downslurred Trill, p. 574.

Carolina Wren, p. 352
House Wren, p. 349

Sage Thrasher, p. 386

BREER
Like Burtt, but longer and distinctly overslurred. Coarser and lower than Vreer.

Dusky-capped Flycatcher,
  p. 279
Brown-crested Flycatcher,
  p. 282

Alder Flycatcher, p. 266
Couch's Kingbird, p. 285
Cassin's Kingbird, p. 286

Short, coarse, plastic
  Olive-sided Flycatcher, p. 265

VEET
Finely burry musical monotone note.

Medium-high
  Canyon Wren, p. 351

Slightly lower
  Veery, p. 372
  Eastern Meadowlark, p. 472

VEW
Like Veet, but sharply downslurred.

Medium pitch
  Lapland Longspur, p. 420
  Chestnut-collared Longspur,
  p. 421

High, sharply downslurred
  Horned Lark, p. 320
  Lawrence's Goldinch, p. 408

VEER (TYPICAL)
Like Vew, but longer.

Hutton's Vireo, p. 297
Yellow-billed Magpie, p. 315

Purple Martin, p. 328
Veery, p. 372

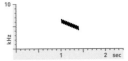

VEER (LONG)
Like Veer, but lower and longer. Medium-low pitch. Occasionally broken into a few similar but shorter whistles in slightly falling series.

Montezuma Quail Male Song, p. 71
Allen's Hummingbird Dive Display (final dive sound), p. 113

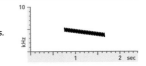

VREER
Like Veet, but overslurred. Finer and higher than Breer.

Shorter, slightly lower
  Gray-crowned Rosy-Finch,
  p. 398

Brown-capped Rosy-Finch,
  p. 399
Black Rosy-Finch, p. 400

VEWY
Like Pewy, but finely burry.

Veery, p. 372

Summer Tanager, p. 513

CHEET
Like Breet, but less musical. Coarser than Dzeet (p. 546).

Medium-high, slightly musical
  Western Sandpiper, p. 147
  Loggerhead Shrike, p. 292
  Northern Shrike, p. 293
  Northern Rough-winged
  Swallow, p. 325

Tree Swallow, p. 326
Violet-green Swallow, p. 327
Brown-headed Cowbird
  (juvenile), p. 480

A BURRY OR BUZZY NOTE, CONTINUED

CHURT
Like Burrt (p. 544), but less musical. Coarser than Dzert. Lower than Cheet (p. 545). Higher than Rasp (p. 560).

Medium-low
Cliff Swallow, p. 323
Bank Swallow, p. 325
Northern Rough-winged
  Swallow, p. 325

Nearly toneless
Dickcissel, p. 515
Snow Bunting, p. 419

Low, coarse, rough
Sedge Wren, p. 356
Gray-crowned Rosy-Finch,
  p. 398

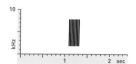

Brown-capped Rosy-Finch,
  p. 399
Black Rosy-Finch, p. 400
Common Yellowthroat, p. 498

CHEWT
Like Churt, but distinctly downslurred.

Low, fine, harsh
Northern Mockingbird,
  p. 387

Higher, longer
Green Kingfisher, p. 232
Verdin, p. 340

Gray-crowned Rosy-Finch,
  p. 398
Brown-capped Rosy-Finch,
  p. 399
Black Rosy-Finch, p. 400

High, short
Pacific Wren, p. 348

SHEET
A semimusical, coarsely burry Chirplike note with a distinctive lisping or sibilant quality; more complex and musical than Cheet (p. 545).

Tree Swallow, p. 326
Northern Rough-winged
  Swallow, p. 325

House Sparrow, p. 393

DZEET
Like Dzit (p. 538), but longer. Lower and more abrupt than Burry Seet (p. 538). Usually nearly monotone. Shorter than Buzz.

Medium-high, fine
Rufous Hummingbird, p. 112
Allen's Hummingbird, p. 113
Swamp Sparrow, p. 453
Lincoln's Sparrow, p. 452

Medium pitch, fairly coarse
Indigo Bunting, p. 520
Lazuli Bunting, p. 521
Painted Bunting, p. 523

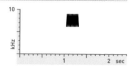

Medium-low, fairly coarse
Blue Grosbeak, p. 524

DZERT
Like Dzeet, but lower.

Bewick's Wren, p. 353

Eastern Meadowlark, p. 472

BUZZ
One long monotone buzzy note. Longer than Dzeet or Dzert.

Loggerhead Shrike, p. 292
Northern Shrike, p. 293
Common Redpoll, p. 405

Longer, sometimes with low consonant at start
Spotted Towhee, p. 425

High, insectlike, with introductory notes that can be inaudible at a distance
Grasshopper Sparrow, p. 462
LeConte's Sparrow, p. 464
Nelson's Sparrow, p. 465

DZIK

Very brief buzzy note, medium in pitch, often slightly nasal. Higher and less musical than Grip (p. 543).

Very brief
Western Grebe, p. 86
Clark's Grebe, p. 85
Common Nighthawk, p. 101

Brief, with simultaneous gulp
Mandarin Duck, p. 44

Longer, variable
Bewick's Wren, p. 353

DZWEET

Like Dzert, but upslurred. Sometimes in series; see Dzik Series (p. 576).

Medium coarse, soft, high
Savannah Sparrow, p. 461
Indigo Bunting, p. 520

Finer, higher
Black-and-white Warbler,
p. 490

Fine, soft, but harsher, more snarling
Vesper Sparrow, p. 440
Lark Sparrow, p. 441
Sagebrush Sparrow, p. 444
Bell's Sparrow, p. 445

Song Sparrow, p. 454
Swamp Sparrow, p. 453
Lincoln's Sparrow, p. 452
Lark Bunting, p. 447

DZREE

Like Dzweet, but longer and less obviously upslurred.

Coarse; upslur slight
Western Wood-Pewee, p. 263

DZEER

Like Dzert, but downslurred.

Usually given singly; very high
Broad-billed Hummingbird,
p. 115
Eastern Kingbird, p. 289

Usually given singly; longer, lower, coarser
Western Wood-Pewee, p. 263

Usually in series or mixed with other calls
Alder Flycatcher, p. 266

Couch's Kingbird, p. 285
Scissor-tailed Flycatcher,
p. 290

PEENT

Medium-pitched, extremely fine buzzy note with some nasal quality.

Downslurred, often 2-syllabled
Common Nighthawk, p. 101

Lower, longer
Royal Tern, p. 178
Forster's Tern, p. 182

A SHORT NASAL NOTE
A brief seminasal or nasal note, broken or unbroken.

EEP

Extremely brief seminasal to nasal notes with almost no change in pitch. Pitch medium-high. Consonant sounds weak or absent. The nasal version of Tip (p. 531).

Soft, often repeated
Black-chinned Hummingbird,
p. 108
Red-breasted Nuthatch,
p. 341

White-breasted Nuthatch (all
populations), pp. 342–344
Wrentit, p. 365

Loud, quite high
Rose-breasted Grosbeak,
p. 518

IP

Extremely brief seminasal to nasal notes. Sharply downslurred, though this can be difficult to hear. The nasal version of Chip (p. 531).

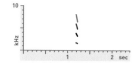

Medium-high
Black-chinned Hummingbird, p. 108

Medium-low
Bronzed Cowbird, p. 481

WINK

Extremely brief seminasal to nasal notes. Sharply upslurred, though this can be difficult to hear. The nasal version of Whit (p. 530).

Sanderling, p. 142

Black-headed Grosbeak, p. 519

WIK (HIGH)

Like Wip (p. 534), but seminasal to fairly nasal. Briefer and sharper than High Reek (p. 550).

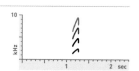

High, clear
Least Grebe, p. 91
Long-eared Owl (juvenile), p. 222
Rosy-faced Lovebird, p. 258

Medium-high, more nasal
Eared Grebe, p. 87

Lower, fairly nasal
Black-necked Stilt, p. 127

Black-throated Magpie-Jay, p. 303

KEW

Like Pew (p. 541), but seminasal to fairly nasal. Briefer and sharper than high Keer.

Medium pitch, sharp
Sora, p. 123
Boreal Owl, p. 224

Higher, less sharp
Semipalmated Plover, p. 130
Arctic Tern, p. 181
Bonaparte's Gull, p. 163

Slightly lower, even less sharp
Forster's Tern, p. 182

Even lower and less sharp
Northern Saw-whet Owl, p. 225
Boreal Owl, p. 224
Elf Owl, p. 229

Fairly low, fairly sharp
Black Tern, p. 183

Quite low, mellow
Black Rail, p. 120

Eastern Screech-Owl, p. 227
Western Screech-Owl, p. 226
Whiskered Screech-Owl, p. 228
Flammulated Owl, p. 229
Northern Cardinal, p. 516

SQUEAK

A brief, broken, seminasal note. Higher than Yelp (see Yelp Series, p. 578); higher and clearer than Screechy Barks (p. 555).

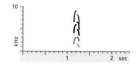

High, brief
Least Bittern, p. 192
Virginia Rail, p. 121
Canyon Towhee, p. 428

Tricolored Blackbird, p. 469

Longer, sharply overslurred
California Quail, p. 68

CHIRP (SQUEALING)

A high, rapid, complex seminasal note, usually broken.

Rose-ringed Parakeet, p. 258
Monk Parakeet, p. 258
Mitred Parakeet, p. 259
Red-masked Parakeet, p. 259

Blue-crowned Parakeet, p. 259
Yellow-chevroned Parakeet, p. 259
Lilac-crowned Parrot, p. 260

KLEE
A brief high whistled or seminasal note, starting at one pitch, then almost immediately breaking upward to a monotone pitch.

American Avocet, p. 128 American Aplomado Falcon, p. 255

KLIP
A brief upslurred nasal note that breaks upward. Lower than Klee, with more abrupt end. Medium pitch.

Burrowing Owl, p. 230 Aplomado Falcon, p. 255

TSOOK
A low, swallowed Bark that begins as a brief high Squeak, the two tone qualities nearly simultaneous.

Northern Shoveler, p. 47 Golden Eagle, p. 210
Blue-winged Teal, p. 46 Peregrine Falcon, p. 257
Sharp-tailed Grouse, pp. 74–75

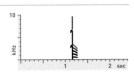

HONK (UNBROKEN)
Brief monotone nasal note, low- to medium-pitched. Listed from highest to lowest pitch within groups.

High, some approaching Keek in pitch
Snow Goose, p. 39
Least Grebe, p. 91
Black-legged Kittiwake, p. 162
Ross's Goose, p. 39

Slightly higher
Cackling Goose, p. 41

High, with metallic quality
Pelagic Cormorant, p. 190

Pitch variable; clear to burry
Common Raven, p. 316

Medium-low pitch
Emperor Goose, p. 38
Trumpeter Swan, p. 43

Low, long; like a truck horn
Greater White-fronted Goose, p. 40
Brant, p. 40

HONK (BROKEN)
Brief nasal note, starting low, then almost immediately breaking upward to a nearly monotone pitch.

Brant, p. 40
Cackling Goose, p. 41
Canada Goose, p. 41
Greater White-fronted Goose, p. 40

Pelagic Cormorant, p. 190

Distinctly 2-syllabled, second note wailing
Indian Peafowl, p. 64

KEEK
High, brief monotone nasal note.

Low to medium high, often harsh
Ridgway's Rail, p. 122

Extremely brief, medium pitch
Common Gallinule, p. 124

Slightly longer, very high
Sora, p. 123

Variably harsh, often in series
Golden-fronted Woodpecker, p. 238

YAP
Higher than Nasal Bark (p. 550); much like Keek, but longer and more noticeably overslurred, with weaker consonant sounds.

Emperor Goose, p. 38
Harlequin Duck, p. 60
Royal Tern, p. 178

Elegant Tern, p. 179
Black Tern, p. 183

YIP
Like Yap (p. 549), but slightly briefer and noisier.

European Starling,
  pp. 388–389

BARK (NASAL)
Brief nasal note of medium pitch.

Very brief, sounding like "Kek"
  Cooper's Hawk, p. 202
  Common Gallinule, p. 124
  Black-necked Stilt, p. 127
  Black Skimmer, p. 175
  Black-throated Magpie-Jay,
    p. 303

Longer, usually clear, highly nasal
  Mandarin Duck, p. 44
  Long-tailed Duck, p. 60
  Harlequin Duck, p. 60
  Cattle Egret, p. 195
  American Coot, p. 125

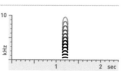

Heermann's Gull, p. 165
Black Skimmer, p. 175
Great Horned Owl, p. 213

WIK (LOW)
A brief upslurred nasal note, lower than High Wik (p. 548); shorter and sharper than Low Reek.

Mandarin Duck, p. 44   Yellow-billed Magpie, p. 315
Black-billed Magpie, p. 314

WIP (NASAL)
A brief overslurred nasal note, shorter than Low Wik. See Whistled Wip, p. 534.

California Quail, p. 68   Gambel's Quail, p. 69

CHWEEK
A brief, sharply underslurred seminasal note.

California Thrasher, p. 382

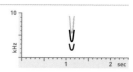

A LONGER NASAL NOTE
A fairly long nasal or seminasal note, broken or unbroken, but lacking a harsh quality.

REEK (HIGH) OR REEP
High, upslurred seminasal note. Longer than High Wik (p. 548); higher than Low Reek. Sometimes in series (Reek-reek).

Wood Duck, p. 45   Surfbird, p. 141
Black Turnstone, p. 140   Marbled Godwit, p. 139
Red Knot, p. 141   Gray Jay, p. 304

KEER (HIGH)
High downslurred seminasal note. Longer than Kew (p. 548); higher than Low Reek. See also Jeer (p. 551) and Single Notes of Gulls (p. 553).

Ruddy Turnstone, p. 140
Marbled Murrelet, p. 159
Bonaparte's Gull, p. 163
Elf Owl, p. 229
Rose-ringed Parakeet, p. 258

White-breasted Nuthatch (all
  populations), pp. 342–344
Blue Jay, pp. 306–307
Gray Jay, p. 304
Rufous-crowned Sparrow,
  p. 427

Higher, longer
Long-eared Owl (juvenile),
  p. 222

WEEW (SEMINASAL)

High, overslurred seminasal note without strong consonant sounds; longer and higher than Yap (p. 549). Higher than Reer.

Black Oystercatcher, p. 128
Pacific Golden-Plover, p. 134

Hudsonian Godwit, p. 139

REER

Medium-high, medium-long, slightly nasal overslurred note. Slightly higher and less nasal than Mew. Lower than Weew.

California Quail Song, p. 68
Gambel's Quail Song, p. 69
Pinyon Jay, p. 312

Clark's Nutcracker, p. 313
Common Raven, p. 316

TEWEE (SEMINASAL)

A medium-low, seminasal underslurred note. See also Whistled Pyoowee or Tewee (p. 542).

Variable and plastic
Brown Thrasher, p. 380

Bendire's Thrasher, p. 381
Curve-billed Thrasher, p. 385

MEW

Medium-high, medium-long, highly nasal notes, variable in pitch pattern. Longer than Yap (p. 549); shorter than Wail (p. 552); higher than Yank (p. 552). See also Single Notes of Gulls (p. 553).

High, usually upslurred
Green-tailed Towhee, p. 426

Slightly lower, whinier, usually upslurred
Loggerhead Shrike, p. 292
Northern Shrike, p. 293

Slightly lower, more nasal, usually downslurred
Red-naped Sapsucker, p. 250
Red-breasted Sapsucker, p. 251
Cooper's Hawk, p. 202

Low, nasal, usually overslurred, sometimes harsh
Gray Catbird, p. 379

California Gnatcatcher, p. 360

Low, nasal, brief, variably harsh
Scott's Oriole, p. 486

Even lower, usually overslurred
Marbled Murrelet, p. 159

REEK (LOW)

A brief upslurred nasal note, lower and more nasal than High Reek. Longer than Low Wik.

Wilson's Phalarope, p. 155
Acorn Woodpecker, p. 236
Cassin's Vireo, p. 298
Plumbeous Vireo, p. 299

Clark's Nutcracker, p. 313
Black-billed Magpie, p. 314
Yellow-billed Magpie, p. 315

KEER (LOW)

High downslurred seminasal note. Longer than Kew (p. 548); higher than Low Reek.

Willet (Western), p. 154

Clark's Nutcracker, p. 313

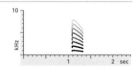

JEER (NASAL)

A medium-long, slightly downslurred nasal note, without any burry quality. Lower and more nasal than High Keer.

Shrill, sometimes metallic
Blue Jay, pp. 306–307

A LONGER NASAL NOTE, CONTINUED

YANK
Low to medium-low, medium-long nasal notes, usually nearly mono-tone, but not as monotone as Honk (p. 549). Lower than Mew (p. 551).

Plastic, nearly monotone, medium in pitch and length
Least Grebe, p. 91
Wild Turkey, p. 65
American Avocet, p. 128

Upland Sandpiper, p. 138
Red-masked Parakeet, p. 259
Mitred Parakeet, p. 259
Red-breasted Nuthatch, p. 341

Low
Clark's Nutcracker, p. 313

SCREAM (NASAL)
A long, medium- to high-pitched seminasal or nasal note. Higher than Wail. Some versions broken.

Monotone, very high, whistled
Sharp-shinned Hawk, p. 201

Downslurred
Gray Hawk, p. 206
Short-tailed Hawk, p. 208

Ferruginous Hawk, p. 209
Rough-legged Hawk, p. 209

Overslurred
Zone-tailed Hawk, p. 206

WAIL
A long, medium-pitched nasal note. Pitch, pattern, and length are plastic in all species, but in general, Wails of owls tend to be overslurred; Wails of hawks tend to be downslurred; and Wails of gulls are often monotone, broken, and/or in series. In many bird species, tremolo versions may be given in agitation.

Pitch quite variable
Ring-billed Gull, p. 166
Herring Gull, p. 170
Glaucous Gull, p. 174
Boreal Owl, p. 224

High, rising, almost whistled
Spotted Owl, p. 221

Medium-high pitch
Franklin's Gull, p. 164
Mew Gull, p. 167
Western Gull, p. 168
Glaucous-winged Gull, p. 169
Ferruginous Hawk, p. 209

Rough-legged Hawk, p. 209
Whiskered Screech-Owl, p. 228
Northern Saw-whet Owl, p. 225
Northern Goshawk, p. 203

Medium pitch
Gambel's Quail, p. 69
Sora, p. 123
American Coot, p. 125
Red-throated Loon, p. 187
Eastern Screech-Owl, p. 227
Flammulated Owl, p. 229

Barred Owl (usually upslurred), p. 220

Low, sometimes nearly moaning and/or slightly hoarse
Ruffed Grouse, p. 80
Pacific Loon, p. 186
Black-legged Kittiwake, p. 162
Heermann's Gull, p. 165
Lesser Black-backed Gull, p. 173
Great Horned Owl, p. 213
Long-eared Owl, p. 222
Crested Caracara, p. 254

MOAN
A very low-pitched nasal sound, usually long and nearly monotone.

Trumpeter Swan, p. 43
Dusky Grouse, p. 82
Sooty Grouse, p. 83
Pelagic Cormorant, p. 190
Snowy Egret, p. 194
Tufted Puffin, p. 159

Yellow-billed Loon, p. 185

Medium-high
Harlequin Duck, p. 60

Low, brief
Brandt's Cormorant, p. 190

Very short, audible at close range
Mountain Plover, p. 133

SINGLE NOTES OF GULLS
Extremely plastic, individually variable Yelps, Squeals, Keers, and broken Wails, usually 1-syllabled.

Very high, squealing
juveniles of many gull species

Medium-high to high, variable
Ring-billed Gull, p. 166
Mew Gull, p. 167
Western Gull, p. 168
Glaucous-winged Gull, p. 169

Herring Gull, p. 170
Glaucous Gull, p. 174

Medium-high, clear and nasal
Franklin's Gull, p. 164

Medium-low, clear and nasal
Heermann's Gull, p. 165

Medium-low, yelping, sometimes hoarse
California Gull, p. 172
Lesser Black-backed Gull, p. 173

KLEER
Like Keer (p. 551), but breaking downward to a lower pitch near start.

Northern Flicker, p. 248

Gilded Flicker, p. 249

SQUEAL (TYPICAL)
A high, long, broken seminasal note. Higher than Broken Wail. See also Single Notes of Gulls.

Rough-legged Hawk, p. 209
Cooper's Hawk (juvenile), p. 202
Northern Goshawk, p. 203
Mississippi Kite, p. 200

Often multisyllabled, with several voice breaks per note
Emperor Goose, p. 38
Greater White-fronted Goose, p. 40

SQUEAL (LONG)
Like Squeal, but longer. Higher than Broken Wail.

Short
many gull species (begging calls; see p. 161)

Averaging longer
Red-tailed Hawk, p. 205
Short-tailed Hawk, p. 208

A RISING BROKEN NASAL NOTE
An upslurred nasal or seminasal note that breaks to a higher pitch at end. See also 2-Syllabled Nasal Note (p. 591) and Longer Upslurred Whistle (p. 540).

Wood Duck Reek, p. 45

Stilt Sandpiper Weep, p. 142

WAIL (BROKEN)
A Wail that breaks at least once to a different pitch. Note that most vocalizations in the Wail category can occasionally break. See also Single Notes of Gulls.

Low pitch
California Gull, p. 172

WAIL (LONG, BROKEN)
A long, mellow, slightly nasal whistle that breaks upward 1–3 times, sometimes downward at end. See also 2-Syllabled Whistles (p. 587).

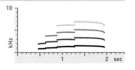

Himalayan Snowcock, p. 72
Common Loon, p. 184

Yellow-billed Loon, p. 185
Pacific Loon, p. 186

A SHORT HARSH NOTE
Very brief noisy notes, nasal notes, or low sharp whistles, with little musical quality. Longer than Clicks or Taps, and often lower in pitch.

PWUT
Very low, brief upslurred whistle, sharp and not very musical; may recall the sound of a drip of water.

Surf Scoter, p. 56
Sharp-tailed Grouse, pp. 74–75
Black-crowned Night-Heron, p. 197

Swainson's Thrush (Russet-backed), p. 374
Swainson's Thrush (Olive-backed), p. 375

KUK
A quick, low, nasal note, lower and shorter than Bark (p. 550). More nasal than Cluck.

Quite nasal
Tundra Swan, p. 43
Scaled Quail, p. 67
California Quail, p. 68
Gambel's Quail, p. 69
Lesser Nighthawk, p. 102

Spruce Grouse, p. 81
Less nasal, less noisy
American Robin, pp. 376–377
Nasal or noisy
Greater Sage-Grouse, p. 78

Gunnison Sage-Grouse, p. 79
White-tailed Ptarmigan, p. 84

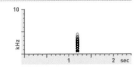

CLUCK
Like Kuk, but harsher, sometimes more complex. Often in series.

Blue-winged Teal, p. 46
Lesser Scaup, p. 54
Greater Scaup, p. 54
Harlequin Duck, p. 60
Domestic Chicken, p. 63

Wild Turkey, p. 65
Buff-collared Nightjar, p. 104
Higher, upslurred
Ruffed Grouse, p. 80
(Red Squirrel)

CHIT
A brief, high, noisy note, sometimes partly nasal. Higher than Chuk.

Gray Partridge, p. 73
Broad-billed Hummingbird, p. 115
Nuttall's Woodpecker, p. 241
Downy Woodpecker, p. 242

Arizona Woodpecker, p. 244
Gray Vireo, p. 296
House Wren, p. 349
Marsh Wren (Eastern), p. 355
Marsh Wren (Western), p. 354

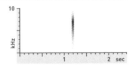

CHUK
A brief, low, noisy note, sometimes partly nasal. Lower than Chit.

Often fairly nasal
Golden-fronted Woodpecker, p. 238
Ladder-backed Woodpecker (rare), p. 240

Usually lower and harsher
Least Bittern, p. 192
Island Scrub-Jay, p. 308
Cactus Wren, p. 357
Brown Thrasher, p. 380
Song Sparrow, p. 454
Bobolink, p. 468

Tricolored Blackbird, p. 469
Red-winged Blackbird, pp. 470–471
Great-tailed Grackle, pp. 474–475
Common Grackle, p. 478
Rusty Blackbird, p. 476
Brewer's Blackbird, p. 477
Brown-headed Cowbird, p. 480
Orchard Oriole, p. 482

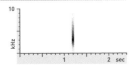

Often slightly longer, grating
Yellow-headed Blackbird, p. 479

Shorter, more Keklike
Yellow-breasted Chat, p. 466
Hooded Oriole, p. 483

Very low and harsh
Ridgway's Rail, p. 122

CHWUT
A brief, low, upslurred noisy note.

White-breasted Nuthatch
(Montane), p. 343

Gray Catbird, p. 379

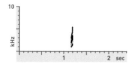

CHUP
Like Chip (p. 531) but lower, slightly nasal or noisy, downslurred.

Medium pitch
Violet-crowned Humming-
bird, p. 116
Common Yellowthroat, p. 498

Medium-low
Mountain Bluebird, p. 367
Western Bluebird, p. 368
Hermit Thrush, p. 373
Bendire's Thrasher, p. 381

Curve-billed Thrasher, p. 385
Sage Thrasher, p. 386
Western Meadowlark, p. 473
Rusty Blackbird, p. 476
Hepatic Tanager, p. 512

Medium-low, clear, almost a Tew
Varied Thrush, p. 371

Rather barking quality
Gyrfalcon, p. 256

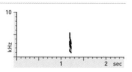

California Thrasher, p. 382
Scott's Oriole, p. 486

Rather noisy, Chuklike
Five-striped Sparrow, p. 442

A LONGER HARSH NOTE

BARK (SCREECHY)
A very brief, high, harsh nasal note. Lower than Cheep (p. 535).

White-winged Scoter, p. 56
Red-necked Grebe, p. 88
Pacific Loon, p. 186

Red-throated Loon, p. 187
Virginia Rail, p. 121
Least Tern, p. 176

BARK (ROUGH)
A brief note that is both nasal and noisy. Pitch medium.

Averaging noisier, less nasal
Bufflehead, p. 57
Greater Prairie-Chicken,
p. 76
Lesser Prairie-Chicken,
p. 77
Short-eared Owl, p. 223
Long-eared Owl, p. 222
Snowy Owl, p. 214

High, noisy, sharply downslurred
Red-bellied Woodpecker,
p. 239

Quite harsh, downslurred, often in flight
Lilac-crowned Parrot, p. 260
Red-crowned Parrot, p. 260
Red-lored Parrot, p. 260

High, clear, almost a Yap
Great Gray Owl, p. 219

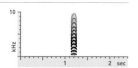

Often short, partly noisy or burry
Ruffed Grouse, p. 80
Spruce Grouse, p. 81
Dusky Grouse, p. 82
Sooty Grouse, p. 83
Common Murre, p. 160

SQUAWK
A brief, medium-pitched nasal and noisy note. Lower and shorter than Screech; harsher than Bark; like a harsh version of Honk (p. 549).

American Bittern, p. 191
Black-crowned Night-Heron,
p. 197

Great Blue Heron, p. 193
Snow Goose, p. 39

GRUNT (TYPICAL)
A low, brief noisy note, usually with a slightly nasal quality. Longer and lower than Chuk. Compare burry versions of Grunt (p. 555).

Slightly nasal
Canada Goose, p. 41
Ross's Goose, p. 39
Common Merganser, p. 61
Double-crested Cormorant,
p. 189

Slightly harsher
Caspian Tern, p. 177

Neotropic Cormorant, p. 189
American White Pelican, p. 188
Brown Pelican, p. 188
American Bittern, p. 191

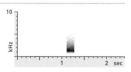

Voiceless, huffing
Black Vulture, p. 198

SCREECH

A high, medium-long nasal and noisy note. Usually plastic, but generally not broken or burry. Higher than Squawk (p. 555); shorter than Shriek.

Green-winged Teal, p. 51
Mountain Quail Song, p. 66
Scaled Quail Song, p. 67
Virginia Rail, p. 121
Ridgway's Rail, p. 122
American Coot, p. 125
Wilson's Snipe, p. 149
Common Murre, p. 160
Whiskered Screech-Owl, p. 228

Long, plastic, variably nasal
Great Horned Owl, p. 213
Great Gray Owl, p. 219
Long-eared Owl, p. 222

Singly, by juveniles
Common Raven, p. 316
Chihuahuan Raven, p. 317

Often in long series, by juveniles
Common Tern, p. 180
Arctic Tern, p. 181
Forster's Tern, p. 182
Least Tern, p. 176
Black Tern, p. 183

CHWEE

A medium-high upslurred seminasal note with some harsh quality.

Violet-green Swallow, p. 327

SKEWCH

Like Screech, but downslurred.

Green Heron, p. 195

Tricolored Blackbird, p. 469

SHRIEK

Long, high, partly nasal harsh note. Longer than Screech.

Usually monotone
Barn Owl, p. 212

Usually upslurred
Great Horned Owl, p. 213
Snowy Owl, p. 214
Great Gray Owl, p. 219
Barn Owl (juvenile), p. 212

Northern Hawk Owl, p. 215
Barred Owl, p. 220
Spotted Owl, p. 221
Short-eared Owl, p. 223

Usually with short Kuk notes
Green Heron, p. 195

Usually overslurred, broken
Wood Duck, p. 45

SCREAM (HARSH)

Long harsh note, at least slightly nasal or seminasal. Harsher than Wail (p. 552); more nasal than Shriek.

Very harsh, snarling, monotone
Harris's Hawk, p. 204

Downslurred
Red-tailed Hawk, p. 205

Swainson's Hawk, p. 208
Cooper's Hawk, p. 202

Overslurred
Common Raven, p. 316

HISS

A rushing of air, a toneless burst of noise without any musical or nasal quality. Usually softer than Snarl (p. 560); not grating.

Canada Goose, p. 41
Turkey Vulture, p. 198
Barn Owl, p. 212
LeConte's Sparrow, p. 464

Red-winged Blackbird, pp. 470–471

Lower, roaring
Black Vulture, p. 198

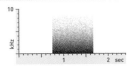

KHOW
A brief, very low, downslurred whispered hiss.

Green Heron, p. 195

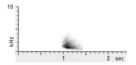

FOOM
A low, downslurred rushing of air, often slightly nasal, with abrupt start.

Common Nighthawk, p. 101

WHOOSH
A swishing sound made by rubbing feathers against one another, or by feathers against the air in fast flight.

Soft, made by opening and closing tail feathers
  Spruce Grouse, p. 81

Loud, at bottom of display dive
  Marbled Murrelet, p. 159

CRACKLE
A burst of noise containing ticking and knocking notes, recalling the sound of a breaking stick. See also Long Songs Punctuated by Crackles (p. 602).

Soft, short
  Long-tailed Duck, p. 60

Longer, fairly loud
  Crested Caracara, p. 254

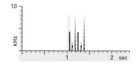

A HARSH BURRY NOTE
A sound with some combination of nasal, noisy, and burry qualities.

VIMP
Brief, finely burry nasal note, sometimes slightly harsh. Can be difficult to distinguish from Jit (p. 563).

Medium-high, clear
  Winter Wren, p. 347

Song Sparrow, p. 454

VRIT
Like Vimp, but lower, harsher, and often coarser.

Gambel's Quail, p. 69
Blue Jay, pp. 306–307

Medium-low, harsh, often Chuklike
  Gray Jay, p. 304

Steller's Jay, p. 305
California Scrub-Jay, p. 309
Woodhouse's Scrub-Jay, p. 310
Mexican Jay, p. 311

KRIK
Like Grate (p. 559) but very brief. Coarser and harsher than Vimp.

Medium-low, coarse, grating
  Mountain Plover, p. 133
  Bonaparte's Gull, p. 163

Royal Tern, p. 178
Elegant Tern, p. 179

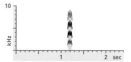

YANK (BURRY)
A burry nasal note. Higher and clearer than Caw.

White-breasted Nuthatch
(Eastern), p. 344

White-breasted Nuthatch
(Pacific), p. 342

ZWOINK
A finely burry upslurred nasal note.

Thick-billed Kingbird, p. 287

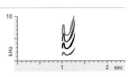

CAW
Brief overslurred nasal note, always at least slightly burry and at least slightly harsh. Pitch low to medium. Finer than Churr.

American Crow, p. 318
Northwestern Crow, p. 319

Deeper, coarser
Common Raven, p. 316
Chihuahuan Raven, p. 317

CHURR (NASAL)
Like Grate, but lower and harsher. More nasal than Snarl (p. 560), and less harsh. Finer than Churring Rattle (p. 565). Longer than burry versions of Grunt (p. 560). See Chatter Duets (p. 579).

Finer, usually harsher
Western Meadowlark, p. 473
Rusty Blackbird, p. 476
Brewer's Blackbird, p. 477

Coarser, more nasal
Black-necked Stilt, p. 127
Pinyon Jay, p. 312
Acorn Woodpecker, p. 236

Williamson's Sapsucker,
p. 252

KRAA
A loud, long, very harsh, coarsely burry, strongly nasal note, usually slightly rising, sometimes monotone.

Clark's Nutcracker, p. 313

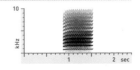

KWIRR
A brief nasal burr, very coarse, almost a tremolo. Higher, coarser, and clearer than Churr. Lower and more musical than Grate.

High, brief, coarse, monotone
Gila Woodpecker, p. 237
Golden-fronted Woodpecker,
p. 238

Red-bellied Woodpecker,
p. 239

Softer, even coarser, less nasal
Northern Flicker, p. 248
Gilded Flicker, p. 249

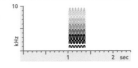

Low, fine, down- or overslurred
Horned Grebe, p. 89

KWEEAH
Like Kwirr, but higher and finer, often harsher. Lower and finer than Grate.

Red-headed Woodpecker,
p. 235

Horned Grebe, p. 89

WHEEZE

A long, finely burry or hoarse nasal note. Easily confused with Whine (p. 563), but not polyphonic (the difference can be hard to hear).

Medium-high, up- or overslurred, slightly harsh
Spotted Towhee, p. 425

Lower, overslurred, finely burry
Eurasian Collared-Dove, p. 94

More monotone, hoarser
White-winged Dove, p. 97

PURR

A very low burry note, usually somewhat nasal. Finer than Bleat (p. 579), higher and coarser than Growl (p. 543).

Sharp-tailed Grouse, pp. 74–75
Ruffed Grouse, p. 80
Spruce Grouse, p. 81
White-tailed Ptarmigan, p. 84

Lower, coarser, slightly longer
Least Bittern Song, p. 192

Very low, almost a Growl
Mexican Whip-poor-will, p. 103

High, nasal, almost a Churr
Wild Turkey, p. 65

GRATE

A high-pitched coarsely burry nasal note, often harsh. Can be considered the burry version of a Screech (p. 556). Higher than Churr; longer than Krik (p. 557).

High, coarse
Mountain Quail, p. 66
Clark's Grebe Song, p. 85
Western Grebe Long Grate, p. 86
Semipalmated Plover, p. 130
Mountain Plover, p. 133
Willet (Western), p. 154
Nanday Parakeet, p. 258

High, fine, almost screeching
White-throated Swift, p. 117
Black Tern, p. 183
Red-necked Phalarope, p. 156
Dunlin, p. 143

Medium pitch
Black-necked Stilt, p. 127
Sabine's Gull, p. 163

Bonaparte's Gull, p. 163
Nanday Parakeet, p. 258
Monk Parakeet, p. 258
Blue-crowned Parakeet, p. 259
Black-throated Magpie-Jay, p. 303

KEER (GRATING)

A distinctly downslurred Grate.

Mountain Plover, p. 133
Black Turnstone, p. 140
Royal Tern, p. 178
Elegant Tern, p. 179
Common Tern, p. 180

Forster's Tern, p. 182
Black Tern, p. 183
Lewis's Woodpecker, p. 234
Rose-ringed Parakeet, p. 258
Black-throated Magpie-Jay, p. 303

Bank Swallow, p. 325

QUACK

A low, nasal note with a finely grating quality.

Rather grating and nasal, variably harsh
Mallard (female), p. 52
Mexican Duck (female), p. 53
Blue-winged Teal, p. 46
Cinnamon Teal, p. 47
Northern Shoveler, p. 47
Northern Pintail, p. 50
Lesser Scaup, p. 54

Tinnier, more grating, less nasal
Mallard (male), p. 52
Mexican Duck (male), p. 53

Higher, squeaky or screechy
Green-winged Teal, p. 51
Snowy Owl, p. 214

Low, nasal, clear
White-faced Ibis, p. 196
Glossy Ibis, p. 196

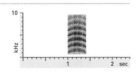

Short, nasal, monotone
Gadwall, p. 48

Nasal to croaking
Red-throated Loon, p. 187

GRUNT (BURRY)
Like Quack (p. 559), but harsher and coarser. Generally lower and harsher than Churr (p. 558). See typical Grunt (p. 555).

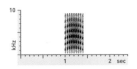

Rather fine and quacking
   Gadwall, p. 48

Harsh, coarse, often churring
   American Wigeon, p. 49
   Eurasian Wigeon, p. 49
   Redhead, p. 55
   Canvasback, p. 55
   Ring-necked Duck, p. 53

Greater Scaup, p. 54
Lesser Scaup, p. 54
Common Goldeneye, p. 58
Barrow's Goldeneye, p. 59
Bufflehead, p. 57
Red-breasted Merganser, p. 61
Hooded Merganser, p. 62

Red-necked Grebe, p. 88
Band-tailed Pigeon, p. 93
Brandt's Cormorant, p. 190

Rather high, nasal, coarse
   Northern Harrier, p. 199

WAIL (GRATING)
A variable, often plastic mix of a Wail (p. 552) and a Grate (p. 559) or a Cranelike Rattle (p. 565).

Grating quality rather fine
   American Coot, p. 125
   Common Gallinule, p. 124

A burry wail or screech
   Horned Grebe Song, p. 89

SNARL
A very harsh noisy note, like a loud Hiss (p. 556), but harsher, generally with a burry or grating quality; like Whine (p. 563) or Jurr (p. 564), but not polyphonic.

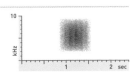

Medium-high
   Rivoli's Hummingbird, p. 106
   Cassin's Kingbird, p. 286
   Tree Swallow, p. 326
   Pacific Wren, p. 348
   Marsh Wren (Eastern), p. 355
   Bendire's Thrasher, p. 381
   Lawrence's Goldfinch, p. 408
   Chipping Sparrow, p. 436
   Vesper Sparrow, p. 440
   Lark Bunting, p. 447
   Snow Bunting, p. 419

Medium-high, often slightly grating
   Common Tern, p. 180
   Arctic Tern, p. 181
   Gull-billed Tern, p. 177

Medium-low
   White-tailed Kite, p. 200
   Red-breasted Nuthatch, p. 341

Black-tailed Gnatcatcher, p. 358
California Gnatcatcher, p. 360
Purple Martin, p. 328
Cactus Wren, p. 357
European Starling, pp. 388–389
California Thrasher, p. 382
Northern Mockingbird, p. 387
Spotted Towhee, p. 425

Sometimes more nasal
   Steller's Jay, p. 305
   Island Scrub-Jay, p. 308
   Bewick's Wren, p. 353
   Black-capped Gnatcatcher, p. 361

Slightly more nasal and churring
   Pectoral Sandpiper, p. 144

Brown Thrasher, p. 380
Black-billed Magpie, p. 314
Yellow-billed Magpie, p. 315

Long, like rattlesnake rattle
   Burrowing Owl, p. 230

Medium-low, short, rather soft
   Ladder-backed Woodpecker, p. 240
   California Scrub-Jay, p. 309
   Woodhouse's Scrub-Jay, p. 310
   Mexican Jay, p. 311

Long, low, loud, sometimes slightly croaking
   Great Blue Heron, p. 193
   Black-crowned Night-Heron, p. 197

RASP
Like Snarl, but coarser and more grating. See also Whine (p. 563).

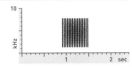

Medium pitch
   Sedge Wren, p. 356

Low and harsh
   Black-capped Vireo, p. 294
   Marsh Wren (Eastern), p. 355

Yellow-breasted Chat, p. 466

Even lower and harsher
   American Crow, p. 318
   Northwestern Crow, p. 319
   Cactus Wren, p. 357

Variable, soft, brief
   Blue-throated Hummingbird, p. 105

SHRIEK (GRATING)
Like Shriek (p. 556), but grating.

Glossy Ibis (juvenile), p. 196

White-faced Ibis (juvenile),
p. 196

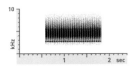

CROAK (MONOTONE)
A note made of very rapid repeated Tocklike or knocking notes,
ranging from toneless to nasal.

Medium to high, resonant,
slightly musical
American Coot, p. 125

Medium-low, harsh, coarse
American White Pelican,
p. 188
Elegant Tern, p. 179

Medium-low, harsh, rather fine
Great Blue Heron, p. 193
Great Egret, p. 193

Snowy Egret, p. 194
Little Blue Heron, p. 194

Very low, resonant, rather
musical
Yellow-headed Parrot, p. 260
Common Raven, p. 316

Very low, like the snorts of a hog
Common Goldeneye, p. 58
Barrow's Goldeneye, p. 59

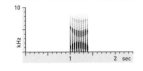

Double-crested Cormorant,
p. 189
Neotropic Cormorant, p. 189

Breathy, slapping or fluttering
Greater Sage-Grouse, p. 78
Gunnison Sage-Grouse,
p. 79

CROAK (SPEECHLIKE)
A sound made of rapid clicklike notes whose loudest (darkest)
frequencies change over the course of the call, creating an effect
vaguely reminiscent of human speech.

Pacific Loon, p. 186

Coarser, more rattling, usually
in series
Crested Caracara, p. 254

CROAK (TICKING)
A short Croak made of soft Ticklike notes rather than knocking or
resonant notes.

Common Goldeneye, p. 58

Barrow's Goldeneye, p. 59

GROAN (CROAKING)
Like low Groan (p. 543), but higher, more nasal, more croaking, and
usually longer.

Great Blue Heron, p. 193
Great Egret, p. 193
Snowy Egret, p. 194
Little Blue Heron, p. 194

Cattle Egret, p. 195

Usually higher, more Cawlike
American Crow, p. 318
Northwestern Crow, p. 319

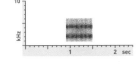

A POLYPHONIC NOTE
A note created by the mixing of two simultaneous sounds, one generated by each side of a bird's syrinx,
resulting in a characteristic dissonant, metallic, or whiny quality.

TEENK
An extremely brief, clear, metallic polyphonic note. Like Pink (p. 532),
but higher, with different consonant sound at start.

Fox Sparrow (Thick-billed), p. 456

A POLYPHONIC NOTE, CONTINUED

ZEET (TYPICAL)
A brief, clear, nearly monotone polyphonic note.

Medium-high
Rose-breasted Grosbeak,
p. 518

Bobolink, p. 468
Orchard Oriole, p. 482
Hooded Oriole, p. 483

ZEET (BURRY)
Like Zeet, but coarsely burry.

Varied Bunting, p. 522

ZREET OR ZOIT
A clear, upslurred polyphonic note.

High to medium-high
Olive Warbler, p. 392
Bobolink, p. 468
Eastern Meadowlark, p. 472
Western Meadowlark, p. 473
Black-headed Grosbeak,
p. 519
Hooded Oriole, p. 483

White-winged Crossbill,
p. 418

Medium-low
Cave Swallow, p. 322
Baltimore Oriole, p. 484
Bullock's Oriole, p. 485
Hepatic Tanager, p. 512

Western Tanager, p. 514

Still lower, shorter (Zoit)
Eastern Towhee, p. 424

ZREE
A medium-long upslurred polyphonic note.

High, clear, long
American Goldfinch, p. 407

Medium-high, fairly long
Brewer's Blackbird, p. 477
Pine Siskin, p. 406
Lawrence's Goldfinch, p. 408

Common Redpoll, p. 405

Medium-low, long
Lesser Goldfinch, p. 409
Eastern Towhee, p. 424

Medium-low, short, almost a Zreet
House Finch, p. 404

ZEER
A high, clear downslurred polyphonic note.

Cliff Swallow, p. 323
Cave Swallow, p. 322
Bank Swallow, p. 325
Bronzed Cowbird (juvenile),
p. 481

Western Tanager, p. 514

Higher, clearer
Bobolink, p. 468

PZEW
Like Zeer, but lower, sharper, with distinct consonant sound at start.

Western Bluebird, p. 368

Summer Tanager, p. 513

ZWEEW OR CHEER
Like seminasal Weew (p. 551), but polyphonic. Cheer higher, clearer, and more downslurred than Zweew.

Rather high, clear
Pine Siskin Zweew, p. 406
Painted Redstart Cheer,
p. 510

Lower, slightly harsher
Black-headed Grosbeak,
p. 519
House Finch, p. 404

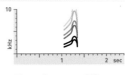

House Sparrow, p. 393

ZEWY OR CHEEREET

A musical, slightly polyphonic underslurred whistle.

Medium-low, soft, whiny (Zewy)
  Eastern Bluebird, p. 369
  Western Tanager, p. 514

Medium-high, mostly clear (Cheereet)
  Painted Redstart, p. 510

JIT

A short polyphonic note, often slightly noisy. Like Vimp (p. 557), but polyphonic instead of burry and nasal (the difference can be hard to hear). See also Finch Flight Call Group (p. 567).

High, loud, metallic
  Gambel's Quail, p. 69

Medium-high, unmusical, downslurred, Chiplike
  Pacific Wren, p. 348
  Wilson's Warbler, p. 508

Lower, monotone, almost a Chit
  Ruby-crowned Kinglet, p. 364

Black-capped Vireo, p. 294

Medium-low, monotone
  Bell's Vireo, p. 295
  Scott's Oriole, p. 486

Low, seminoisy, upslurred
  Barn Swallow, p. 324
  Baltimore Oriole, p. 484
  Bullock's Oriole, p. 485

Similar but clearer
  House Finch, p. 404

Similar but more sharply upslurred
  Hutton's Vireo, p. 297
  Red-eyed Vireo, p. 302
  Warbling Vireo, pp. 300–301

JIRP

A polyphonic version of Chirp (p. 535). More complex and usually slightly harsher than Jit. See also Finch Flight Call Group (p. 567).

House Finch, p. 404

House Sparrow, p. 393

WHINE

A long, harsh, polyphonic note, often finely burry. Medium pitch. Like a polyphonic Snarl (p. 560). Compare Wheeze (p. 559).

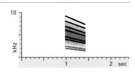

High, long
  Greater Roadrunner, p. 100

Usually not burry to finely burry
  Loggerhead Shrike, p. 292
  Northern Shrike, p. 293
  Hermit Thrush, p. 373
  Townsend's Solitaire, p. 370
  Gray Catbird, p. 379
  House Wren, p. 349
  Cactus Wren, p. 357
  Island Scrub-Jay, p. 308

California Scrub-Jay, p. 309
Woodhouse's Scrub-Jay, p. 310
Mexican Jay, p. 311
Gray Jay, p. 304
Violet-green Swallow, p. 327
Black-capped Gnatcatcher, p. 361
Bobolink, p. 468
Yellow-headed Blackbird, p. 479

Often coarsely burry. See also Rasp (p. 560).
  Warbling Vireo, pp. 300–301
  Gray Vireo, p. 296
  Hutton's Vireo, p. 297
  Cassin's Vireo, p. 298
  Plumbeous Vireo, p. 299

SHEER

Like Whine, but higher and distinctly downslurred.

High, whiny
  Blue-gray Gnatcatcher, p. 359
  Black-capped Gnatcatcher, p. 361

Lower, harsher
  Black-tailed Gnatcatcher, p. 358

Even lower, quite brief
  California Gnatcatcher, p. 360

JENK

A brief, monotone burry polyphonic note, rather harsh, with a metallic ring.

Louder
  Red-winged Blackbird, pp. 470–471

Common Grackle, p. 478
Brewer's Blackbird, p. 477

JURR
A harsh, metallic, monotone polyphonic snarl. Longer than Jenk (p. 563); more monotone and more metallic than a burry Whine (p. 563).

Phainopepla, p. 391
Common Grackle, p. 478

Yellow-headed Blackbird, p. 479

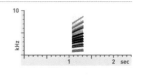

ZREEK
An upslurred, slightly harsh polyphonic note. Always at least slightly burry. Often in series.

Even shorter and finer
Mexican Jay, p. 311

Medium pitch, long, coarse
California Scrub-Jay, p. 309
Woodhouse's Scrub-Jay, p. 310

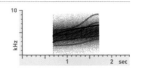

ZHREE
A long, upslurred polyphonic snarl.

Pine Siskin, p. 406

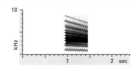

ZHEER
A medium-long downslurred burry polyphonic note. Variably harsh.

Medium-high, fine
Red-eyed Vireo, p. 302

Yellow-breasted Chat, p. 466

JEER (BURRY)
Like Zheer but lower, coarser, and more metallic.

Phainopepla, p. 391
Common Grackle, p. 478

European Starling, pp. 388–389

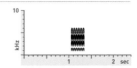

CHURR (POLYPHONIC)
A low, polyphonic coarse burr or short trill.

Medium-low, sometimes slightly noisy
Veery, p. 372

Lower, clearer, audible only at close range
Wood Duck, p. 45

Loud, harsh
Common Grackle, p. 478

INDEX PART II: A SOUND MADE OF REPEATED SIMILAR NOTES

The same note repeated, without a significant pause. Includes series that change in speed or pitch, and irregular Twitters.

A TICKING OR TAPPING SERIES

A series or trill of Ticklike vocalizations, or of tapping or clapping sounds, including woodpecker drums and other mechanical sounds made by the bill or wings.

RATTLE (TICKING)

A trill or rapid series of Ticklike or clicking notes, coarser than Croaks (p. 561).

Voiceless, coarse
Mountain Plover, p. 133
Pelagic Cormorant, p. 190
American Crow, p. 318
Northwestern Crow, p. 319

Often slower, almost a series
California Scrub-Jay, p. 309
Woodhouse's Scrub-Jay, p. 310

Mexican Jay, p. 311
Pinyon Jay, p. 312
Clark's Nutcracker, p. 313

Fast, high
Steller's Jay, p. 305
Blue Jay, pp. 306–307
Island Scrub-Jay, p. 308
Common Goldeneye, p. 58

High, brief
LeConte's Thrasher, p. 383

Plastic, irregular in rhythm
Western Grebe, p. 86

RATTLE (CROAKING)

Like a Croak (p. 561), but coarser; like a Ticking Rattle, but lower, with longer, noisier notes. Coarser, harsher, and less resonant than Cranelike Rattle and Knocking Rattle.

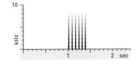

Plastic, sometimes crackling
Crested Caracara, p. 254

Short, monotone
Cinnamon Teal, p. 47

RATTLE (CRANELIKE)

A grating note with a distinctive resonant knocking quality. Lower and more knocking than Grate (p. 559).

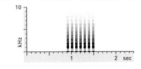

Sandhill Crane, p. 126

RATTLE (KNOCKING)

A short series of very low knocking notes. Lower and less grating than Cranelike Rattle. See also Croaking Rattle and Knocking or Purring Phrase (p. 589).

Common Raven, p. 316
Chihuahuan Raven, p. 317
Woodhouse's Scrub-Jay, p. 310

Slower, longer, notes more Keklike
Yellow-billed Cuckoo, p. 98

RATTLE (CHURRING)

More nasal than Croaking or Cranelike Rattles; coarser than Churr (p. 558).

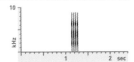

Helmeted Guineafowl, p. 64

Wilson's Snipe, p. 149

A SERIES OF SNAPS OR CLAPS
Made by the wings or bill.

Wing claps in flight
most dove species (see
p. 96)
Barn Owl, p. 212
Mexican Whip-poor-will,
p. 103

With rattles between each clap
Buff-collared Nightjar, p. 104

Usually only 2 claps
Spruce Grouse (Franklin's),
p. 81

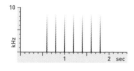

A STEADY MECHANICAL DRUM OR RATTLE
A trill of taps, snaps, or claps made by wings or bill.

Drum of bill against hard surface
most woodpecker species
(see p. 233)

Short quick rattle of bill snaps
Greater Roadrunner, p. 100
Phainopepla, p. 391

Short quick rattle of wing claps
Short-eared Owl, p. 223

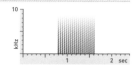

A FLUTTER OR WHIRR
Rapid flutter of wings or tail, often audible only at close range.

Without any musical quality
Spruce Grouse, p. 81
Sharp-tailed Grouse,
pp. 74–75
Inca Dove, p. 95
Common Ground-Dove,
p. 95

(other dove species
occasionally)
Vermilion Flycatcher, p. 278

*With a whirring hum, during
flight display*
Scissor-tailed Flycatcher,
p. 290

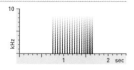

Lucifer Hummingbird, p. 107

In 2 distinct bursts
Dusky Grouse, p. 82

AN ACCELERATING MECHANICAL DRUM OR RATTLE
Made by the wings or bill.

*Drum of bill against hard
surface; acceleration subtle*
American Three-toed
Woodpecker, p. 246
Black-backed Woodpecker,
p. 247
Pileated Woodpecker, p. 253

Bill snaps
Vermilion Flycatcher, p. 278

*Clicklike notes, then a quick low
buzz*
Ruddy Duck Bubble Display,
p. 62

Extremely low, beating thumps
Ruffed Grouse, p. 80

A DECELERATING MECHANICAL DRUM OR RATTLE
Made by the wings or bill.

*Drum of bill against hard
surface; deceleration obvious*
Red-naped Sapsucker,
p. 250
Red-breasted Sapsucker
p. 251

*Similar, but drum broken into
short bursts, the first longest*
Williamson's Sapsucker,
p. 252

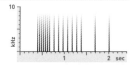

A PLASTIC, IRREGULAR MECHANICAL DRUM OR RATTLE
Made by the wings or bill.

*Bill clatter, given at nesting
colonies*
several heron species

*Wing claps, single or in irregular
series*
Long-eared Owl, p. 222

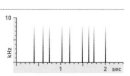

OTHER WINGBEAT SOUNDS

Short snapping rattle
Emperor Goose, p. 38
Trumpeter Swan, p. 43
Tundra Swan, p. 43

Nasal or polyphonic hum
Black Vulture, p. 198
Rock Pigeon, p. 92

Whooshing beats like helicopter
Marbled Murrelet, p. 159
See also dove wing sounds,
p. 96, and Wing Whistles,
p. 574

A CHIP SERIES OR UNMUSICAL TRILL

CHIP SERIES
A series of Chips or similar notes. See also Finch Flight Call Group.

Mountain Quail, p. 66
White-tailed Kite, p. 200
Verdin, p. 340
Wilson's Warbler, p. 508

Pyrrhuloxia, p. 517
With harsh notes interspersed
Black-tailed Gnatcatcher
Complex Song, p. 358

SPIT SERIES
A series of Spits or similar notes. See also Finch Flight Call Group.

California Quail, p. 68

FINCH FLIGHT CALL GROUP
"Flight calls" of finches are variable but stereotyped short Chiplike, Piplike, or polyphonic notes, usually given in series of 3–4 (not necessarily in flight). Red Crossbill call types are indexed here for reference; field identification can be difficult (see p. 412).

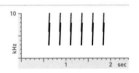

Chip Chip Chip
Red Crossbill Type 3
Flight Call, p. 414
Red Crossbill Type 7
Flight Call, p. 416

Jeet Jeet Jeet (polyphonic)
Red Crossbill Type 3
Excitement Call, p. 414
Red Crossbill Type 7
Excitement Call, p. 416
Lesser Goldfinch, p. 409

Jip Jip Jip (polyphonic)
Red Crossbill Type 5
Excitement Call, p. 415
Red Crossbill Type 6
Excitement Call, p. 416
Cassia Crossbill Excitement
Call, p. 417

Jirp Jirp Jirp (polyphonic)
Common Redpoll, p. 405
Pine Siskin, p. 406

Jit Jit Jit (polyphonic)
White-winged Crossbill,
p. 418

Kip Kip Kip
Red Crossbill Type 1
Flight Call, p. 413
Red Crossbill Type 2
Flight Call, p. 414

Klip Klip Klip
Red Crossbill Type 5
Flight Call, p. 415

Kwit Kwit Kwit
Red Crossbill Type 4
Flight Call, p. 415

Peep Peep Peep
Red Crossbill Type 1
Excitement Call, p. 413

Peer Peer Peer
Red Crossbill Type 10
Excitement Call, p. 417

Pik Pik Pik
Cassia Crossbill Flight Call,
p. 417

Pip Pip Pip
Greater Pewee, p. 264
Olive-sided Flycatcher,
p. 265
Red Crossbill Type 2
Excitement Call, p. 414

**Pitter Pitter Pitter (see also
Pitter, p. 590)**
Olive-sided Flycatcher,
p. 265
Red Crossbill Type 10
Excitement Call, p. 417

Pwip Pwip Pwip
Red Crossbill Type 4
Excitement Call, p. 415

Tew Tew Tew
American Goldfinch, p. 407

Tyit Tyit Tyit
Red Crossbill Type 6 Flight
Call, p. 416

Whit Whit Whit
Red Crossbill Type 10 Flight
Call, p. 417

TSWEET SERIES
A series of rising, complex Chiplike notes.

Pacific-slope Flycatcher,
p. 272

Baird's Sparrow, p. 463

RATCHET
A short series of doubled Smacklike notes, each pair of notes so close together that they sound like a single note. Very abrupt and harsh.

Gray Catbird, p. 379
Brown Thrasher, p. 380

Curve-billed Thrasher, p. 385

AN ACCELERATING SERIES OF CHIPS
A series of Chiplike notes that accelerates, sometimes into a trill. See also Accelerating Series of Whistles or Chirps, p. 573.

Rufous-winged Sparrow, p. 431

TRILL (CHIPPING)
A single unmusical trill of Chiplike or Tsitlike notes.

Little change in volume or speed
Northern Pygmy-Owl
(juvenile), p. 216
Chipping Sparrow, p. 436

Gets louder, sometimes higher
Grace's Warbler, p. 504

RATTLE (SMACKING)
A rapid trill of Smacklike notes; faster and more ticking than classic Chatter (p. 582).

Long
Carolina Wren, p. 352
Common Yellowthroat, p. 498

Short, often mixed with single notes
Winter Wren, p. 347
Pacific Wren, p. 348

TWITTER (CHIPPING)
A highly plastic series of Chiplike or Spitlike notes, often accelerating briefly into a trill.

Almost always from high overhead
Black Swift, p. 117
Chimney Swift, p. 118
Vaux's Swift, p. 118

In agitation, usually from perch
Anna's Hummingbird, p. 109
Verdin, p. 340
Song Sparrow, p. 454
Fox Sparrow (Slate-colored),
p. 457
LeConte's Sparrow, p. 464

Savannah Sparrow, p. 461
Rusty Blackbird, p. 476
Rufous-capped Warbler,
p. 499

Notes Spitlike or slightly complex
Costa's Hummingbird, p. 110
Pin-tailed Whydah, p. 394
Indigo Bunting, p. 520
Botteri's Sparrow, p. 433
Cassin's Sparrow, p. 432

Often squeaky at start
Rivoli's Hummingbird, p. 106
Violet-crowned Hummingbird, p. 116

Often squeaky at end
Broad-billed Hummingbird, p. 115

Squeaky throughout
Sulphur-bellied Flycatcher,
p. 283

AN UNSTEADY TRILL OR SERIES
A slow trill or rapid series of semimusical, often Chiplike notes that changes gradually in pitch and/or quality; even stereotyped versions may sound plastic.

Orange-crowned Warbler, p. 492

Colima Warbler, p. 493
Wilson's Warbler, p. 508

TWITTER (OF TSITLIKE NOTES)
A highly plastic series of medium- to high-pitched Tsitlike or Spitlike notes, often accelerating briefly into a trill. Can be tough to distinguish from Chipping Twitters. Less musical than Twitters of Tiplike notes.

Notes musical, almost Seetlike
Chipping Sparrow, p. 436
Field Sparrow, p. 439
Lesser Goldfinch, p. 409
Botteri's Sparrow, p. 433

Notes very sharp, Spitlike
Northern Cardinal, p. 516
Pyrrhuloxia, p. 517

Notes very sharp, Tsitlike
Green-tailed Towhee, p. 426
Black-chinned Sparrow, p. 435

Lark Sparrow, p. 441
Clay-colored Sparrow, p. 437
Brewer's Sparrow, p. 438
Five-striped Sparrow, p. 442
Black-throated Sparrow, p. 443
Sagebrush Sparrow, p. 444
Bell's Sparrow, p. 445
Dark-eyed Junco, pp. 458–459
Yellow-eyed Junco, p. 460

White-throated Sparrow, p. 448
White-crowned Sparrow, p. 450
Golden-crowned Sparrow, p. 451
Grasshopper Sparrow, p. 462
LeConte's Sparrow, p. 464
Painted Redstart, p. 510

TWITTER (OF TIPLIKE NOTES)
A highly plastic series of semimusical, medium- to high-pitched Tiplike notes, often accelerating briefly into a trill. See also Kree Series (p. 576).

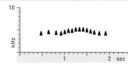

Medium-pitch, rather slow
Vermilion Flycatcher, p. 278
Brown-crested Flycatcher, p. 282

Medium-pitch, rapid, trilled
Eared Grebe, p. 87
Killdeer, p. 132
Greater Pewee, p. 264
Alder Flycatcher, p. 266
Willow Flycatcher, p. 267
Hammond's Flycatcher, p. 268
Dusky Flycatcher, p. 269
Gray Flycatcher, p. 270

Pacific-slope Flycatcher, p. 272
Cordilleran Flycatcher, p. 273
Buff-breasted Flycatcher, p. 274
Least Flycatcher, p. 271
Say's Phoebe, p. 275
Marsh Wren (Eastern), p. 355

High, musical, sometimes polyphonic; compare Twitter of Psip or Seetlike notes (p. 572)
Black-throated Sparrow Interaction Calls, p. 443

Eastern Towhee, p. 424
Spotted Towhee, p. 425
Indigo Bunting, p. 520

High, musical, notes often Pinklike
Red-winged Blackbird, pp. 470–471

High, less musical
White-rumped Sandpiper, p. 145
Eastern Kingbird, p. 289
Tropical Kingbird, p. 284

CHITTER (SCREECHY OR COMPLEX)
Variable, usually plastic twitters of high-pitched, usually unmusical notes, often with a screechy quality. See Noisy Chitter, p. 582.

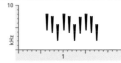

Long, loud, extremely plastic
White-throated Swift, p. 117
Lewis's Woodpecker, p. 234

Similar but shorter
Rosy-faced Lovebird, p. 258

Softer, slower
Tricolored Blackbird Female Song, p. 469

Shorter, notes more clicking
Eastern Screech-Owl, p. 227

Western Screech-Owl, p. 226
Whiskered Screech-Owl, p. 228
Elf Owl, p. 229
Snowy Owl, p. 214
Great Horned Owl, p. 213
Spotted Owl, p. 221
Great Gray Owl, p. 219
Northern Hawk Owl, p. 215
Northern Saw-whet Owl, p. 225

Boreal Owl, p. 224
Barn Owl, p. 212
Ferruginous Pygmy-Owl, p. 218
Northern Pygmy-Owl, p. 216
Short-eared Owl, p. 223
Peregrine Falcon, p. 257
(likely other falcon species)

A SOUND MADE OF REPEATED SIMILAR NOTES

A PEEPING SERIES OR RATTLE

RATTLE (TYPICAL)
A coarse trill of sharp, unmusical Piklike or Pwiklike notes.

Higher, longer, more sputtering
Brown-headed Cowbird,
p. 480
Bronzed Cowbird, p. 481
Black-and white Warbler,
p. 490
several warbler species
(see p. 511)

Lower, shorter
Lapland Longspur, p. 420
Chestnut-collared Longspur,
p. 421
McCown's Longspur, p. 422
Snow Bunting, p. 419

*Lower, longer, less musical,
sometimes slightly buzzy*
Eastern Meadowlark, p. 472
Western Meadowlark, p. 473

Low, clucking, downslurred
Summer Tanager, p. 513

Short, peeping
Long-billed Dowitcher, p. 148
Pygmy Nuthatch, p. 345

*High, slow, slightly more musical
or metallic at start*
Red-winged Blackbird,
pp. 470–471
Black Phoebe, p. 276

High, fast, slightly screechy
Belted Kingfisher High Rattle,
p. 232
Nuttall's Woodpecker, p. 241
American Kestrel, p. 254

*Medium-low, harsh, slightly
screechy*
Belted Kingfisher Low Rattle,
p. 232
Hairy Woodpecker, p. 243
Arizona Woodpecker, p. 244
White-headed Woodpecker,
p. 245
American Three-toed
Woodpecker, p. 246
Black-backed Woodpecker,
p. 247

PSEEP SERIES
Long strings of Pseeplike notes, usually from begging juvenile birds.

Fast
most grebe species (see
pp. 85–91)
American Coot, p. 125
Scripps's Murrelet, p. 158

Scaly-breasted Munia, p. 395

Slow
Sulphur-bellied Flycatcher,
p. 283

PIP OR PEEP SERIES
*A series of Piplike or Peeplike notes; slower and more stereotyped
than Peeping Twitters. See also Finch Flight Call Group (p. 567).*

Slow series of 2–4 Pee notes
Spotted Sandpiper, p. 150

Fast series of 4–8 Pee notes
Wandering Tattler, p. 155

Long series of Pips
Spotted Sandpiper, p. 150

Short series of 5–7 Pips
Whimbrel, p. 137
Ash-throated Flycatcher,
p. 280

Long plastic series of Peeps
Royal Tern, p. 178

Elegant Tern, p. 179
Arctic Tern, p. 181
Least Tern, p. 176

PITTIK
Like a Rattle of Piklike notes, but shorter, often just 2 notes.

Nuttall's Woodpecker, p. 241
White-headed Woodpecker, p. 245

A SHORT TRILL OF PIPLIKE NOTES
Like a Pipping Trill, but usually shorter, often just 2–3 notes.

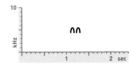

Usually 2 identical sharp Pips
Red-necked Phalarope
Piddip, p. 156
Gray Flycatcher Pittit, p. 270
Pacific-slope Flycatcher Pittit,
p. 272

Pygmy Nuthatch Song, p. 345

Usually 3 rather mellow Pips
Upland Sandpiper Quiddyquit,
p. 138

TRILL (PIPPING)
A trill of medium-low Piplike notes. Often plastic. See Peeping Twitter.

Least Sandpiper, p. 146
Western Sandpiper, p. 147
Solitary Sandpiper, p. 151
Northern Pygmy-Owl, p. 216
Ferruginous Pygmy-Owl,
  p. 218
Gray Vireo, p. 296

Black-capped Vireo, p. 294
Cassin's Vireo, p. 298
Plumbeous Vireo, p. 299
Pine Grosbeak, p. 401
(antelope ground squirrel
  species)

TWITTER (PEEPING)
A highly plastic series of medium- to high-pitched Pip- or Peeplike notes, often accelerating briefly into a trill.

Often slightly chattering, with some noisy notes
  Red Phalarope, p. 156
  Red-necked Phalarope, p. 156

Long, with large changes in pitch
  Black Oystercatcher, p. 128

High, notes sharp, plastic
  Orange-cheeked Waxbill,
    p. 394

High, notes sharp, usually from high overhead
  Black Swift, p. 117

High, notes sharp, during copulation
  Northern Pygmy-Owl, p. 216

Most notes near same pitch
  Fulvous Whistling-Duck,
    p. 36
  Black-bellied Whistling-
    Duck, p. 37
  Mandarin Duck, p. 44
  Long-billed Dowitcher, p. 148
  Short-billed Dowitcher,
    p. 148
  Least Sandpiper, p. 146
  Semipalmated Sandpiper,
    p. 147

Western Sandpiper, p. 147
Baird's Sandpiper, p. 145
Sanderling, p. 142
Dunlin, p. 143
Pygmy Nuthatch, p. 345

Rather soft, medium-low; notes musical Pips
  Baltimore Oriole, p. 484
  Bullock's Oriole, p. 485

Short, low, slightly nasal; first note usually highest
  Chestnut-collared Longspur
    Kiddle, p. 421

A VERY HIGH SERIES OR TRILL

DZIT OR DZEET SERIES
A series of high burry or buzzy notes.

Very high
  Bushtit, p. 339
  Chipping Sparrow, p. 436
  Clay-colored Sparrow, p. 437

Brewer's Sparrow, p. 438

Medium-high
  American Dipper, p. 362

SEET SERIES
A high to very high monotone series of Seetlike notes.

Stereotyped, regularly repeated
  Golden-crowned Kinglet,
    p. 363

Plastic, given irregularly
  Black-capped Chickadee,
    p. 334
  Mountain Chickadee, p. 335
  Boreal Chickadee, p. 336
  Chestnut-backed Chickadee,
    p. 337
  Mexican Chickadee, p. 338

Black-crested Titmouse,
    p. 330
  Bridled Titmouse, p. 331
  Oak Titmouse, p. 332
  Juniper Titmouse, p. 333

Plastic, series often quite long
  House Wren, p. 349

Plastic, with trills or series of Pseeps interspersed
  Pigeon Guillemot, p. 157

Notes very brief, well spaced
  Blue-throated Hummingbird
    Seet Song, p. 105

4-6 notes with consonant at start
  Brown Creeper Psee Series,
    p. 346

A VERY HIGH TRILL
A fairly musical trill of Psip- or Seetlike notes. Compare Burry Seet (p. 538).

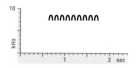

Fairly musical
Cedar Waxwing, p. 390

Less musical; trills often downslurred
Guadalupe Murrelet, p. 158
Craveri's Murrelet, p. 158

Bohemian Waxwing, p. 390

Musical, extremely coarse; notes sometimes slow enough to count
Bushtit Trill, p. 339
(some ground squirrel species)

Notes sharp, trill cricketlike
Ferruginous Pygmy-Owl (juvenile), p. 218

TWITTER (OF PSIP- OR SEETLIKE NOTES)
A highly plastic series of very high-pitched notes, often accelerating briefly into a trill.

Golden-crowned Kinglet, p. 363

Cassin's Sparrow, p. 432

A MUSICAL SERIES OR TRILL

A SERIES OF VARIABLE WHISTLES
These species give a huge variety of whistled notes in series. The notes in any given series are identical, but the series vary individually and geographically.

Short series, often repeated; notes medium-high
Black-crested Titmouse Song, p. 330
Bridled Titmouse Song, p. 331
Oak Titmouse Song, p. 332
Juniper Titmouse Song, p. 333
Olive Warbler Song, p. 392
Dark-eyed Junco (Red-backed), p. 459

Yellow-eyed Junco, p. 460

Short series, fast or slow, often slightly plastic
Gray Jay, p. 304

Short series; notes high, musical
Canyon Towhee, p. 428
American Redstart, p. 500

Short series; notes musical to metallic
Great-tailed Grackle Series Calls, pp. 474–475

Rather long series; notes medium-high, chirping
Wrentit Female Song, p. 365

Long series; notes high, musical to chirping
American Pipit Song, p. 396

Long series; notes shrill, usually nearly monotone
Killdeer, p. 132

A SERIES OF MONOTONE WHISTLES
A series of high Pee or Hee notes.

5–15 medium-high monotone whistles
Montezuma Quail Female Song, p. 71

Higher, longer, much faster
Wandering Tattler Song, p. 155

Highly plastic, unsteady in pitch and rhythm
Western Wood-Pewee, p. 263

A SERIES OF UPSLURRED WHISTLES
Slower than Sputter. See also Northern Cardinal Song Group (p. 605).

Notes medium-low, variable, short
Spotted Sandpiper Pee-pwee-pwee, p. 150
Solitary Sandpiper Pwee-pwee-pwee, p. 151

Cordilleran Flycatcher Pweet Series, p. 273

Notes higher, Weeplike, long
Baird's Sandpiper Weep Series, p. 145
See also Weep, p. 540

Notes high, Weeplike, short
Olive-sided Flycatcher Weep Series, p. 265

A SERIES OF DOWNSLURRED WHISTLES
See also *Northern Cardinal Song Group (p. 605)* and *Finch Flight Call Group (p. 567)*.

3–5 medium-low Pewlike notes, often in flight
Lesser Yellowlegs Pew
Series, p. 153
Greater Yellowlegs Pew
Series, p. 152
Short-billed Dowitcher Pew
Series, p. 148

3–5 medium-high Peerlike notes
Black-crested Titmouse
Song, p. 330

5–7 high Peerlike notes, series downslurred
Northern Beardless-
Tyrannulet Day Song, p. 262

Rapid series of high Tewlike notes
Chestnut-backed Chickadee
Trill Song, p. 337

Longer series of Peerlike notes
Lesser Yellowlegs Peer
Series, p. 153

Thick-billed Kingbird Peer
Series, p. 287

Longer series of high Teerlike notes
Violet-crowned Hummingbird
Simple Song, p. 116

A DOWNSLURRED SERIES OF WHISTLES
A downslurred series of whistled notes, sometimes changing in speed and quality.

Notes nearly monotone
Montezuma Quail Female
Song, p. 71
Northern Beardless-
Tyrannulet Day Song, p. 262

Notes variable; series slowing
Canyon Wren Male Song,
p. 351

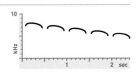

AN ACCELERATING SERIES OF WHISTLES OR CHIRPS
A series of similar whistled or chirping notes that increases in speed. See also Accelerating Series of Chiplike Notes, p. 568.

All notes musical whistles
Field Sparrow, p. 439

Starts with musical whistles, becomes unmusical buzz
Black-chinned Sparrow, p. 435

All notes semimusical, Piplike
Wrentit Male Song, p. 365

All notes high, musical Teets
California Towhee, p. 429

All notes complex, semimusical
Dickcissel, p. 515

TRILL (MUSICAL, WHISTLED)
A single musical to semimusical trill or rapid series.

High, whistled; notes mostly sharp or complex
Chipping Sparrow, p. 436
Dark-eyed Junco,
pp. 458–459
Dark-eyed Junco (Red-
backed), p. 459
Swamp Sparrow, p. 453

Usually slightly unsteady in pitch
Orange-crowned Warbler,
p. 492
Wilson's Warbler, p. 508

Medium-low, musical to semimusical
Black-crested Titmouse Song,
p. 330

Bridled Titmouse Song, p. 331
Oak Titmouse Song, p. 332
Juniper Titmouse Song, p. 333
Medium-low, plastic, semimusical
Rock Wren, p. 350

SPUTTER OR WEET SERIES
High, semimusical slow trills or fast series, the individual notes usually upslurred Weetlike notes. Less nasal than Laugh (p. 577).

Slightly slower (Weet Series)
Curve-billed Thrasher
(Eastern), p. 378

Slightly faster (Sputters), high and squeaky
Rivoli's Hummingbird, p. 106

Northern Beardless-
Tyrannulet, p. 262

Fast (Sputters), rather mellow
Brown-crested Flycatcher,
p. 282

Western Kingbird, p. 288

A MUSICAL SERIES OR TRILL, CONTINUED

WING WHISTLES
A trill or rapid series of breathy, musical, nearly monotone whistles. Sometimes sounds slightly polyphonic, if left and right wing are not precisely in tune. For unmusical wing sounds, see pp. 566–567.

Short (given mostly at takeoff)
  Mourning Dove, p. 96
  sometimes other doves
  (see p. 92)

Short, repeated (given at bottom of display dives)
  Black-chinned Hummingbird,
  p. 108

Long (given throughout flight)
  Common Goldeneye, p. 58
  Barrow's Goldeneye, p. 59
  Surf Scoter, p. 56
  White-winged Scoter, p. 56
  Hooded Merganser, p. 62

Long or short semimusical trill
  Broad-tailed Hummingbird,
  p. 114

TRILL (OVERSLURRED)
A musical to semimusical trill that rises and then falls in pitch.

High, recalling referee whistle
  Broad-tailed Hummingbird Dive Display, p. 114

TRILL (DOWNSLURRED)
A downslurred musical to semimusical trill. See also Beert (p. 544).

Very high, not very musical
  Bohemian Waxwing, p. 390

Medium-low, musical to semimusical
  Pinyon Jay, p. 312
  Carolina Wren Beert, p. 352
  House Wren Beert, p. 349

A LOW-PITCHED SERIES OR TRILL

A SERIES OF MELLOW MONOTONE WHISTLES
Lower than Series of Monotone Whistles (p. 572).

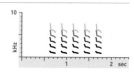

3–5 notes, series often repeated
  Black-billed Cuckoo Song,
  p. 99

Long steady series
  Northern Saw-whet Owl
  Song, p. 225
  Northern Pygmy-Owl Song
  (all populations),
  pp. 216–217

Ferruginous Pygmy-Owl
Song, p. 218

A SERIES OF MELLOW DOWNSLURRED WHISTLES
Series of rather low, mellow downslurred notes.

  Northern Pygmy-Owl (all populations), pp. 216–217

KLEW SERIES
Series of rather low, mellow downslurred notes, each of which breaks to a lower pitch.

  Black-billed Cuckoo, p. 99

MELLOW WAIL SERIES
Series of long overslurred mellow whistles, sometimes slightly nasal, sometimes broken. See also Longer Overslurred Whistle (p. 541).

Gray Hawk, p. 206

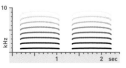

HOOTING OR COOING SERIES
A steady series of single hooting or cooing notes.

Extremely low, sometimes syncopated
Dusky Grouse Song, p. 82
Sooty Grouse Song, p. 83

Slow; individual notes upslurred
Common Ground-Dove Song, p. 95

Slow; individual notes monotone or overslurred
Rock Pigeon Song, p. 92
Great Gray Owl Song, p. 219

Snowy Owl Song, p. 214

Slow; series slightly downslurred
Yellow-billed Cuckoo Song, p. 98
Greater Roadrunner Song, p. 100

Medium-fast, series steady or slowing
Spotted Owl Series Song, p. 221

Fast; individual notes monotone
Short-eared Owl Song, p. 223
Whiskered Screech-Owl Song, p. 228

Series very long
Whiskered Screech-Owl Long Song, p. 228
Boreal Owl Long Song, p. 224

TRILL (HOOTING)
One low, hooting trill.

Long (10 seconds or more)
Lesser Nighthawk Trill, p. 102
Boreal Owl Song, p. 224
Northern Hawk Owl Song, p. 215

(trills of some toad species)

Shorter, rhythm often slightly irregular
Eastern Screech-Owl Trill, p. 227

WILSON'S SNIPE GROUP
A rapid, upslurred or barely overslurred series of low, almost hooting whistles.

Wilson's Snipe Winnow, p. 149   Boreal Owl Song, p. 224

AN ACCELERATING SERIES OF HOOTS
An accelerating series of hooting notes or very low-pitched whistles.

Western Screech-Owl Song, p. 226

AN UNSTEADY SERIES OF HOOTS
An irregular or syncopated series of hooting notes or very low-pitched whistles.

Whiskered Screech-Owl Morse Code Song, p. 228

GROWL SERIES
A series of very low-pitched burry notes. Not very loud.

Band-tailed Pigeon Growl Song, p. 93

Wilson's Phalarope, p. 155
Mexican Whip-poor-will, p. 103
Greater Roadrunner, p. 100

GROAN SERIES
A series of low noisy notes. Not very loud.

Accelerating
White-tailed Ptarmigan, p. 84

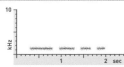

A SERIES OF BURRS OR BUZZES

A SERIES OF CHEERLIKE NOTES
A series of coarsely burry, rather unmusical notes.

Red-winged Blackbird Female Song, pp. 470–471

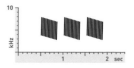

BREET SERIES
A series of medium-high, coarsely burry, nearly monotone musical notes.

White-headed Woodpecker, p. 245

Cassia Crossbill Song, p. 417

DZIK SERIES
A series of high, brief, sharply upslurred buzzes. See Dzweet (p. 547).

Eastern Meadowlark, p. 472
American Goldfinch, p. 407

Lesser Goldfinch, p. 409

KREE SERIES
A series of soft, medium-pitched upslurred burry notes. Some Twitters of Tiplike notes (p. 569) can sound similar. See also Screechy Chitters (p. 569).

Greater Pewee, p. 264

Baltimore Oriole, p. 484
Bullock's Oriole, p. 485

A MONOTONE SERIES OF TRILLS
A series of 2 or more fairly musical trills or burry notes, all similar in pitch and quality.

Rapid, about 3–4 trills/second
Rufous Hummingbird Shuttle Display, p. 112

Allen's Hummingbird Shuttle Display, p. 113
Broad-tailed Hummingbird Shuttle Display, p. 114

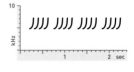

A MONOTONE SERIES OF BUZZES
A series of 2 or more fairly long buzzes, all similar in pitch and quality.

Slower, usually 2–6 notes
Clay-colored Sparrow, p. 437

Faster, sometimes almost a trill, sometimes a couplet series
Brewer's Sparrow, p. 438

Buzzes low, made by wings
Lucifer Hummingbird Shuttle Display, p. 107
Calliope Hummingbird Shuttle Display, p. 111

AN ACCELERATING BUZZY SERIES
A series of buzzy notes, getting faster, usually all on the same pitch.

Bank Swallow Song, p. 325

A DECELERATING BUZZY SERIES
A series of buzzy notes, slowing and usually falling in pitch at end.

Series usually overslurred
Canyon Wren Female Song, p. 351

Series usually falling
Mountain Plover Song, p. 133
Dunlin Song, p. 143

A SERIES OF SHORT NASAL NOTES

KEEK SERIES
A series of high, short, abrupt, monotone nasal notes. Like Whinny but more monotone.

Northern Harrier, p. 199
Whimbrel, p. 137
Long-billed Curlew, p. 136
Northern Hawk Owl, p. 215
Northern Saw-whet Owl,
   p. 225
Aplomado Falcon, p. 255

Higher, notes downslurred
Common Black Hawk, p. 204
Sharp-shinned Hawk, p. 201
Sora, p. 123

High, series often overslurred
Common Black Hawk, p. 204
Bald Eagle, p. 210

Notes slightly Piklike
American Three-toed
   Woodpecker, p. 246

Notes short; series long, slow
Willet (Western), p. 154

Notes longer, lower, series slow
Red-shouldered Hawk,
   p. 207

Notes nasal, Keklike; series long
Wood Duck, p. 45
Mountain Quail, p. 66
Gila Woodpecker, p. 237
Northern Flicker, p. 248
Gilded Flicker, p. 249
Pileated Woodpecker, p. 253

LAUGH
A rapid series or slow trill of nasal, often upslurred notes. Pitch medium. More nasal than Sputter (p. 573).

Medium-high
Semipalmated Sandpiper,
   p. 147
LeConte's Thrasher, p. 383

Fairly high, often quite slow
Willet (Western), p. 154

Medium-low
American Avocet, p. 128
African Collared-Dove, p. 94

Hutton's Vireo, p. 297

A SERIES OF SQUEEPS OR PWEEWS
A series of high complex Cheeplike notes, each usually upslurred.

Medium-high, usually 2–3 notes
American Robin Squee-
   squeep, p. 377

Medium-high, first note Whitlike
Least Flycatcher Pweew
   Series, p. 271

Medium-low, plastic, sometimes noisy
Greater Pewee Pweew
   Series, p. 264
White-throated Sparrow
   Squeep Series, p. 448
Harris's Sparrow Squeep
   Series, p. 449

White-crowned Sparrow
   Squeep Series, p. 450
Golden-crowned Sparrow
   Squeep Series, p. 451

WIK SERIES (HIGH)
A series of fairly high, seminasal upslurred Wik notes.

Northern Flicker, p. 248
Gilded Flicker, p. 249
Hairy Woodpecker, p. 243

White-headed Woodpecker,
   p. 245
Pileated Woodpecker, p. 253

TEW SERIES
Series of high, sharply downslurred seminasal notes.

American Pipit, p. 396
Dark-eyed Junco, pp. 458–459

Yellow-eyed Junco, p. 460

YAP SERIES
A series of high nasal notes, higher than Nasal Bark Series (p. 578), the notes sometimes running together. See also 2- or 3-Syllabled Nasal Note (p. 591).

Fast, usually just 2 or 3 notes
Bar-tailed Godwit Pup-pup,
   p. 138
Franklin's Gull Pup-pup,
   p. 164

Mew Gull Pup-pup, p. 167
Royal Tern, p. 178

Fast, often longer
Gull-billed Tern, p. 177

Notes slower, high, Keklike
Gila Woodpecker, p. 237

CHWEEK SERIES
Series of high, sharply underslurred seminasal notes. Compare Klee Series.

Slow
Ferruginous Pygmy-Owl, p. 218

Fast
Arizona Woodpecker, p. 244
American Three-toed Woodpecker, p. 246

Green Kingfisher, p. 232

BARK SERIES (NASAL)
A series of brief medium-low nasal notes. Lower than Yap Series (p. 577); more nasal than Kek Series (p. 581), and notes often longer.

Notes brief, Keklike; series fast
Cooper's Hawk, p. 202
Bald Eagle, p. 210

Notes brief, Keklike; series slow
Black-legged Kittiwake, p. 162

Series monotone, often long
Common Goldeneye, p. 58
Barrow's Goldeneye, p. 59
Black-necked Stilt, p. 127
Wilson's Snipe, p. 149
Wilson's Phalarope, p. 155
Heermann's Gull, p. 165
Black Tern, p. 183

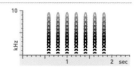

Western Screech-Owl, p. 226

Notes monotone or downslurred
Northern Harrier, p. 199

Notes medium to low, mellow
Greater Roadrunner, p. 100
Spotted Owl, p. 221

KLEE SERIES, KWEEK SERIES, AND SIMILAR SOUNDS
Like Whinny, but with broken notes. Compare Chweek Series.

Long, slow series, only some notes broken
Red-tailed Hawk, p. 205
Swainson's Hawk, p. 208
American Avocet, p. 128

Shorter, faster series, most notes broken
Golden Eagle, p. 210
Osprey, p. 211

Solitary Sandpiper Song, p. 151
Greater Yellowlegs, p. 152
Least Tern, p. 176
American Kestrel, p. 254

Series fast, notes slightly noisy, Peeklike (Kweek Series)
Ladder-backed Woodpecker, p. 240

Nuttall's Woodpecker, p. 241
Hairy Woodpecker, p. 243
Downy Woodpecker, p. 242
Arizona Woodpecker, p. 244
White-headed Woodpecker, p. 245

YELP SERIES
A series of brief, medium-pitched broken nasal notes. See also Large Gull Long Calls (p. 613).

Wild Turkey, p. 65

Northern Bobwhite, p. 70

WHINNY (OVERSLURRED)
A rapid series or slow trill of Keek- or Keklike notes, the first and last 1–2 notes usually slightly lower than the rest. Higher and coarser than Bleats. Compare Screechy Whinny (p. 581) and Keek Series (p. 577).

High, piping, some notes broken
Common Black Hawk, p. 204
Bald Eagle Keek Series, p. 210

High, semimusical
American Robin, pp. 376–377

Medium-high, nasal
Pileated Woodpecker, p. 253

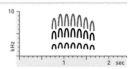

Medium-high, seminasal, slow
Elf Owl, p. 229

WHINNY (DOWNSLURRED)
A rapid series or slow trill of Keek-, Kek-, or Piklike notes, the series generally falling and sometimes slowing.

High, downslurred; notes Piklike
Downy Woodpecker, p. 242
Ladder-backed Woodpecker, p. 240

High, decelerating; notes Keklike
Sora, p. 123

Lower, slowing; notes Keklike
Common Gallinule, p. 124

WHINNY-RASP
A whinny of Keeklike notes that accelerate into a harsh Rasp (p. 560).

Black-backed Woodpecker, p. 247

CHATTER (NASAL)
Rapid, highly plastic rattle of seminasal, semimusical Piplike notes. Higher and more plastic than Nasal Trill. **See also Chatter Duet.**

Medium-high
Rufous-crowned Sparrow, p. 427

Low
White-tailed Ptarmigan, p. 84

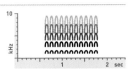

CHATTER DUET
A rapid series of nasal Piklike notes, usually by 2 birds in highly synchronized chorus, often sounding like a single bird. **See also Churr (p. 558) and Nasal Chatter.**

Slower, more laughing
Pied-billed Grebe, p. 90

Faster, churring
Least Grebe, p. 91
Black Rail, p. 120

TRILL (NASAL)
A rapid trill of brief, highly nasal notes. Faster and more monotone than Whinnies; more stereotyped than nasal Chatter.

Red-breasted Nuthatch Fast Song, p. 341

White-breasted Nuthatch (all populations), pp. 342–344

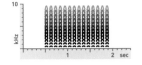

TWITTER (NASAL)
A highly plastic rapid series or trill of very brief nasal notes, often irregular in rhythm.

Soft, often trilled
Red-breasted Nuthatch, p. 341
White-breasted Nuthatch (Pacific), p. 342

White-breasted Nuthatch (Montane), p. 343

Usually irregular single notes
White-breasted Nuthatch (Eastern), p. 344

BLEAT (HIGH)
A fairly high burry note, trill, or tremolo, with a nasal to screechy quality.

Extremely coarse, screechy
Northern Hawk Owl, p. 215

A long, high, coarse nasal burr
Horned Grebe, p. 89

Similar but finer
Eared Grebe, p. 87

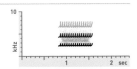

GARGLE
Comical nasal gibberish, like the language of cartoon aliens.

Snowy Egret, p. 194
Yellow-headed Parrot, p. 260

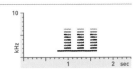

A SERIES OF SHORT NASAL NOTES, CONTINUED

BLEAT (LOW)
A burry wail or moan. Coarser than Churr (p. 558).

High, rather clear
Tundra Swan, p. 43

Medium pitch, very nasal
Lesser Nighthawk, p. 102

Medium pitch, nasal, breathy
Rufous Hummingbird Dive
Display, p. 112

Medium pitch, noisier
Brant, p. 40
Sandhill Crane, p. 126
California Gull, p. 172
Common Murre, p. 160

Low, rather fine
Glossy Ibis, p. 196

White-faced Ibis (probably),
p. 196

Even lower, coarser
Least Bittern Song, p. 192

GOBBLE
A short rapid series of Kuklike notes. Coarser and higher than Purr (p. 559).

Short, monotone
Sharp-tailed Grouse,
pp. 74–75

Longer, downslurred
Wild Turkey, p. 65

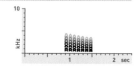

A SERIES OF LONGER NASAL NOTES

REEK OR YANK SERIES
A series of rising, medium- to low-pitched nasal notes.

Series of 5–15 notes
Cooper's Hawk, p. 202
White-breasted Nuthatch
(all populations),
pp. 342–344

*Longer series of single
well-spaced notes*
Red-breasted Nuthatch
Slow Song, p. 341

Similar but higher
Pinyon Jay Reek Series,
p. 312

DEER SERIES
A series of falling, medium- to low-pitched nasal notes.

Cassin's Kingbird, p. 286
Thick-billed Kingbird, p. 287

Western Kingbird, p. 288

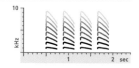

KLEER SERIES
A series of high downslurred seminasal or nasal notes that break to a lower pitch.

Quite slow, nasal
Red-shouldered Hawk,
p. 207

Quite fast, seminasal
Greater Yellowlegs, p. 152
Rose-ringed Parakeet,
p. 258

TEWEE SERIES
A series of high underslurred seminasal notes.

White-breasted Nuthatch (Pacific), p. 342

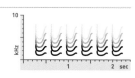

SCREAM SERIES
A series of long, medium- to high-pitched seminasal or nasal notes.

Screams very high, monotone or upslurred
  Sharp-shinned Hawk, p. 201
  American Kestrel, p. 254

Screams upslurred, high, seminasal to nasal
  Wood Duck, p. 45

Northern Goshawk, p. 203
Bald Eagle, p. 210
Osprey, p. 211
Merlin, p. 255
Aplomado Falcon, p. 255
Peregrine Falcon, p. 257
Prairie Falcon, p. 256

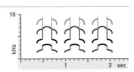

Gyrfalcon, p. 256
Crested Caracara, p. 254

SQUEAL SERIES
A series of high seminasal notes that break to a higher pitch and back down, quality clear to harsh.

3–5 harsh, hoarse Squeals
  Red-naped Sapsucker, p. 250
  Red-breasted Sapsucker, p. 251
  Williamson's Sapsucker, p. 252

Longer series, notes mostly clear
  Red-tailed Hawk, p. 205
  Red-shouldered Hawk, p. 207
  Zone-tailed Hawk, p. 206
  Short-tailed Hawk, p. 208
  Harris's Hawk, p. 204

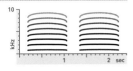

many raptor species in copulation (see p. 197)

Squeals short, variable
  Black-throated Magpie-Jay, p. 303

WAIL SERIES
A series of long overslurred wails.

Medium-low
  Indian Peafowl, p. 64

MOAN SERIES
A series of low nasal moans.

  Rhinoceros Auklet, p. 159

A SERIES OF SHORT HARSH NOTES

WHINNY (SCREECHY)
A series of high, variably screechy Keeklike notes, the first and last 1–2 notes in the series often slightly lower. Compare Keek Series (p. 577) and Overslurred Whinny (p. 578).

  Red-necked Grebe, p. 88   Merlin, p. 255

KEK SERIES
A series of very brief noisy Chuklike notes.

  Green Heron, p. 195   Common Gallinule, p. 124
  Ridgway's Rail, p. 122   Marsh Wren (Western), p. 354

BARK SERIES (ROUGH)
A series of rather harsh nasal notes. Noisier than Nasal Bark Series (p. 578).

  Gila Woodpecker, p. 237

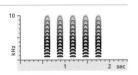

KUK OR CLUCK SERIES
A series of very brief low nasal notes, sometimes harsh. Lower and more nasal than Kek Series (p. 581).

White-tailed Ptarmigan, p. 84

CHITTIT AND SIMILAR SOUNDS
A rapid pair of identical Chiplike, Chitlike, or Chuklike notes.

High, slightly nasal Jiddit
Ruby-crowned Kinglet, p. 364
Black-capped Vireo, p. 294

Higher, notes Chitlike
Broad-billed Hummingbird, p. 115

High, notes Chiplike
Rivoli's Hummingbird Chittip, p. 106

Harsher
White-winged Crossbill Chittit, p. 418

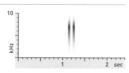

Snow Bunting Chip-it, p. 419

Low, notes Chuklike
Yellow-headed Blackbird Chuttuk, p. 479

CHITTER (NOISY)
Like Chatter, but higher-pitched, made of Chitlike, not Chuklike notes. Usually plastic. See Screechy Chitter, p. 569.

Anna's Hummingbird, p. 109
Broad-billed Hummingbird, p. 115
Downy Woodpecker, p. 242
Nuttall's Woodpecker, p. 241
Arizona Woodpecker, p. 244
Chimney Swift (juv.), p. 118
Vaux's Swift (juv.), p. 118
European Starling, pp. 388–389
Barn Swallow, p. 324

Notes rather nasal
Wood Duck, p. 45
Fox Sparrow (all forms), pp. 455–457

Notes rather chipping
Canyon Wren, p. 351

Short, rather sputtering
Boreal Chickadee, p. 336
Black-capped Chickadee, p. 334
Mountain Chickadee, p. 335
Chestnut-backed Chickadee, p. 337
Mexican Chickadee, p. 338

Soft, plastic, often polyphonic
Cassin's Finch, p. 403

Notes usually nasal, slightly buzzy
Ruddy Turnstone, p. 140
Bewick's Wren, p. 353

Eastern Phoebe, p. 277
Western Kingbird, p. 288
Scissor-tailed Flycatcher, p. 290

Slightly screechy
Black Turnstone, p. 140
Rock Sandpiper, p. 143
Pectoral Sandpiper, p. 144
Stilt Sandpiper, p. 142
Ash-throated Flycatcher, p. 280
Thick-billed Kingbird, p. 287

Quite high
Tropical Kingbird, p. 284

CHATTER (NOISY)
A rapid series or slow trill of Chuklike noisy notes. Faster than Snarl Series.

Lesser Yellowlegs, p. 153
Red-headed Woodpecker, p. 235
Red-naped Sapsucker, p. 250
Red-breasted Sapsucker, p. 251
Williamson's Sapsucker, p. 252
Green Kingfisher, p. 232
House Wren, p. 349
Sedge Wren, p. 356
Cactus Wren, p. 357
Black-tailed Gnatcatcher, p. 358
Western Bluebird, p. 368
Eastern Bluebird, p. 369
Varied Thrush, p. 371
Bendire's Thrasher, p. 381
Curve-billed Thrasher, p. 385

Hooded Oriole, p. 483
Orchard Oriole, p. 482
Scott's Oriole, p. 486
Baltimore Oriole, p. 484
Bullock's Oriole, p. 485
Great-tailed Grackle, pp. 474–475
Brewer's Blackbird, p. 477
House Sparrow, p. 393
Bronze Mannikin, p. 394
Five-striped Sparrow, p. 442

Short, almost always 3 notes
Western Tanager Chippituk, p. 514

Fast to slow
Gray Jay, p. 304

Slow, plastic
Steller's Jay, p. 305

Usually in middle of long barking series
Egyptian Goose, p. 42

Rather slow and nasal
Black-billed Magpie, p. 314
Yellow-billed Magpie, p. 315

Fast, almost rasping or churring
Yellow-chevroned Parakeet, p. 259
Winter Wren, p. 347
Wrentit, p. 365
Black-throated Sparrow, p. 443

CHUCKLE

A rapid series of grunts, low barks, or Kuklike notes. Lower and more nasal than Chatters, and often slower. Less croaking than Kruk Series (p. 584). Faster than Grunt Series (p. 584).

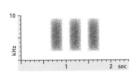

Short series of rather harsh grunts
  Snow Goose, p. 39
  Ross's Goose, p. 39
  Greater White-fronted Goose, p. 40
  Mallard, p. 52
  Mexican Duck, p. 53
  Gadwall, p. 48
  Lesser Scaup, p. 54
  Common Merganser, p. 61
  Snowy Owl, p. 214
  Elegant Trogon, p. 231

Series of low, swallowed Kuks
  Greater Sage-Grouse, p. 78
  Gunnison Sage-Grouse, p. 79
  Dusky Grouse, p. 82
  Sooty Grouse, p. 83
  Least Bittern Song, p. 192

Short series of grunting to screeching Kuklike notes
  Heermann's Gull, p. 165
  California Gull, p. 172

Ring-billed Gull, p. 166
Western Gull, p. 168
Glaucous-winged Gull, p. 169
Herring Gull, p. 170
Lesser Black-backed Gull, p. 173
Glaucous Gull, p. 174

Short series of very nasal honks
  Trumpeter Swan, p. 43

Short series, harsh to metallic
  Steller's Jay, p. 305

Short fast series, almost churring
  Northern Pintail, p. 50
  Green-winged Teal, p. 51
  Gull-billed Tern, p. 177

Medium-long series of nasal Kuklike notes
  Sharp-tailed Grouse, pp. 74–75
  Greater Prairie-Chicken, p. 76
  Lesser Prairie-Chicken, p. 77

Rather long series of harsh downslurred barks
  Red-bellied Woodpecker, p. 239
  Williamson's Sapsucker, p. 252

Rather long series of low Chuklike notes
  Cactus Wren Song, p. 357

Low, harsh; rhythm irregular
  Red-necked Grebe, p. 88

Notes grating
  Sabine's Gull, p. 163

Notes slightly screechy
  Burrowing Owl, p. 230

Notes variable, semimusical
  Piping Plover, p. 131
  Upland Sandpiper, p. 138
  Western Bluebird, p. 368
  Western Meadowlark, p. 473

A SERIES OF LONGER HARSH NOTES

SNARL SERIES
A series of very harsh noisy notes. Slower than Noisy Chatter.

Egyptian Goose, p. 42
Steller's Jay, p. 305
Island Scrub-Jay, p. 308
Carolina Wren, p. 352
Cactus Wren, p. 357

Black-tailed Gnatcatcher Simple Song, p. 358
California Gnatcatcher Simple Song, p. 360
Black-capped Gnatcatcher Simple Song, p. 361

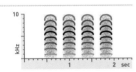

Tricolored Blackbird Female Song, p. 469

SCREECH SERIES
A series of medium-long screechy notes.

Horned Grebe, p. 89
Least Bittern, p. 192
Northern Goshawk, p. 203
juveniles of some tern species, pp. 176–183
Northern Hawk Owl, p. 215
Peregrine Falcon, p. 257
Prairie Falcon, p. 256
Gyrfalcon, p. 256

Highly plastic; sometimes in screechy couplets
  Red-naped Sapsucker, p. 250
  Red-breasted Sapsucker, p. 251
  Williamson's Sapsucker, p. 252

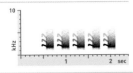

Harsh, notes often overslurred, slightly metallic
  Island Scrub-Jay, p. 308
  California Scrub-Jay, p. 309
  Woodhouse's Scrub-Jay, p. 310

BARK SERIES (YELPING)
A series of swallowed, harsh nasal Barks (p. 550), each note breaking downward at the start.

Elegant Trogon Song, p. 231

A SERIES OF LONGER HARSH NOTES, CONTINUED

GRUNT SERIES
A series of low, fairly brief harsh notes, like Chuckle (p. 583) but slower. See also Rail Long Calls (p. 614).

Egyptian Goose, p. 42   Elegant Trogon, p. 231

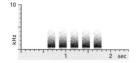

A HARSH BURRY SERIES

DZEER SERIES
A series of finely burry harsh or nasal downslurred notes. See also Dzeet, p. 546.

Cassin's Kingbird, p. 286

CHURR SERIES OR KRAA SERIES
A series of low, harsh nasal burry notes.

Notes usually monotone (Churr Series)
Black Rail, p. 120
Red-headed Woodpecker, p. 235
Acorn Woodpecker, p. 236

Gila Woodpecker, p. 237
Red-bellied Woodpecker, p. 239
Golden-fronted Woodpecker, p. 238

Notes more nasal, often rising
Clark's Nutcracker, p. 313

KRIK OR KRUK SERIES
Rapid, often plastic series of very brief harsh croaking notes.

Medium low, croaking (Krik Series)
Elegant Tern, p. 179

Medium-low, nasal (Kruk Series)
Great Egret, p. 193

Low, grating (Kruk Series)
Great Blue Heron, p. 193
Snowy Egret, p. 194
Little Blue Heron, p. 194
Cattle Egret, p. 195

RASP SERIES
A series of coarse harsh grating notes.

Black-backed Woodpecker, p. 247

American Dipper, p. 362

CROAK SERIES
A steady series of low Croaks.

Fast to slow, rather low, harsh
Greater Sage-Grouse, p. 78
Gunnison Sage-Grouse, p. 79

Glossy Ibis, p. 196
White-faced Ibis (probably), p. 196

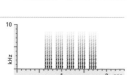

AN ACCELERATING SERIES OF CROAKS
A series of low, harsh croaking notes that increases in speed.

Spruce Grouse, p. 81
Dusky Grouse, p. 82

Sooty Grouse, p. 83

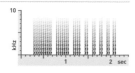

A POLYPHONIC SERIES
See Finch Flight Call Group, p. 567.

WHINE SERIES (HARSH)
A series of noisy polyphonic notes. Slower than Whiny Chatter.

Each note rising
Black-crested Titmouse Dee
Series, p. 330

Notes rather long
Northern Shrike, p. 293
Loggerhead Shrike, p. 292

Black-tailed Gnatcatcher,
p. 358
Yellow-headed Blackbird,
p. 479

Plastic, grading into a chatter
Belted Kingfisher, p. 232

Bell's Vireo, p. 295
Gray Jay, p. 304

CHATTER (WHINY)
A trill of noisy polyphonic notes. Faster than Harsh Whine Series.

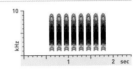

Notes short, high, series rapid
Red-eyed Vireo, p. 302
Bridled Titmouse, p. 331
Oak Titmouse, p. 332
Juniper Titmouse, p. 333

Notes longer, series slower
Bell's Vireo, p. 295

First note often longest
Black-capped Vireo, p. 294
Gray Vireo, p. 296
Cassin's Vireo, p. 298
Plumbeous Vireo, p. 299
Rock Wren, p. 350

Last note often longest
Warbling Vireo, pp. 300–301

METALLIC SERIES
A series of brief Jenklike notes, usually rather harsh, with a metallic ring. See also Zreek Series.

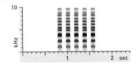

Often fairly harsh
Steller's Jay Chuckle, p. 305
Steller's Jay Pumphandle
Calls, p. 305

Common Grackle Jenk
Series, p. 478

Often less harsh
Great-tailed Grackle Series Calls, pp. 474–475

ZREEK SERIES
A series of upslurred, finely burry polyphonic notes, with at least some metallic quality. See also Metallic Series.

California Scrub-Jay, p. 309
Woodhouse's Scrub-Jay, p. 310

Notes screechier, series slower
Scaled Quail, p. 67

WHINE SERIES (HIGH)
A plastic, usually falling series of high polyphonic notes.

Vesper Sparrow, p. 440
Savannah Sparrow, p. 461

Baird's Sparrow, p. 463
LeConte's Sparrow, p. 464

WHINE SERIES (LOW)
A plastic series of often underslurred polyphonic notes. Pitch medium.

Black-capped Vireo, p. 294
Song Sparrow, p. 454

Lincoln's Sparrow, p. 452
Swamp Sparrow, p. 453

TRILL (POLYPHONIC)
A polyphonic trill.

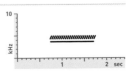

Varied Thrush, p. 371

INDEX PART III: A PHRASE OF 2–3 SYLLABLES

A phrase of 2 or 3 different notes, or a note of 2 or 3 syllables, by itself, or repeated after a pause.

A 2- OR 3-SYLLABLED PHRASE CONTAINING CHIPLIKE NOTES

A SHORT SHARP NOTE AND A WHISTLE
A phrase of 2 different notes, the one Chiplike or Piplike, the other whistled.

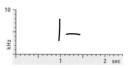

Second note monotone
Wandering Tattler Pi-tee, p. 155
Broad-winged Hawk Pee-tee, p. 207
Rock Wren Spit-tee, p. 350
Orchard Oriole Songlike Calls, p. 482

Second note downslurred
Scaled Quail Chip-seer, p. 67

Calliope Hummingbird Dive Display, p. 111
Mississippi Kite Keek-keer, p. 200
Gray Flycatcher Pi-teep, p. 270
Buff-breasted Flycatcher Pi-teer, p. 274
Say's Phoebe Pi-deer, p. 275
Eastern Phoebe Pi-teer, p. 277
Eastern Kingbird Pi-teer, p. 289

Second note a Pip or Kip
Northern Bobwhite Yoy-kip, p. 70
Olive-sided Flycatcher Peer-pip, p. 265
Dusky Flycatcher Peer-pip, p. 269
Western Kingbird Weer-dip, p. 288

2 OR 3 SHORT SHARP NOTES (PITCHIT GROUP)
A phrase of 2 different Chiplike or Piplike notes.

2 notes, rather peeping
Pygmy Nuthatch Song, p. 345

2 notes, at least one Chiplike
Least Flycatcher Song, p. 271
Eastern Phoebe Pi-teep, p. 277
Black Phoebe Pi-teep, p. 276
Dusky-capped Flycatcher Blee-bit, p. 279
Barn Swallow, p. 324
Carolina Wren P-dit, p. 352
Horned Lark Pitchit, p. 320

Eurasian Skylark Pitchit, p. 321
Snow Bunting, p. 419
Summer Tanager Pittuk, p. 513

3 notes
Summer Tanager Pittituk, p. 513

2 or 3 notes, last one sometimes musical
Mountain Bluebird Chuppa-tew, p. 367

Western Bluebird Chuppa-tew, p. 368
Pine Grosbeak Flight Call, p. 401
Summer Tanager Pitti-tew, p. 513
Western Tanager Chippi-tee, p. 514

A VERY HIGH 2- OR 3-SYLLABLED PHRASE

A VERY HIGH 2- OR 3-SYLLABLED PHRASE
A phrase of 2 or 3 very high Seetlike notes, sometimes burry.

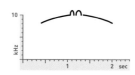

One long, high, rising whistle, then a falling one
Costa's Hummingbird Song, p. 110

A 2- OR 3-SYLLABLED WHISTLED PHRASE

HIGHLY VARIABLE SHORT WHISTLED OR BURRY PHRASES
These species give a huge variety of 2- or 3-syllabled phrases that vary individually and geographically.

Whistled, buzzy, burry, or any combination
Loggerhead Shrike Song, p. 292
Northern Shrike Song, p. 293
Hutton's Vireo Song, p. 297
Steller's Jay Pumphandle Calls, p. 305
Blue Jay Pumphandle Calls, pp. 306–307
Phainopepla Song, p. 391
Red-winged Blackbird Alert Calls, p. 471

Variable 2- or 3-syllabled whistle
American Golden-Plover Pleewee, p. 134

Pacific Golden-Plover Pi-wee-er, p. 134
Caspian Tern (juvenile), p. 177
Purple Finch Tidilip, p. 402
Cassin's Finch Tidilip, p. 403
American Tree Sparrow Tee-yup Calls, p. 434
Painted Redstart Cheer, p. 510

Variable 2- to 3-syllabled phrase of separate whistles
Ferruginous Pygmy-Owl Weet-weet, p. 218
Carolina Wren Peeder, p. 352
Verdin Song, p. 340
Snow Bunting Teer (variant), p. 419

Lapland Longspur Whistled Calls, p. 420
Pine Grosbeak Flight Calls, p. 401
Lesser Goldfinch Tea-cup (juvenile), p. 409
American Goldfinch Tew-weet (juvenile), p. 407
See also Mostly Single 1-syllabled Notes, p. 596

A 2-SYLLABLED WHISTLE
A whistle that breaks once to a different pitch. See also Long, Broken Wails (p. 553), and Pitter (p. 590).

Long, breaking upward; sometimes slightly nasal
Eared Grebe Song, p. 87
Whimbrel Cur-lew, p. 137
Long-billed Curlew Cur-lee, p. 136
Semipalmated Plover Peweep, p. 130
Snowy Plover Ter-weet, p. 129

Short, breaking downward.
Anna's Hummingbird Dive Display, p. 109
Piping Plover Pee-lo, p. 131
Pacific Golden-Plover Pwee-er, p. 134
Red-crowned Parrot Pee-loo, p. 260
Northern Beardless-Tyrannulet Pee-uk, p. 262

Gray Flycatcher Pee-uk, p. 270

Highly plastic, often slightly screechy
Common Murre, p. 160

2 LONG MONOTONE WHISTLES
A pair of high monotone whistles that differ from one another in pitch or length.

High; second note distinctly lower
Black-capped Chickadee, p. 334
Mountain Chickadee, p. 335

High; notes on same pitch
Broad-winged Hawk Pee-tee, p. 207
Harris's Sparrow, p. 449

Low, mellow; notes on nearly same pitch
Black Scoter Song, p. 57

2 SHORT SEPARATE MUSICAL WHISTLES
A pair of different whistled notes.

Notes 1 second apart, second upslurred
Northern Bobwhite Bob-white, p. 70

Notes closer together
Northern Bobwhite Hoyp-woo, p. 70
Killdeer Dee-dit, p. 132

Spotted Sandpiper Pee-pwee, p. 150
Hammond's Flycatcher Peer-pewit, p. 268
Cordilleran Flycatcher Wee-seet, p. 273
Horned Lark Tee-ter, p. 320
Curve-billed Thrasher (Western) Wit-weet, p. 378

California Thrasher Weep-it, p. 382
Lawrence's Goldfinch Teet-tip, p. 408

2–4 short high notes, variable
Verdin Song, p. 340

3 SHORT SEPARATE WHISTLES
A trio of different whistled notes.

2–4 short high notes, variable
Verdin Song, p. 340
Lawrence's Goldfinch
Teeter-tip, p. 408

All notes upslurred
Spotted Sandpiper
Pee-pwee-pwee, p. 150
Curve-billed Thrasher
(Chiricahua dialect)
Weet-wit-weet, p. 378

1 Pip, then 2 long slurred whistles
Olive-sided Flycatcher
Song, p. 265

2 Pips, then 1 long overslurred whistle
Black Oystercatcher
Pitti-weew, p. 128

Variable, shrill, first note longest and highest
Killdeer Dee-dit-dit,
p. 132

A 2- OR 3-SYLLABLED BREATHY PHRASE
Phrase of 2–3 high, breathy, Wheewlike notes. See also Breathy Phrase (p. 605).

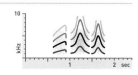

2-noted, high
Fulvous Whistling-Duck
Ki-wheer, p. 36
Black-bellied Whistling-Duck
Ki-whew, p. 37

Eurasian Wigeon Wheew,
p. 49

2-noted, lower, more nasal
Red-breasted Merganser
Whiddew, p. 61

Usually 3- to 4-noted
American Wigeon
Whi-whee-whew, p. 49

A 2- OR 3-SYLLABLED LOW-PITCHED PHRASE

WIDDOO
Like Wow (p. 543), but 2- to 3-syllabled.

Lesser Prairie-Chicken, p. 77
Canvasback, p. 55

Lesser Scaup, p. 54
Greater Scaup, p. 54

A 2- TO 3-SYLLABLED HOOTING OR COOING PHRASE
A phrase of 2 or 3 low-pitched hooting or cooing notes.

2-syllabled
Band-tailed Pigeon Song,
p. 93
Inca Dove Song, p. 95
Common Poorwill Song, p. 104

3-syllabled
Greater Prairie-Chicken
Ooh-loo-woo, p. 76

Eurasian Collared-Dove
Song, p. 94

Usually longer, but sometimes 3-syllabled, first note broken
Mourning Dove Song, p. 96
Flammulated Owl Complex
Song, p. 229

A DOUBLE HOOTING TRILL OR GROWL
A 2-parted low, hooting trill, burry coo, purr, or growl.

Low, growling
Hooded Merganser Growl
Display, p. 62

Slightly higher and slower
Western Screech-Owl
Double-trill, p. 226

A GROWLING OR GROANING PHRASE
A low 2- or 3-syllabled phrase, at least partly constructed of growling notes.

Vaguely 2-syllabled growls, often repeated
Common Merganser G'daa, p. 61

2 notes, both low growls
Hooded Merganser Growl Display, p. 62

2 notes, second long, overslurred, and burry
African Collared-Dove Song, p. 94

2–4 growling or groaning notes
Spotted Dove Song, p. 93

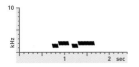

Eurasian Collared-Dove Growl Song, p. 94
Inca Dove Growl Song, p. 95
Common Ground-Dove Growl Song, p. 95

A KNOCKING OR PURRING PHRASE
A brief phrase containing a rattle of very low knocking or clicking notes. See also Knocking Rattle (p. 565).

Knocking
Lesser Nighthawk Chortle, p. 102
Black-billed Cuckoo Chortle, p. 99
Yellow-billed Cuckoo Chortle, p. 98

Common Raven, p. 316
Chihuahuan Raven, p. 317

Purring
Rock Pigeon Growl Song, p. 92

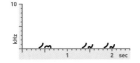

Plastic mix of Clucks, Growls, and clicking phrases
Buff-collared Nightjar Growl-Cluck Phrase, p. 104

A BUBBLING OR POPPING PHRASE
An odd sound like bubbling water or popping bubbles.

Greater Sage-Grouse, p. 78
Gunnison Sage-Grouse, p. 79

A 2- OR 3-SYLLABLED BURRY OR BUZZY PHRASE

VARIABLE BURRY OR BUZZY 2-NOTE PHRASES
These species give a huge variety of 2-syllabled phrases with at least one burry or buzzy note or syllable. These phrases vary individually and geographically.

Bewick's Wren Pewik, Greevit, etc., p. 353
Rose-breasted Grosbeak Zwee-er, p. 518
Black-headed Grosbeak Zwee-er, p. 519
Eastern Towhee Zoo-ee, p. 424
Eastern Meadowlark Veet, p. 472

Whistled, buzzy, burry, or any combination
Loggerhead Shrike Song, p. 292
Northern Shrike Song, p. 293
Hutton's Vireo Song, p. 297
Steller's Jay Pumphandle Calls, p. 305
Blue Jay Pumphandle Calls, p. 306

Gray-crowned Rosy-Finch Chee-er, p. 398
Brown-capped Rosy-Finch Chee-er, p. 399
Black Rosy-Finch Chee-er, p. 400
Red-winged Blackbird Alert Calls, p. 471

A 2-NOTE PEENTING PHRASE
Like Peent, but usually distinctly 2-noted

2-noted, second note longer
Common Goldeneye Pi-peent, p. 58

2-noted, second note shorter
Barrow's Goldeneye Peent-ip, p. 59

A 2- OR 3-SYLLABLED BURRY OR BUZZY PHRASE, CONTINUED

A 2-NOTE BURRY OR BUZZY PHRASE

A 2-noted phrase with at least one part burry or buzzy.

2 different buzzes
Brewer's Sparrow, p. 438

2 different burry notes, or a Peer and then a burry note
Evening Grosbeak Song
(Types 1 and 2), pp. 410–411

Only first note burry
Eastern Phoebe Breep-it,
p. 277
Gray Flycatcher Song
(A phrase only), p. 270

First note brief, not burry; second overslurred
Alder Flycatcher Pip-pweer,
p. 266
Ash-throated Flycatcher
Pip-breer, p. 280
Couch's Kingbird Pip-breer,
p. 285
Thick-billed Kingbird
Chi-kveer, p. 287
Cassin's Kingbird Pip-breer,
p. 286

First note brief, not burry; second monotone or downslurred
Snowy Plover Peer-purr,
p. 129
Hammond's Flycatcher Song
(C phrase only), p. 268
Dusky Flycatcher Song
(A phrase only), p. 269
Gray Flycatcher Tsi-burrt,
p. 270
Least Flycatcher Song, p. 271
Eastern Phoebe Fee-burr,
p. 277

Like above group, but coarser
Eastern Phoebe Fee-blitty,
p. 277
Ash-throated Flycatcher
Ka-brik, p. 280

First note upslurred, variably burry
Himalayan Snowcock Bree-a,
p. 72

Brown-crested Flycatcher
Bree-burr, p. 282

First note a burry upslur, second a burry downslur
Willow Flycatcher Zree-beer,
p. 267

First note a low Pwit, second a nasal monotone burr
Swainson's Thrush (Olive-backed) Pwit-burr, p. 375

First note variable, second a polyphonic Churr
Swainson's Thrush (Russet-backed) Wee-churr, p. 374

First note Chiplike, second a monotone burr
Rock Wren Spit-tee, p. 350

A 3-NOTE BURRY OR BUZZY PHRASE

A 3-noted phrase with at least one part burry or buzzy.

Alder Flycatcher Song, p. 266
Evening Grosbeak Song
(Type 2), p. 411

Plastic, sometimes 2 or 4 notes
Purple Martin Veer Phrase,
p. 328
Verdin Chewt Phrase, p. 340

Townsend's Solitaire
Zweer-zweer-zree, p. 370

Very soft, high, usually repeated in series
Lucifer Hummingbird Song,
p. 107

A 2- OR 3-SYLLABLED NASAL PHRASE

PITTER

Very brief medium-pitched whistled to nasal note, broken into 2 distinct syllables; second may be higher or lower. Often in series. See also Finch Flight Call Group (p. 567).

Medium-high, whistled
Harris's Sparrow, p. 449

Medium-high, seminasal
Bar-tailed Godwit Song,
p. 138

Phainopepla, p. 391
Ruby-crowned Kinglet,
p. 364

Medium-low, nasal
White-breasted Nuthatch
(Montane), p. 343

2- OR 3-SYLLABLED NASAL NOTE
A single short nasal note that breaks or changes pitch, sounding 2- or 3-syllabled. See also Rising Broken Nasal Note (p. 553) and Yap Series (p. 577).

Very high, squeaky, like a dog toy
Sulphur-bellied Flycatcher Squeal, p. 283

High, usually 3-syllabled
Gull-billed Tern Kee-erik, p. 177

Pinyon Jay Ree-er-er, p. 312

Medium-high, usually 2-syllabled
see Marbled Godwit Gur-rik Series, p. 139

Low, mellow, seminasal
Western Screech-Owl Kew-du-du, p. 226

2- OR 3-NOTED NASAL PHRASE
A 2- or 3-noted phrase with at least one part nasal. See also 2- or 3-Syllabled Breathy Phrase (p. 588) and Tewee (p. 542).

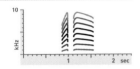

2 notes on similar pitch, second longer
Western Grebe Song, p. 86
Burrowing Owl Song, p. 230
Scaly-breasted Munia Pi-peep, p. 395

First note a Pip
Sulphur-bellied Flycatcher Pip-William, p. 283

First note upslurred, second Tseetlike
Green-tailed Towhee Kwoy-tseet, p. 426

2 or 3 unbroken honks in stereotyped pattern
Common Raven, p. 316

2 or 3 longer, Keerlike notes
Willet (Western) Keer-wee-wee, p. 154

3- or 4-note phrase, often repeated, second note longest
California Quail Hu-haa-how, p. 68
Gambel's Quail Hu-haa-ha-how, p. 69

Variable, wailing or honking
Indian Peafowl, p. 64

A 2- OR 3-SYLLABLED HARSH PHRASE

2- TO 3-SYLLABLED SCREECHY PHRASE
A short phrase of screechy notes.

Of 2 notes; high, wheezy
Fulvous Whistling-Duck Ki-wheer, p. 36

Of 2 notes; medium-low
Ring-necked Pheasant Keek-kuk, p. 73
Gray Partridge Kee-raw, p. 73

Of 2–4 syllables; medium-high
Caspian Tern Kee-kareer, p. 177
Red-lored Parrot Keep-away, p. 260
Tricolored Blackbird Gur-eek, p. 469

Of 3–5 notes; medium-low
Domestic Chicken Crow, p. 63

A 2- OR 3-SYLLABLED HARSH BURRY PHRASE

2- OR 3-SYLLABLED GRATING, CROAKING, OR CHURRING PHRASE
A high-pitched harsh grating or croaking sound of 2 or 3 syllables.

High, only first syllable grating
Mountain Quail Kree-urk, p. 66

High, 2-syllabled, variably grating
Black Tern Chivik, p. 183
Least Tern Ki-deek, p. 176

2-syllabled, harshly grating
Common Tern Kee-arr, p. 180
Arctic Tern Keeyer, p. 181

Elegant Tern Keerik, p. 179

2-syllabled Grate or 2 grating notes
Western Grebe Song, p. 86

3-syllabled Grate
Black Tern Chividik, p. 183

Deep 2-syllabled Croak
Common Raven Kro-uk, p. 316

Nasal Churr, usually 2-syllabled
Acorn Woodpecker Kreek-kut, p. 236

A WHISTLE AND A HARSH NOTE OR QUACK
A 2-noted phrase, one note a clear high musical whistle, the other a Snarl, Quack, or Grunt.

A high rising whistle, then a noisy to grating note
    White-tailed Kite Wee-Snarl, p. 200

A high whistled Hee, then a nasal quack
    Gadwall Hee-Quack, p. 48

A high whistled Hee, then a low soft grunt
    Mallard, p. 52

A low soft grunt, then a high whistled Hee
    Northern Pintail Grunt-hee, p. 50

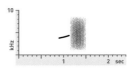

A wail, then a high squeaky Peep
Boreal Owl Wail-peep, p. 224

A 2- OR 3-SYLLABLED POLYPHONIC PHRASE

A 2-NOTED POLYPHONIC PHRASE
A 2-noted phrase with at least one part polyphonic.

Bobolink See-yur, p. 468
American Goldfinch Bee-bee, p. 407

Red Crossbill Chit-too, p. 413
Scaly-breasted Munia Pi-peep, p. 395

A 2- OR 3-SYLLABLED POLYPHONIC NOTE
A single polyphonic note that changes in pitch, creating the impression of 2 or 3 syllables.

Sharp-tailed Grouse Hoo-wee-urr, p. 74–75
Mountain Bluebird Vee-er-er, p. 367
Eastern Bluebird Zewy, p. 369

Summer Tanager Vewy, p. 513
Rose-breasted Grosbeak Zwee-er, p. 518
Black-headed Grosbeak Zwee-er, p. 519

TWANGING OR METALLIC PHRASE
A phrase that includes at least one nearly monotone note with a bizarre, metallic nasal tone quality, like the sound of a harmonica.

Indian Peafowl, p. 64
Steller's Jay Pumphandle Calls, p. 305

Blue Jay Pumphandle Calls, pp. 306–307

INDEX PART IV: THE SAME PHRASE REPEATED (A COMPLEX SERIES)

The same phrase repeated without a significant pause. Includes series of couplets, triplets, and more complex phrases, as well as series of series.

A COMPLEX SERIES OF CLICKING OR MECHANICAL NOTES

COMPLEX TICKING SERIES
Ticklike notes in rhythmically complex series.

A COMPLEX SERIES OF CHIPLIKE NOTES

COMPLEX CHIPLIKE SERIES
A repeated 2- or 3-syllabled phrase of high, sharp Chiplike notes.

A VERY HIGH COMPLEX SERIES

COMPLEX SEET SERIES
Very high couplet or triplet series of Seetlike notes.

A COMPLEX SERIES OF WHISTLED NOTES

WHISTLED COUPLET SERIES
A repeated 2-syllabled phrase made of whistled notes. **See also** *Northern Cardinal Song Group, p. 605, and Pitter, p. 590.*

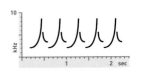

A COMPLEX SERIES OF WHISTLED NOTES, CONTINUED

YELLOWTHROAT SONG GROUP
A triplet or quadruplet series of sharp but musical whistles, usually medium in pitch.

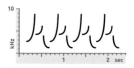

Common Yellowthroat, p. 498

Triplet series, often semimusical
    Bridled Titmouse, p. 331
    Oak Titmouse, p. 332
    Juniper Titmouse, p. 333

Triplet series, musical
    Ruby-crowned Kinglet, p. 364
    Painted Redstart, p. 510

Triplet or quadruplet series
    Carolina Wren, p. 352

WILLET SONG GROUP
A triplet or quadruplet series of medium-low clear whistled notes, often continuing for long periods. Frequently given in flight.

Willet (Western), p. 154

Triplet series
    Greater Yellowlegs, p. 152

Quadruplet series
    Lesser Yellowlegs, p. 153

SERIES OF SERIES
A short series of 3–5 whistled or semimusical notes, themselves repeated in a stereotyped fashion.

Variable, but usually not very musical
    Juniper Titmouse, p. 333
    Great-tailed Grackle Series Calls, pp. 474–475

A LOW-PITCHED COMPLEX SERIES

COMPLEX HOOTING OR COOING SERIES
A couplet, triplet, or quadruplet series of hooting or cooing notes.

Couplet series
    Band-tailed Pigeon Song, p. 93

Triplet series
    Eurasian Collared-Dove Song, p. 94

Burry couplet or triplet series
    Spotted Dove Song, p. 93
    African Collared-Dove Song, p. 94

GULPING COMPLEX SERIES
A couplet or triplet series of very low-pitched gulping notes.

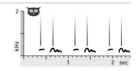

    American Bittern Song, p. 191
    Little Blue Heron Chuk-a Series, p. 194

A COMPLEX SERIES OF BURRS OR BUZZES

BURRY COUPLET SERIES
A repeated 2-syllabled phrase of burry notes.

Medium-high, very finely burry
    Black-throated Gray Warbler, p. 505

Medium-low, coarser, musical
    Mexican Chickadee Whistled Song, p. 338
    Olive Warbler, p. 392
    MacGillivray's Warbler, p. 497

Lower, coarser, and slower
    Spotted Sandpiper, p. 150

BUZZY COUPLET OR TRIPLET SERIES
A repeated 2- or 3-syllabled phrase of buzzy notes.

High, soft
Lucifer Hummingbird Song,
p. 107
Broad-billed Hummingbird
Series Song, p. 115

Medium-low, coarse, musical
Mexican Chickadee Whistled
Song, p. 338
Black-throated Gray Warbler,
p. 505

A COMPLEX SERIES OF NASAL NOTES

NASAL COUPLET SERIES
A repeated 2-syllabled phrase of nasal notes.

*Longer notes in each phrase
downslurred*
Cassin's Kingbird Pi-deer
Series, p. 286

*Longer notes in each phrase
upslurred*
American Coot Long Call,
p. 125
Hairy Woodpecker Wik-a
Series, p. 243
Northern Flicker Wik-a
Series, p. 248
Gilded Flicker Wik-a Series,
p. 249

Acorn Woodpecker Wik-a
Series, p. 236
Gila Woodpecker Wik-a
Series, p. 237
Red-bellied Woodpecker
Wik-a Series, p. 239

One note squeaky, one Chiplike
Peregrine Falcon Ee-chup
Series, p. 257
Prairie Falcon Ee-chup
Series, p. 256

All notes short, clipped
Gambel's Quail Wit-Wut
Series, p. 69

Willet (Western) Ka-lip
Series, p. 154
Wilson's Snipe Kik-a Series,
p. 149

Series of 2-syllabled notes
Marbled Godwit Gur-rik
Series, p. 139
Hudsonian Godwit Poowit
Series, p. 139
Bar-tailed Godwit Pi-weer
Series, p. 138

NASAL TRIPLET SERIES
A repeated 3-syllabled phrase of nasal notes.

Sulphur-bellied Flycatcher Pip-William, p. 283

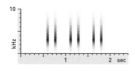

A COMPLEX SERIES OF HARSH NOTES

COUPLET SERIES OF KEKS
A couplet series of brief harsh noisy notes.

Virginia Rail Song, p. 121

COUPLET SERIES OF GRUNTS OR SCREECHES
A repeated 2-syllabled phrase of barking, grunting, or screechy notes.

*Noisy, barking to grunting
(Chuk-a Series)*
Black-crowned Night-Heron,
p. 197
Cattle Egret, p. 195
Golden-fronted Woodpecker,
p. 238
Red-naped Sapsucker,
p. 250

Red-breasted Sapsucker,
p. 251
Williamson's Sapsucker,
p. 252

Low, gulping (Chuk-a Series)
Little Blue Heron, p. 194

High, slow, nasal to screechy
Scaled Quail Kuk-curr Series,
p. 67

Helmeted Guineafowl
Buck-wheat Series, p. 64

High, fast, screechy
Ring-necked Pheasant
Kuttuk, p. 73

INDEX PART V: DIFFERENT SHORT NOTES, SERIES, OR PHRASES

A complex song made of different short notes, series, or phrases, all separated by pauses.

A SONG OF MOSTLY DIFFERENT SEPARATE SINGLE NOTES

MOSTLY SINGLE 1-SYLLABLED NOTES

Songs consisting mostly of well-spaced single 1-syllabled notes, with consecutive notes often different. See also *Variable Single Whistles, p. 538, in which consecutive notes are usually the same.*

All notes long and monotone, burry or polyphonic
  Varied Thrush Song, p. 371

Notes downslurred, burry, sometimes mixed with Seets
  Gray-crowned Rosy-Finch, p. 398
  Brown-capped Rosy-Finch, p. 399
  Black Rosy-Finch, p. 400

Musical, many notes whistled and/or polyphonic
  Osprey, p. 211
  Lapland Longspur Whistled Calls, p. 420
  McCown's Longspur Whistled Calls, p. 422

Orchard Oriole Songlike Calls, p. 482
Baltimore Oriole Songlike Calls, p. 484
Bullock's Oriole Songlike Calls, p. 485

Unmusical, most notes resembling Chips or Spits, sometimes with Seets
  Pin-tailed Whydah Display Song, p. 394
  Eastern Towhee Complex Song, p. 424
  Spotted Towhee Complex Song, p. 425
  Green-tailed Towhee Complex Song, p. 426

Botteri's Sparrow Complex Song, p. 433
Fox Sparrow (all forms) Complex Song, pp. 455–457

Most notes Chiplike, often in short series, often with occasional burry phrases
  Red Crossbill (all types), pp. 412–417
  Cassia Crossbill, p. 417

All notes high brief Seets on slightly different pitches
  Blue-throated Hummingbird Seet Song, p. 105

SINGLE NOTES WITH SOME RATTLES OR CHATTERS

Songs consisting mostly of single musical whistles mixed with short Rattles or Chatters, all sounds well spaced. See also *Yellow-breasted Chat (p. 466).*

Whistles and short Rattles or Chatters
  Chestnut-collared Longspur Whistled Calls, p. 421

McCown's Longspur Whistled Calls, p. 422

Hooded Oriole Songlike Calls, p. 483

MOSTLY SINGLE CHIRPS

Songs consisting mostly of well-spaced single complex 1- to 2-syllabled chirping notes, with consecutive notes usually different.

Tree Swallow Dawn Song, p. 326

Violet-green Swallow Dawn Song, p. 327

House Finch, p. 404
House Sparrow, p. 393

A SONG OF MOSTLY DIFFERENT SEPARATE SERIES OR TRILLS

BROWN THRASHER SONG GROUP
Extremely varied short doubled phrases, separated by pauses.   Brown Thrasher, p. 380

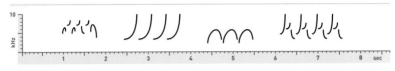

MOCKINGBIRD/CHAT SONG GROUP
Widely varied series, single notes, and/or trills, separated by pauses. Some versions include chatters.

*Mostly series and trills,
semimusical; pauses short*
  Common Redpoll, p. 405

*Mostly series, often including
imitations; pauses short to long*
  Northern Mockingbird, p. 387

*Soft, with single notes, series,
and sometimes imitations*
  Phainopepla Quiet Song,
  p. 391

Canyon Towhee Complex
  Song, p. 428
Abert's Towhee Complex
  Song, p. 430

*Mostly 1- or 2-noted series,
some trills; all musical*
  Rock Wren, p. 350

Series, trills, chatters, and
single notes; pauses long
  Yellow-breasted Chat, p. 466

(*see also* Songlike Calls of
orioles)

*Mostly harsh Snarls, with some
Chips, Tinks, or Mews*
  Black-tailed Gnatcatcher,
  p. 358
California Gnatcatcher
  Simple Song, p. 360
Black-capped Gnatcatcher
  Simple Song, p. 361

A SONG OF MOSTLY DIFFERENT SEPARATE SINGLE PHRASES

WIDELY VARIED PHRASES AND SERIES
*A mix of phrases and series, often of varied lengths, sometimes with trills, single notes, or more complex
patterns.*

*Often including polyphony, or long
slurred whistles*
  European Starling, pp. 388–389

Crissal Thrasher, p. 384
Great-tailed Grackle Short
  Song, pp. 474–475

*Phrases usually contain 1–2
short semimusical chatters*
  Black-capped Vireo, p. 294

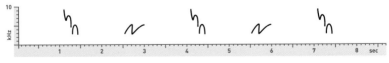

2 DIFFERENT SHORT PHRASES
Songs consisting of 2 different 1- to 3-syllabled phrases, consecutive phrases usually different.

Faster, usually partly burry
  Gray Flycatcher Song, p. 270
  Buff-breasted Flycatcher
  Song, p. 274
  Eastern Phoebe Song, p. 277

Black Phoebe Song, p. 276
Western Wood-Pewee Dawn
  Song, p. 263
Ash-throated Flycatcher
  Dawn Song, p. 280

Great Crested Flycatcher
  Dawn Song, p. 281
Brown-crested Flycatcher
  Dawn Song, p. 282
Cassin's Kingbird Dawn Song,
  p. 286

2–5 OF 1 NOTE OR PHRASE, THEN A DIFFERENT PHRASE

Songs that prominently feature an AAAB pattern of well-spaced notes or phrases, sometimes with other single notes or phrases in between instances of the pattern.

Introductory notes include twitters of nasal Pips
Western Kingbird Dawn Song, p. 288
Scissor-tailed Flycatcher Dawn Song, p. 290

Another note type sometimes between AAAB patterns
Northern Beardless-Tyrannulet Dawn Song, p. 262

Greater Pewee Dawn Song, p. 264
Couch's Kingbird Dawn Song, p. 285
Cassin's Kingbird Dawn Song, p. 286

3 DIFFERENT SHORT PHRASES

Songs consisting of 3 different 1- to 3-syllabled phrases, consecutive phrases usually different.

All strongly burry
Willow Flycatcher, p. 267

High, one phrase a simple "Teet"
Pacific-slope Flycatcher, p. 272
Cordilleran Flycatcher, p. 273

Whistled to slightly burry
Say's Phoebe Dawn Song, p. 275
Dusky Flycatcher, p. 269
Hammond's Flycatcher, p. 268
Buff-breasted Flycatcher, p. 274

Nasal, screechy, twittering
Thick-billed Kingbird Dawn Song, p. 287

MULTIPLE SHORT DIFFERENT PHRASES

Songs consisting mostly of well-spaced single 2- to 4-syllabled phrases, usually without any trills or series, consecutive phrases usually different.

Mostly series, some phrases
Five-striped Sparrow, p. 442

Phrases high, tinkling, complex
Horned Lark Long Song, p. 320

Phrases medium-low, soft, polyphonic
Mountain Bluebird, p. 367
Western Bluebird, p. 368

Phrases medium-high, usually whistled, not burry
Red-eyed Vireo, p. 302
Gray Vireo, p. 296
Townsend's Solitaire Disjunct Song, p. 370
Purple Finch Disjunct Song, p. 402
Cassin's Finch Disjunct Song, p. 403

Hepatic Tanager Dawn Song, p. 512

Phrases medium-high, both whistled and burry
Dusky-capped Flycatcher Song, p. 279
Summer Tanager Dawn Song, p. 513

Phrases mostly burry
Hutton's Vireo, p. 297
Cassin's Vireo, p. 298
Plumbeous Vireo, p. 299
Western Tanager Dawn Song, p. 514

Phrases medium-high, whistled, burry, or polyphonic
Loggerhead Shrike, p. 292
Northern Shrike, p. 293
Phainopepla, p. 391

Many phrases high, polyphonic
American Robin Complex Song, pp. 376–377
American Goldfinch Disjunct Song, p. 407
Lawrence's Goldfinch Disjunct Song, p. 408
Lesser Goldfinch Disjunct Song, p. 409
Pine Siskin Disjunct Song, p. 406

Phrases varying widely in pitch and tone quality
Gray Catbird, p. 379

Some phrases high slurred whistles, others gurgling
Brown-headed Cowbird, p. 480
Bronzed Cowbird, p. 481

A SONG OF MOSTLY DIFFERENT SEPARATE PHRASES IN CLUSTERS

A ROBINLIKE SONG
Songs consisting of 1- to 3-syllabled phrases grouped into clusters separated by longer pauses.

Phrases high, musical, tinkling
McCown's Longspur, p. 422

Phrases medium-high, musical, not burry
American Robin, pp. 376–377

Clusters long; phrases mostly slurred whistles
Hepatic Tanager, p. 512
Rose-breasted Grosbeak, p. 518
Black-headed Grosbeak, p. 519

Phrases finely burry
Purple Martin Dawn Song, p. 328
Scarlet Tanager, p.
Summer Tanager, p. 513
Western Tanager, p. 514

Phrases of mostly rising, polyphonic Jitlike notes
Barn Swallow Dawn Song, p. 324

Phrases of soft, hoarse polyphonic notes
Northern Rough-winged Swallow Song, p. 325
Mountain Bluebird, p. 367
Western Bluebird, p. 368
Hepatic Tanager Quiet Song, p. 512

Phrases include Chirps, often with whistles and gurgles
Tree Swallow Day Song, p. 326
Violet-green Swallow Day Song, p. 327

INDEX PART VI: A LONG COMPLEX SONG

A very long continuous song that lasts at least 4 seconds without a significant pause, including a variety of different note types or patterns. For long songs that consist of the same note repeated, see Index Part II (p. 565). For long songs that consist of the same phrase repeated, see Index Part IV (p. 593).

A LONG SONG WITH FEW REPEATED NOTES

A VERY LONG PHRASE OR WARBLE

A long (usually 4 or more seconds), mostly musical warble or phrase, without significant pauses or repetition.

A LONG SONG WITH SOME REPEATED NOTES

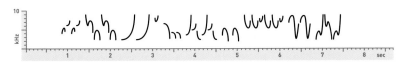

CURVE-BILLED THRASHER SONG GROUP

A long (usually 4 or more seconds) mostly musical phrase without significant pauses, but with at least some notes repeated once.

A MIX OF SINGLE NOTES, CHATTERS, AND TRILLS
A long mix of single musical notes, brief chatters, and trills or short series, without significant pauses.

Black-capped Vireo Complex
Song, p. 294
Bullock's Oriole Complex
Song, p. 485

Baltimore Oriole Complex
Song, p. 484
Hooded Oriole, p. 483

Black-throated Sparrow,
p. 443
Lark Sparrow, p. 441

A LONG SONG OF MOSTLY SERIES OR TRILLS

MULTIPLE CONSECUTIVE SERIES
At least 3 series of mostly musical notes, sometimes with trills and/or single notes; no significant pauses.

*Medium-low, made of complex
burry, nasal phrases*
Bendire's Thrasher, p. 381

*Medium-pitch; all notes highly
musical, whistled*
Field Sparrow Complex Song,
p. 439

Slow, high, musical, burry
American Dipper, p. 362

*Mostly high, musical, some notes
polyphonic*
American Goldfinch, p. 407

High, musical, with imitations
Lawrence's Goldfinch, p. 408
Lesser Goldfinch, p. 409

Notes extremely varied
California Thrasher, p. 382
Northern Mockingbird, p. 387

Notes unmusical, chirping
Dickcissel Complex Song,
p. 515

With long, high, slurred whistles
Red-winged Blackbird Flight
Song, pp. 470–471

*Notes mostly squeaky, harsh, or
nasal*
Tricolored Blackbird Flight
Song, p. 469

*With long, medium-low slurred
whistles, often trills*
Northern Cardinal, p. 516
Pyrrhuloxia, p. 517
Lark Bunting, p. 447

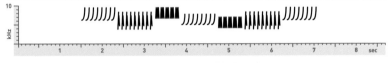

MULTIPLE CONSECUTIVE TRILLS, CHATTERS, OR BUZZY SERIES
3 or more trills or fast series, some often chattery or buzzy, without pauses in between.

Only 3 semimusical trills
Colima Warbler, p. 493

Series and trills, mostly buzzy
Brewer's Sparrow, p. 438

*Mostly trills, semimusical to
buzzy*
White-winged Crossbill,
p. 418
Common Redpoll, p. 405

*Mostly musical, sometimes with
harsh buzzes and single notes*
Vesper Sparrow, p. 440
Lark Sparrow, p. 441

SERIES OF CHIPLIKE NOTES AND BURRY PHRASES
1 to several series of Chiplike notes, then a simple or complex musical series that usually includes burry notes. Some versions more complex, culminating in multiple consecutive burry series.

Red Crossbill (all types), pp. 412–417

Cassia Crossbill, p. 417

A LONG SONG THAT MIXES WARBLES AND TRILLS, OR OTHER PATTERNS

WINTER WREN SONG GROUP
A long, high, musical tinkling warble, with trills interspersed.

Entire song very high, musical
Winter Wren, p. 347
Pacific Wren, p. 348

Grasshopper Sparrow
Complex Song, p. 462

Note types varied; song often plastic, with irregular pauses
Dark-eyed Junco Complex Song, pp. 458–459

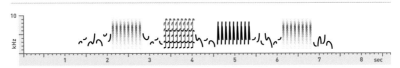

MARSH WREN LONG SONGS
Chatters or complex semimusical trills, often with some gurgling or tinkling warbles interspersed.

Marsh Wren (Eastern), p. 355   Marsh Wren (Western), p. 354

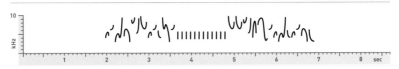

PUNCTUATED BY TICKING TRILLS
A complex, often very long song of warbles, phrases, and sometimes series, with occasional ticking trills interspersed.

Barn Swallow Song, p. 324
Cliff Swallow Song, p. 323

Cave Swallow Song, p. 322
Purple Martin Day Song, p. 328

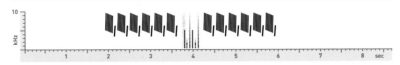

PUNCTUATED BY CRACKLES
A complex, often very long song of whistles, burrs, or other notes, with occasional crackles or "stick-breaking sounds" interspersed.

Song more varied, sometimes with sirenlike whistles
Great-tailed Grackle, pp. 474–475

INDEX PART VII: A COMPLEX SONG

A complex song of more than 3 notes or syllables, less than 4 seconds long. Includes all songs not indexed in the prior categories.

A VERY HIGH-PITCHED COMPLEX SONG

A VERY HIGH COMPLEX SONG

A complex, often plastic phrase of very high long whistles and trills. Often preceded by lower, complex notes or phrases. See also Complex Songs of Hummingbirds, p. 612.

Often quite long; often with both whistles and trills
European Starling, pp. 388–389
Bronzed Cowbird Whistled
Song, p. 481

Often quite long, with Seets, Pseeps, and/or Trills
Pigeon Guillemot, p. 157

Usually 4–6 notes, none very long
Brown Creeper, p. 346

Usually 4 notes or fewer, usually including trills
Brown-headed Cowbird
Whistled Song, p. 480

CHIPS AND TRILLS

A mix of Chiplike notes and repeating trills, given during flight displays.

Chips nasal, trills rattling
Black-chinned Hummingbird
Shuttle Display, p. 108

A COMPLEX SONG WITH SEETS OR TINKS AT START, LOWER NOTES AT END

TITMOUSE TSEET-SONGS

Highly variable short (2- to 4-note) song; typically starts with a high Tseetlike note, followed by much lower, musical whistles.

Black-crested Titmouse,
p. 330

Oak Titmouse, p. 332
Juniper Titmouse, p. 333

SPARROW PAIR REUNION CALLS

Usually a series of high Seets transitioning gradually or suddenly into a chipping Twitter or harsh Chatter. Highly plastic, often given by 2 or more birds at once.

Usually starts with Seets
Rufous-crowned Sparrow,
p. 427
Canyon Towhee, p. 428
California Towhee, p. 429

Abert's Towhee, p. 430

Usually a complex series without Seets at start
Rufous-winged Sparrow, p. 431

Five-striped Sparrow, p. 442

DZEET-CHIPPITY

1–3 quick high buzzes, then a short chipping trill, or another combination of similar elements. Highly plastic.

Lucifer Hummingbird, p. 107
Black-chinned Hummingbird,
p. 108
Costa's Hummingbird, p. 110

Calliope Hummingbird,
p. 111
Rufous Hummingbird, p. 112
Allen's Hummingbird, p. 113

Broad-tailed Hummingbird,
p. 114

A 3-PART MUSICAL SONG STARTING WITH SEET SERIES

A unique 3-part song: an accelerating series of Seetlike notes, then a lower seminasal couplet series, then a variable complex series of musical whistles.

Ruby-crowned Kinglet, p. 364

CHICK-A-DEE CALL GROUP

Starts with high Pseeplike notes, transitioning gradually or suddenly into a series of nasal notes or whines. Plastic; "dee" notes at end may be given separately.

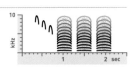

"Dee" notes rather slow, nasal
Black-capped Chickadee, p. 334

"Dee" notes rather slow, hoarse or whiny
Mountain Chickadee, p. 335
Boreal Chickadee, p. 336
Mexican Chickadee, p. 338

"Dee" notes rather rapid
Bridled Titmouse, p. 331

Black-crested Titmouse, p. 330
Oak Titmouse, p. 332
Juniper Titmouse, p. 333

"Dee" notes very high and nasal
Chestnut-backed Chickadee, p. 337

Ends in a semimusical Sputter instead of nasal notes
Golden-crowned Kinglet Song, p. 363

Starts with sharp Tsits or Chips, ends with musical whines
Song Sparrow Whine Series, p. 454
Lincoln's Sparrow Whine Series, p. 452

COMPLEX SONGS OF WARBLERS AND SPARROWS

A catch-all category of generally quiet, infrequent songs that often start with high-pitched call-like notes, followed by a complex, varied phrase that may include buzzes, trills, and single notes. Sometimes given in flight.

Very few repeated notes
Ovenbird, p. 488
Common Yellowthroat, p. 498

Some repeated notes
Northern Waterthrush, p. 489
Cassin's Sparrow, p. 432
MacGillivray's Warbler, p. 497

Mostly series and trills
Vesper Sparrow, p. 440

Rather short: a few Tinks, a short phrase, then a trill or buzz
Nelson's Sparrow, p. 465
LeConte's Sparrow, p. 464
Swamp Sparrow, p. 453

A COMPLEX SONG OF MOSTLY LONG MUSICAL WHISTLES

CHICKADEE AND *ZONOTRICHIA* SONG GROUP

Songs consisting almost entirely of high clear, monotone whistles.

Usually 2 notes, the second one lower
Black-capped Chickadee, p. 334

2–3 notes on the same pitch (next song on different pitch)
Harris's Sparrow, p. 449

Usually 3–4 notes on 1–3 different pitches
Mountain Chickadee, p. 335
Brown-headed Cowbird Whistled Song, p. 480
Golden-crowned Sparrow, p. 451

Usually 5 or more notes, ending in a monotone triplet series
White-throated Sparrow, p. 448

A simple or couplet series, then 1 note or a phrase
Hermit Warbler, p. 507

NORTHERN CARDINAL SONG GROUP
Usually 1–3 consecutive whistled series; pitch medium-low and most notes highly musical. Series of very long slurred whistles distinctive.

Northern Cardinal, p. 516   Pyrrhuloxia, p. 517

A WHISTLED PHRASE
A phrase of 4 or more musical whistled notes that are not monotone.

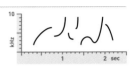

Very high and loud
European Starling,
pp. 388–389

2–4 short high notes
Verdin Song, p. 340

Variable, shrill, first note longest and highest
Killdeer Dee-dit-dit, p. 132

High, with few or no sharp notes
Eastern Meadowlark, p. 472

High, some notes sharp
Brown Creeper, p. 346

Medium-high, some notes sharp
Sulphur-bellied Flycatcher
Dawn Song, p. 283

Medium-low, musical
Mexican Chickadee Whistled
Song, p. 338

Medium-low, musical, usually ending in gurgle
Western Meadowlark, p. 473

Medium-low, musical, with long slurred whistles
Greater Pewee Day Song,
p. 264

Fox Sparrow (all forms),
pp. 455–457

Medium-low, musical; second half often repeats first half
Scott's Oriole, p. 486
Baltimore Oriole, p. 484

Lower, mellower, plastic
American Golden-Plover,
p. 134
Pacific Golden-Plover, p. 134

A MELLOW WHISTLED PHRASE OR CHORUS
A phrase, duet, or chorus of short, mellow whistled phrases, some broken. Most notes well spaced.

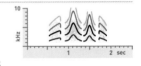

Repeated over and over at night
Mexican Whip-poor-will
Song, p. 103

Duet or chorus; notes well spaced, some broken
Northern Bobwhite Duet,
p. 70

A BREATHY PHRASE
A phrase of 4 or more high, breathy, Wheewlike notes. See also 2- or 3-Syllabled Breathy Phrase (p. 588).

Usually 3- to 4-noted
American Wigeon
Whi-whee-whew, p. 49

Usually 4 or more notes
Black-bellied Whistling-Duck
Kip-whee-witter, p. 37

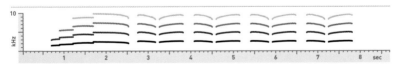

LOON SONG GROUP
A long phrase of musical whistles, some broken. Sometimes in chorus.

Common Loon, p. 184
Yellow-billed Loon, p. 185
Pacific Loon, p. 186

CURLEW SONG GROUP
A long phrase of nearly monotone musical whistles and trills, usually culminating in a long overslurred whistle that breaks downward and then upward.

Black-bellied Plover, p. 135
Long-billed Curlew, p. 136   Whimbrel, p. 137   Upland Sandpiper, p. 138

A COMPLEX SONG OF MOSTLY MUSICAL SERIES OR TRILLS
Including most nonbuzzy warbler and sparrow songs.

ONE NOTE, SERIES, OR SHORT PHRASE, THEN A SERIES
1–3 single notes or a 2- to 3-note phrase, then a series.

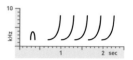

1 sharply downslurred whistle, then a lower series of same
   American Goldfinch
   Pee-tee-tee-tee, p. 407
   Mississippi Kite Keek-tew-tew-tew, p. 200

First note often a Cheep, the rest lower Kuks
   American Robin Cheep/Kuk, pp. 376–377

First note a Whit or a Pip
   Hammond's Flycatcher
   Peer-Pewit, p. 268

Least Flycatcher Pweew
   Series, p. 271

First note a Squeak
   Canyon Towhee, p. 428

Series highly musical, of single notes or couplets
   Cassin's Vireo Trill, p. 298
   Plumbeous Vireo Trill, p. 299
   Mexican Chickadee Whistled
   Song, p. 338
   Olive Warbler, p. 392
   Harris's Sparrow, p. 449
   Eastern Towhee, p. 424

Yellow-eyed Junco, p. 460
Northern Cardinal, p. 516
Pyrrhuloxia, p. 517

All notes semi- to unmusical
   Sedge Wren, p. 356
   Dickcissel, p. 515
   Eastern Towhee, p. 424
   Spotted Towhee, p. 425
   Rufous-capped Warbler,
   p. 499

ONE NOTE, SERIES, OR SHORT PHRASE, THEN A TRILL
1–3 single notes or a 2- to 3-note phrase, then a trill. Some songs may end with an additional single note after the trill. Compare the above category.

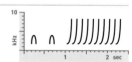

1–2 whistled, polyphonic, or buzzy notes, then a musical trill
   Dusky-capped Flycatcher
   Wee-beer, p. 279
   Bewick's Wren, p. 353
   Eastern Towhee, p. 424
   Rufous-winged Sparrow, p. 431
   Black-throated Sparrow, p. 443

Baird's Sparrow, p. 463
Yellow-eyed Junco, p. 460

Trill slightly metallic, sometimes with 1 note at end
   Red-winged Blackbird Male
   Song, pp. 470–471

A few notes, then a semimusical or unmusical trill
   Chestnut-backed Chickadee
   Trill Song, p. 337
   Island Scrub-Jay Trill, p. 308
   Marsh Wren (Eastern), p. 355
   Marsh Wren (Western), p. 354
   Spotted Towhee, p. 425

A TRILL OR SERIES, THEN 1 NOTE
A short trill or series followed by a single note; all notes high sharp whistles.

Starts with trill
   Colima Warbler, p. 493

Starts with high series
   American Redstart, p. 500

Starts with medium-high series, often of couplets
   Dark-eyed Junco (Red-backed), p. 459

Yellow-eyed Junco, p. 460
Grace's Warbler, p. 504

A SERIES, THEN A PHRASE
A short series followed by a short phrase; all notes medium-high, fairly sharp whistles.

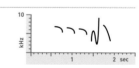

Medium-low, often starting with couplet series
   Yellow-eyed Junco, p. 460

2 TRILLS, OR A SERIES AND A TRILL
2 consecutive musical to unmusical whistled simple trills, or 1 fast series and 1 trill.

First part usually higher; all notes fairly Chiplike
   Wilson's Warbler, p. 508
   Orange-crowned Warbler,
   p. 492

Second part faster and higher; all notes rather unmusical
   Spotted Towhee, p. 425

Variable, usually unmusical
   Rufous-capped Warbler,
   p. 499

Both parts fairly musical
   Dark-eyed Junco,
   pp. 458–459

Dark-eyed Junco (Red-backed), p. 459
Yellow-eyed Junco, p. 460
Colima Warbler, p. 493
Lucy's Warbler, p. 494

2 SIMPLE SERIES
2 consecutive series of unmusical to musical 1-syllabled whistles.

All notes Chiplike, rather unmusical
   Rufous-capped Warbler, p. 499
   Wilson's Warbler, p. 508

Start often musical; final series Chiplike, unmusical
   Grace's Warbler, p. 504

All notes medium-high, musical
   Virginia's Warbler, p. 496
   MacGillivray's Warbler, p. 497

All notes high, musical, simple
   Yellow Warbler, p. 501

All notes high, musical, usually complex
   Yellow-rumped Warbler (Myrtle), p. 502

Yellow-rumped Warbler (Audubon's), p. 503
Dark-eyed Junco (Red-backed), p. 459
Yellow-eyed Junco, p. 460

Musical Teets, then a Chatter
   Abert's Towhee, p. 430

A COUPLET SERIES, THEN A SIMPLE SERIES OR TRILL
A whistled couplet series, then a musical or unmusical simple series or trill.

Second part an unmusical, chipping trill
   Nashville Warbler, p. 495
   Tennessee Warbler, p. 491
   Grace's Warbler, p. 504

Trill and series in either order
   Dark-eyed Junco (Red-backed), p. 459
   Yellow-eyed Junco, p. 460
   Painted Redstart, p. 510

Second part a series, not a trill
   Virginia's Warbler, p. 496

Second part a very musical trill
   Baird's Sparrow, p. 463

2 COUPLET SERIES
2 consecutive couplet series of unmusical to musical whistles.

Medium-low, usually burry, musical
   Olive Warbler, p. 392
   Dark-eyed Junco (Red-backed), p. 459
   Yellow-eyed Junco, p. 460
   MacGillivray's Warbler, p. 497

Medium-low, musical
   Painted Redstart, p. 510

Medium-high, musical
   Virginia's Warbler, p. 496

High, musical, notes often only vaguely 2-syllabled
   Yellow-rumped Warbler (Myrtle), p. 502
   Yellow-rumped Warbler (Audubon's), p. 503

Medium-low, rather unmusical, becoming louder
   Ovenbird, p. 488

Very high
   Black-and-white Warbler, p. 490

2 SIMPLE SERIES, THEN 1 NOTE OR PHRASE
2 quick series, then a final note or 2-note phrase, all notes high clear sharp whistles.

Medium-high to high; all notes clear, whistled
   Virginia's Warbler, p. 496
   Yellow Warbler, p. 501
   American Redstart, p. 500

Usually lower, most notes complex
   Yellow-eyed Junco, p. 460
   Yellow-rumped Warbler (Audubon's), p. 503
   Yellow-rumped Warbler (Myrtle), p. 502

Faster, at least the first series trilled
   Colima Warbler, p. 493

3 OR 4 CONSECUTIVE SERIES
Usually 3, rarely 4 consecutive series of clear sharp whistles. Some series may be of couplets.

Medium-low, last part usually a series of Tews
   Northern Waterthrush, p. 489

Medium-high, last part usually a chipping trill
   Tennessee Warbler, p. 491

Medium to high, all notes clear and musical
   Lucy's Warbler, p. 494

Virginia's Warbler, p. 496
Yellow Warbler, p. 501
Painted Redstart, p. 510

High, notes complex and musical, first series longest
   Yellow-rumped Warbler (Audubon's), p. 503
   Yellow-rumped Warbler (Myrtle), p. 502

3–4 very high series, some of couplets
   Black-and-white Warbler, p. 490

Variable, notes complex or burry
   MacGillivray's Warbler, p. 497
   Rufous-capped Warbler, p. 499

MULTIPLE CONSECUTIVE SHORT SERIES
4–7 consecutive series of musical notes, most series only 2 notes.

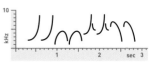

Indigo Bunting, p. 520
Lazuli Bunting, p. 521
Rufous-crowned Sparrow,
p. 427
Yellow-eyed Junco, p. 460

Painted Redstart, p. 510

*Often including single notes
or short phrases*
Red-faced Warbler, p. 509

COMPLEX ACCELERATING SONGS
*High, generally musical songs that prominently feature an
accelerating series, with 1 or more notes before and/or after it.*

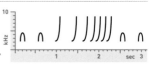

*High, upslurred, accelerating
series ending on 1 lower note*
Vermilion Flycatcher, p. 278

*Low, knocking, accelerating
series ending on 1 lower note*
Buff-collared Nightjar, p. 104

*Chiplike notes before the
accelerating series*
Botteri's Sparrow, p. 433

*An accelerating musical trill,
then 2–4 monotone whistles*
Cassin's Sparrow, p. 432

*An accelerating musical trill, at
or near the start of a complex
musical song*
Song Sparrow, p. 454

*1–2 accelerating series that start
musical and end buzzy, sometimes
with 1–2 other notes*
Black-chinned Sparrow, p. 435

*Accelerating Teetlike notes, then a
semimusical series or phrase*
California Towhee, p. 429
Abert's Towhee, p. 430

A PHRASE WITH SOME TRILLS OR BUZZES
*A musical song of mostly single notes, with at least some buzzes,
trills, or whistled series; some include polyphonic notes.*

Highly variable
Yellow-eyed Junco, p. 460

*Usually 3 clear whistles, then
a trill*
Golden-crowned Sparrow,
p. 451

*An accelerating musical trill,
then 2–4 monotone whistles*
Cassin's Sparrow, p. 432

*Medium to medium-high, with
long slurred musical whistles*
American Tree Sparrow,
p. 434
Fox Sparrow (all forms),
pp. 455–457

*Medium-high, most notes sharp;
sometimes buzzes or series*
Chestnut-collared Longspur,
p. 421

Snow Bunting, p. 419

*Medium-high; notes can vary
widely in quality*
Orchard Oriole, p. 482
Bullock's Oriole, p. 485

*Medium-high, many notes
polyphonic*
Lapland Longspur, p. 420

SONG SPARROW SONG GROUP
*A musical mix of buzzes, trills, and whistled series, sometimes with
single notes.*

*Usually starts with a single note
or short phrase*
Black-throated Sparrow,
p. 443
Bewick's Wren, p. 353
Lark Sparrow, p. 441
Green-tailed Towhee, p. 426

Fox Sparrow (Thick-billed),
p. 456

*Usually starts with a musical
series*
Song Sparrow, p. 454

*2–3 downslurred whistles, then
only series or trills*
Vesper Sparrow, p. 440

Short, not very musical
Dickcissel, p. 515

A COMPLEX SONG CONTAINING MUSICAL WARBLES

AN ACCELERATING PHRASE
*A musical whistled phrase that speeds up, the last few notes running
together into a warble.*

*Medium-low, first 1–2 notes
nearly monotone whistles*
Western Meadowlark, p. 473

High, tinkling, rising
Horned Lark, p. 320

A 1- TO 4-SECOND WARBLE

Short (1–4 seconds) musical warbles of whistled, burry, and/or polyphonic notes. See also Gurgle (p. 611).

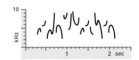

High, mostly sharp whistles
Pin-tailed Whydah, p. 394
Red-faced Warbler, p. 509
Rufous-crowned Sparrow, p. 427

Medium-low, whistled, with a few short burry notes
Warbling Vireo (Western), p. 300
Warbling Vireo (Eastern), p. 301
Varied Bunting, p. 522
Painted Bunting, p. 523
Blue Grosbeak, p. 524

Similar but with squeaky notes, whines, or Jits
Warbling Vireo Complex Song (Western), p. 300
Warbling Vireo Complex Song (Eastern), p. 301

Whistled, with burry notes, often with repeated phrases
Snow Bunting, p. 419
Sagebrush Sparrow, p. 444
Bell's Sparrow, p. 445

Medium-low, whistled, usually with 1–2 long slurred buzzes
House Finch, p. 404

Lower, richly musical, usually lacking buzzes
Pine Grosbeak, p. 401
Purple Finch, p. 402
Cassin's Finch, p. 403

Includes polyphonic and burry notes
Bell's Vireo, p. 295
Eastern Bluebird, p. 369

WARBLE PLUS SERIES OR TRILLS

A highly musical song that starts with rapid notes, most in short series; usually culminates in high trill, then a lower trill and sometimes other notes.

Starts fast, usually culminates in a high trill near the end
House Wren, p. 349

Similar but slower and more musical at start
Lincoln's Sparrow, p. 452

Variable, often lacking trills
Purple Finch, p. 402
Cassin's Finch, p. 403

Higher, with sharper notes, no trills
Rufous-crowned Sparrow, p. 427

A HOOTING, COOING, OR GROWLING COMPLEX SONG

PIED-BILLED GREBE SONG

2–3 consecutive series of low, almost hooting whistles, the later series slower, of couplets or triplets

Pied-billed Grebe, p. 90

A HOOTING OR COOING PHRASE

A phrase of 4 or more very low-pitched hooting or cooing notes.

4-note phrase "WHO COOKS for YOU?"
Barred Owl Song, p. 220
White-winged Dove Short Song, p. 97

4-note phrase with 2 long pauses
Spotted Owl Song, p. 221

Usually 3–4 notes, the first one broken
Mourning Dove Song, p. 96

Usually 4–6 notes, none broken
Great Horned Owl Song, p. 213
Flammulated Owl Complex Song, p. 229

Usually 6 or more notes
Dusky Grouse Song, p. 82
Sooty Grouse Song, p. 83
Barred Owl Series Song, p. 220

Spotted Owl Series Song, p. 221
White-winged Dove Long Song, p. 97

Long irregular series
Whiskered Screech-Owl Morse Code Song, p. 228

A HOOTING AND CACKLING CHORUS
A distinctive "caterwauling" duet or chorus of hooting phrases and barking nasal cackles.

Barred Owl, p. 220

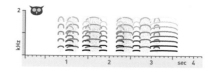

GROWLING OR GROANING PHRASES
See p. 589.

A BUZZY COMPLEX SONG

ALL-BUZZY PHRASE
A phrase of buzzy notes, buzzy series, and/or buzzy trills, at least some of which differ in pitch or quality.

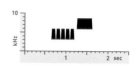

2–5 different buzzes, variable
  Brewer's Sparrow, p. 438

1- to 3-parted, quite variable
  Black-throated Gray Warbler, p. 505

Townsend's Warbler, p. 506
Hermit Warbler, p. 507

A PHRASE OF WHISTLES AND BUZZES
A phrase of 3 or more fairly long notes, mostly buzzes and musical whistles, occasionally with musical trills or short series.

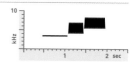

A series of notes starting whistled, becoming buzzy
  Harris's Sparrow, p. 449

1–2 musical series, then a buzz
  American Redstart, p. 500

5 or more whistles and buzzes mixed in various patterns
  Black-throated Gray Warbler, p. 505
  Townsend's Warbler, p. 506
  Hermit Warbler, p. 507

First note a monotone whistle; song complex, often including musical trills or short series
  White-crowned Sparrow, p. 450

SAVANNAH SPARROW SONG GROUP
Phrases of a few high Tinklike notes, then a high, insectlike buzz, sometimes with a few different notes right before or after the buzz.

Only 3–4 notes before the buzz, 0–1 notes after
  Grasshopper Sparrow, p. 462
  LeConte's Sparrow, p. 464

More complex
  Savannah Sparrow, p. 461

A BURRY COMPLEX SONG

A HIGH TWITTERING OR BURRY PHRASE
A high short burry or twittering phrase, introduced by an accelerating twitter.

Given almost exclusively prior to sunrise
  Tropical Kingbird Dawn Song, p. 284

Eastern Kingbird Dawn Song, p. 289

BURRY COMPLEX SHOREBIRD/ FLYCATCHER SONG GROUP

Usually long, a mix of whistled notes and burry phrases, many notes and phrases repeated. Often rather plastic.

Long and fairly stereotyped
Short-billed Dowitcher Song, p. 148
Long-billed Dowitcher Song, p. 148
Semipalmated Plover Song, p. 130
Snowy Plover Ter-weet/ Peer-purr, p. 129
Eastern Phoebe Flight Song, p. 277

Say's Phoebe Flight Song, p. 275

Averaging higher, less musical, often with Whits or Pips
Least Flycatcher Flight Song, p. 271
Willow Flycatcher Flight Song, p. 267
Hammond's Flycatcher Flight Song, p. 268
Dusky Flycatcher Flight Song, p. 269

Highly plastic interaction calls
Northern Beardless- Tyrannulet Interaction Calls, p. 262
Western Wood-Pewee Interaction Calls, p. 263
Say's Phoebe Interaction Calls, p. 275

Mostly coarse, harsh Churrs, with few musical notes
Yellow-headed Blackbird Flight Song, p. 479

A COMPLEX SONG FEATURING GURGLES, CREAKS, CHATTERS, AND/OR IMITATIONS

A GURGLE

A very rapid brief jumble of fairly musical notes, usually 1–3 syllables total. Generally shorter and faster than Warbles (p. 609).

Ending in short trill, or else second half repeating first half
Black-capped Chickadee Gurgle Song, p. 335
Mountain Chickadee Gurgle Song, p. 334
Boreal Chickadee Gurgle Song, p. 336

Chestnut-backed Chickadee Gurgle Song, p. 337
Mexican Chickadee Gurgle, p. 338

All notes different, no trills
Blue Jay Pumphandle Calls, pp. 306–307

Brewer's Blackbird Gurgle Song, p. 477
Rusty Blackbird Gurgle Song, p. 476

Usually with Chirps and whistles
Tree Swallow, p. 326

SQUEAKY GATE HINGE GROUP

1–2 noisy or complex gurgling notes, then a high monotone whistle or creak.

Usually starts with noisy and/or polyphonic notes
Common Grackle Song, p. 478

Usually starts with 1 medium-low warble or gurgle
Brewer's Blackbird Creak Song, p. 477
Rusty Blackbird Gurgle- Creak Song, p. 476

Usually starts with 1 or more low gurgles; final whistles variable
Brown-headed Cowbird Gurgle Song, p. 480
Bronzed Cowbird Gurgle Song, p. 481

Usually starts with 1 low note
Loggerhead Shrike Song, p. 292
Northern Shrike Song, p. 293

Rusty Blackbird Creak Song, p. 476

Highly variable
Steller's Jay Pumphandle Calls, p. 305
Blue Jay Pumphandle Calls, pp. 306–307

A PHRASE OF MUSICAL NOTES AND RATTLES

A soft phrase containing 1-3 musical notes with a ticking Rattle or Snarl.

Musical notes low, mellow
American Crow Gurgle Phrase, p. 318

Northwestern Crow Gurgle Phrase, p. 319

Musical notes seminasal; plastic, given on wing
Cliff Swallow Songlike Calls, p. 323

A COMPLEX SONG FEATURING GURGLES, CREAKS, CHATTERS, AND/OR IMITATIONS, CONTINUED

A COMPLEX HISSING GURGLE
A complex phrase of rapid notes superimposed on a soft hiss.

Nelson's Sparrow Song,
p. 465
Brewer's Blackbird Gurgle
Song, p. 477

Rusty Blackbird Gurgle Song,
p. 476

See also
Great-tailed Grackle, pp. 474–475

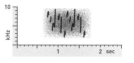

COMPLEX SONGS OF HUMMINGBIRDS
A complex song, usually soft, combining rapid chitters, semimusical series, gurgles, and/or ticking or squeaking notes.

Blue-throated Hummingbird,
p. 105
Rivoli's Hummingbird, p. 106
Anna's Hummingbird, p. 109

Violet-crowned Hummingbird,
p. 116
Broad-billed Hummingbird,
p. 115

A CHATTER, THEN A PHRASE
A short musical phrase introduced by 1–3 quick low noisy notes, usually in a syncopated pattern.

Phrase of polyphonic, burry notes
Eastern Bluebird Agitated
Song, p. 369

Phrase short, highly varied in tone quality
Hooded Oriole Song, p. 483
Bullock's Oriole Song, p. 485

A NOTE OR 2, THEN A CHATTER
A single note or 2- to 3-note series or phrase, followed by a Chatter or unmusical series.

Sedge Wren, p. 356
Marsh Wren (Western), p. 354

Marsh Wren (Eastern), p. 355

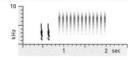

A NASAL, HARSH, HARSH BURRY, OR KNOCKING COMPLEX SONG (INCLUDING "LONG CALLS")

A NASAL PHRASE
A phrase of 4 or more different nasal notes.

Stereotyped rising "How, how-are-you?"
Long-tailed Duck Song, p. 60

High, 4-syllabled, squeaky phrase
Least Tern Kideer-kiddik,
p. 176

Soft, plastic sounds like a radio being tuned
Northern Bobwhite, p. 70

A NASAL AND CHITTERING PHRASE
A phrase with both clear nasal notes and harsh Chitters, Churrs, or Rattles.

Plastic, in interactions, often in chorus
Thick-billed Kingbird Deer
Series, p. 287

Western Kingbird Deer
Series, p. 288
Scissor-tailed Flycatcher
Peer-churr, p. 290

A CROAKING PHRASE
A phrase of low, harsh croaking notes.

Crested Caracara, p. 254

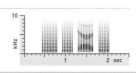

GODWIT SONG GROUP

1–2 consecutive, usually complex series of high seminasal notes, often with various notes before and/or after the series.

High, with Klip or Klee notes
Solitary Sandpiper, p. 151

Built around a high, seminasal couplet series
Bar-tailed Godwit, p. 138

Built around a high, seminasal triplet series
Hudsonian Godwit, p. 139

Built around a medium-low, nasal triplet series
Marbled Godwit, p. 139

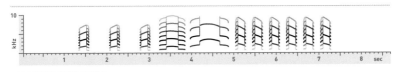

LARGE GULL LONG CALLS

1–3 high, long broken Wails or Squeals, followed by a series of shorter, lower broken notes, the whole performance often preceded by a number of introductory notes. All versions variable and plastic.

High, clear, yelping to squealing
Western Gull, p. 168
Glaucous-winged Gull, p. 169
Herring Gull, p. 170
Glaucous Gull, p. 174
Iceland Gull, p. 171

High, final series rather slow
Ring-billed Gull, p. 166

Mew Gull, p. 167

Slightly lower, nasal and hoarse
California Gull, p. 172
Lesser Black-backed Gull, p. 173

Low, clear, and simple
Yellow-footed Gull, p. 171

Clear and nasal, fast notes absent or at end
Franklin's Gull, p. 164
Heermann's Gull, p. 165

Low, clear and nasal, mostly a triplet series
Black-legged Kittiwake, p. 162

COOT AND GALLINULE LONG CALLS

Nasal to partly grating notes in 1–3 consecutive series. Many patterns possible; usually fastest at start. Toward end, notes often longer and series complex.

High, downslurred, decelerating series of high Keeklike notes
Sora, p. 123

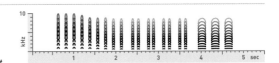

Medium-low downslurred, decelerating series becoming complex or grating
Common Gallinule, p. 124

Medium-low clear nasal couplet series, second note of each couplet longer and upslurred
American Coot, p. 125

GROUSE AND ALCID LONG CALLS

Low nasal or harsh notes in 1–3 consecutive series. Many patterns possible; usually fastest in middle or at end. Some versions can include high Squeals.

Accelerating series of Kuks
Dusky Grouse, p. 82
Sooty Grouse, p. 83

All notes Chuklike, rhythm often irregular
Chukar, p. 72

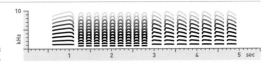

Alternating squeals and Chuks
Chukar Squeal-Chuk Series, p. 72

Accelerating Clucks, with a whistle near the climax
Himalayan Snowcock, p. 72

Nasal, often 2- or 3-parted, fastest in first or second part
Greater Prairie-Chicken, p. 76

Spruce Grouse, p. 81

Nasal, often 1- or 2-parted, fastest near end
Sharp-tailed Grouse, pp. 74–75
Lesser Prairie-Chicken, p. 77

Lower, harsher, bleating and groaning
Common Murre, p. 160

A NASAL, HARSH, HARSH BURRY, OR KNOCKING COMPLEX SONG (INCLUDING "LONG CALLS"), CONTINUED

COMPLEX KNOCKING SERIES
A knocking series or Rattle, either becoming a complex series or changing in pitch or tone quality.

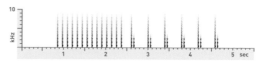

Rising, accelerating knocking series, becoming semimusical
  Buff-collared Nightjar Song,
  p. 104

All notes knocking; second series slower, made of couplets
  Yellow-billed Cuckoo, p. 98

Knocking notes change gradually into clear notes
  Black-billed Cuckoo, p. 99

DUCK LONG CALLS
A series of Quacks, notes usually loudest and longest near start.

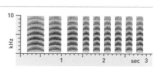

Usually long, loud, the Quacks slightly harsh
  Mallard, p. 52
  Mexican Duck (likely), p. 53
  Northern Shoveler, p. 47

Rather high, screechy, short
  Green-winged Teal, p. 51

Nasal, rather short
  Blue-winged Teal, p. 46
  Cinnamon Teal, p. 47

Nasal, usually 2–4 notes, not accelerating
  Northern Pintail, p. 50

Little-known, not often heard
  Gadwall, p. 48
  American Wigeon, p. 49
  Eurasian Wigeon, p. 49

A SCREECHY PHRASE
A phrase of 4 or more screechy notes. See 2- to 3-Syllabled Screechy Phrase, p. 591.

Of 4-5 syllables, medium-high
  White-tailed Ptarmigan Song,
  p. 84

Of 2–4 syllables; medium-high
  Caspian Tern Kee-kareer,
  p. 177

Of 3–5 notes; medium-low
  Domestic Chicken Crow, p. 63

A SQUEALING CHORUS
A chorus of high-pitched nasal squeaky phrases given by multiple birds at once.

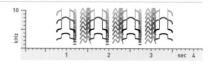

  Sulphur-bellied Flycatcher, p. 283

RAIL LONG CALLS
A series of screechy Grunts, usually falling in pitch and often slightly accelerating. See also Grunt Series (p. 584).

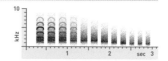

  Ridgway's Rail, p. 122

  Virginia Rail, p. 121

SMALL GULL AND TERN LONG CALLS
High, rather harsh grating notes in 1–3 consecutive series. Many patterns possible; some include short Keks or Kriks. All versions variable and plastic.

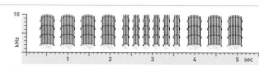

Most notes down- or under-slurred, about the same length
  Common Tern, p. 180

Longest notes upslurred
  Sabine's Gull, p. 163

1–5 rapid Keks between Grates
  Arctic Tern, p. 181
  Forster's Tern, p. 182

Variable mix of long and short Grates
  Bonaparte's Gull, p. 163
  Forster's Tern, p. 182

Fairly low, first notes vaguely 2-syllabled
  American Coot, p. 125

Mix of grating notes and 2-syllabled Kideeks
  Least Tern, p. 176

GRATING, WAILING LONG CALLS
A series of Groans or Wails with burry or tremolo sections, often in plastic, unsynchronized chorus.

Red-necked Grebe Song, p. 88
Horned Grebe Song, p. 89
Red-throated Loon Song, p. 187

Sandhill Crane Long Call, p. 126

Less grating, more whooping
Tundra Swan Long Call, p. 43

KUK-KUK-SCREECH AND SIMILAR PATTERNS
1 to many short Kuklike or Clucklike notes, then a longer, harsher, higher Screech or Shriek.

Kek-kek-shriek
Green Heron, p. 195

Grating Krik-krik-keer
Royal Tern, p. 178

Kuk-kuk-quack
Red-throated Loon, p. 187
(courting ducks may reproduce this pattern incidentally)

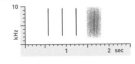

Cluck-cluck-screech
Domestic Chicken, p. 63
White-tailed Ptarmigan, p. 84

KEK-KEK-CHURR AND SIMILAR PATTERNS
1 to many short Keklike notes, then a longer, coarse harsh burry note, usually downslurred.

Noisy to screechy, unmusical
Helmeted Guineafowl, p. 64
Ridgway's Rail, p. 122
Wilson's Snipe Rattle, p. 149

First notes high, semimusical, more like Keeks
Virginia Rail Kee-kee-burr, p. 121
Black Rail Song, p. 120
Red-winged Blackbird Female Song, pp. 470–471

In repeated grating series, often with other calls
Arctic Tern, p. 181
Forster's Tern, p. 182

TICK-TICK-SNARL
1 to many Ticklike notes, then a longer Snarl, given while swooping on potential predators.

Loud, often long
Common Tern, p. 180
Arctic Tern, p. 181

Rather soft
Tree Swallow, p. 326

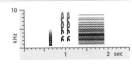

A POLYPHONIC COMPLEX SONG

A POLYPHONIC PHRASE
A phrase of 4 or more different polyphonic notes.

Blue-gray Gnatcatcher
Simple Song, p. 359
California Gnatcatcher
Complex Song, p. 360

Black-capped Gnatcatcher
Simple Song, p. 361
Eastern Bluebird Song, p. 369

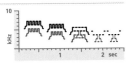

"Eastern" Blue-gray Gnatcatcher, p. 359

A NASAL OR METALLIC SNARLING SONG
A short song ending in a loud harsh nasal or polyphonic Snarl or Groan.

Red-winged Blackbird (Bicolored) Male Song, p. 467
Tricolored Blackbird Male Song, p. 469
Yellow-headed Blackbird, p. 479

THRUSH SONG GROUP
Intricate burry polyphonic phrases, with a jingling metallic quality.

Spiraling upward (overslurred phrases in rising series)
Swainson's Thrush (Russet-backed), p. 374
Swainson's Thrush (Olive-backed), p. 375

Spiraling downward (overslurred phrases in falling series)
Sprague's Pipit, p. 397
Veery, p. 372

Starts with monotone whistle
Hermit Thrush, p. 373

ACKNOWLEDGMENTS

This book was over a dozen years in the making, and countless people helped bring it to fruition. Unfortunately, I only have space to mention a few of them here.

I thank all those who contributed audio recordings. They are acknowledged on the website, www.petersonbirdsounds.com. Some, like Andrew Spencer, Tayler Brooks, Ian Cruickshank, and Bob McGuire, went out of their way to make recordings specifically for this project. Special thanks to Angelika Nelson of the Borror Laboratory of Bioacoustics, Carla Cicero of the Museum of Vertebrate Zoology at Berkeley, and Tom Webber of the Florida Museum of Natural History.

Many people at the Cornell Laboratory of Ornithology deserve thanks, including Jessie Barry, Tammy Bishop, Greg Budney, Greg Delisle, Jay McGowan, Matt Schloss, Mya Thompson, Brad Walker, and Michael Webster. Matt Medler arranged the digitization of old audiotapes, and helped me navigate the immense collection. Matt Young edited hundreds of sounds, hunted down hundreds more, and gave much expertise and encouragement. He is also responsible for the Red Crossbill maps.

I could not have completed the research for this project without the Xeno-Canto website, and many of the recordings come from Xeno-Canto contributors. I owe much to Willem-Pier Vellinga, Bob Planque, Jonathon Jongsma, and the rest of the Xeno-Canto team and recordist community.

Andrew Spencer contributed over 700 recordings, many hours of editing, and vast amounts of expertise and feedback from the very beginning of the project. Walter Szeliga volunteered his time to automate the spectrographic analysis of thousands of sounds. The contributions of Bill Evans, Andrew Farnsworth, Michael Lanzone, and Michael O'Brien were essential to the treatment of nocturnal flight calls. Christopher Clark contributed essential recordings and knowledge to the treatment of hummingbirds. For technical review, I am indebted to Andrew Spencer, Tayler Brooks, Paul Driver, Ted Floyd, Don Kroodsma, Arch McCallum, Michael O'Brien, Richard Webster, and Chris Wood.

Others who helped with obtaining recordings or information include Patrik Aberg, Peter Adriaens, Chuck Aid, Walter Ammann, Nicholas Anich, Kat Avila, Rick Baetsen, Mary Banker, Suzanne Beauchaine, Chris Benesh, Lance Benner, Magnus Bergsson, Peter Boesman, Kelly Bryan, Robin Carter, Ken Chamberlain,

Kevin Colver, Ian Davies, Jeff Dowler, Marco Dragonetti, Caroline Eastman, Elias Elias, Jesse Fagan, Bruce Falls, Matthias Feuersenger, Owen Fitzsimmons, Gerald Fuller, Pat Gonzalez, Walter Graul, Thomas Graves, Phil Green, Larry Gregg, Manuel Grosselet, Wayne Hall, Steve Hampton, Lauren Harter, David Hof, Eric Hopps, Steve N. G. Howell, Rich Hoyer, Paul Hurtado, Marshall Iliff, Diana Iriarte, Alvaro Jaramillo, Stephanie Jones, Richard Kern, Nick Komar, Chrissy Kondrat-Smith, Niels Krabbe, Frank Lambert, Dan Lane, Albert Lastukhin, Michael Lester, Tony Leukering, Ross Lein, Tim Manolis, Paul Marvin, Jarek Matusiak, Kevin McGowan, Martin Muller, Kristie Nelson, Mike Nelson, Stein Nilsen, Stephen Nowicki, Ryan P. O'Donnell, Shawn O'Grady, Scott Olmstead, Brent Ortego, Leroy Overstreet, Ed Pandolfino, Chris Parrish, Gail Patricelli, Iliana Pena, Maria Pereyra, Susan Peters, Bill Pranty, Carolyn Pytte, Will Richardson, Bruce Rideout, Micah Riegner, KT Rusch, Andrew Rush, Michael Schroeder, David Sibley, Dave Slager, Tim Spahr, Martin St. Michel, Ha-Cheol Sung, Jesse Svejcar, Patrick Turgeon, Brad Walker, Denise Wight, Bobby Wilcox, Todd Wilson, Paige Warren, Richard Webster, Russ Wigh, Rene Valdes, Benjamin Van Doren, David Vander Pluym, Maarten van Kleinwee, Gus van Vliet, David Yee, and Jessica Young. My apologies and sincere thanks to all the others who helped out, but are not mentioned here.

My agent, Regina Ryan, was indefatigable and indispensable. My editor, Lisa White, was a pleasure to work with, as were many other people at Houghton Mifflin Harcourt, including Mary Dalton-Hoffman, Beth Burleigh Fuller, and Brian Moore. Donna Riggs worked on the indexing. Eugenie Delaney's excellent work is evident in the layout and graphic design.

My family and my students workshopped parts of this book, from the initial proposal to the species account layouts. Thanks to all, especially my grandparents, my mother, my sister, my brother, and my sister-in-law. Last, a million thanks and a lifetime of love to Molly, for the many sacrifices she made for this project over the years I spent working on it.

I am very proud of this book and I could not have done it without you.

ART CREDITS

p. i: Peter Burke (Wilson's Warbler)

pp. ii–iii: Andrew Spencer (Steller's Jay)

p. vi: Andrew Spencer (White-tailed Ptarmigan)

p. viii: Andrew Spencer (Pacific Wren)

p. ix: Nathan Pieplow (Canada Goose)

p. x: Peter Burke (Lazuli Bunting)

p. 1: Peter Burke (American Avocet)

pp. 34–35: Andrew Spencer (Baird's Sparrow)

p. 525: Peter Burke (Common Grackle)

pp. 526–527: Andrew Spencer (Bohemian Waxwing)

p. 528: Peter Burke (Bald Eagle)

p. 529: Andrew Spencer (Mew Gull)

p. 616: Nathan Pieplow (Orchard Oriole)

p. 618: Peter Burke (Forster's Tern)

p. 620: Peter Burke (Common Raven)

Paintings by Roger Tory Peterson, except:

Michael DiGiorgio: Egyptian Goose, p. 42, Mandarin Duck, p. 44, Domestic Chicken, p. 63, Indian Peafowl, p. 64, Helmeted Guineafowl, p. 64, Red-masked Parakeet, Himalayan Snowcock, p 72, Greater Sage-Grouse, p. 78, Calliope Hummingbird, p. 111, flying Common Loon, p. 184, flying Yellow-billed Loon, p. 185, flying Pacific Loon, p. 186, California Condor, p. 199, Common Black-Hawk, p. 204, Blue-crowned Parakeet, p. 259, Pacific-slope Flycatcher, p. 272, Cordilleran Flycatcher, p. 273, Buff-breasted Flycatcher, p. 274, Black-throated Magpie-Jay, p. 303, flying American Crow, p. 318, Black-tailed Gnatcatcher, p. 358, California Gnatcatcher, p. 360, Pin-tailed Whydah, p. 394, Bronze Mannikin, p. 394, Cassin's Finch, p. 403, MacGillivray's Warbler, p. 497

INDEX

Purchase Peterson Field Guide titles wherever books are sold.
For more information on Peterson Field Guides, visit **www.petersonfieldguides.com.**

PETERSON FIELD GUIDES®

Roger Tory Peterson's innovative format uses accurate, detailed drawings to pinpoint key field marks for quick recognition of species and easy comparison of confusing look-alikes.

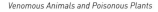

PETERSON FIRST GUIDES®

The first books the beginning naturalist needs, whether young or old. Simplified versions of the full-size guides, they make it easy to get started in the field, and feature the most commonly seen natural life.

Astronomy

Birds

Butterflies and Moths

Caterpillars

Clouds and Weather

Fishes

Insects

Mammals

Reptiles and Amphibians

Rocks and Minerals

Seashores

Shells

Trees

Urban Wildlife

Wildflowers

PETERSON FIELD GUIDES
FOR YOUNG NATURALISTS

This series is designed with young readers ages eight to twelve in mind, featuring the original artwork of the celebrated naturalist Roger Tory Peterson.

Backyard Birds

Birds of Prey

Songbirds

Butterflies

Caterpillars

PETERSON FIELD GUIDES® COLORING BOOKS®

Fun for kids ages eight to twelve, these color-your-own field guides include color stickers and are suitable for use with pencils or paint.

Birds

Butterflies

Reptiles and Amphibians

Wildflowers

Shells

Mammals

PETERSON REFERENCE GUIDES®

Reference Guides provide in-depth information on groups of birds and topics beyond identification.

Behavior of North American Mammals

Birding by Impression

Woodpeckers of North America

Sparrows of North America

PETERSON AUDIO GUIDES

Birding by Ear: Eastern/Central

Bird Songs: Eastern/Central

PETERSON FIELD GUIDE / *BIRD WATCHER'S DIGEST* BACKYARD BIRD GUIDES

Identifying and Feeding Birds

Bird Homes and Habitats

The Young Birder's Guide to Birds of North America

The New Birder's Guide to Birds of North America

DIGITAL

App available for Apple and Android.

Peterson Mammals of North America

Peterson Birds of North America

E-books

Birds of Arizona

Birds of California

Birds of Florida

Birds of Massachusetts

Birds of Minnesota

Birds of New Jersey

Birds of New York

Birds of Ohio

Birds of Pennsylvania

Birds of Texas